EXPLORING ANIMAL BEHAVIOR

Readings from *American Scientist* THIRD EDITION

EDITED BY

Paul W. Sherman
Cornell University

John Alcock
Arizona State University

Sinauer Associates, Inc. *Publishers*
Sunderland, Massachusetts

The Cover

In a honey bee colony, a single queen bee manages to influence the social behavior of thousands of her daughters. Much of the communication between the queen and her worker daughters is based on chemical messages (pheromones). Here, a queen has attracted workers that are acquiring pheromones from their mother. The workers will pass them on to their hivemates, enabling the queen to communicate with bees she never encounters personally. Photograph by Kenneth Lorenzen, University of California, Davis.

Exploring Animal Behavior: Readings from *American Scientist*, Third Edition

Copyright © 2001 Sinauer Associates, Inc.

For information contact Sinauer Associates, Inc., P.O. Box 407, Sunderland, Massachusetts, 01375-0407 U.S.A. FAX: 413-549-1118.
Internet: publish@sinauer.com; http://www.sinauer.com

Illustrations for the articles reprinted in this collection were drawn by artists on the *American Scientist* staff, with the following exceptions: All figures in "Conduct, misconduct, and the structure of science" by Tom Sullivan. Figs. 2 and 7 in "Shaping brain sexuality" by Margaret Nelson and Linda Huff. Figs. 4, 8–10, 13, and 14 in "Aerial defense tactics of flying insects" by Virge Kask. Figs. 2, 4, and 6 in "Why do bowerbirds build bowers?" by Albert Uy. Figs. 3, 7, and 8 in "Early Canid Domestication by Linda Huff. Fig. 1 in "Prairie-vole partnerships" by Virge Kask. Fig. 7 in "Why ravens share" by Linda Huff and Bernd Heinrich. Fig. 1 in "Naked mole-rats" by Sally Black; Fig. 12 by Sally Black and Linda Huff. Figs 2, 8, 10, and 12 in "The essence of royalty" by Elizabeth Carefoot.

"Infanticide as a Primate Reproductive Strategy" reprinted by permission of the publishers of *The Langurs of Abu: Female and Male Strategies of Reproduction*, by Sarah Blaffer Hrdy, Cambridge, MA: Harvard University Press, Copyright © 1977 by the President and Fellows of Harvard College.

"Animal Thinking" by Donald R. Griffin, Copyright © 1984 by Donald R. Griffin.

Library of Congress Cataloging-in-Publication Data
Exploring animal behavior: readings from American scientist / edited by Paul W. Sherman, John Alcock.—3rd ed.
     p.   cm.
    Includes bibliographical references.
    ISBN 0-87893-814-1 (pbk.)
    1. Animal behavior. I. Sherman, Paul W., 1949- II. Alcock, John, 1942- III. American scientist.
QL751.6.E96   2001
591.5—dc21   2001020701

Printed in Hong Kong
8 7 6 5 4 3 2 1

CONTENTS

PREFACE

The first edition of *Exploring Animal Behavior* was published in 1993. It contained 25 articles originally written for *American Scientist*, the journal of the scientific society Sigma Xi. The second edition, which appeared in 1998, included 30 articles—indicating the growth of behavioral research—and our unwillingness to dispense with many excellent contributions from the first edition. In the past three years, the field has continued to flourish, as illustrated by the many new articles that have appeared in *American Scientist*. We have selected some of these for inclusion in this third edition, integrating the new choices with articles from previous editions to create 33 chapters that cover the entire field of behavioral biology.

We have assembled this collection as supplementary reading for students in animal behavior courses. We believe the articles can be profitably employed in classrooms in several ways. First, they provide material for discussion and debate on how science is conducted and on key concepts in behavioral biology. Second, they cover topics and organisms in greater depth than is generally possible in textbook accounts of the same work. Third, they provide personal introductions to some of the leading figures in behavioral biology. Through their articles, these men and women explain how they conduct their research and interpret their results, while also illustrating why they find studying behavior so rewarding.

This anthology should be particularly useful for classes that use John Alcock's textbook, *Animal Behavior: An Evolutionary Approach*, because the articles are organized in a sequence complementary to that text. However, the collection can also enrich courses based on other textbooks. Indeed, this reader can easily stand alone as a sampler of the diversity of questions and research approaches that constitute the modern study of behavior.

We have grouped the articles into five sections. The first section examines practical and ethical aspects of conducting and explaining science, including but not limited to behavioral biology; the second focuses on investigations into the immediate causes or proximate mechanisms that underlie animal behavior; the third shifts to studies that attempt to trace the evolutionary histories of selected behaviors; the fourth examines the adaptive significance of various reproductive behaviors; and the final section illustrates the fitness consequences of social behavior. We hope that students and teachers alike will benefit from this collection of instructive and enjoyable articles.

ACKNOWLEDGMENTS

This book is a reality because of the good will and hard work of many individuals, most notably the authors of the articles, who graciously provided permission to reprint their work, and the many photographers and illustrators whose images brighten virtually every page. Michelle Press and Rosalind Reid, former and current editors of *American Scientist*, respectively, deserve our gratitude for their keen interest in the field of behavior and their efforts to see that it features prominently in the journal. The small size of this volume belies the number of details that had to be attended to in its production. At *American Scientist*, Rosalind Reid and Lil Chappell made sure everything was in order before we went to press. At Sinauer Associates, Peter Farley, Marie Scavotto, Christopher Small, Chelsea Holabird, and Mara Silver facilitated production of a handsome book at a modest price. At work, our colleagues and students who used the two previous editions encouraged us to keep the tradition going. And, at home our families cheerfully supported us, but never let us take ourselves too seriously. Thanks to all.

Paul W. Sherman
John Alcock

PART I
Doing Science

We begin our collection with a series of six articles that examine what it means to be a scientist. In the first of these, Edward O. Wilson defines science as "the organized systematic enterprise that gathers knowledge about the world and condenses the knowledge into testable laws and principles." The goal of science is original discovery and so, according to Wilson, "the true and final test of a scientific career is how well the following declarative sentence can be completed: *He (or she) discovered that...*"

How do scientists make discoveries? Wilson believes it takes a combination of training, creativity, confidence, luck, and being "bright enough to see what needs to be done but not so bright as to suffer boredom doing it." 'Doing it' involves using something called the scientific method. According to Percy Bridgman (a noted physicist) "the scientific method is doing your damnedest, no holds barred." This makes science sound like a synonym for hard work, but it is much more than that. The scientific method is a set of procedures that come into play when attempting to explain how or why a particular phenomenon occurs. It involves developing one or more potential explanations or hypotheses about what causes the phenomenon. Each hypothesis is then examined to see what predictions or expected results can be derived from it. These predictions are checked against reality through observation and experimentation.

Since this book is about behavior, we will illustrate the scientific method with reference to a behavior—one that is spectacularly unpleasant. As described by Sarah Blaffer Hrdy, when male gray langurs attempt to join groups of females, they sometimes kill the females' infants. Hrdy puts forth one possible evolutionary explanation for infanticidal behavior, namely that it enables the killer male to eliminate the offspring of a rival male, thereby gaining sexual access to females who have lost their offspring. If this explanation is correct, we can make some predictions about when males will practice infanticide in nature, predictions that may be tested by further observation. Thus, infanticide "should" (is predicted to) occur only at times when a new male langur ousts the previous resident and acquires control of a band of females, some of whom will have dependent youngsters sired by the previous male. The discovery that newcomer males attack only the infants of other males and not their own offspring (when these come on the scene months later) is consistent with the prediction, and so constitutes evidence for the hypothesis that infanticide is adaptive—a behavior that males use to raise their reproductive success in competition with other males.

However, Richard Curtin and Phyllis Dolhinow offer an alternative, nonadaptive explanation for langur infanticide. They suggest that recent human population expansion, accompanied by habitat destruction and elimination of natural predators, have raised langur population densities and increased frequencies of inter-troop encounters. Groups of bachelor males attempt to take over female bands by attacking the resident male en masse. Chases and fights erupt and, in the ensuing social chaos, infants can become mortally injured or permanently separated from their mothers. Thus, infanticide can be interpreted as a by-product of male-male competition and unnatural crowding, not a reproductive "strategy." Under this hypothesis infanticide "should" be rare in areas where langurs (and humans) occur at lower densities, and unmated males reside in bisexual troops instead of as roving groups of bachelors. Curtin and Dolhinow present some evidence that this is in fact the case.

These competing views of the same behavior, langur infanticide, are juxtaposed not to force you to decide whether infanticide is adapative or nonadaptive (although a class could debate this issue, after searching the more recent literature for relevant information). Instead they are used to illustrate the workings of science. The ability of scientists to logically deduce what they "should" see in nature, if a particular hypothesis is correct, enables them to test their predictions by collecting actual data to be matched against the expected outcomes. Hypotheses whose predictions are apparently incorrect are discarded, whereas hypotheses that yield predictions that are supported by the results of experiments or additional observations, or comparisons among populations and related species, can be accepted.

The approach of developing competing alternative hypotheses, deriving predictions from each, and gathering data that exclude one or more of the alternatives is known as "strong inference" (J. Platt, *Science 146*: 347-353, 1964). It is applicable to all scientific endeavors, including the analysis of behaviors, as illustrated by many of the articles in this volume. By winnowing through alternative ideas, keeping the "good" ones and getting rid of those that produce inaccurate predictions, scientists claim they make progress, and get closer to the truth.

But, as Lewis Branscomb points out, scientists are like everyone else in wanting economic and social rewards in addition to the satisfaction that comes from doing their jobs well. These rewards go disproportionately to the first persons to publish important new findings, which creates pressures that can lead researchers to deceive themselves into thinking that they have done an adequate job of testing their favorite hypothesis, when they have not. Branscomb believes that this kind of self-deception is probably common in science, much more so than outright fraud, and that the "honest mistakes" of scientists harm the scientific community at large because this community has no way "to protect itself from sloppy or deceptive literature." He argues that scientists, therefore, have an ethical responsibility to be aware of the risks of self-deception and to check and double-check their results, while also considering all possible alternative explanations, before reporting their discoveries in the literature. Thus, "quality control in original research is the responsibility of the individual, part of the duty, if not the honor, expected from each of us."

In contrast, James Woodward and David Goodstein come to very different conclusions about what comprises ethical scientific behavior. They note that prescriptions similar to the ones Branscomb advocates are widely accepted in theory, but not in practice. They argue further that many ethical principles that sound good on paper would actually do harm if followed. They believe, for example, that the pressure on scientists to get their novel results out quickly has the important benefit of distributing information promptly and efficiently. Likewise, if one scientist strongly advocates his or her pet idea, no substantative harm is done, according to Woodward and Goodstein, since other scientists are likely to remain skeptical and may soon publish findings that demonstrate the defects associated with a rival's argument.

Woodward and Goodstein would probably also claim that even the "honest mistakes" that Branscomb says are common are unlikely to be numerous, thanks to competition among scientists for recognition and high status, which do not go to those whose conclusions are overthrown by others. Do you agree? Or are Woodward and Goodstein, and others like them, engaged in a form of self-deception when they speak of science as a self-correcting enterprise? How would Branscomb on the one hand, and Woodward and Goodstein on the other, explain the apparent rarity with which researchers simply invent data and publish totally fraudulent papers? Which approach do you find the more persuasive (i.e., the "carrot" or the "stick") in accounting for the infrequency of outright cheating?

In the light of the these two articles, revisit the langur infanticide controversy. What kind of approach was taken by Hrdy on the one hand and Curtin and Dolhinow on

the other? Are they committed advocates for a particular hypothesis, or do they appear to be impartial and objective about their competing explanations? Would controversy exist in science if Branscomb's views were not only widely accepted but actually practiced? Would the elimination of controversy in favor of reaching "genteel agreement" be a good thing in terms of getting closer to the truth about infanticide or anything else of interest? Or does progress in science arise from controversies of this sort? Do you agree with Woodward and Goodstein that "Popperian falsification" (i.e., strong inference) is the best tool for scientists to use in this process and that there is no place for "Baconian inductivism" in science? Or is it possible that in answering a scientific question, one proceeds through sequential stages, from inductivism to falsification?

As noted by several authors, publishing one's findings is a central element of doing science. As Wilson puts it, bluntly, "a discovery does not exist until it is safely reviewed and in print," a point also captured in the academic admonition *"Publish or Perish!"* Writing about one's conclusions and the means by which they were reached enables scientists to communicate with a broad audience, who can then evaluate the message and learn from it. Yet, the training that most scientists receive rarely includes any formal instruction in how to communicate effectively, which may explain why scientific articles often are turgid, incomprehensible, and no fun at all to read. Happily, there are numerous exceptions to this "rule," as the articles in this anthology demonstrate. Moreover, useful formal instruction on how to write scientifically does exist, both in the article by George Gopen and Judith Swan and in a marvelous little book called *The Elements of Style* (Macmillan, New York, reprinted: 1998) by William Strunk, Jr. and E. B. White (author of *Charlotte's Web*). Gopen and Swan's suggestions strike us as superbly helpful—not just for student scientists but for anyone who wants to write in ways that readers will appreciate. Read their article, then analyze and dissect a paragraph or two from any other article in this collection. After pulling the writing apart, see if you can put it back together in improved form, taking advantage of what you have learned about effective writing. Then apply Gopen and Swan's suggestions to your own writing projects. We hope you will discover, as we have, that part of the fun of doing science is explaining your results to others in writing that is clear and understandable.

SCIENTISTS, SCHOLARS, KNAVES AND FOOLS

Edward O. Wilson

Science, its imperfections notwithstanding, is the sword in the stone that humanity finally pulled. The question it poses, of ultimately lawful materialism, is the most important that can be asked in philosophy and religion. Its procedures are not easy to master, even to conceptualize; that is why it took so long to get started, and then mostly in one place, which happened to be western Europe.

Science, to put its warrant as concisely as possible, is the organized systematic enterprise that gathers knowledge about the world and condenses the knowledge into testable laws and principles. The diagnostic features of science that distinguish it from pseudoscience are first, repeatability: The same phenomenon is sought again, preferably by independent investigation, and the interpretation given to it is confirmed or discarded by means of novel analysis and experimentation. Second, economy: Scientists attempt to abstract the information into the form that is both simplest and aesthetically most pleasing—the combination called elegance—while yielding the largest amount of information with the least amount of effort. Third, mensuration: If something can be properly measured, using universally accepted scales, generalizations about it are rendered unambiguous. Fourth, heuristics: The best science stimulates further discovery, often in unpredictable new directions; and the new knowledge provides an additional test of the original principles that led to its discovery. Fifth and finally, consilience: The explanations of different phenomena most likely to survive are those that can be connected and proved consistent with one another.

Astronomy, biomedicine and physiological psychology possess all these criteria. Astrology, ufology, creation science and Christian Science, sadly, possess none. And it should not go unnoticed that the true natural sciences lock together in theory and evidence to form the ineradicable technical base of modern civilization. The pseudosciences satisfy personal psychological needs, but lack the ideas or the means to contribute to the technical base.

Edward O. Wilson is Pellegrino University Professor Emeritus and Honorary Curator in Entomology at Harvard University. In 1997 he received Sigma Xi's William Procter Prize for Scientific Achievement. He is currently working on a series of projects in biodiversity and conservation. This article was adapted from his book, Consilience: The Unity of Knowledge, *to be published in April by Alfred A. Knopf. © 1998 Edward O. Wilson. Address: The Museum of Comparative Zoology, Harvard University, 26 Oxford Street, Cambridge, MA 02138.*

The work of real science is hard and often for long intervals frustrating. You have to be a bit compulsive to be a productive scientist. Keep in mind that new ideas are commonplace, and almost always wrong. Most flashes of insight lead nowhere; statistically, they have a half-life of hours or maybe days. Most experiments to follow up the surviving insights are tedious and consume large amounts of time, only to yield negative or (worse!) ambiguous results.

Over the years I have been presumptuous enough to counsel new Ph.D.'s in biology as follows: If you choose an academic career you will need 40 hours a week to perform teaching and administrative duties, another 20 hours on top of that to conduct respectable research and still another 20 hours to accomplish really important research. This formula is not boot-camp rhetoric. More than half of the Ph.D.'s in science are stillborn, dropping out of original research after at most one or two publications. Percy Bridgman, the founder of high-pressure physics—no pun intended—put the guideline another way: "The scientific method is doing your damnedest, no holds barred."

Original discovery is everything. Scientists as a rule do not discover in order to know but rather, as the philosopher Alfred North Whitehead observed, they know in order to discover. They learn what they need to know, often remaining poorly informed about the rest of the world, including most of science for that matter, in order to move speedily to some part of the frontier of science where discoveries are made. There they spread out like foragers on a picket line, each alone or in small groups probing a carefully chosen, narrow sector. When two scientists meet for the first time the usual conversation entry is, "What do you work on?" They already know what generally bonds them. They are fellow prospectors pressing deeper into an abstracted world, content most of the time to pick up an occasional nugget but dreaming of the mother lode. They come to work each day thinking subconsciously, *It's there, I'm close, this could be the day.*

They know the first rule of the professional game book: Make an important discovery, and you are a successful scientist in the true, elitist sense in a profession where elitism is practiced without shame. You go into the textbooks. Nothing can take that away; you may rest on your laurels the rest of your life. But of course you won't. Almost no one driven enough to make an important discovery ever rests. And any discovery at all is thrilling. There is no feeling more pleasant, no drug more addictive, than setting foot on virgin soil.

Fail to discover, and you are little or nothing in the culture of science, no matter how much you learn and write about science. Scholars in the humanities also make discoveries, of course, but their most original and valuable scholarship is the interpretation and explanation of already existing factual knowledge. When a scientist begins to sort out knowledge in order to sift for meaning, and especially when he carries that knowledge outside the circle of discoverers, he is classified as a scholar in the humanities. Without scientific discoveries of his own, he may be a veritable archangel among intellectuals, his broad wings spread above science, and still not be in the circle. The true and final test of a scientific career is how well the following declarative sentence can be completed: *He (or she) discovered that....* A fundamental distinction thus exists in the natural sciences between process and product. The difference explains why so many accomplished scientists are narrow, foolish people, and why so many wise scholars in the field are considered weak scientists.

Yet, oddly, there is very little science culture, at least in the strict tribal sense. Few rites are performed to speak of. There is at most only a scattering of icons. One does, however, hear a great deal of bickering over territory and status. The social organization of science most resembles a loose confederation of petty fiefdoms. In religious belief individual scientists vary from born-again Christians, admittedly rare, to hard-core atheists, very common. Few are philosophers. Most are intellectual journeymen, exploring locally, hoping for a strike, living for the present. They are content to work at discovery, often teaching science at the college level, pleased to be relatively well-paid members of one of the least conspiratorial of professions.

In character they are as variable as the population at large. Take any random sample of a thousand and you will find the near full human range on every axis of measurement—generous to predatory, well adjusted to psychopathic, casual to driven, gregarious to reclusive. Some are as stolid as tax accounts in April, while a few are clinically certifiable as manic-depressives (or bipolars, to use the ambiguous new term).

In motivation they run from venal to noble. Einstein classified scientists very well during the celebration of Max Planck's 60th birthday in 1918. In the temple of science, he said, are three kinds of people. Many take to science out of a joyful sense of their superior intellectual power. For them research is a kind of sport that satisfies personal ambition. A second class of researchers engage in science to achieve purely utilitarian ends. But of the third: If "the angel of the Lord were to come and drive all the people belonging to these two categories out of the temple, a few people would be left, including Planck, and that is why we love him."

Scientific research is an art form in this sense: It does not matter how you make a discovery, only if your claim is true and convincingly validated. The ideal scientist thinks like a poet and works like a bookkeeper, and I suppose that if gifted with a full quiver, he also writes like a journalist. As a painter stands before canvas or a novelist recycles old emotion with eyes closed, he searches his imagination for subjects as much as for conclusions, for questions as much as for answers. Even if his highest achievement is only to perceive the need for a new instrument or theory, that may be enough to open the door to a new industry of research.

This level of creativity in science, as in art, depends as much on self-image as on talent. To be highly successful the scientist must be confident enough to steer for blue water, abandoning sight of land for a while. He values risk for its own sake. He keeps in mind that the footnotes of forgotten treatises are strewn with the names of the gifted but timid. If on the other hand he chooses like the vast majority of his colleagues to hug the coast, he must be fortunate enough to possess what I like to define as optimum intelligence for normal science: bright enough to see what needs to be done but not so bright as to suffer boredom doing it.

Advice to the novice scientist: There is no fixed way to make and establish a scientific discovery. Throw everything you can at the subject, so long as the procedures can be duplicated by others. Consider repeated observations of a physical event under varying circumstances, experiments in different modes and styles, correlation of supposed causes and effects, statistical analyses to reject null hypotheses (those deliberately raised to threaten the conclusion), logical argument, and attention to detail and consistency with the results published by others. All these actions, singly and in combination, are part of the tested and true armamentarium of science. As the work comes together, also think about the audience to whom it will be reported. Plan to publish in a reputable, peer-reviewed journal. One of the strictures of the scientific ethos is that a discovery does not exist until it is safely reviewed and in print.

Integrity in Science

Lewis M. Branscomb

Much of the problem of honor—or lack of honor—in science stems not from malice but from self-deception

In 1945 a physics graduate student at Harvard began a Ph.D. thesis project involving the use of molecular spectroscopy to determine the temperature of the atmosphere 1,000 km above the earth, at that time quite unknown. The Schumann-Runge bands of molecular oxygen had been observed as very weak emissions from the upper atmosphere. It was thought that they could be used as a thermometer, subject to verification in laboratory studies. But the bands had been observed only in absorption at very high pressures. Then in 1948 there appeared in *Nature* a report that stated that the Schumann-Runge bands had been observed in emission, excited at low pressure in a high-frequency discharge. The author also analyzed the molecular constants of the states involved (1).

Delighted to find from the literature that his thesis problem could be successfully attacked, the student set about reproducing the experiment described in *Nature*. After months of fruitless effort, he became suspicious that the results reported were in error and even that the photograph published with the text was not a picture of the Schumann-Runge spectrum at all. Indeed, it appeared that the results might have been fabricated from the proverbial whole cloth. In any case, six months of a predoctoral fellowship were lost, and another way to tackle the thesis problem had to be found.

I was that graduate student, and I have always felt sorry for the author of the article in *Nature*, who must have been under terrible pressure to show something for his efforts. I doubt that he had any intent to injure anyone, certainly not an unknown student thousands of miles away.

I believe that there are very few scientists who deliberately falsify their work, cheat on their colleagues, or steal from their students. On the other hand, I am afraid that a great many scientists deceive themselves from time to time in their treatment of data, gloss over problems involving systematic errors, or understate the contributions of others. These are the "honest mistakes" of science, the scientific equivalent of the "little white lies" of social discourse. But unlike polite society, which

Lewis M. Branscomb is Vice President and Chief Scientist of the IBM Corporation, President of Sigma Xi, and a past-president of the American Physical Society. He joined IBM in 1972 after a 21-year career at the National Bureau of Standards, of which he became director in 1969. He taught physics at several universities, was editor of Reviews of Modern Physics, *and conceived and chaired for several years the Joint Institute for Laboratory Astrophysics at the University of Colorado. Dr. Branscomb was appointed by President Carter to the National Science Board in 1979 and was elected chairman the following year, serving until 1984. Address: IBM Corporation, Armonk, NY 10504.*

easily interprets those white lies, the scientific community has no way to protect itself from sloppy or deceptive literature except to learn whose work to suspect as unreliable. This is a tough sentence to pass on an otherwise talented scientist.

The pressures on young science faculty are often fierce, not so much from tenure committees or even from peers, but from within. A young untenured scientist has all his emotional eggs in one basket. He picks a research problem and invests a year or more in its pursuit. Getting a successful start is important to the opportunity to do research. A lifetime career hinges on nature's cooperation as well as his own diligence and ingenuity. As we are reminded on television, it is dangerous to trifle with Mother Nature. Scientists run that risk every day. It takes a very self-confident young scientist to laugh at Tom Lehrer's "Lobachevsky" without a twinge of fear.

The Sigma Xi project on Honor in Science must deal with the broader question of the integrity of scientists' behavior, not just with the morality of what is admittedly the more serious evil, deliberate cheating (2–5). Unless science students are thoroughly inculcated with the discipline of correct scientific process, they are in serious danger of being damaged by the temptation to take the easy road to apparent success. And outright cheating can best be contained if the standards in all disciplines are held at high levels.

When is an experiment complete?

The reader may feel that the rules are simple and easy to follow for those who care about the integrity of their work. That is not necessarily so. Take, for example, the problem of knowing when an experiment is finished and the results are ready to publish. In 1953, building on the work of Wade Fite and profiting from his help at a critical time, I succeeded in making the first laboratory measurement of the photodetachment cross section for a negatively charged atomic ion in vacuum (6). The absorption of light by the negative ion of hydrogen (H−) was believed by Rupert Wildt to dominate the opacity of the solar photosphere. Simply put, the temperature of the sun, and thus the wavelengths to which human eyes are sensitive, is determined by this cross section. No one knew how accurate the quantum calculation of this three-body problem might be.

In order to test the calculation, Stephen J. Smith and I undertook an experiment requiring an absolute measurement in a very complex crossed-beam apparatus. After several years of preparation, the experiment began to yield data, and we made a reasonably diligent search for sources of systematic errors. The results differed from the quantum calculations by about 15%, a not unreasonable percentage considering the challenge of the three-body problem at the time. Stephen Smith and I were writing up the paper and making some final tests on the radiometric calibration system when the apparatus gave us a hint that something was amiss. We put the paper aside, tore the experiment down, and started over again on the calibrations. Three months later we had done everything necessary to quantify the limits of systematic error. Only then did we convert the results to cross-section units. We discovered to our utter amazement that the corrections we had introduced measured exactly 15%, bringing the experiment and the theory into an agreement so exact as to be clearly fortuitous.

At that point we were faced with a tough decision. What to do now? The experiment was finished. But we had thought it was finished once before. Were we in danger of stopping when we liked the answer? I realized then, as I have often said since, that Nature does not "know" what experiment a scientist is trying to do. "God loves the noise as much as the signal" (7). I decided to spend another three months looking for more sources of systematic error—a time exactly equal to the time we had spent on the last effort, which resulted in bringing experiment into agreement with theory. Fortunately, no additional sources could be found, and Steve and I felt we were ready to publish (8).

Perhaps this degree of conservatism is not necessary in every case, but it is certainly crucial in the case of absolute as opposed to relative measurements. The most severe requirement for such care is in the measurement of the fundamental constants of nature, a major interest of scientists at the National Bureau of Standards. Ever since the 1960s, scientists measuring atomic constants have adopted the policy of never reducing their data to final form (permitting comparison with the work of others) until all error analysis has been completed and the experiment is over.

Why, if the scientists are both honest and disciplined, is this necessary? Because the temptation to get a "good" (i.e., "safe" or "significant") result by stopping when the data pass through the desired coordinates is ever present. Some excellent scientists may have succumbed to the temptation. Back in the 1930s, for example, there was a long series of measurements of the universal constant of nature c, the speed of light in a vacuum. Following the pioneering measurement of Michelson and his co-workers, who used a rotating polygonal mirror to chop a beam of light passing between Mt. Wilson and Mt. Baldy in California, subsequent experimenters found more precise results using better equipment. In 1941, Birge's review of all the work concluded that the best weighted average of all the prior work was $c = 299,776 \pm 4$ km/sec (9).

Then came World War II. New technology and new people came into science. Very low frequency radio navigation (Loran) had been developed for military use, and electrical engineers realized that this system could be used to measure the speed of propagation of those 16 KHz waves in ways totally independent of the prewar optical methods. Within a few years, microwave cavity methods and free-space microwave interferometry gave consistent values with much higher precision.

Froome found $c = 299,793 \pm 0.3$ km/sec (10). There had been a shift of 17 km/sec, yet the stated accuracy of most of the previous measurements was 4 km/sec or better.

In their review of this mystery, Cohen and DuMond concluded that "two things contributed strongly to mislead [Birge in 1941] and would have misled anyone else in the same circumstances. These were the great prestige of Michelson's name as an expert in the field, and the fact that . . . two measurements . . . in 1937 and 1941 agreed quite well with the Michelson-Pease-Pearson result" (11). Writing in 1957, Birge said: "In any highly precise experimental arrangement there are initially many instrumental difficulties that lead to numerical results far from the accepted value of the quantity being measured. . . . Accordingly, the investigator searches for the source or sources of such error, and continues searching until he gets a result close to the accepted value. Then he stops! . . . In this way one can account for the close agreement of several different results and also for the possibility that all of them are in error by an unexpectedly large amount" (12). Cohen and DuMond credit Peter Franken with labeling this tendency "intellectual phase locking."

Commitment to quality

What might be done to reduce these "honest mistakes," to support the quality and thus the integrity of science? It takes the concerted efforts of teachers and research mentors, of promotion and tenure committees, of journal editors and referees. Above all it takes renewed commitment by the working scientist.

Young scientists should understand all the subtle ways in which they can delude themselves in the design of observations and the interpretation of data and statistics. They should understand metrology and should know what tendencies to manipulate information are built into their digital signal processors. They should also get to know the algorithms used in their favorite computers, which may under certain circumstances give strange results. Above all they should be trained in the detection and control of systematic errors.

The responsibility of the gatekeepers of scientific careers, the tenure committees, deans, and laboratory directors, is a heavy one. To reward people solely on the basis of numbers of papers published is destructive of the quality of science. Publication is of course the conventional method of making one's work available for critical appraisal by one's peers, but it is not the only way. And while perhaps even a necessary way, it is most emphatically not sufficient.

Journal editors and referees are, of course, the stewards of scientific quality, and they face a very difficult task. No journal can afford to publish all the evidence required to support an author's experimental conclusions. But how can a referee approve publication, when information necessary to proof is missing? The traditional answer is that authors use a certain shorthand to refer to procedures used which are either common practice or documented elsewhere. The reader has to trust the author to invoke those procedures properly. Thus one's reputation for trustworthiness, call it intellectual integrity if not honesty, is crucial to a scientific career. Are young people entering the world of scientific research as aware of this as they should be?

The quality of science places another burden on the scientist: not only to ensure that his own work meets the highest standards, but to participate in both the peer review of primary literature and the authorship of reviews of areas of work in which he is competent. Maurice Goldhaber, when he was director of the Brookhaven National Laboratory, encouraged his staff to write scholarly reviews. He felt that the review literature was a special responsibility of scientists at national laboratories; his motto was, "A good review is the moral equivalent of teaching."

During the last two decades substantial organized efforts at professional reviews of the literature of physical science have been undertaken. Groups of research experts have undertaken critical evaluations of original literature, usually dealing with properties of matter and materials. The goal of such reviews is to increase the density of useful information in the literature. Information that is wrong is not useful. And information lacking evidence revealing whether it is right or wrong is scarcely more so. Quality control in original research is the responsibility of the individual, part of the duty, if not the honor, expected from each of us.

To make the literature worth reviewing, authors of original papers must give the reader quantitative estimates of the amount by which the values given may be in error, and scientific justification for their conclusions. Scientists must demand of others and of themselves a revival of sound scholarship, instead of the cream-skimming and large numbers of hastily written papers with which we are all too familiar.

Commitment to integrity

The broader view of honor in science that I have discussed here should help everyone understand that this is not someone else's problem and is not just the problem of fraud in science. Most of us will never encounter a piece of truly fraudulent research. But concerns about scientific integrity permeate every piece of research we do, every talk we hear, every paper we read. A revitalization of interest in scientific honesty and integrity could have an enormous benefit both to science and to the society we serve.

First of all, integrity is essential for the realization of the joy that exploring the world of science can and should bring to each of us. Beyond that, the integrity of science affects the way the public looks at the pronouncements of scientists and the seriousness with which it takes our warnings, whether they relate to acid rain, the loss of genetic materials from endangered species, or the possibilities for science to help solve the global problems facing mankind. The users of our results, the decision-makers who need our advice, will always press us to be more sure of ourselves than our data permit, for it would make their jobs easier. The pressures to take shortcuts in science come from outside, as well as inside, the community.

We must help the public understand the rules of scientific evidence, just as we insist on rules of judicial evidence in our courts. A precondition for success in this endeavor is to refine and apply those rules with great rigor in our own work and literature. The future of mankind hangs in no small measure on the integrity, and thus the credibility, of science.

References

1. L. Lal. 1948. *Nature* 161:477.
2. C. I. Jackson and J. W. Prados. 1983. *Am. Sci.* 71:462.
3. *Honor in Science.* 1984. Sigma Xi.
4. R. N. Hall. 1968. Gen. Elec. rep. no. 68-C-035.
5. R. P. Feynman. 1974. *Engineering and Science* 37(7):10.
6. L. M. Branscomb and W. L. Fite. 1954. *Phys. Rev.* 93:651.
7. L. M. Branscomb. 1980. *Phys. Today* 33(4):42.
8. L. M. Branscomb and S. J. Smith. 1955. *Phys. Rev.* 98:1028.
9. R. T. Birge. 1941. *Rep. Progr. Phys.* 8:90.
10. K. D. Froome. 1954. *Proc. Royal Soc. London* A223:195.
11. E. R. Cohen and J. DuMond. 1965. *Rev. Modern Phys.* 37:537.
12. R. T. Birge. 1957. *Nuovo Cimento,* supp. 6:39.

Conduct, Misconduct and the Structure of Science

Many plausible-sounding rules for defining ethical conduct might be destructive to the aims of scientific inquiry

James Woodward and David Goodstein

In recent years the difficult question "what constitutes scientific misconduct?" has troubled prominent ethicists and scientists and tied many a blue-ribbon panel in knots. In teaching an ethics class for graduate and undergraduate students over the past few years, we have identified what seems to be a necessary starting point for this debate: the clearest possible understanding of *how science actually works*. Without such an understanding, we believe, one can easily imagine formulating plausible-sounding ethical principles that would be unworkable or even damaging to the scientific enterprise.

Our approach may sound so obvious as to be simplistic, but actually it uncovers a fundamental problem, which we shall try to explore in this article. The nature of the problem can be glimpsed by considering the ethical implications of the earliest theory of the scientific method. Sir Francis Bacon, a contemporary of Galileo, thought the scientist must be a disinterested observer of nature, whose mind was cleansed of prejudices and preconceptions. As we shall see, the reality of science is radically different from this ideal. If we expect to find scientists who are disinterested observers of nature we are bound to be disappointed, not because scientists have failed to measure up to the appropriate standard of behavior, but because we have tried to apply the wrong standard of behavior. It can be worse: Rules or standards of conduct that seem intuitively appealing can turn out to have results that are both unexpected and destructive to the aims of scientific inquiry.

In drafting this article, we set out to examine the question of scientific ethics in light of what we know about science as a system and about the motivations of the scientists who take part in it. The reader will find that this exercise unearths contradictions that may be especially unpleasant for those who believe clear ethical principles derive directly from the principles of scientific practice. In fact, one can construct a wonderful list of plausible-sounding ethical principles, each of which might be damaging or unworkable according to our analysis of how science works.

Ideals and Realities

We can begin where Sir Francis left off. Here is a hypothetical set of principles, beginning with the Baconian ideal, for the conduct of science:

1. Scientists should always be disinterested, impartial and totally objective when gathering data.
2. A scientist should never be motivated to do science for personal gain, advancement or other rewards.
3. Every observation or experiment must be designed to falsify an hypothesis.
4. When an experiment or an observation gives a result contrary to the prediction of a certain theory, all ethical scientists must abandon that theory.
5. Scientists must never believe dogmatically in an idea nor use rhetorical exaggeration in promoting it.
6. Scientists must "lean over backwards" (in the words of the late physicist Richard Feynman) to point out evidence that is contrary to their own hypotheses or that might weaken acceptance of their experimental results.
7. Conduct that seriously departs from that commonly accepted in the scientific community is unethical.
8. Scientists must report what they have done so fully that any other scientist can reproduce the experiment or calculation. Science must be an open book, not an acquired skill.

James Woodward and David Goodstein are professors at Caltech, where, in addition to the Research Ethics course they teach together, Woodward teaches philosophy and Goodstein teaches physics. Woodward has served Caltech as Executive Officer for the Humanities; Goodstein serves as vice provost and is the Frank J. Gilloon Distinguished Teaching and Service Professor. Address for both: California Institute of Technology, Pasadena, CA 91125.

9. Scientists should never permit their judgments to be affected by authority. For example, the reputation of a scientist making a given claim is irrelevant to the validity of the claim.

10. Each author of a multiple-author paper is fully responsible for every part of the paper.

11. The choice and order of authors on a multiple-author publication must strictly reflect the contributions of the authors to the work in question.

12. Financial support for doing science and access to scientific facilities should be shared democratically, not concentrated in the hands of a favored few.

13. There can never be too many scientists in the world.

14. No misleading or deceptive statement should ever appear in a scientific paper.

15. Decisions about the distribution of resources and publication of results must be guided by the judgment of scientific peers who are protected by anonymity.

Should the behavior of scientists be governed by rules of this sort? We shall argue that it should not. We first consider the general problems of motivation and the logical structure of science, then the question of how the community of scientists actually does its work, showing along the way why each of these principles is defective. At the end, we offer a positive suggestion of how scientific misconduct might be recognized.

Behavior that may seem at first glance morally unattractive can, in a properly functioning system, produce results that are generally beneficial.

Motives and Consequences

Many of the provocative statements we have just made raise general questions of motivation related to the issue explicitly raised in principle 2, and it is worth dealing with these up front. We might begin with a parallel: the challenge of devising institutions, rules and standards to govern commerce. In economic life well-intentioned attempts to reduce the role of greed or speculation can turn out to have disastrous consequences. In fact, behavior that may seem at first glance morally unattractive, such as the aggressive pursuit of economic self-interest, can, in a properly functioning system, produce results that are generally beneficial.

In the same way it might appear morally attractive to demand that scientists take no interest in obtaining credit for their achievements. Most scientists are motivated by the desire to discover important truths about nature and to help others to do so. But they also prefer that they (rather than their competitors) be the ones to make discoveries, and they want the recognition and the advantages that normally reward success in science. It is tempting to think that tolerating a desire for recognition is a concession to human frailty; ideally, scientists should be interested only in truth or other purely epistemic goals. But this way of looking at matters misses a number of crucial points.

For one thing, as the philosopher Philip Kitcher has noted, the fact that the first person

to make a scientific discovery usually gets nearly all the credit encourages investigators to pursue a range of different lines of inquiry, including lines that are thought by most in the community to have a small probability of success. From the point of view of making scientific discoveries as quickly and efficiently as possible, this sort of diversification is extremely desirable; majority opinion turns out to be wrong with a fairly high frequency in science.

Another beneficial feature of the reward system is that it encourages scientists to make their discoveries public. As Noretta Koertge has observed, there have been many episodes in the history of early modern science in which scientists made important discoveries and kept them private, recording them only in notebooks or correspondence, or in cryptic announcements designed to be unintelligible to others. The numerous examples include Galileo, Newton, Cavendish and Lavoisier. It is easy to see how such behavior can lead to wasteful repetition of effort. The problem is solved by a system of rewards that appeals to scientists' self-interest. Finally, in a world of limited scientific resources, it makes sense to give more resources to those who are better at making important discoveries.

We need to be extremely careful, in designing institutions and regulations to discourage scientific misconduct, that we not introduce changes that disrupt the beneficial effects that competition and a concern for credit and reputation bring with them. It is frequently claimed that an important motive in a number of recent cases of data fabrication has been the desire to establish priority and to receive credit for a discovery, or that a great deal of fraud can be traced to the highly competitive nature of modern science. If these claims are correct, the question becomes, how can we reduce the incidence of fraud without removing the beneficial effects of competition and reward?

The Logical Structure of Science

The question of how science works tends to be discussed in terms of two particularly influential theories of scientific method, *Baconian inductivism* and *Popperian falsification,* each of which yields a separate set of assumptions.

According to Bacon's view, scientific investigation begins with the careful recording of observations. These should be, insofar as is humanly possible, uninfluenced by any prior prejudice or theoretical preconception. When a large enough body of observations is accumulated, the investigator generalizes from these, via a process of induction, to some hypothesis or theory describing a pattern present in the observations. Thus, for example, an investigator might inductively infer, after observing a large number of black ravens, that all ravens are black. According to this theory good scientific conduct consists in recording all that one observes and not just some selected part of it, and in asserting only hypotheses that are strongly inductively supported by the evidence. The guiding ideal is to avoid any error that may slip in as a result of prejudice or preconception.

How can we reduce the incidence of fraud without removing the beneficial effects of competition and reward?

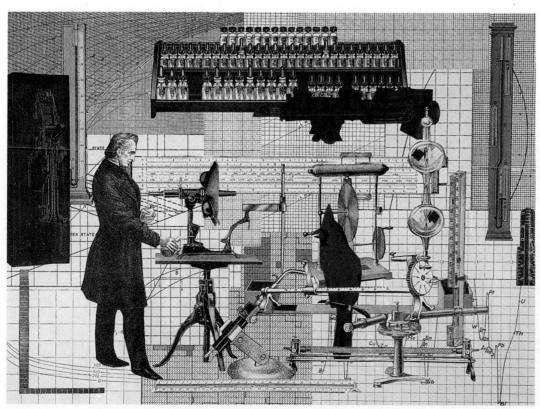

Historians, philosophers and those scientists who care are virtually unanimous in rejecting Baconian inductivism as a general characterization of good scientific method. The advice to record all that one observes is obviously unworkable if taken literally; some principle of selection or relevance is required. But decisions about what is relevant inevitably will be influenced heavily by background assumptions, and these, as many recent historical studies show, are often highly theoretical in character. The vocabulary we use to describe the results of measurements, and even the instruments we use to make the measurements, are highly dependent on theory. This point is sometimes expressed by saying that all observation in science is "theory-laden" and that a "theoretically neutral" language for recording observations is impossible.

The idea that science proceeds only and always via inductive generalization from what is directly observed is also misguided. Theories in many different areas of science have to do with entities whose existence or function cannot be directly observed: forces, fields, subatomic particles, proteins and other large organic molecules, and so on. For this and many other reasons, no one has been able to formulate a defensible theory of inductive inference of the sort demanded by inductivist theories of science.

The difficulties facing inductivism as a general conception of scientific method are so well known that it is surprising to find authoritative characterizations of scientific misconduct that appear to be influenced by this conception. Consider the following remarks by Suzanne Hadley, at one time acting head of what used to be called the Office of Scientific Integrity (now the Office of Research Integrity),

the arm of the U.S. Public Health Service charged with investigating allegations of scientific misconduct. In a paper presented at the University of California, San Diego, in October 1991, Hadley wrote: "…it is essential that observation, data recording, and data interpretation and reporting be veridical with the phenomena of interest, *i.e.*, be as free as humanly possible of 'taint' due to the scientist's hopes, beliefs, ambitions, or desires." Elsewhere she writes: "Anything that impinges on the verdical perception, recording and reporting of scientific phenomena is antithetical to the very nature of science." She also says, "…it is the human mind, which based on trained observations, is able to form higher-order conceptions about phenomena."

Hadley's view may not be as rigidly inductivist as these remarks imply. She adds, "I hasten to say that I am not suggesting that a scientist can or should be relegated to a mechanistic recording device." But this more nuanced view is not allowed to temper excessively her oversight responsibility as a government official: "The really tough cases to deal with are the cases closest to the average scientist: those in which 'fraud' is not clearly evident, but 'out of bounds' conduct is: data selection, failure to report discrepant data, over-interpretation of data…."

The idea that data selection and overinterpretation of data are forms of misconduct seem natural if one begins with Hadley's view of scientific method. A less restrictive view would lead to a different set of conclusions about what activities constitute misconduct.

Although relatively few contemporary scientists espouse inductivism, there are many scientists who have been influenced by the falsificationist ideas of Karl Popper. According to falsificationists, we test a hypothesis by deducing from it a testable prediction. If this prediction turns out to be false, the hypothesis from which it is deduced is said to be falsified and must be rejected. For example, the observation of a single nonblack raven will falsify the hypothesis H: "All ravens are black." But if we set out to test H and observe a black raven or even a large number of such ravens we cannot, according to Popper, conclude that H is true or verified or even that it is more probable than it was before. All that we can conclude is that H has been tested and has not yet been falsified. There is thus an important asymmetry between the possibility of falsification and the possibility of verification; we can show conclusively that an hypothesis is false, but not that it is true.

Because of this asymmetry it is a mistake to think, as the inductivist does, that good science consists of hypotheses that are proved or made probable by observation, whereas bad science does not. Instead, according to Popper, good

science requires hypotheses that might be falsified by some conceivable observation. For example, the general theory of relativity predicts that starlight passing sufficiently close to the sun will be deflected by a certain measurable amount E. General relativity is a falsifiable theory because observations of starlight deflection that differ substantially from E are certainly conceivable and had they been made, would have served to falsify general relativity. By contrast, writes Popper, Freudian psychology is unfalsifiable and hence unscientific. If, for example, a son behaves in a loving way toward his mother, this will be attributed to his Oedipus complex. If, on the contrary, he behaves in an hostile and destructive way, this will be attributed to the same Oedipus complex. No possible empirical observation constitutes a refutation of the hypothesis that the son's behavior is motivated by an Oedipus complex.

According to Popper, bad scientific behavior consists in refusing to announce in advance what sorts of evidence would lead one to give up an hypothesis, in ignoring or discarding evidence contrary to one's hypothesis or in introducing ad hoc, content-decreasing modifications in one's theories in order to protect them against refutation. Good scientific method consists in putting forward highly falsifiable hypotheses, specifying in advance what sorts of evidence would falsify these hypotheses, testing the hypotheses at exactly those points at which they seem most likely to break down and then giving them up should such evidence be observed. More generally (and moving somewhat beyond the letter of Popper's theory) we can say that to do science in a Popperian spirit is to hold to one's hypothesis in a tentative, nondogmatic fashion, to explore and draw to the attention of others various ways in which one's hypothesis might break down or one's experimental result may be invalid, to give up one's hypothesis willingly in the face of contrary evidence, to take seriously rather than to ignore or discard evidence that is contrary to it, and in general not to exaggerate or overstate the evidence for it or suppress problems that it faces. Richard Feynman, in a commencement address at Caltech some years ago, recommended a recognizably Popperian attitude in the following remarks:

> [There is an] idea that we all hope you have learned in studying science in school—we never explicitly say what this is, but just hope that you catch on by all the examples of scientific investigation.… It's a kind of scientific integrity, a principle of scientific thought that corresponds to a kind of utter honesty—a kind of leaning over backwards. For example, if you're doing an experiment, you should report everything that you think might make it invalid—not only what you

think is right about it; other causes that could possibly explain your results; and things you thought of that you've eliminated by some other experiment, and how they worked—to make sure the other fellow can tell they have been eliminated.

> .…In summary, the idea is to try to give all the information to help others to judge the value of your contribution, not just the information that leads to judgment in one particular direction or another.

These views form the basis of principles 3–6 above.

Although falsificationism has many limitations (see below), it introduces several corrections to inductivism that are useful in understanding how science works and how to characterize misconduct. To begin with, falsificationism rejects the idea that good scientific behavior consists in making observations without theoretical preconceptions. For Popper, scientific activity consists in attempting to falsify. Such testing requires that one have in mind a hypothesis that will indicate which observations are relevant or worth making. Rather than something to be avoided, theoretical preconceptions are essential to doing science.

Inductivists attach a great deal of weight to the complete avoidance of error. By contrast, falsificationists claim that the history of science shows us that all hypotheses are falsified sooner or later. In view of this fact, our aim should be to detect our errors quickly and to learn as

For science to advance, scientists must be free to be wrong.

efficiently as possible from them. Error in science thus plays a constructive role. Indeed, according to falsificationists, putting forward a speculative "bold conjecture" that goes well beyond available evidence and then trying vigorously to falsify it will be the strategy that enables us to progress as efficiently as possible. For science to advance, scientists must be free to be wrong.

Despite these advantages, there are also serious deficiencies in falsificationism, when it is taken as a general theory of method. One of the most important of these is sometimes called the Duhem-Quine problem. We claimed above that testing a hypothesis H involved deriving from it some observational consequence O. But in most realistic cases such observational consequences will not be derivable from H alone, but only from H in conjunction with a great many other assumptions A (auxiliary assumptions, as philosophers sometimes call them). For example, to derive an observational claim from a hypothesis about rates of evolution, one may need auxiliary assumptions about the processes by which the fossil record is laid down. Suppose one hypothesizes that a certain organism has undergone slow and continuous evolution and derives from this that one should see numerous intermediate forms in the fossil record. If such forms are absent it may mean H is false, but it may also be the case that H is true but fossils were preserved only in geological deposits that were laid down at widely separated times. It is possible that H is true and that the reason that O is false is that A is false.

One immediate result of this simple logical fact is that the logical asymmetry between falsification and verification disappears. It may be true, as Popper claims, that we cannot conclusively verify a hypothesis, but we cannot conclusively falsify it either. Thus, as a matter of method, it is sometimes a good strategy to hold onto a hypothesis even when it seems to imply an observational consequence that looks to be false. In fact, the history of science is full of examples in which such anti-Popperian behavior has succeeded in finding out important truths about nature when it looks as though more purely Popperian strategies would have been less successful.

Anti-Popperian strategies seem particularly prevalent in experiments. In doing an experiment one's concern is often to find or demonstrate an effect or to create conditions that will allow the effect to appear, rather than to refute the claim that the effect is real. Suppose a novel theory predicts some previously unobserved effect, and an experiment is undertaken to detect it. The experiment requires the construction of new instruments, perhaps operating at the very edge of what is technically possible, and the use of a novel experimental design,

It is sometimes a good strategy to hold onto a hypothesis even when it seems to imply an observational consequence that looks to be false.

which will be infected with various unsuspected and difficult-to-detect sources of error. As historical studies have shown, in this kind of situation there will be a strong tendency on the part of many experimentalists to conclude that these problems have been overcome if and when the experiment produces results that the theory predicted. Such behavior certainly exhibits anti-Popperian dogmatism and theoretical "bias," but it may be the best way to discover a difficult-to-detect signal. Here again, it would be unwise to have codes of scientific conduct or systems of incentives that discourage such behavior.

Social Structure

Inductivism, falsificationism and many other traditional accounts of method are inadequate as theories of science. At bottom this is because they neglect the psychology of individual scientists and the social structure of science. These points are of crucial importance in understanding how science works and in characterizing scientific misconduct.

Let us begin with what Philip Kitcher has called the division of cognitive labor and the role of social interactions in scientific investigation. Both inductivism and falsificationism envision an individual investigator encountering nature and constructing and assessing hypotheses all alone. But science is carried out by a community of investigators. This fact has important implications for how we should think about the responsibilities of individual scientists.

Suppose a scientist who has invested a great deal of time and effort in developing a theory is faced with a decision about whether to continue to hold onto it given some body of evidence. As we have seen, good Popperian method requires that scientists act as skeptical critics of their own theories. But the social character of science suggests another possibility. Suppose that our scientist has a rival who has invested time and resources in developing an alternative theory. If additional resources, credit and other rewards will flow to the winner, perhaps we can reasonably expect that the rival will act as a severe Popperian critic of the theory, and vice versa. As long as others in the community will perform this function, failure to behave like a good Popperian need not be regarded as a violation of some canon of method.

There are also psychological facts to consider. In many areas of science it turns out to be very difficult, and to require a long-term commitment of time and resources, to develop even one hypothesis that respects most available theoretical constraints and is consistent with most available evidence. Scientists, like other human beings, find it difficult to sustain commitments to arduous, long-term projects if they spend too much time contemplating the various ways in which the project might turn out to be unsuccessful.

A certain tendency to exaggerate the merits of one's approach, and to neglect or play down, particularly in the early stages of a project, contrary evidence and other difficulties, may be a necessary condition for the success of many scientific projects. When people work very hard on something over a long period of time, they tend to become committed or attached to it; they strongly want it to be correct and find it increasingly difficult to envision the possibility that it might be false, a phenomenon related to what psychologists call *belief-perseverance*. Moreover, scientists like other people like to be right and to get credit and recognition from others for being right: The satisfaction of demolishing a theory one has laboriously constructed may be small in comparison with the satisfaction of seeing it vindicated. All things considered, it is extremely hard for most people to adopt a consistently Popperian attitude toward their own ideas.

Given these realistic observations about the psychology of scientists, an implicit code of conduct that encourages scientists to be a bit dogmatic and permits a certain measure of rhetorical exaggeration regarding the merits of their work, and that does not require an exhaustive discussion of its deficiencies, may be perfectly sensible. In many areas of science, if a scientist submits a paper that describes all of the various ways in which an idea or result might be defective, and draws detailed attention to the contrary results obtained by others, the paper is likely to be rejected. In fact, part of the intellectual responsibility of a scientist is to provide the best possible case for important ideas, leaving it to others to publicize their defects and limitations. Studies of both historical and contemporary science seem to show that this is just what most scientists do.

If this analysis is correct, there is a real danger that by following proposals (like that advocated by Hadley) to include within the category of "out-of-bounds conduct" behavior such as overinterpretation of data, exaggeration of available evidence that supports one's conclusion or failure to report contrary data, one may be proscribing behavior that plays a functional role in science and that, for reasons rooted deep in human psychology, will be hard to eliminate. Moreover, such proscriptions may be unnecessary, because interactions between scientists and criticisms by rivals may by themselves be sufficient to remove the bad consequences at which the proscriptions are aimed. Standards that might be optimal for single, perfectly rational beings encountering nature all by themselves may be radically deficient when applied to actual scientific communities.

Rewarding Useful Behavior

From a Popperian perspective, discovering evidence that merely supports a hypothesis is easy to do and has little methodological value; therefore one might think it doesn't deserve much credit. It is striking that the actual distribution of reward and credit in science reflects a very different view. Scientists receive Nobel prizes for finding new effects predicted by theories or for proposing important theories that are subsequently verified. It is only when a hypothesis or theory has become very well established that one receives significant credit for refuting it. Unquestionably, rewarding confirmations over refutations provides scientists with incentives to confirm theories rather than refute them and thus discourages giving up too quickly in the face of contrary experimental results. But, as we have been arguing, this is not necessarily bad for science.

Conventional accounts of scientific method (of which there are many examples in the philosophical literature) share the implicit assumption that all scientists in a community should adopt the same strategies. In fact, a number of government agencies now have rules that define as scientific misconduct "practices that seriously deviate from those that are commonly accepted within the scientific community…" (see principle 7). But rapid progress will be more likely if different scientists have quite different attitudes toward appropriate methodology. As noted above, one important consequence of the winner-takes-all (or nearly all) system by which credit and reward are allocated in science is that it encourages a variety of research programs and approaches. Other features of human cognitive psychology—such as the belief-perseverance phenomenon described above—probably have a similar consequence. It follows that attempts to characterize misconduct in terms of departures from practices or methods commonly accepted within the scientific community will be doubly misguided: Not only will such commonly accepted practices fail to exist in many cases, but it will be undesirable to try to enforce the uniformity of practice that such a characterization of misconduct would require. More generally, we can see why the classical methodologists have failed to discover "the" method by which science works. There are deep, systematic reasons why all scientists should not follow some single uniform method.

Our remarks so far have emphasized the undesirability of a set of rules that demand that all scientists believe the same things or behave in the same way, given a common body of evidence. This is not to say, however, that "anything goes." One very important distinction has to do with the difference between claims and behavior that are open to public assess-

A certain tendency to exaggerate the merits of one's approach, and to neglect or play down contrary evidence and other difficulties, may be a necessary condition for the success of many scientific projects.

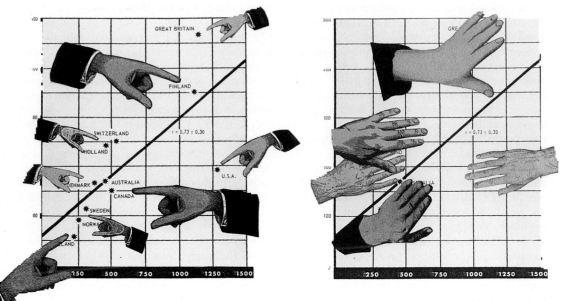

Standards that might be optimal for single, perfectly rational beings encountering nature all by themselves may be radically deficient when applied to actual scientific communities.

ment and those that are not. Exaggerations, omissions and misrepresentations that cannot be checked by other scientists should be regarded much more harshly than those that can, because they subvert the processes of public assessment and intellectual competition on which science rests. Thus, for example, a scientist who fabricates data must be judged far more harshly than one who does a series of experiments and accurately records the results but then extrapolates beyond the recorded data or insists on fitting some favored function to them. The difference is the fact that in the case in which there is no fabrication nothing has been done to obstruct the critical scrutiny of the work by peers; they can look at the data themselves and decide whether there is support for the conclusions. By contrast, other scientists will not be able to examine firsthand the process by which the data have been produced. They must take it on trust that the data resulted from an experiment of the sort described. Fabrication should thus be viewed as much more potentially damaging to the process of inquiry and should be more harshly punished than other forms of misrepresentation.

Science as Craft

Contemporary scientific knowledge is so vast and complex that even a very talented and hardworking scientist will be able to master only an extremely small fragment well enough to expect to make contributions to it. In part for this reason scientists must rely heavily on the authority of other scientists who are experts in domains in which they are not. A striking example of this is provided by the sociologist Trevor Pinch's recent book, *Confronting Nature*, which is a study of a series of experiments that discovered in the solar-neutrino flux far fewer neutrinos than seemed to be predicted by accepted theory. Pinch found that what

he called the "personal warrant" of the experimenters involved in this project played a large role in how other scientists assessed the experimental results. According to Pinch, other scientists often place at least as much weight on an experimentalist's general reputation for careful, painstaking work as on the technical details of the experiment in assessing whether the data constitute reliable evidence.

One reason why such appeals to personal warrant play a large role in science has to do with the specialized character of scientific knowledge. There is, however, another related reason which is of considerable importance in understanding how science works and how one should think about misconduct. This has to do with the fact that science in general—and especially experimentation—has a large "skill" or "craft" component.

Conducting an experiment in a way that produces reliable results is not a matter of following algorithmic rules that specify exactly what is to be done at each step. As Pinch put it, experimenters possess skills that "often enable the experimenter to get the apparatus to work without being able to formulate exactly or completely what has been done." For the same reason, assessing whether another investigator has produced reliable results requires a judgment of whether the experimenter has demonstrated the necessary skills in the past. These facts about the role of craft knowledge may be another reason why the general rules of method sought by the classical methodologists have proved so elusive.

The importance of craft in science is supported by empirical studies. For example, in a well-known study, Harry Collins investigated a number of experimental groups working in Britain to recreate a new kind of laser that had been successfully constructed elsewhere. Collins found that no group was able to reproduce a working laser simply on the basis of detailed written instructions. By far the most reliable method was to have someone from the original laboratory who had actually built a functioning laser go to the other laboratories and participate in the construction. The skills needed to make a working device could be acquired by practice *"without necessarily formulating, enumerating or understanding them."* Remarks on experimental work by working scientists themselves often express similar claims, not withstanding principle 8 above. If claims of this sort are correct, it often will be very difficult for those who lack highly specific skills and knowledge to assess a particular line of experimental work. A better strategy may be to be guided at least in large part by the experimenter's general reputation for reliability.

These facts about specialization, skill and authority have a number of interesting consequences for understanding what is proper scientific conduct. For example, a substantial amount of conduct that may look to an outsider like nonrational deference to authority may have a serious epistemological rationale. When an experimentalist discards certain data on the basis of subtle clues in the behavior of the apparatus, and other scientists accept the experimentalist's judgment in this matter, we should not automatically attribute this to the operation of power relationships, as is implied by principle 9 in our list above.

A second important consequence has to do with the responsibility of scientists for the misconduct or sloppy research practices of collaborators. It is sometimes suggested that authors should not sign their names to joint papers unless they have personally examined the evidence and are prepared to vouch for the correctness of every claim in the paper (principle 10). However, many collaborations bring together scientists from quite different specializations who lack the expertise to evaluate one another's work directly. This is exactly why collaboration is necessary. Requiring that scientists not collaborate unless they are able to check the work of collaborators or setting up a general policy of holding scientists responsible for the misconduct of coauthors would discourage a great deal of valuable collaboration.

Understanding the social structure of science and the operation of the reward system within science also has important ethical implications. We consider three examples: the Matthew effect, the Ortega hypothesis and scientific publication.

Matthew *vs.* Ortega

The sociologist Robert K. Merton has observed that credit tends to go to those who are already famous, at the expense of those who are not. A paper signed by Nobody, No-

body and Somebody often will be casually referred to as "work done in Somebody's lab," and even sometimes cited (incorrectly) in the literature as due to "Somebody *et al.*" Does this practice serve and accurately depict science, or does the tendency to elitism distort and undermine the conduct of science?

It is arguable that what Merton called the Matthew Effect plays a useful role in the organization of science; there are so many papers in so many journals that no scientist has time to read more than a tiny fraction of those in even a restricted area of science. Famous names tend to identify those works that are more likely to be worth noticing. In certain fields, particularly biomedical fields, it has become customary to make the head of the laboratory a coauthor, even if the head did not participate in the research. One reason for this practice is that by including the name of the famous head on the paper, chances are greatly improved that the paper will be accepted by a prestigious journal and noticed by its readers. Some people refer to this practice as "guest authorship" and regard it as unethical (as would be implied by principle 11 above). However, the practice may be functionally useful and may involve little deception, since conventions regarding authorship may be well understood by those who participate in a given area of science.

"In the cathedral of science," a famous scientist once said, "every brick is equally important." The remark (heard by one of the authors at a gathering at the speaker's Pasadena, California home) evokes a vivid metaphor of swarms of scientific workers under the guidance, perhaps, of a few master builders erecting a grand monument to scientific faith. The speaker was Max Delbrück, a Nobel laureate often called the father of molecular biology. The remark captures with some precision the scientists' ambivalent view of their craft. Delbrück never for an instant thought the bricks

A substantial amount of conduct that may look to an outsider like nonrational deference to authority may have a serious epistemological rationale.

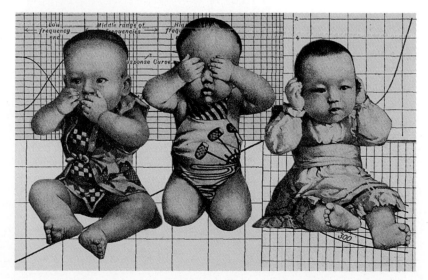

If the elitist view is right, then since science is largely financed by the public purse, it is best for science and best for society to restrict our production to fewer and better scientists.

he laid were no better than anyone else's. If anything, he regarded himself as the keeper of the blueprints, and he had the fame and prestige to prove it. It was exactly his exalted position that made it obligatory that he make a ceremonial bow to the democratic ideal that many scientists espouse and few believe. In fact it is precisely the kind of recognition that Delbrück enjoyed that propels the scientific enterprise forward.

The view expressed by Delbrück has been called the Ortega Hypothesis. It is named after Jose Ortega y Gasset, who wrote in his classic book, *The Revolt of the Masses*, that

> ...it is necessary to insist upon this extraordinary but undeniable fact: experimental science has progressed thanks in great part to the work of men astoundingly mediocre. That is to say, modern science, the root and symbol of our actual civilization, finds a place for the intellectually commonplace man and allows him to work therein with success. In this way the majority of scientists help the general advance of science while shut up in the narrow cell of their laboratory like the bee in the cell of its hive, or the turnspit of its wheel.

This view (see principle 12) is probably based on the empirical observation that there are indeed, in each field of science, many ordinary scientists doing more or less routine work. It is also supported by the theoretical view that knowledge of the universe is a kind

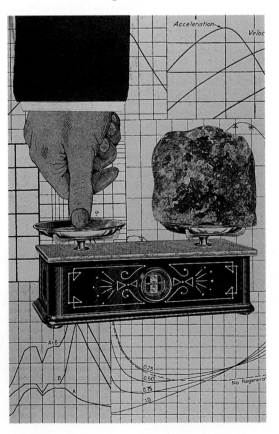

of limitless wilderness to be conquered by relentless hacking away of underbrush by many hands. An idea that is supported by both theory and observation always has a very firm standing in science.

The Ortega Hypothesis was named by Jonathan and Steven Cole when they set out to demolish it, an objective they pursued by tracing citations in physics journals. They concluded that the hypothesis is incorrect, stating:

> It seems, rather, that a relatively small number of physicists produce work that becomes the base for future discoveries in physics. We have found that even papers of relatively minor significance have used to a disproportionate degree the work of the eminent scientists....

In other words, a small number of elite scientists produce the vast majority of scientific progress. Seen in this light, the reward system in science is a mechanism evolved for the purpose of identifying, promoting and rewarding the star performers.

One's view of the Ortega Hypothesis has important implications concerning how science ought to be organized. If the Ortega Hypothesis is correct, science is best served by producing as many scientists as possible, even if they are not all of the highest quality (principle 13). On the other hand, if the elitist view is right, then since science is largely financed by the public purse, it is best for science and best for society to restrict our production to fewer and better scientists. In any case, the question of whether to produce more or fewer scientists involves ethical issues (what is best for the common good?) as well as policy issues (how to reach the desired goal).

Peers and Publication

In a classic paper called "Is the scientific paper a fraud?" Peter Medawar has argued that typical experimental papers intentionally misrepresent the actual sequence of events involved in the conduct of an experiment, the process of reasoning by which the experimenter reached various conclusions and so on. In general, experimentalists will make it look as if they had a much clearer idea of the ultimate result than was actually the case. Misunderstandings, blind alleys and mistakes of various sorts will fail to appear in the final written account.

Papers written this way are undoubtedly deceptive, at least to the uninitiated, and they certainly stand in contrast to Feynman's exhortation to "lean over backward." They also violate principle 14. Nevertheless, the practice is virtually universal, because it is a much more efficient means of transmitting results than an accurate historical account of the scientist's activities would be. Thus it is a simple fact that, contrary to normal belief, there are types of misrepresentation that are condoned and accepted in scientific publications, whereas other types are harshly condemned.

Nevertheless, scientific papers have an exalted reputation for integrity. That may be because the integrity of the scientific record is protected, above all, by the institution of peer review. Peer review has an almost mystical role in the community of scientists. Published results are considered dependable because they have been reviewed by peers, and unpublished data are considered not dependable because they have not been. Many regard peer review to be (as principle 15 would suggest) the ethical fulcrum of the whole scientific enterprise.

Peer review is used to help determine whether journals should publish articles submitted to them, and whether agencies should grant financial support to research projects. For most small projects and for nearly all journal articles, peer review is accomplished by sending the manuscript or proposal to referees whose identities will not be revealed to the authors.

Peer review conducted in this way is extremely unlikely to detect instances of intentional misconduct. But the process is very good at separating valid science from nonsense. Referees know the current thinking in a field, are aware of its laws, rules and conventions, and will quickly detect any unjustified attempt to depart from them. Of course, for precisely this reason peer review can occasionally delay a truly visionary or revolutionary idea, but that may be a price that we pay for conducting science in an orderly way.

Peer review is less useful for adjudicating an intense competition for scarce resources. The pages of prestigious journals and the funds distributed by government agencies have become very scarce resources in recent times. The fundamental problem in using peer review to decide how these resources are to be allocated is obvious enough: There is an intrinsic conflict of interest. The referees, chosen because they are among the few experts in the author's field, are often competitors with the author for those same resources.

A referee who receives a proposal or manuscript to judge is being asked to do an unpaid professional service for the editor or project officer. The editor or officer thus has a responsibility to protect the referee, both by protecting the referee's anonymity and by making sure that the referee is never held to account for what is written in the report. Without complete confidence in that protection, referees cannot be expected to perform their task. Moreover, editors and project officers are never held to account for their choice of referees, and they can be confident that, should anybody ask, their referees will have the proper credentials to withstand scrutiny.

Referees would have to have high ethical standards to fail to take personal advantage of

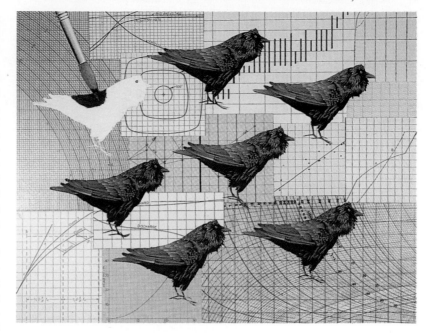

their privileged anonymity and to make peer review function properly in spite of these conditions. Undoubtedly, most referees in most circumstances do manage to accomplish that. However, the fact is that many referees have themselves been victims of unfair reviews and this must sometimes influence their ability to judge competing proposals or papers fairly. Thus the institution of peer review seems to be suffering genuine distress.

Once again, this analysis shows that science is a complex enterprise that must be understood in some detail before ethical principles can be formulated to help guide it.

Conclusions

We have put forth arguments in this article that indicate why each of the principles listed above may be defective as a guide to the behavior of scientists. However, our repeated admonition that there are no universal rules of scientific conduct does not mean that it is impossible to recognize distinctive scientific misconduct. We would like to conclude with some thoughts on how scientific misconduct might be distinguished from other kinds of misconduct.

We propose that distinctively scientific forms of misconduct are those that require the expert judgment of a panel of scientists in order to be understood and assessed. Other forms of misconduct may take place in science, but they should not constitute scientific misconduct. For example, fabricating experimental data is scientific misconduct, but stealing scientific instruments is not. Similarly, misappropriation of scientific ideas is scientific misconduct, but plagiarism (copying someone else's words) is not. Stealing and plagiarism are serious misdeeds, but there are other well-established means for dealing with them, even

There are types of misrepresentation that are condoned and accepted in scientific publications, whereas other types are harshly condemned.

Fabrication or covert and unwarranted manipulation of data is an example of the kind of deceptive practice that cannot be tolerated because it undermines the mutual trust essential to the system of science.

when they are associated with science or committed by scientists. No special knowledge is required to recognize them.

On the other hand, only a panel of scientists can deal with matters such as data fabrication that require a detailed understanding of the nature of the experiments, the instruments used, accepted norms for presenting data and so on, to say nothing of the unique importance of experimental data in science. In a dispute over an allegation that a scientific idea has been misappropriated, the issues are likely to be so complex that it is difficult to imagine a lay judge or jury coming to understand the problem from testimony by expert witnesses or any other plausible means. Similarly, expert judgment will usually be required to determine whether an experimenter's procedures in selecting or discarding data constitute misconduct—the conventions governing this vary so much across different areas of science that judgments about what is reasonable will require a great deal of expert knowledge, rather than simply the application of some general rule that might be employed by nonscientists.

In the section on the logical structure of science, we drew a sharp distinction between advocacy, which is permitted or encouraged in science, and deception that is not open to public assessment, which is judged very harshly in science. Fabrication or covert and unwarranted manipulation of data is an example of the kind of deceptive practice that cannot be tolerated because it undermines the mutual trust essential to the system of science that we have described. Similarly, misappropriation of ideas undermines the reward system that helps motivate scientific progress. In both cases, a panel of scientists will be required to determine whether the deed occurred and, if so, whether it was done with intent to deceive or with reckless disregard for the

truth. Should these latter conditions be true, the act may be judged to be not merely scientific misconduct but, in fact, scientific fraud.

Acknowledgment

The authors wish to thank Kathy Cooke, Ph.D., for her valuable assistance in thinking through the problems discussed in this article.

Bibliography

Cole, J., and S. Cole. 1972. The Ortega hypothesis. *Science* 178:368–375.

Collins, H. 1974. The TEA-set: Tacit knowledge and scientific networks. *Science Studies* 4:165–186.

Collins, H. 1975. The seven sexes: A study in the sociology of a phenomenon, or the replication of experiments in physics. *Sociology* 9:205–224.

Duhem, P. 1962. *The Aim and Structure of Physical Theory.* New York: Athenaeum.

Feyerabend, P. 1975. *Against Method: Outline of an Anarchistic Theory of Knowledge.* London: New Left Books.

Feynman, R. 1985. *Surely You're Joking, Mr. Feynman.* New York: W. W. Norton.

Galison, P. 1987. *How Experiments End.* Chicago: University of Chicago Press.

Hadley, S. "Can Science Survive Scientific Integrity?" Paper presented on October 17, 1991, at the University of California, San Diego. Unpublished.

Kitcher, P. 1990. The division of cognitive labor. *Journal of Philosophy* 87:5–22.

Koertge, N. 1991. The function of credit in Hull's evolutionary model of science. *Proceedings of the Philosophy of Science Association* 2:237–244.

Kuhn, T. 1970. *The Structure of Scientific Revolutions.* Chicago: University of Chicago Press.

Lakatos, I. 1974. Falsification and the methodology of scientific research programmes. In *Criticism and the Growth of Knowledge,* ed. I. Lakatos and A. Musgrave. Cambridge, U.K.: Cambridge University Press. Pp. 91–196.

Medawar, P. 1963. Is the scientific paper a fraud? *The Listener* (12 September) 70:377–378.

Merton, R. 1968. The Matthew Effect in science. *Science* 159:56–63.

Ortega y Gasset, J. 1932. *The Revolt of the Masses.* New York: W. W. Norton.

Pinch, T. 1986. *Confronting Nature.* Dordrecht, Holland: P. Reidel.

Popper, K. 1968. *The Logic of Scientific Discovery.* New York: Harper and Row.

Popper, K. 1969. *Conjectures and Refutations.* London: Routledge and Kegan Paul.

Quine, W. V. O. 1961. *From a Logical Point of View.* New York: Harper and Row.

The Science of Scientific Writing

If the reader is to grasp what the writer means,
the writer must understand what the reader needs

George D. Gopen and Judith A. Swan

Science is often hard to read. Most people assume that its difficulties are born out of necessity, out of the extreme complexity of scientific concepts, data and analysis. We argue here that complexity of thought need not lead to impenetrability of expression; we demonstrate a number of rhetorical principles that can produce clarity in communication without oversimplifying scientific issues. The results are substantive, not merely cosmetic: Improving the quality of writing actually improves the quality of thought.

The fundamental purpose of scientific discourse is not the mere presentation of information and thought, but rather its actual communication. It does not matter how pleased an author might be to have converted all the right data into sentences and paragraphs; it matters only whether a large majority of the reading audience accurately perceives what the author had in mind. Therefore, in order to understand how best to improve writing, we would do well to understand better how readers go about reading. Such an understanding has recently become available through work done in the fields of rhetoric, linguistics and cognitive psychology. It has helped to produce a methodology based on the concept of reader expectations.

Writing with the Reader in Mind: Expectation and Context
Readers do not simply read; they interpret. Any piece of prose, no matter how short, may "mean" in 10 (or more) different ways to 10 different readers. This methodology of reader expectations is founded on the recognition that readers make many of their most important interpretive decisions about the substance of prose based on clues they receive from its structure.

This interplay between substance and structure can be demonstrated by something as basic as a simple table. Let us say that in tracking the temperature of a liquid over a period of time, an investigator takes measurements every

George D. Gopen is associate professor of English and Director of Writing Programs at Duke University. He holds a Ph.D. in English from Harvard University and a J.D. from Harvard Law School. Judith A. Swan teaches scientific writing at Princeton University. Her Ph.D., which is in biochemistry, was earned at the Massachusetts Institute of Technology. Address for Gopen: 307 Allen Building, Duke University, Durham, NC 27706.

three minutes and records a list of temperatures. Those data could be presented by a number of written structures. Here are two possibilities:

t (time) = 15', T (temperature) = 32°; $t = 0'$, $T = 25°$; $t = 6'$, $T = 29°$; $t = 3'$, $T = 27°$; $t = 12'$, $T = 32°$; $t = 9'$, $T = 31°$

time (min)	temperature (°C)
0	25
3	27
6	29
9	31
12	32
15	32

Precisely the same information appears in both formats, yet most readers find the second easier to interpret. It may be that the very familiarity of the tabular structure makes it easier to use. But, more significantly, the structure of the second table provides the reader with an easily perceived context (time) in which the significant piece of information (temperature) can be interpreted. The contextual material appears on the left in a pattern that produces an expectation of regularity; the interesting results appear on the right in a less obvious pattern, the discovery of which is the point of the table.

If the two sides of this simple table are reversed, it becomes much harder to read.

temperature (°C)	time (min)
25	0
27	3
29	6
31	9
32	12
32	15

Since we read from left to right, we prefer the context on the left, where it can more effectively familiarize the reader. We prefer the new, important information on the right, since its job is to intrigue the reader.

Information is interpreted more easily and more uniformly if it is placed where most readers expect to find it. These needs and expectations of readers affect the interpretation not only of tables and illustrations but also of prose itself. Readers have relatively fixed expectations about where in the structure of prose they will encounter particular items of its substance. If writers can become consciously aware of these locations, they can better control the degrees of recognition and emphasis a reader will give to the various pieces of information being presented. Good writers are intuitively aware of these expectations; that is why their prose has what we call "shape."

This underlying concept of reader expectation is perhaps most immediately evident at the level of the largest units of discourse. (A unit of discourse is defined as anything with a beginning and an end: a clause, a sentence, a section, an article, etc.) A research article, for example, is generally divided into recognizable sections, sometimes labeled Introduction, Experimental Methods, Results and Discussion. When the sections are confused—when too much experimental detail is found in the Results section, or when discussion and results intermingle—readers are often equally confused. In smaller units of discourse the functional divisions are not so explicitly labeled, but readers have definite expectations all the same, and they search for certain information in particular places. If these structural expectations are continually violated, readers are forced to divert energy from understanding the content of a passage to unraveling its structure. As the complexity of the content increases moderately, the possibility of misinterpretation or noninterpretation increases dramatically.

We present here some results of applying this methodology to research reports in the scientific literature. We have taken several passages from research articles (either published or accepted for publication) and have suggested ways of rewriting them by applying principles derived from the study of reader expectations. We have not sought to transform the passages into "plain English" for the use of the general public; we have neither decreased the jargon nor diluted the science. We have striven not for simplification but for clarification.

Reader Expectations for the Structure of Prose

Here is our first example of scientific prose, in its original form:

The smallest of the URF's (URFA6L), a 207-nucleotide (nt) reading frame overlapping out of phase the NH_2-terminal portion of the adenosinetriphosphatase (ATPase) subunit 6 gene has been identified as the animal equivalent of the recently discovered yeast H^+-ATPase subunit 8 gene. The functional significance of the other URF's has been, on the contrary, elusive. Recently, however, immunoprecipitation experiments with antibodies to purified, rotenone-sensitive NADH-ubiquinone oxido-reductase [hereafter referred to as respiratory chain NADH dehydrogenase or complex I] from bovine heart, as well as enzyme fractionation studies, have indicated that six human URF's (that is, URF1, URF2, URF3, URF4, URF4L, and URF5, hereafter referred to as ND1, ND2, ND3, ND4, ND4L, and ND5) encode subunits of complex I. This is a large complex that also contains many subunits synthesized in the cytoplasm.*

Ask any ten people why this paragraph is hard to read, and nine are sure to mention the technical vocabulary; sev-

eral will also suggest that it requires specialized background knowledge. Those problems turn out to be only a small part of the difficulty. Here is the passage again, with the difficult words temporarily lifted:

The smallest of the URF's, an [A], has been identified as a [B] subunit 8 gene. The functional significance of the other URF's has been, on the contrary, elusive. Recently, however, [C] experiments, as well as [D] studies, have indicated that six human URF's [1-6] encode subunits of Complex I. This is a large complex that also contains many subunits synthesized in the cytoplasm.

It may now be easier to survive the journey through the prose, but the passage is still difficult. Any number of questions present themselves: What has the first sentence of the passage to do with the last sentence? Does the third sentence contradict what we have been told in the second

---

*I*nformation is interpreted more easily and more uniformly if it is placed where most readers expect to find it.

---

sentence? Is the functional significance of URF's still "elusive"? Will this passage lead us to further discussion about URF's, or about Complex I, or both?

Knowing a little about the subject matter does not clear up all the confusion. The intended audience of this passage would probably possess at least two items of essential technical information: first, "URF" stands for "Uninterrupted Reading Frame," which describes a segment of DNA organized in such a way that it could encode a protein, although no such protein product has yet been identified; second, both ATPase and NADH oxido-reductase are enzyme complexes central to energy metabolism. Although this information may provide some sense of comfort, it does little to answer the interpretive questions that need answering. It seems the reader is hindered by more than just the scientific jargon.

To get at the problem, we need to articulate something about how readers go about reading. We proceed to the first of several reader expectations.

Subject-Verb Separation

Look again at the first sentence of the passage cited above. It is relatively long, 42 words; but that turns out not to be the main cause of its burdensome complexity. Long sentences need not be difficult to read; they are only difficult to write. We have seen sentences of over 100 words that flow

---

*The full paragraph includes one more sentence: "Support for such functional identification of the URF products has come from the finding that the purified rotenone-sensitive NADH dehydrogenase from *Neurospora crassa* contains several subunits synthesized within the mitochondria, and from the observation that the stopper mutant of *Neurospora crassa*, whose mtDNA lacks two genes homologous to URF2 and URF3, has no functional complex I." We have omitted this sentence both because the passage is long enough as is and because it raises no additional structural issues.

easily and persuasively toward their clearly demarcated destination. Those well-wrought serpents all had something in common: Their structure presented information to readers in the order the readers needed and expected it.

The first sentence of our example passage does just the opposite: it burdens and obstructs the reader, because of an all-too-common structural defect. Note that the grammatical subject ("the smallest") is separated from its verb ("has been identified") by 23 words, more than half the sentence.

---

*B*eginning with the exciting material and ending with a lack of luster often leaves us disappointed and destroys our sense of momentum.

Readers expect a grammatical subject to be followed immediately by the verb. Anything of length that intervenes between subject and verb is read as an interruption, and therefore as something of lesser importance.

The reader's expectation stems from a pressing need for syntactic resolution, fulfilled only by the arrival of the verb. Without the verb, we do not know what the subject is doing, or what the sentence is all about. As a result, the reader focuses attention on the arrival of the verb and resists recognizing anything in the interrupting material as being of primary importance. The longer the interruption lasts, the more likely it becomes that the "interruptive" material actually contains important information; but its structural location will continue to brand it as merely interruptive. Unfortunately, the reader will not discover its true value until too late—until the sentence has ended without having produced anything of much value outside of that subject-verb interruption.

In this first sentence of the paragraph, the relative importance of the intervening material is difficult to evaluate. The material might conceivably be quite significant, in which case the writer should have positioned it to reveal that importance. Here is one way to incorporate it into the sentence structure:

> The smallest of the URF's is URFA6L, a 207-nucleotide (nt) reading frame overlapping out of phase the NH$_2$-terminal portion of the adenosinetriphosphatase (ATPase) subunit 6 gene; it has been identified as the animal equivalent of the recently discovered yeast H$^+$-ATPase subunit 8 gene.

On the other hand, the intervening material might be a mere aside that diverts attention from more important ideas; in that case the writer should have deleted it, allowing the prose to drive more directly toward its significant point:

> The smallest of the URF's (URFA6L) has been identified as the animal equivalent of the recently discovered yeast H$^+$-ATPase subunit 8 gene.

Only the author could tell us which of these revisions more accurately reflects his intentions.

These revisions lead us to a second set of reader expectations. Each unit of discourse, no matter what the size, is expected to serve a single function, to make a single point. In the case of a sentence, the point is expected to appear in a specific place reserved for emphasis.

The Stress Position

It is a linguistic commonplace that readers naturally emphasize the material that arrives at the end of a sentence. We refer to that location as a "stress position." If a writer is consciously aware of this tendency, she can arrange for the emphatic information to appear at the moment the reader is naturally exerting the greatest reading emphasis. As a result, the chances greatly increase that reader and writer will perceive the same material as being worthy of primary emphasis. The very structure of the sentence thus helps persuade the reader of the relative values of the sentence's contents.

The inclination to direct more energy to that which arrives last in a sentence seems to correspond to the way we work at tasks through time. We tend to take something like a "mental breath" as we begin to read each new sentence, thereby summoning the tension with which we pay attention to the unfolding of the syntax. As we recognize that the sentence is drawing toward its conclusion, we begin to exhale that mental breath. The exhalation produces a sense of emphasis. Moreover, we delight in being rewarded at the end of a labor with something that makes the ongoing effort worthwhile. Beginning with the exciting material and ending with a lack of luster often leaves us disappointed and destroys our sense of momentum. We do not start with the strawberry shortcake and work our way up to the broccoli.

When the writer puts the emphatic material of a sentence in any place other than the stress position, one of two things can happen; both are bad. First, the reader might find the stress position occupied by material that clearly is not worthy of emphasis. In this case, the reader must discern, without any additional structural clue, what else in the sentence may be the most likely candidate for emphasis. There are no secondary structural indications to fall back upon. In sentences that are long, dense or sophisticated, chances soar that the reader will not interpret the prose precisely as the writer intended. The second possibility is even worse: The reader may find the stress position occupied by something that does appear capable of receiving emphasis, even though the writer did not intend to give it any stress. In that case, the reader is highly likely to emphasize this imposter material, and the writer will have lost an important opportunity to influence the reader's interpretive process.

The stress position can change in size from sentence to sentence. Sometimes it consists of a single word; sometimes it extends to several lines. The definitive factor is this: The stress position coincides with the moment of syntactic closure. A reader has reached the beginning of the stress position when she knows there is nothing left in the clause or sentence but the material presently being read. Thus a whole list, numbered and indented, can occupy the stress position of a sentence if it has been clearly announced as being all that remains of that sentence. Each member of that list, in turn, may have its own internal stress position, since each member may produce its own syntactic closure.

Within a sentence, secondary stress positions can be

formed by the appearance of a properly used colon or semicolon; by grammatical convention, the material preceding these punctuation marks must be able to stand by itself as a complete sentence. Thus, sentences can be extended effortlessly to dozens of words, as long as there is a medial syntactic closure for every piece of new, stress-worthy information along the way. One of our revisions of the initial sentence can serve as an example:

> The smallest of the URF's is URFA6L, a 207-nucleotide (nt) reading frame overlapping out of phase the NH$_2$-terminal portion of the adenosinetriphosphatase (ATPase) subunit 6 gene; it has been identified as the animal equivalent of the recently discovered yeast H$^+$-ATPase subunit 8 gene.

By using a semicolon, we created a second stress position to accommodate a second piece of information that seemed to require emphasis.

We now have three rhetorical principles based on reader expectations: First, grammatical subjects should be followed as soon as possible by their verbs; second, every unit of discourse, no matter the size, should serve a single function or make a single point; and, third, information intended to be emphasized should appear at points of syntactic closure. Using these principles, we can begin to unravel the problems of our example prose.

Note the subject-verb separation in the 62-word third sentence of the original passage:

> Recently, however, immunoprecipitation experiments with antibodies to purified, rotenone-sensitive NADH-ubiquinone oxido-reductase [hereafter referred to as respiratory chain NADH dehydrogenase or complex I] from bovine heart, as well as enzyme fractionation studies, have indicated that six human URF's (that is, URF1, URF2, URF3, URF4, URF4L, and URF5, hereafter referred to as ND1, ND2, ND3, ND4, ND4L, and ND5) encode subunits of complex I.

After encountering the subject ("experiments"), the reader must wade through 27 words (including three hyphenated compound words, a parenthetical interruption and an "as well as" phrase) before alighting on the highly uninformative and disappointingly anticlimactic verb ("have indicated"). Without a moment to recover, the reader is handed a "that" clause in which the new subject ("six human URF's") is separated from its verb ("encode") by yet another 20 words.

If we applied the three principles we have developed to the rest of the sentences of the example, we could generate a great many revised versions of each. These revisions might differ significantly from one another in the way their structures indicate to the reader the various weights and balances to be given to the information. Had the author placed all stress-worthy material in stress positions, we as a reading community would have been far more likely to interpret these sentences uniformly.

We couch this discussion in terms of "likelihood" because we believe that meaning is not inherent in discourse by itself; "meaning" requires the combined participation of text and reader. All sentences are infinitely interpretable, given an infinite number of interpreters. As communities of readers, however, we tend to work out tacit agreements as to what kinds of meaning are most likely to be extracted

We cannot succeed in making even a single sentence mean one and only one thing; we can only increase the odds that a large majority of readers will tend to interpret our discourse according to our intentions.

from certain articulations. We cannot succeed in making even a single sentence mean one and only one thing; we can only increase the odds that a large majority of readers will tend to interpret our discourse according to our intentions. Such success will follow from authors becoming more consciously aware of the various reader expectations presented here.

Here is one set of revisionary decisions we made for the example:

> The smallest of the URF's, URFA6L, has been identified as the animal equivalent of the recently discovered yeast H$^+$-ATPase subunit 8 gene; but the functional significance of other URF's has been more elusive. Recently, however, several human URF's have been shown to encode subunits of rotenone-sensitive NADH-ubiquinone oxido-reductase. This is a large complex that also contains many subunits synthesized in the cytoplasm; it will be referred to hereafter as respiratory chain NADH dehydrogenase or complex I. Six subunits of Complex I were shown by enzyme fractionation studies and immunoprecipitation experiments to be encoded by six human URF's (URF1, URF2, URF3, URF4, URF4L, and URF5); these URF's will be referred to subsequently as ND1, ND2, ND3, ND4, ND4L, and ND5.

Sheer length was neither the problem nor the solution. The revised version is not noticeably shorter than the original; nevertheless, it is significantly easier to interpret. We have indeed deleted certain words, but not on the basis of wordiness or excess length. (See especially the last sentence of our revision.)

When is a sentence too long? The creators of readability formulas would have us believe there exists some fixed number of words (the favorite is 29) past which a sentence is too hard to read. We disagree. We have seen 10-word sentences that are virtually impenetrable and, as we mentioned above, 100-word sentences that flow effortlessly to their points of resolution. In place of the word-limit concept, we offer the following definition: A sentence is too long when it has more viable candidates for stress positions than there are stress positions available. Without the stress position's locational clue that its material is intended to be emphasized, readers are left too much to their own devices in deciding just what else in a sentence might be considered important.

In revising the example passage, we made certain decisions about what to omit and what to emphasize. We put subjects and verbs together to lessen the reader's syntactic burdens; we put the material we believed worthy of emphasis in stress positions; and we discarded material for which we could not discern significant connections. In doing so, we have produced a clearer passage—but not one that necessarily reflects the author's intentions; it reflects only our interpretation of the author's intentions. The more problematic the structure, the less likely it becomes that a grand majority of readers will perceive the discourse in exactly the way the author intended.

It is probable that many of our readers—and perhaps even the authors—will disagree with some of our choices. If so, that disagreement underscores our point: The origi-

---

*T*he information that begins a sentence establishes for the reader a perspective for viewing the sentence as a unit.

---

nal failed to communicate its ideas and their connections clearly. If we happened to have interpreted the passage as you did, then we can make a different point: No one should have to work as hard as we did to unearth the content of a single passage of this length.

The Topic Position

To summarize the principles connected with the stress position, we have the proverbial wisdom, "Save the best for last." To summarize the principles connected with the other end of the sentence, which we will call the topic position, we have its proverbial contradiction, "First things first." In the stress position the reader needs and expects closure and fulfillment; in the topic position the reader needs and expects perspective and context. With so much of reading comprehension affected by what shows up in the topic position, it behooves a writer to control what appears at the beginning of sentences with great care.

The information that begins a sentence establishes for the reader a perspective for viewing the sentence as a unit: Readers expect a unit of discourse to be a story about whoever shows up first. "Bees disperse pollen" and "Pollen is dispersed by bees" are two different but equally respectable sentences about the same facts. The first tells us something about bees; the second tells us something about pollen. The passivity of the second sentence does not by itself impair its quality; in fact, "Pollen is dispersed by bees" is the superior sentence if it appears in a paragraph that intends to tell us a continuing story about pollen. Pollen's story at that moment is a passive one.

Readers also expect the material occupying the topic position to provide them with linkage (looking backward) and context (looking forward). The information in the topic position prepares the reader for upcoming material by connecting it backward to the previous discussion. Although linkage and context can derive from several sources, they stem primarily from material that the reader has already encountered within this particular piece of discourse. We refer to this familiar, previously introduced ma-

terial as "old information." Conversely, material making its first appearance in a discourse is "new information." When new information is important enough to receive emphasis, it functions best in the stress position.

When old information consistently arrives in the topic position, it helps readers to construct the logical flow of the argument: It focuses attention on one particular strand of the discussion, both harkening backward and leaning forward. In contrast, if the topic position is constantly occupied by material that fails to establish linkage and context, readers will have difficulty perceiving both the connection to the previous sentence and the projected role of the new sentence in the development of the paragraph as a whole.

Here is a second example of scientific prose that we shall attempt to improve in subsequent discussion:

> Large earthquakes along a given fault segment do not occur at random intervals because it takes time to accumulate the strain energy for the rupture. The rates at which tectonic plates move and accumulate strain at their boundaries are approximately uniform. Therefore, in first approximation, one may expect that large ruptures of the same fault segment will occur at approximately constant time intervals. If subsequent mainshocks have different amounts of slip across the fault, then the recurrence time may vary, and the basic idea of periodic mainshocks must be modified. For great plate boundary ruptures the length and slip often vary by a factor of 2. Along the southern segment of the San Andreas fault the recurrence interval is 145 years with variations of several decades. The smaller the standard deviation of the average recurrence interval, the more specific could be the long term prediction of a future mainshock.

This is the kind of passage that in subtle ways can make readers feel badly about themselves. The individual sentences give the impression of being intelligently fashioned: They are not especially long or convoluted; their vocabulary is appropriately professional but not beyond the ken of educated general readers; and they are free of grammatical and dictional errors. On first reading, however, many of us arrive at the paragraph's end without a clear sense of where we have been or where we are going. When that happens, we tend to berate ourselves for not having paid close enough attention. In reality, the fault lies not with us, but with the author.

We can distill the problem by looking closely at the information in each sentence's topic position:

> Large earthquakes
> The rates
> Therefore... one
> subsequent mainshocks
> great plate boundary ruptures
> the southern segment of the San Andreas fault
> the smaller the standard deviation...

Much of this information is making its first appearance in this paragraph—in precisely the spot where the reader looks for old, familiar information. As a result, the focus of the story constantly shifts. Given just the material in the topic positions, no two readers would be likely to construct exactly the same story for the paragraph as a whole.

If we try to piece together the relationship of each sentence to its neighbors, we notice that certain bits of old information keep reappearing. We hear a good deal about the recurrence time between earthquakes: The first sentence introduces the concept of nonrandom intervals between earthquakes; the second sentence tells us that recurrence rates due to the movement of tectonic plates are more or less uniform; the third sentence adds that the recurrence rate of major earthquakes should also be somewhat predictable; the fourth sentence adds that recurrence rates vary with some conditions; the fifth sentence adds information about one particular variation; the sixth sentence adds a recurrence-rate example from California; and the last sentence tells us something about how recurrence rates can be described statistically. This refrain of "recurrence intervals" constitutes the major string of old information in the paragraph. Unfortunately, it rarely appears at the beginning of sentences, where it would help us maintain our focus on its continuing story.

In reading, as in most experiences, we appreciate the opportunity to become familiar with a new environment before having to function in it. Writing that continually begins sentences with new information and ends with old information forbids both the sense of comfort and orientation at the start and the sense of fulfilling arrival at the end. It misleads the reader as to whose story is being told; it burdens the reader with new information that must be carried further into the sentence before it can be connected to the discussion; and it creates ambiguity as to which material the writer intended the reader to emphasize. All of these distractions require that readers expend a disproportionate amount of energy to unravel the structure of the prose, leaving less energy available for perceiving content.

We can begin to revise the example by ensuring the following for each sentence:

1. The backward-linking old information appears in the topic position.

2. The person, thing or concept whose story it is appears in the topic position.

3. The new, emphasis-worthy information appears in the stress position.

Once again, if our decisions concerning the relative values of specific information differ from yours, we can all blame the author, who failed to make his intentions apparent. Here first is a list of what we perceived to be the new, emphatic material in each sentence:

time to accumulate strain energy along a fault
approximately uniform
large ruptures of the same fault
different amounts of slip
vary by a factor of 2
variations of several decades
predictions of future mainshock

Now, based on these assumptions about what deserves stress, here is our proposed revision:

Large earthquakes along a given fault segment do not occur at random intervals because it takes time to accumulate the strain energy for the rupture. The rates at which tectonic plates move and accumulate strain at their boundaries are roughly uniform. Therefore, nearly constant time intervals (at first approximation) would be expected between large ruptures of the same fault segment. [However?], the recurrence time may vary; the basic idea of periodic mainshocks may need to be modi-

fied if subsequent mainshocks have different amounts of slip across the fault. [Indeed?], the length and slip of great plate boundary ruptures often vary by a factor of 2. [For example?], the recurrence interval along the southern segment of the San Andreas fault is 145 years with variations of several decades. The smaller the standard deviation of the average recurrence interval, the more specific could be the long term prediction of a future mainshock.

Many problems that had existed in the original have now surfaced for the first time. Is the reason earthquakes do not occur at random intervals stated in the first sentence or in the second? Are the suggested choices of "however," "indeed," and "for example" the right ones to express the connections at those points? (All these connections were left unarticulated in the original paragraph.) If "for example" is an inaccurate transitional phrase, then exactly how does the San Andreas fault example connect to ruptures that "vary by a factor of 2"? Is the author arguing that recurrence rates must vary because fault movements often vary? Or is the author preparing us for a discussion of how in spite of such variance we might still be able to predict earthquakes? This last question remains unanswered because the final sentence leaves behind earthquakes that recur at variable intervals and switches instead to earthquakes that recur regularly. Given that this is the first paragraph of the article, which type of earthquake

---

*I*n our experience, the misplacement of old and new information turns out to be the No. 1 problem in American professional writing today.

---

will the article most likely proceed to discuss? In sum, we are now aware of how much the paragraph had not communicated to us on first reading. We can see that most of our difficulty was owing not to any deficiency in our reading skills but rather to the author's lack of comprehension of our structural needs as readers.

In our experience, the misplacement of old and new information turns out to be the No. 1 problem in American professional writing today. The source of the problem is not hard to discover: Most writers produce prose linearly (from left to right) and through time. As they begin to formulate a sentence, often their primary anxiety is to capture the important new thought before it escapes. Quite naturally they rush to record that new information on paper, after which they can produce at their leisure the contextualizing material that links back to the previous discourse. Writers who do this consistently are attending more to their own need for unburdening themselves of their information than to the reader's need for receiving the material. The methodology of reader expectations articulates the reader's needs explicitly, thereby making writers consciously aware of structural problems and ways to solve them.

*P*ut in the topic position the old information that links backward; put in the stress position the new information you want the reader to emphasize.

A note of clarification: Many people hearing this structural advice tend to oversimplify it to the following rule: "Put the old information in the topic position and the new information in the stress position." No such rule is possible. Since by definition all information is either old or new, the space between the topic position and the stress position must also be filled with old and new information. Therefore the principle (not rule) should be stated as follows: "Put in the topic position the old information that links backward; put in the stress position the new information you want the reader to emphasize."

Perceiving Logical Gaps
When old information does not appear at all in a sentence, whether in the topic position or elsewhere, readers are left to construct the logical linkage by themselves. Often this happens when the connections are so clear in the writer's mind that they seem unnecessary to state; at those moments, writers underestimate the difficulties and ambiguities inherent in the reading process. Our third example attempts to demonstrate how paying attention to the placement of old and new information can reveal where a writer has neglected to articulate essential connections.

The enthalpy of hydrogen bond formation between the nucleoside bases 2'deoxyguanosine (dG) and 2'deoxycytidine (dC) has been determined by direct measurement. dG and dC were derivatized at the 5' and 3' hydroxyls with triisopropylsilyl groups to obtain solubility of the nucleosides in non-aqueous solvents and to prevent the ribose hydroxyls from forming hydrogen bonds. From isoperibolic titration measurements, the enthalpy of dC:dG base pair formation is –6.65 ± 0.32 kcal/mol.

Although part of the difficulty of reading this passage may stem from its abundance of specialized technical terms, a great deal more of the difficulty can be attributed to its structural problems. These problems are now familiar: We are not sure at all times whose story is being told; in the first sentence the subject and verb are widely separated; the second sentence has only one stress position but two or three pieces of information that are probably worthy of emphasis—"solubility... solvents," "prevent... from forming hydrogen bonds" and perhaps "triisopropylsilyl groups." These perceptions suggest the following revision tactics:

1. Invert the first sentence, so that (*a*) the subject-verb-complement connection is unbroken, and (*b*) "dG" and "dC" are introduced in the stress position as new and interesting information. (Note that inverting the sentence requires stating who made the measurement; since the au-

thors performed the first direct measurement, recognizing their agency in the topic position may well be appropriate.)

2. Since "dG" and "dC" become the old information in the second sentence, keep them up front in the topic position.

3. Since "triisopropylsilyl groups" is new and important information here, create for it a stress position.

4. "Triisopropylsilyl groups" then becomes the old information of the clause in which its effects are described; place it in the topic position of this clause.

5. Alert the reader to expect the arrival of two distinct effects by using the flag word "both." "Both" notifies the reader that two pieces of new information will arrive in a single stress position.

Here is a partial revision based on these decisions:

We have directly measured the enthalpy of hydrogen bond formation between the nucleoside bases 2'deoxyguanosine (dG) and 2'deoxycytidine (dC). dG and dC were derivatized at the 5' and 3' hydroxyls with triisopropylsilyl groups; these groups serve both to solubilize the nucleosides in non-aqueous solvents and to prevent the ribose hydroxyls from forming hydrogen bonds. From isoperibolic titration measurements, the enthalpy of dC:dG base pair formation is –6.65 ± 0.32 kcal/mol.

The outlines of the experiment are now becoming visible, but there is still a major logical gap. After reading the second sentence, we expect to hear more about the two effects that were important enough to merit placement in its stress position. Our expectations are frustrated, however, when those effects are not mentioned in the next sentence: "From isoperibolic titration measurements, the enthalpy of dC:dG base pair formation is –6.65 ± 0.32 kcal/mol." The authors have neglected to explain the relationship between the derivatization they performed (in the second sentence) and the measurements they made (in the third sentence). Ironically, that is the point they most wished to make here.

At this juncture, particularly astute readers who are chemists might draw upon their specialized knowledge, silently supplying the missing connection. Other readers are left in the dark. Here is one version of what we think the authors meant to say, with two additional sentences supplied from a knowledge of nucleic acid chemistry:

We have directly measured the enthalpy of hydrogen bond formation between the nucleoside bases 2'deoxyguanosine (dG) and 2'deoxycytidine (dC). dG and dC were derivatized at the 5' and 3' hydroxyls with triisopropylsilyl groups; these groups serve both to solubilize the nucleosides in non-aqueous solvents and to prevent the ribose hydroxyls from forming hydrogen bonds. Consequently, when the derivatized nucleosides are dissolved in non-aqueous solvents, hydrogen bonds form almost exclusively between the bases. Since the interbase hydrogen bonds are the only bonds to form upon mixing, their enthalpy of formation can be determined directly by measuring the enthalpy of mixing. From our isoperibolic titration measurements, the enthalpy of dG:dC base pair formation is –6.65 ± 0.32 kcal/mol.

Each sentence now proceeds logically from its predecessor. We never have to wander too far into a sentence without being told where we are and what former strands of discourse are being continued. And the "measurements" of the last sentence has now become old information, reaching back to the "measured directly" of the preceding sentence. (It also fulfills the promise of the "we have directly measured" with which the paragraph began.) By following our knowledge of reader expectations, we have been able to spot discontinuities, to suggest strategies for bridging gaps, and to rearrange the structure of the prose, thereby increasing the accessibility of the scientific content.

Locating the Action

Our final example adds another major reader expectation to the list.

> Transcription of the 5S RNA genes in the egg extract is TFIIIA-dependent. This is surprising, because the concentration of TFIIIA is the same as in the oocyte nuclear extract. The other transcription factors and RNA polymerase III are presumed to be in excess over available TFIIIA, because tRNA genes are transcribed in the egg extract. The addition of egg extract to the oocyte nuclear extract has two effects on transcription efficiency. First, there is a general inhibition of transcription that can be alleviated in part by supplementation with high concentrations of RNA polymerase III. Second, egg extract destabilizes transcription complexes formed with oocyte but not somatic 5S RNA genes.

The barriers to comprehension in this passage are so many that it may appear difficult to know where to start revising. Fortunately, it does not matter where we start, since attending to any one structural problem eventually leads us to all the others.

We can spot one source of difficulty by looking at the topic positions of the sentences: We cannot tell whose story the passage is. The story's focus (that is, the occupant of the topic position) changes in every sentence. If we search for repeated old information in hope of settling on a good candidate for several of the topic positions, we find all too much of it: egg extract, TFIIIA, oocyte extract, RNA polymerase III, 5S RNA, and transcription. All of these reappear at various points, but none announces itself clearly as our primary focus. It appears that the passage is trying to tell several stories simultaneously, allowing none to dominate.

We are unable to decide among these stories because the author has not told us what to do with all this information. We know who the players are, but we are ignorant of the actions they are presumed to perform. This violates yet another important reader expectation: Readers expect the action of a sentence to be articulated by the verb.

Here is a list of the verbs in the example paragraph:

is
is... is
are presumed to be
are transcribed
has
is... can be alleviated
destabilizes

The list gives us too few clues as to what actions actually take place in the passage. If the actions are not to be found in the verbs, then we as readers have no secondary structural clues for where to locate them. Each of us has to make a personal interpretive guess; the writer no longer controls the reader's interpretive act.

Worse still, in this passage the important actions never

As critical scientific readers, we would like to concentrate our energy on whether the experiments prove the hypotheses.

---

appear. Based on our best understanding of this material, the verbs that connect these players are "limit" and "inhibit." If we express those actions as verbs and place the most frequently occurring information—"egg extract" and "TFIIIA"—in the topic position whenever possible,* we can generate the following revision:

> In the egg extract, the availability of TFIIIA limits transcription of the 5S RNA genes. This is surprising because the same concentration of TFIIIA does not limit transcription in the oocyte nuclear extract. In the egg extract, transcription is not limited by RNA polymerase or other factors because transcription of tRNA genes indicates that these factors are in excess over available TFIIIA. When added to the nuclear extract, the egg extract affected the efficiency of transcription in two ways. First, it inhibited transcription generally; this inhibition could be alleviated in part by supplementing the mixture with high concentrations of RNA polymerase III. Second, the egg extract destabilized transcription complexes formed by oocyte but not by somatic 5S genes.

As a story about "egg extract," this passage still leaves something to be desired. But at least now we can recognize that the author has not explained the connection between "limit" and "inhibit." This unarticulated connection seems to us to contain both of her hypotheses: First, that the limitation on transcription is caused by an inhibitor of TFIIIA present in the egg extract; and, second, that the action of that inhibitor can be detected by adding the egg extract to the oocyte extract and examining the effects on transcription. As critical scientific readers, we would like to concentrate our energy on whether the experiments prove the hypotheses. We cannot begin to do so if we are left in doubt as to what those hypotheses might be—and if we are using most of our energy to discern the structure of the prose rather than its substance.

Writing and the Scientific Process

We began this article by arguing that complex thoughts expressed in impenetrable prose can be rendered accessible and clear without minimizing any of their complexity. Our

---

*We have chosen these two pieces of old information as the controlling contexts for the passage. That choice was neither arbitrary nor born of logical necessity; it was simply an act of interpretation. All readers make exactly that kind of choice in the reading of every sentence. The fewer the structural clues to interpretation given by the author, the more variable the resulting interpretations will tend to be.

examples of scientific writing have ranged from the merely cloudy to the virtually opaque; yet all of them could be made significantly more comprehensible by observing the following structural principles:

1. Follow a grammatical subject as soon as possible with its verb.

2. Place in the stress position the "new information" you want the reader to emphasize.

3. Place the person or thing whose "story" a sentence is telling at the beginning of the sentence, in the topic position.

4. Place appropriate "old information" (material already stated in the discourse) in the topic position for linkage backward and contextualization forward.

5. Articulate the action of every clause or sentence in its verb.

6. In general, provide context for your reader before asking that reader to consider anything new.

7. In general, try to ensure that the relative emphases of the substance coincide with the relative expectations for emphasis raised by the structure.

None of these reader-expectation principles should be

---

*I*t may seem obvious that a scientific document is incomplete without the interpretation of the writer; it may not be so obvious that the document cannot "exist" without the interpretation of each reader.

---

considered "rules." Slavish adherence to them will succeed no better than has slavish adherence to avoiding split infinitives or to using the active voice instead of the passive. There can be no fixed algorithm for good writing, for two reasons. First, too many reader expectations are functioning at any given moment for structural decisions to remain clear and easily activated. Second, any reader expectation can be violated to good effect. Our best stylists turn out to be our most skillful violators; but in order to carry this off, they must fulfill expectations most of the time, causing the violations to be perceived as exceptional moments, worthy of note.

A writer's personal style is the sum of all the structural choices that person tends to make when facing the challenges of creating discourse. Writers who fail to put new information in the stress position of many sentences in one document are likely to repeat that unhelpful structural pattern in all other documents. But for the very reason that writers tend to be consistent in making such choices, they can learn to improve their writing style; they can permanently reverse those habitual structural decisions that mislead or burden readers.

We have argued that the substance of thought and the expression of thought are so inextricably intertwined that changes in either will affect the quality of the other. Note that only the first of our examples (the paragraph about URF's) could be revised on the basis of the methodology to reveal a nearly finished passage. In all the other examples, revision revealed existing conceptual gaps and other problems that had been submerged in the originals by dysfunctional structures. Filling the gaps required the addition of extra material. In revising each of these examples, we arrived at a point where we could proceed no further without either supplying connections between ideas or eliminating some existing material altogether. (Writers who use reader-expectation principles on their own prose will not have to conjecture or infer; they know what the prose is intended to convey.) Having begun by analyzing the structure of the prose, we were led eventually to reinvestigate the substance of the science.

The substance of science comprises more than the discovery and recording of data; it extends crucially to include the act of interpretation. It may seem obvious that a scientific document is incomplete without the interpretation of the writer; it may not be so obvious that the document cannot "exist" without the interpretation of each reader. In other words, writers cannot "merely" record data, even if they try. In any recording or articulation, no matter how haphazard or confused, each word resides in one or more distinct structural locations. The resulting structure, even more than the meanings of individual words, significantly influences the reader during the act of interpretation. The question then becomes whether the structure created by the writer (intentionally or not) helps or hinders the reader in the process of interpreting the scientific writing.

The writing principles we have suggested here make conscious for the writer some of the interpretive clues readers derive from structures. Armed with this awareness, the writer can achieve far greater control (although never complete control) of the reader's interpretive process. As a concomitant function, the principles simultaneously offer the writer a fresh re-entry to the thought process that produced the science. In real and important ways, the structure of the prose becomes the structure of the scientific argument. Improving either one will improve the other.

The methodology described in this article originated in the linguistic work of Joseph M. Williams of the University of Chicago, Gregory G. Colomb of the Georgia Institute of Technology and George D. Gopen. Some of the materials presented here were discussed and developed in faculty writing workshops held at the Duke University Medical School.

Bibliography

Williams, Joseph M. 1988. *Style: Ten Lessons in Clarity and Grace.* Scott, Foresman, & Co.

Colomb, Gregory G., and Joseph M. Williams. 1985. Perceiving structure in professional prose: a multiply determined experience. In *Writing in Non-Academic Settings*, eds. Lee Odell and Dixie Goswami. Guilford Press, pp. 87–128.

Gopen, George D. 1987. Let the buyer in ordinary course of business beware: suggestions for revising the language of the Uniform Commercial Code. *University of Chicago Law Review* 54:1178–1214.

Gopen, George D. 1990. *The Common Sense of Writing: Teaching Writing from the Reader's Perspective.* To be published.

Infanticide as a Primate Reproductive Strategy

Sarah Blaffer Hrdy

Conflict is basic to all creatures that reproduce sexually, because the genotypes, and hence self-interests, of consorts are necessarily nonidentical. Infanticide among langurs illustrates an extreme form of this conflict

The Hanuman langur, *Presbytis entellus,* is the most versatile member of a far-flung subfamily of African and Asian leaf-eating monkeys known as Colobines. Langurs are traditionally classified as arboreal, but these elegant monkeys are built like greyhounds and can cover distances on the ground with speed and agility. Far more omnivorous than "leaf-eater" implies, Hanuman langurs feed on fully mature leaves, leaf flush, seeds, sap, fruit, insect pupae, and whatever delicacies might be fed them or left unguarded by local people. In forests, langurs spend much of their days in trees, but near open areas the adaptable Hanuman descends to the ground to feed and groom and may spend as much as 80 percent of daytime there. Monkeys are considered sacred by Hindus. This tolerance and their flexibility of diet and locomotion combine to make the Hanuman langur the most widespread primate other than man on the vast subcontinent of India. Ranging from as high

Sarah Blaffer Hrdy received her Ph.D. from Harvard in 1975 and was appointed a lecturer in biological anthropology there. Five years of research on langurs are chronicled in her forthcoming book, The Langurs of Abu: Male and Female Strategies of Reproduction *(Harvard Univ. Press). Currently she is doing research on monogamous primates.*
Dr. Hrdy wishes to acknowledge her debt to the community of langur fieldworkers, most especially to P. (Jay) Dolhinow, S. M. Mohnot, and Y. Sugiyama, and to other primatologists, J. Fleagle, D. Fossey, G. Hausfater, S. Kitchener, J. Oates, T. Struhsaker, R. Tilson, and K. Wolf, who allowed her access to unpublished findings. D. Hrdy, J. Seger, and R. Trivers made valuable comments on the manuscript. Dr. Blaffer Hrdy is also author of The Black-man of Zinacantan *(Univ. of Texas Press, 1972), an analysis of myths of Maya-speaking people. Address: Department of Anthropology, Harvard University, Cambridge, MA 02138.*

as 400 meters in the Himalayas down to sea level, and living in habitats that grade from moist montane forest to semidesert, this flexible Colobine occurs in pockets and in connected swaths from Nepal, down through India, to the island country of Sri Lanka.

The stable core of langur social organization is overlapping generations of close female relatives who spend their entire lives in the same matrilineally inherited 40 hectare plot of land. Troops have an average of 25 individuals, including as many as three or more adult males, but more often only one fully adult male is present. Whereas females remain in the same range and in the company of the same other females throughout their lives, males typically leave their natal troop or are driven out by other males prior to maturity. Loose males join with other males (in some cases brothers or cousins) in a nomadic existence. These all-male bands, containing anywhere from two to 60 or more juvenile and fully adult males, traverse the ranges of a number of female lineages. They will not return again to troop life unless as adults they are successful in invading a bisexual troop and usurping resident males.

With the exception of male invasions, langur troops are closed social units. Troops are spaced out in separate ranges with some areas of overlap between them. When troops meet at the borders of their ranges, both males and females participate in defending their territory. Males are especially active, relying on a wide repertoire of impressive audiovisual displays, such as whooping, canine grinding, and daring leaps that create a swaying turmoil in the treetops. Despite chases and lunges, the ap-

parent aggressiveness of intertroop encounters is largely bravado and almost never results in injuries. Serious fighting among langurs is largely confined to the business of defending troops against invading males; invasions are the only encounters in which males have actually been seen to inflict injuries on one another.

Because of the close association between man and langurs in a part of the world where monkeys are considered sacred, the earliest published accounts of their behavior date back before the time of Darwin and provide us with extraordinary descriptions of langur males battling among themselves for access to females and of females going to great lengths to defend their own destinies. In the 1836 issue of the *Bengal Sporting Magazine,* for example, we are told that in langur society, males compete for females and "the strongest usurps the sole office of perpetuating his species" (Hughes 1884). Another account (see also Hughes 1884) was written by a Victorian naturalist who witnessed invading males attack and kill a resident male followed by a counterattack against the invaders by resident females, who—if we are to believe the account—castrated and mortally wounded one of the invaders:

In April 1882, when encamped at the village of Singpur . . . my attention was attracted to a restless gathering of Hanumans. . . . Two opposing troops [were] engaged in demonstrations of an unfriendly character. Two males of one troop . . . and one of another—a splendid looking fellow of stalwart proportions—were walking round and displaying their teeth. . . . It was some time—at least a quarter of an hour—before actual hostilities took place, when, having got within striking distance, the two monkeys made a rush at their adversary. I saw their arms

Figure 1. Aggressive behavior in encounters between troops of Hanuman langurs is largely bluff. Here, a spread-eagled, open-mouthed male hurtles dangerously through space. In fact, he landed nowhere near his opponent, the familiar resident male from the neighboring troop. (Photos by D. B. Hrdy and author.)

and teeth going viciously, and then the throat of one of the aggressors was ripped right open and he lay dying.

He had done some damage however before going under having wounded his opponent in the shoulder. . . . I fancy the tide of victory would have been in [this male's favor] had the odds against him not been reinforced by the advance of two females. . . . Each flung herself upon him, and though he fought his enemies gallantly, one of the females succeeded in seizing him in the most sacred portion of his person, and depriving him of his most essential appendages. This stayed all power of defense, and the poor fellow hurried to the shelter of a tree where leaning against the trunk, he moaned occasionally, hung his head, and gave every sign that his course was nearly run. . . . Before the morning he was dead.

Social pathology hypothesis

Despite the vivid accounts of langur aggression set down by early naturalists, one of the first steps of modern primatology was to put aside these anecdotes so that the fledgling science of primatology could be laid on a purely factual foundation. By the late 1950s the modern era of primate studies, launched primarily by social scientists, had begun. The early workers were profoundly influenced by current social theory and in particular by the work of Radcliffe-Brown, who believed that any healthy society had to be a "fundamentally integrated social structure" and that in such a society every class of individuals would have a role to play in the life of the group in order to ensure its survival.

In 1959, Phyllis Jay went out from the University of Chicago to the Indian forest of Orcha (Fig. 2) and to Kaukori, a village on the heavily cultivated Gangetic plain. Jay found among North Indian langurs a remarkably peaceful society. She reported that relations among adult male langurs were relaxed, domi-

nance relatively unimportant, and aggressive threats and fighting exceedingly uncommon (1963 diss., 1965). All troop members, she wrote, were functioning so as to maintain the fabric of the social structure. Because of the overriding conviction that primates behave as they do for the good of their group, the early naturalists' descriptions were dismissed as "anecdotal, often bizarre, certainly not typical behavior" (Jay 1963 diss.).

Nevertheless, a second study turned up findings that forced reconsideration of the question of langur aggressiveness. In 1963, a team of Japanese primatologists led by Yukimaru Sugiyama were tracking langurs through the teak forests near Dharwar, South India, when they witnessed a band of seven langur males drive out the leader of a bisexual troop, after which one male from among the invaders usurped control and remained in sole possession of the troop. Within days of this takeover, all six infants in the troop were bitten to death by the new male. Curiously, and contrary to all previous reports concerning the solicitude of langur mothers (who have been known to carry the corpse of a dead infant for days), mothers whose infants were wounded by the usurping male abandoned them (Sugiyama 1967).

It was difficult to explain such behavior in terms of group survival and of a "fundamentally integrated" social structure. To circumvent this problem, it was suggested that there was something abnormal about the langurs of Dharwar and that their extreme aggressiveness was somehow pathological. In fact, if Jay's Kaukori study—the only other one available at that time—was taken as the norm, there *was* something unusual about Dharwar: langurs there were living in an area of rapid deforestation and of environmental disruption. Population densities (84–133 langurs per km²) were some 30 times higher than the very low density recorded at Kaukori (3 per km²).

Almost concurrently, John Calhoun (1962), at the National Institutes of Health, was studying the effects of crowding on the behavior of rats. He demonstrated that when the animals were crowded, normal rat social conventions broke down. The rats sank into a "pathological" state characterized by excessively high infant mortality due to inadequate maternal care, infanticide, and cannibalism. Comparisons between Calhoun's rats and the langurs of Dharwar were inevitable. A number of explanations were offered as to why langur infants were killed, and the social pathology hypothesis figured prominently among them. It was suggested that infanticide was a product of crowding (Sugiyama 1967; Eisenberg et al. 1972) and as such a mechanism for population control (Rudran 1973; Kummer et al. 1974). Alternatively, it was suggested that the behavior had no adaptive value (Bygott 1972) or that it was "dysgenic" (Warren 1967). Functional explanations for infanticide included the idea that males were somehow displacing aggression built up by the "simultaneous sexual excitement and enragement" of the new leader (Mohnot 1971) or that the male attacked infants in order to strengthen his "social bonds" with females in his new troop (Sugiyama 1965). All of these explanations derived from the basic assumption that under normal conditions animals act so as to maintain, not disrupt, the prevailing social structure.

Only one of the early explanations focused on the possible advantages of infanticide for the animal actually responsible for the act—the male. In 1967, Sugiyama suggested that the male attacked infants to avoid the two- to three-year delay in female sexual receptivity while she continued to nurse her offspring. This argument has been expanded into the more general sexual selection hypothesis that will be offered here.

It was to find out whether crowding really was at issue in infant-killing and desertion that in 1971 I first went to India. By the time I arrived, there was a new report of infanticide, this time from the desert region near Jodhpur, far to the north of Dharwar at a location where the Indian primatologist S. M. Mohnot had been studying langurs for several years. Already it seemed possible that infanticide was a more widespread and normal behavior than the social pathology hypothesis suggested.

From Jodhpur, I traveled southwestward to Mt. Abu. For 1,503 hours during five annual two- to three-month study periods between 1971 and 1975 I monitored political changes in five troops of langurs in and around the town. In the following section I will summarize the evidence—based largely on work at Abu, but drawing also on the detailed observations of Y. Sugiyama and S. M.

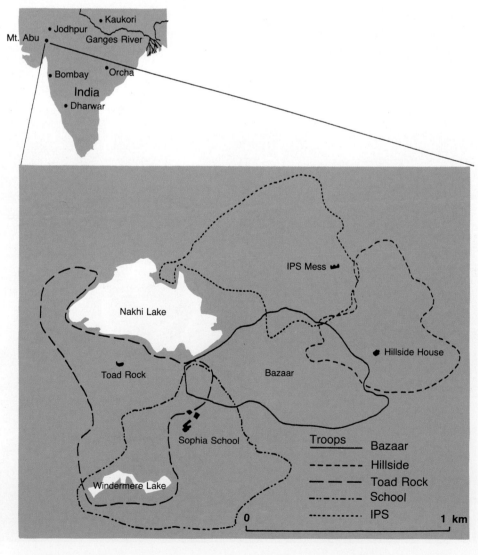

Figure 2. Over a 5-year period, the author studied five troops of Hanuman langurs in and around Mt. Abu, one of several sites on the Indian subcontinent where these widespread monkeys have been investigated.

Mohnot—that led me to reject my initial crowding hypothesis in favor of the theory that infanticide is adaptive behavior, extremely advantageous for the males who succeed at it.

The langurs of Abu

The forested hillsides of Mt. Abu rise steeply from the parched Rajasthani plains. The town itself is an Indian pilgrimage and tourist center 1,300 m above sea level. My study concentrated on five troops in the vicinity of the town, but I will focus here on just two of these: the small Hillside troop and its neighbor, the Bazaar troop, whose name derives from the fact that these langurs spent a portion of almost every day scavenging in the bazaar (see Fig. 2).

In June 1971 the Hillside troop contained one adult male, seven adult females, six infants, and one juvenile male. In August of that year, Mug was replaced by a new male, Shifty Leftless—named for a bite-sized chunk missing from his left ear. At the time of the takeover, one adult female and all six infants disappeared from the troop. Soon after, mothers who had lost infants came into estrus and solicited the new male. Local inhabitants witnessed the killing of two infants by an adult male. Each killing took place at a site well within the range of the Hillside troop; in fact, one occurred at a location used exclusively by that group. It seemed highly probable that the missing infants had been killed, and that the usurping male Shifty was the culprit. (These events are discussed in greater detail in Blaffer Hrdy 1974 and 1975 diss.)

On my return to Abu in June 1972, I was surprised to find that the same male, Shifty, had now transferred to the neighboring Bazaar troop. In 1971, Bazaar troop had contained three adult males, ten subadult and adult females, five infants, and four juveniles. Three of these infants were

now missing. The killing of one had been observed by a local amateur ornithologist who lived beside the bazaar. The three Bazaar troop males remained in the vicinity of their former troop; the second-ranking of these bore a deep wound in his right shoulder.

During 1972, Mug took advantage of Shifty's absence to return to his former troop. At this time Hillside troop consisted of the same six adult females and their four new infants. Two females, an older, one-armed female called Pawless and a very old female named Sol, had no infants. Although Mug was able to return to his troop for extended visits, whenever Shifty left Bazaar troop on reconnaissance to Hillside troop, Mug fled. On at least eight occasions, Mug left the troop abruptly just as the more dominant Shifty arrived, or else the "interloper" was actually chased by Shifty. Typically, Shifty's visits to

Hillside troop were brief, but if one of the Hillside females was in estrus he might remain for as long as eight hours before returning to Bazaar troop.

During the periods Mug was able to spend with his former harem he made repeated attacks on infants that had been born since his loss of control. On at least nine occasions in 1972, Mug actually assaulted the infants he was stalking. Each time one or both childless females intervened to thwart his attack. Despite their heroic intervention, on three occasions the infant was wounded. During this same period, other animals in the troop were never wounded by the male. When the same male, Mug, had been present in the troop in 1971, he had not attacked infants. Similarly, during Shifty's visits to the Hillside troop in 1971, his demeanor toward infants was aloof but never hostile. Whereas Hillside mothers were very

Figure 3. This Hillside infant was conceived in 1971, during the time that Shifty was the troop's resident male, and was later killed by an adult male langur, probably Mug. The age of langur infants can be determined with some precision: between the third and fifth months of life, the all-black natal coat changes to cream color, starting with the top of the head and a little white goatee.

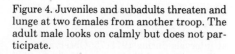

Figure 4. Juveniles and subadults threaten and lunge at two females from another troop. The adult male looks on calmly but does not participate.

Figure 4. Juveniles and subadults threaten and lunge at two females from another troop. The adult male looks on calmly but does not participate.

restrictive with their infants when Mug was present, gathering them up and moving away whenever he approached, these same mothers were quite casual around Shifty. Infants could be seen clambering about and playing within inches of Shifty without their mothers' taking notice.

In 1973, Mug was joined by a band of five males. Nevertheless, the double usurper Shifty could still chase out all six males whenever he visited the Hillside troop. A daughter born to Pawless during the period when both Shifty and Mug were vying for control of Hillside troop was assaulted on several occasions by the five newer invaders; the infant eventually disappeared and was presumed dead.

By 1974, Mug was once again in sole possession of the Hillside troop and holding his own against Shifty. When the Hillside and Bazaar troops met, Mug remained with his harem. On several occasions, the newly staunch Mug confronted Shifty and in one instance grappled with him briefly before retreating behind females in the Hillside troop. Mug resolutely chased away members of a male band who attempted to enter his troop. By 1975, Mug's star had risen.

When I returned to Abu in March of that year, Shifty was no longer with

the Bazaar troop. In his place was Mug. It was not known what had become of the extraordinary old male with the bite out of his left ear. Perhaps he died or moved on to another troop, or perhaps he was at last usurped by his longtime antagonist Mug.

Mug's former position in the Hillside troop was filled by a young adult male called Righty Ear. Righty (with a missing half-moon out of his right ear) was one of the five males who had joined Mug in the Hillside troop two years previously. Since that time, Righty had passed in and out of the troop's range, traveling with other males but not (so far as I knew) attempting to enter the troop. Righty's "waiting game" apparently paid off that March, when he came into sole possession of the Hillside troop. But, as in the case of his predecessors, Hillside troop was only a stepping stone: in April 1975, Righty replaced Mug as the leader of Bazaar troop.

The first indication I had of Righty's arrival in Bazaar troop was a report from local inhabitants that an adult male langur had killed an infant. On the following day when I investigated this report, the young adult male with the unmistakable half-moon out of his right ear was present in Bazaar troop; Mug was nowhere to be found.

An elderly langur mother still carried about the mauled corpse of her infant; by the following day, she had abandoned it. Righty subsequently made more than 50 different assaults on mothers carrying infants. Nevertheless, only one other infant disappeared. Five infants in the Bazaar troop remained unharmed when my observations terminated on June 20.

After Righty switched from Hillside to Bazaar troop, there followed some nine or more weeks during which the Hillside females had no resident male except for brief visits from Righty. Whenever the two troops met at their common border, Hillside females sought out Righty Ear and lingered beside him. These females were fiercely rebuffed by resident females in the Bazaar troop. Hostility of Bazaar troop females toward "trespassers" from Righty's previous harem prevented a merger of the two. The troops were still separate when Harvard biologist James Malcolm visited Abu in October 1975, but the vacuum in Hillside troop had been filled by a new male, christened Slash-neck for the deep gash in his neck.

The evolution of infanticide

Over a period of five years, then, political histories of the Hillside and Bazaar troops were linked by a succession of shared usurpers. First Shifty, then Mug, and finally Righty switched from the small and apparently rather vulnerable Hillside troop to the larger Bazaar troop (Fig. 5). Possibly the shifts were motivated by the greater number of reproductively active females in Bazaar troop. Between 1971 and 1975, at least four different males usurped control of Hillside troop. Infant mortality in this troop between 1971 and 1974 reached 83 percent, and extinction of the troop loomed as a real possibility. In contrast, during the same period, another troop at Abu, the School troop, was exceedingly stable, retaining the same male throughout.

Figure 5. The vicissitudes of male tenure in two troops of langurs are charted during the months the author spent observing the troops at intervals during 1971–75. Observations of infants missing, killed, or assaulted coincided with tenure shifts (as shown in italics in the chart).

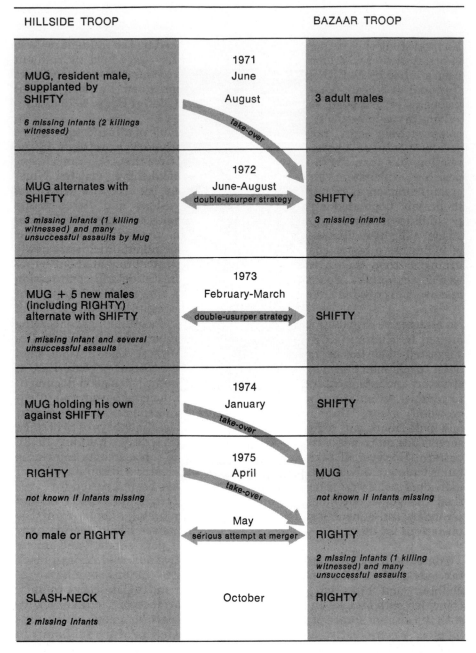

Combining all troop studies, the average male tenure at Abu was 27.5 months, a figure astonishingly close to the average tenure of 27 months calculated by Sugiyama for Dharwar (1967).

The short average duration of male tenure among langurs underlies the most crucial counterargument against the social pathology hypothesis: the extent to which adult males appear to gain from killing infants. Given that the tenure of a usurper is likely to be short, he would benefit from telescoping as much of his females' reproductive career as possible into the brief period during which he has access to them. By eliminating infants sired by a competitor, the usurping male hastens the mothers' return to sexual receptivity; on average, a mother whose infant is killed will become sexually receptive again within eight days of the death. In other words, infanticide permits an incoming male to use his short reign more efficiently than if he allowed unweaned infants present in the troop at his entrance to survive, to continue to suckle, and thus to delay the mother's next conception.

In three troops at Dharwar and Abu for which we have reliable information on subsequent births, 70 percent of the females who lost infants gave birth again within 6 to 8 months of the death of their infants, on average—just over one langur gestation period later. In the harsh desert environment of Jodhpur, however, the postinfanticide birth interval was much longer, up to 27 months.

Once infant-killing began, a usurper would be penalized for *not* committing infanticide. If a male failed to kill infants upon taking over a troop, and instead waited for those infants already in the troop to be weaned before he inseminated their mothers, then his infants would still be unweaned and hence vulnerable when the next usurper (presumably an infanticidal male) entered.

Other variations in the social system might likewise be expected to select for changes in male behavior. For example, if the rate of takeovers were speeded up and then held constant over time, male tolerance toward weaned immatures might be drastically altered. With a faster rate of takeover, it would be unlikely that one male could remain in control of a troop long enough for immature females to reach menarche and to give birth to an infant that would in turn grow old enough to survive the next takeover. Immature females, then, would be worth no more to the usurper than young males would be, and they might compete with the productive females of his harem for resources. Under these circumstances, it would behoove a usurper to drive out immatures of both sexes. This is precisely what occurs among a related langur species, *Presbytis senex*, living at very high densities (as high as 215 animals per km²) at Horton Plains in Sri Lanka (Rudran 1973). The ousted females travel with former male troopmates in mixed-sex bands.

Up to this point, I have not dealt with the apparent correlation between male takeovers and high population density. At both Dharwar (84–133 langurs per km²) and Abu (50 per km²), population densities are relatively high. In the desert region near Jodhpur, langurs have vast open areas available to them but tend to cluster about waterholes and garden spots. Infanticide has been reported at all three locations, but it has been recorded for none of the areas with low densities (at Jay's Orcha and Kaukori study sites or at any of three Himalayan sites where langurs have been studied by N. Bishop, the Curtins, and C. Vogel). This finding is inconclusive, however, since observations in the low-density areas were comparatively short, ranging from several months to a year. If the correlation does turn out to be valid, a possible explanation may be the greater numbers of extratroop males in heavily populated areas. If the possibilities for male recruitment are greater at high densities, and if a band of males has a better chance of usurping a troop than a single male would, then there would be more takeovers in crowded areas.

An alternative explanation has been offered by Rudran (1973), who has suggested that takeovers occur in order to maintain the one-male troop structure and infanticide occurs so as to curtail population growth in crowded areas. Unquestionably, one-male troops and reduced infant survival are outcomes of the takeover pattern. However, if takeovers and infanticide are advantageous to the individual males who engage in them, then the above outcomes are only secondary consequences and not explanations for them.

To date, we have information on 15 takeovers, 5 at Dharwar, one at Jodhpur, and 9 at Abu. At least 9 coincided with attacks on infants or with the disappearance of unweaned infants. A conservative estimate of the number of infants who have disappeared at the time of takeovers is 39. The important point (and this is the second line of evidence against the social pathology hypothesis) is that attacks on infants have been observed *only* when males enter the breeding system from outside—even if, as in the case of Mug, they have been only temporarily outside it. Such males are unlikely to be the progenitors of their victims. In contrast to what is considered "pathological" behavior, attacks on infants were highly goal-directed. An important area of future research will be learning exactly what means a langur male has at his disposal for discriminating infants probably his own from those probably sired by some other male. Quite possibly, males are evaluating past consort relations with the

mother (Blaffer Hrdy 1976). Interestingly, infants kidnapped by females from neighboring troops were not attacked by the resident male so long as they were held by resident females from his own troop and were not accompanied by their (alien) mothers (Blaffer Hrdy, 1975 diss.).

The third line of evidence against the social pathology hypothesis is the length of time that conditions favoring infanticide have persisted. Nineteenth-century accounts describing male invasions and fierce fights among males for access to females undermine the position that langur aggression and infanticide are newly acquired traits brought about by recent deforestation and compression of langur ranges. More important (and this constitutes the fourth line of evidence), recent findings concerning other members of the subfamily Colobinae suggest that a time span much longer than a few centuries is at issue. In addition to good documentation for male takeovers and infanticide among the closely related purple-faced leaf-monkeys of Sri Lanka (*Presbytis senex*) (Rudran 1973), adult male replacements coinciding with the death or disappearance of infants have been reported for *Presbytis cristata* of Malaysia (Wolf and Fleagle, in press); *P. potenziani,* the rare Mentawei Island leaf-monkey (R. Tilson, pers. comm.); and among both captive and wild African black and white colobus monkeys (S. Kitchener and J. Oates, pers. comm.). This recurrence of the takeover/infanticide pattern among widely separated members of the subfamily in Africa, India, and Southeast Asia argues strongly for its antiquity. Though the possibility of environmental convergence cannot be ruled out, the case of phylogenetic inheritance of these traits among geographically disparate relatives is a compelling one. Far from being recent responses to crowded conditions, it appears that a predisposition to male takeovers and infanticide has been part of the colobine repertoire since Pliocene times, some ten million or more years ago, when the split between the African and Asian forms occurred.

Beyond the Colobines

But the tale of infanticide does not stop with the Colobines. In what may be the most startling finding by pri-

matologists in recent years, we are discovering that the gentle souls we claim as our near relatives in the animal world are by and large an extraordinarily murderous lot. It is apparent now that the events witnessed at Abu and Dharwar are not aberrations. Increased observation of primates had led to an increase in the number of species in which adult males are known to attack and kill infants—and, occasionally, each other. Although murder is uncommon, cases of adults fighting to the death have been reported for rhesus, pig-tailed, and Japanese macaques, baboons, and chimpanzees, as well as Hanuman langurs.

At the time of this writing, infanticide, either observed or inferred from the disappearance of infants at times when males have usurped new females, has been reported for more than a dozen species of primates. Every major group of primates, including the prosimians, the New and Old World monkeys, apes, and man, is represented.

Not all these reports parallel the pattern of events recorded for Hanuman langurs, but many are disturbingly similar: males attack infants when they come into possession of females who are accompanied by offspring sired by another male. Typically, these are unfamiliar females. Perhaps the clearest illustration of the potential importance of previous acquaintance is provided by an experiment with caged crab-eating macaques (Thompson 1967). Here, infanticide was the unexpected outcome in a cage study on the effects of familiarity or lack of it in relations between male and female *Macaca fascicularis.* When paired with his accustomed companion and her infant, the adult male displayed typical behavior, mounting the female briefly and then casually exploring his surroundings. He entirely ignored the infant. Paired with an unfamiliar mother-infant pair, the male responded quite differently. After a brief attempt at mounting, the male attacked the infant as it lay clutched to its mother's belly. When the mother tried to escape, the male pinned her to the ground and gnawed the infant, making three different punctures in its brain with his canines.

Two suspected cases of infanticide

Figure 6. A female langur holding a newborn infant takes food from a priest of Shiva who lives in one of the sacred caves in the hillsides surrounding Mt. Abu.

among wild hamadryas baboons were occasioned by human manipulation. In the course of capture-and-release experimentation on the process of harem formation among *Papio hamadryas* of Ethiopia, two mothers were switched to new one-male units. In one case the infant was missing a day later; in the other, the infant was seen dead, "its skull pierced and its thighs lacerated by large canine teeth." The witnessed killing of two hamadryas infants at the Zurich Zoo just after their mothers changed "owners" adds plausibility to the inference that the wild infants were similarly murdered (Kummer et al. 1974).

Less contrived perhaps is the following account of chimpanzees from the Gombe Stream Reserve in Tanzania, where a young British researcher, David Bygott, happened to be following a band of five male chimpanzees when they encountered a strange female whom "in hundreds of hours of field observation," Bygott had never seen before. This female and her infant were immediately and intensely attacked by the males. For a few moments, the screaming mass of chimps disappeared from Bygott's view. When he relocated them, the strange female had disappeared and one of the males held a struggling infant. "Its nose was bleeding as though from a blow, and [the male], holding the infant's legs, intermittently beat its head against a branch. After 3 minutes, he began to eat the flesh from the thighs of the infant which stopped struggling and calling" (Bygott 1972). In contrast with normal chimp predation, this cannibalized corpse was nibbled by several males but never consumed.

Dian Fossey's remarkable decade-long study of wild mountain gorillas in central Africa provides what may be the most dramatic instances of adult male invaders mauling infants. For several days, a lone "silverback" (or fully mature) male had been following a harem of gorillas, presumably in quest of females. At last, he made his move, penetrating the group with a "violent charging run." A pri-

miparous female who had given birth to an infant on the previous night countered his charge by running at him. Halting within arm's reach of the male, she stood bipedally to beat her chest. The male struck her ventrally exposed body region where her newly born infant was clinging. Immediately following this blow, a "thin wail" was heard from the dying infant. On two other occasions, Fossey witnessed silverbacks kill infants belonging to primiparous mothers. In the best documented of these cases, the mother subsequently copulated with the male who had killed her infant (Fossey 1974; pers. comm.). To date, of the killings witnessed, only first-born gorilla infants have been seen to be victims. This could be owing to maternal inexperience, or, as I believe is more probable, to Fossey's finding that in gorilla society only *young* females routinely change social units. Since an older mother would in all

likelihood not join a usurper anyway, he would rarely benefit from killing her infant.

Isolated instances of infanticide by adult males have also been reported for various prosimians: among free-ranging Barbary (*Macaca sylvana*) and rhesus macaques (*M. mulatta*) (Burton 1972; Carpenter 1942) and among wild *Cercopithecus ascanius,* the red-tailed monkeys of Africa (T. Struhsaker, in press). Infanticide is suspected among wild chacma baboons (*Papio ursinus*) (Saayman 1971); wild howler monkeys (*Alouatta*) of South America (Collias and Southwick 1952); and among caged squirrel monkeys (*Saimiri*) (Bowden et al. 1967).

The explanation for infanticide need not be the same in every case, but the parallels with the well-documented langur pattern are striking. According to the explanatory hypothesis offered here for langurs, infant-killing is a reproductive strategy whereby the usurping male increases his own reproductive success at the expense of the former leader (presumably the father of the infant killed), the mother, and the infant. If this model applies, the primatewide phenomenon of infanticide might be viewed as yet another outcome of the process Darwin termed *sexual selection:* any struggle between individuals of one sex (typically males) for reproductive access to the other sex, in which the result is not death to the unsuccessful competitor, but few or no offspring (Trivers 1972). Crucial to the evolution of infanticide are, first, a nonseasonal and flexible female reproductive physiology such that it is both feasible and advantageous for a mother to ovulate again soon after the death of her infant and, second, competition between males such that tenure of access to females is on average short.

Female counterstrategies

Confronted with a population of males competing among themselves, often with adverse consequences for females and their offspring, one would expect natural selection to favor those females most inclined and best able to protect their interests. When an alien langur male invades a troop, he may be chased away and harassed by resident females as well as by the resident male. After a new male takes over, females may form temporary alliances to prevent him from killing their infants (e.g. Sol and Pawless's combined front against the infanticidal Mug).

Females are often able to delay infanticide. Less often are they able to prevent it. Pitted against a male who has the option to try again and again until he finally succeeds, females have poor odds. For this reason, one of the best counterinfanticide tactics may be a peculiar form of female deceit. Almost invariably, langur males have attacked infants sired by some other male; a male who attacked his own offspring would rapidly be selected against. It may be significant then that at Dharwar, Jodhpur, and Abu, pregnant females confronted with a usurper displayed the traditional langur estrous signals: the female presents her rump to the male and

frenetically shudders her head. These females mated with the usurper even though they could not possibly have been ovulating at the time. Postconception estrus in this context may serve to confuse the issue of paternity.

After birth, an infant's survival is best ensured if its mother is able to associate with the father, or at least with a male who "considers" himself the father or who acts like one—in short, a male who tolerates her infant. In at least three instances at Abu, females with unweaned infants left recently usurped troops to spend time in the vicinity of males that on the basis of other evidence I suspected of having fathered their offspring.

If all else fails and her infant is attacked and wounded, a mother may continue to care for it, or abandon it. In several cases at Dharwar and Jodhpur, mothers abandoned their murdered infants soon after or even before death (Sugiyama 1967; Mohnot 1971). Rudran has suggested that the mother abandons her infant for fear of injury to herself and "because an adult female is presumably more valuable than an infant to the troop" (1973). It is far more likely, however, that desertion reflects a practical evaluation of what *this* infant's chances are weighed against the probability that her next infant will survive.

Under some circumstances a mother may opt to abandon an unwounded infant. In a single case from Abu, a female in a recently usurped troop who had been traveling apart from the troop (presumably to avoid the new male's assaults) left her partially weaned infant in the company of another mother and returned to the main body of the troop alone. If this was in fact an attempt to save her infant by deserting it, the ploy failed when the babysitter herself returned to the troop, some time later, bringing both infants with her. Nevertheless, both infants did survive the takeover.

Despite the various tactics that a female may employ to counter males, infanticide was the single greatest source of infant mortality at Abu. The plight of these females raises a perplexing question: How has this situation come about? Langur males contribute little to the rearing of off-

spring; apart from insemination, females have little use for males *except* to protect them from other langur males who might otherwise invade the troop and kill infants. Why then should females tolerate males at all, suffering subjection to the tyranny of warring polygynists? On the vast time scale of evolution, alternatives have been open to the female since the dawn of Colobines. Large body size, muscle mass, and saber-sharp canines might just as well have been selected among females as among males. Why should females weigh only 12 kg, on average, and not the 18 kg that males routinely do? Alternatively, female relatives could ally themselves to a much greater extent than they do. The combined 36 kg of three females operating as a united front against an infanticidal male surely should prevail. Infanticide depends for its evolutionary feasibility on the prior female adaptation of conceiving again as soon as possible after the death of an infant. If females failed to ovulate after a male killed their infants, or if they "refused" to copulate with an infanticide, the trait would be eliminated from the population.

The facts that females do not grow so large as males, that they do not selflessly ally themselves to one another, and that they do not boycott infanticides, suggest that counterselection is at work. Once again, the pitfall is intrasexual competition—this time competition among females themselves for representation in the next generation's gene pool. Whereas head-on competition between males for access to females selects for males who are as big and as strong (or stronger) than their opponents, a female who "opted" for large size in order to fight off males might not be so well-adapted for her dual role of ecological survivor and childbearer. An over-sized female might produce fewer offspring than her smaller cousin. In time, the smaller cousin's progeny would prevail.

Intrasexual competition is mitigated by the close genetic relatedness between female troop members, but it is by no means eliminated. A female in her reproductive prime who altruistically defended her kin, in spite of the cost to herself, might be less fit than her cousin who sat on the sidelines. Finally, if infanticide really is advantageous behavior for males, a female who sexually boycotted infanticides would do so to the detriment of her male progeny. Her sons would suffer in competition with the offspring of nondiscriminating mothers.

For generations langur females have possessed the means to control their own destinies. Caught in an evolutionary trap, they have never been able to use them.

References

Blaffer Hrdy, S. 1974. Male-male competition and infanticide among the langurs (*Presbytis entellus*) of Abu, Rajasthan. *Folia. Primat.* 22:19–58.

———. Male and female strategies of reproduction among the langurs of Abu. 1975 diss., Harvard University.

———. 1976. The care and exploitation of nonhuman primate infants by conspecifics other than the mother. In *Advances in the Study of Behavior 6*, ed. J. Rosenblatt, R. Hinde, C. Beer, and E. Shaw. Academic Press.

———. In press. *The Langurs of Abu.* Harvard Univ. Press.

Bowden, D., P. Winter, and D. Ploog. 1967. Pregnancy and delivery behavior of the squirrel monkey (*Saimiri sciureus*) and other primates. *Folia Primat.* 5:1–42.

Burton, F. 1972. The integration of biology and behavior in the socialization of *Macaca sylvana* of Gibraltar. In *Primate Socialization,* ed. F. Poirier. Random House.

Bygott, D. 1972. Cannibalism among wild chimpanzees. *Nature* 238:410–11.

Calhoun, J. 1962. Population density and social pathology. *Sci. Am.* 206:139–48.

Carpenter, C. R. 1942. Societies of monkeys and apes. *Biol. Symposia* 8:177–204.

Collias, N., and C. H. Southwick. 1952. A field study of population density and social organization in howling monkeys. *Proc. of the Amer. Phil. Soc.* 96:143–56.

Eisenberg, J. F., N. A. Muckenhirn, and R. Rudran. 1972. The relation between ecology and social structure in primates. *Science* 176:863–74.

Fossey, D. 1974. Development of the mountain gorilla (*Gorilla gorilla beringei*) through the first thirty-six months. Paper presented at Berg Wartenstein symposium no. 62, The Behavior of the Great Apes, Wenner-Gren Foundation for Anthropological Research.

Hughes, T. H. 1884. An incident in the habits of *Semnopithecus entellus,* the common Indian Hanuman monkey. *Proc. Asiatic Soc. of Bengal,* pp. 147–50.

Jay, P. The social behavior of the langur monkey. 1963 diss., University of Chicago.

———. 1965. The common langur of North India. In *Primate Behavior,* ed. I. DeVore. Holt, Rinehart and Winston.

Kummer, H., W. Gotz, and W. Angst. 1974. Triadic differentiation: An inhibitory process protecting pairbonds in baboons. *Behaviour* 49:62.

Mohnot, S. M. 1971. Some aspects of social change and infant-killing in the Hanuman langur, *Presbytis entellus* (Primates: Cercopithecinae) in Western India. *Mammalia* 35:175–98.

Rudran, R. 1973. Adult male replacement in one-male troops of purple-faced langurs (*Presbytis senex senex*) and its effects on population structure. *Folia Primat.* 19: 166–92.

Saayman, G. S. 1971. Behaviour of the adult males in a troop of free-ranging chacma baboons (*Papio ursinus*). *Folia Primat.* 15: 36–57.

Struhsaker, T. In press. Infanticide in the redtail monkey (*Cereopithecus ascanius schmidti*). In *Proceedings of the Sixth Congress of International Primatological Society.* Academic Press.

Sugiyama, Y. 1965. On the social change of Hanuman langurs (*Presbytis entellus*) in their natural conditions. *Primates* 6:381–417.

———. 1967. Social organization of Hanuman langurs. In *Social Communication among Primates,* ed. S. Altmann. Univ. of Chicago Press.

Thompson, N. S. 1967. Primate infanticide: A note and request for information. *Laboratory Primate Newsletter* 6(3):18–19.

Trivers, R. L. 1972. Parental investment and sexual selection. In *Sexual Selection and the Descent of Man 1871–1971,* ed. B. Campbell, pp. 136–79. Aldine.

Warren, J. M. 1967. Discussion of social dynamics. In *Social Communication Among Primates,* ed. S. Altmann. Univ. of Chicago Press.

Wolf, K., and J. Fleagle. In press. Adult male replacement in a group of silvered leaf-monkeys (*Presbytis cristata*) at Kuala Selangor, Malaysia. *Primates.*

Richard Curtin
Phyllis Dolhinow

Primate Social Behavior in a Changing World

Human alteration of the environment may be pushing the gray langur monkey of India beyond the limits of its adaptability

The gray langur monkey (*Presbytis entellus*) finds itself at present in the midst of heated controversy. Infant mortality in one of the langur troops at Mt. Abu, in northwest India (Fig. 1), rose to an astonishing 83% over a period of four years, and the causes for this striking figure immediately commanded the attention of students of primate behavior. It was assumed by some researchers that the Abu monkeys, and other langurs living at crowded sites, were typical of the species (Mohnot 1971; Sugiyama 1965, 1966, 1967; Hrdy 1974, 1977), and Hrdy has proposed an evolutionary model that includes males routinely killing infants as an adaptation to competition for females. Our observations of langurs at the less crowded sites of Orcha and Kaukori (Dolhinow 1972) and at Junbesi (Curtin 1975 diss.) have forced us to question this interpretation and to ask instead: How could such levels of infant death arise from the more widespread, generally peaceful pattern of behavior associated with the

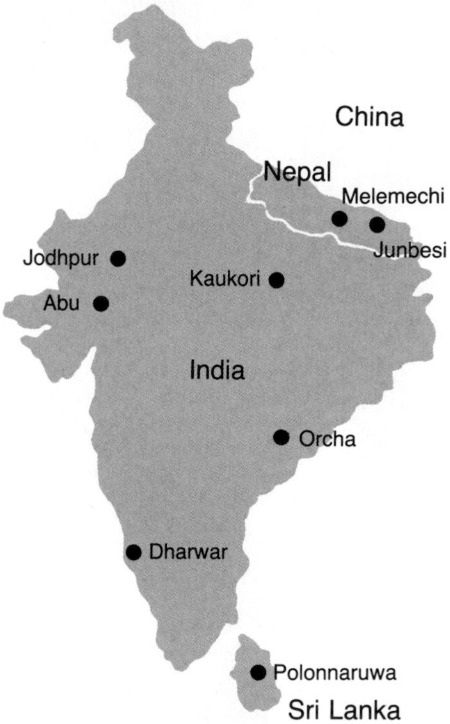

Figure 1. Gray langurs have been studied at a variety of sites in India, Nepal, and Sri Lanka, ranging from heavily populated areas to relatively unspoiled forest.

Richard Curtin, who has studied langur monkeys in the Nepal Himalaya for twenty months, received his Ph.D. in 1975 from the University of California, Berkeley, where he is currently a Research Associate and Lecturer and does research on captive langurs. During his field work he was supported by an NIH Traineeship.
Phyllis Dolhinow, Professor of Anthropology at Berkeley, spent three and one half years in India studying langur monkeys and rhesus macaques. She has also worked in East Africa. Since 1972 she has done research on langur monkeys in a captive colony at the University of California. Her recent research concentrates on langur development, attachment, caretaking and adoption, and the effects of mother-loss. Address: Department of Anthropology, University of California, Berkeley, CA 94720.

gray langurs' notable ecological and evolutionary success?

Langurs are members of the subfamily Colobinae, the Old World monkeys that possess digestive tracts which allow them to eat mature leaves. Primate studies have historically focused on the other subfamily—the Cercopithecinae, which includes such animals as baboons and rhesus macaques. Perhaps more important, colobines have been neglected because most of them are forest dwellers, and early studies concentrated on animals that could

be easily seen. However, field researchers of the 1960s and 1970s have been less intimidated by the difficulties of working in forest habitats, and several excellent studies of forest leaf-eaters have been completed or are under way.

Gray langurs live from Nepal to Sri Lanka in an array of habitats ranging from tropical jungle to desert margin, and from undisturbed forest to bustling village. Their behavior varies greatly over this range, but they have now been well studied at a number of sites scattered over the Indian subcontinent, and the limits of their behavioral variation have emerged clearly. The behavior of langurs at Orcha and Junbesi appears to represent a pattern of social organization that probably occurs over the species' whole range and persists under all but the most stressful environmental conditions.

Typical of the gray langur social pattern are multiple infant caretakers and male rivalry that can be dramatic but that does not normally disrupt the orderly rearing of the young. Langur troops include relatively stable cores of adult females plus adult and subadult males whose membership in the group is somewhat less constant. Males can and do move from troop to troop, and although relations among adult males differ from one site to another, competition for females is intense enough in some areas to bring about changes in troop composition. At Orcha, serious fighting among males was never observed during a year's study, but at Junbesi, fighting among males, particularly during the peak mating season, led to animals being driven from the troop, at least temporarily. These populations can be said to have

41

a "multi-male troop pattern," and are to be distinguished from populations in which troops typically have only one adult male, and a very high percentage of males are not members of reproductive troops.

The multi-male troop pattern is as widespread as the species itself. It has been found in all three continuous long-term studies of gray langurs living where human influence on ecological conditions has been slight: at Orcha, where leopards and tigers still prowled the virgin forest; in minimally disturbed oak forest in central Nepal (Bishop 1975 diss.); and at Junbesi, where a marginal habitat and substantial predation helped limit population density to a single langur every square kilometer. The pattern has also persisted under quite different circumstances: Ripley (1967) reported it from Sri Lanka where langurs lived at a population density fifty times that at Junbesi, and the langurs at Kaukori maintained their stable multi-male troop in an area where 98% of the land was occupied by villages or under cultivation (see Fig. 2). This remarkable adaptability of langur social organization carries with it a high degree of behavioral variability, but the wealth of studies now complete allows us to describe, in general terms, the social life a langur is likely to lead.

Normal social life

From the time a newborn presence is noticed in a group, it becomes a focus of great interest for young and old females alike (Fig. 3). The infant is conspicuous because the bare pink skin of its hands, feet, face, and disproportionately large and convoluted ears contrasts sharply with its sparse dark brown fur. The newborn may be passed among waiting females more than fifty times in four hours on the first day of life. It may be away from its mother for hours, although normally she can retrieve it when she desires, and most transfers are calm and without peril to the infant. Many females allow the infant to nurse, even though they are not lactating or may never have had an infant.

This high frequency of passing the young infant from female to female lasts only a few weeks. By the end of the second month of life the infant is very active and ably goes where and when it wishes. In the meanwhile, it

Figure 2. A juvenile gray langur begs a bite of mango from her mother, who ignores her. Although the Kaukori troop, of which they are members, takes a toll of both unripe and ripe fruit from the mango orchards, an abundance of fruit remains on the large trees.

has had many hours of contact with most of the females in the troop and has developed strong preferences for certain of them.

Weaning, at the end of the first year, is seldom a harsh or traumatic experience. Inconsistent maternal rejection often prolongs weaning, and occasionally an adult with a neonate is observed being followed closely by her last infant. At the time of weaning, a yearling may adopt another adult female to serve as its mother. Although adoptions seldom have been witnessed in the field, probably because of difficulties of observation, they occur frequently in captivity and may be stimulated by even brief removals of the mother from the group. This is, of course, assuming that substitute caretakers are available—as they normally are.

Male and female paths of development diverge soon after the first year of life. Play is important for both sexes, but the immature female also shows interest in neonates. By the time she is two, she takes an active part in attempting to take, hold, and carry new infants. At first, her efforts are hopelessly clumsy, and her ineptness, as reflected in the infant's struggles and cries, stimulates older females to retrieve it. She gradually acquires skill and becomes a fair or even a competent caretaker well before she becomes a mother. There is a great individual variability in motivation to perform caretaking activities among females of all ages. A woefully inadequate juvenile may become a splendidly able mother!

Furthermore, adult females may treat successive infants in very different manners. The degree of interest in the infants of other females varies greatly among adults, and definite preferences are shown for certain infants by some females. The reasons for these preferences and for changes in caretaking styles of individual females are being investigated, because to the human observer it is not at all apparent why the variations exist. The vast majority of attentions directed toward infants can best be described as adequate and protective. Infants too are far from a homogeneous lot: some are strugglers and squealers, while others seem passive, seldom showing discomfort.

Adult females live an orderly existence. There are social changes as, for example, when males enter and leave the group, but the paramount markers of the passage of time are a succession of infants. A female has her first infant when she is 4 to 5 years old, and her ability to produce viable offspring improves as she ages. In the wild, females usually give birth at intervals of 18 to 24 months. There is variability in peak birth seasons in different areas of the Indian subcontinent, but most areas are characterized by seasonal regularity of births. When a female is in estrus, she solicits and mates with several males if the group is large enough, often showing a preference for one, who may not always be the alpha male. Complete copulations may occur during pregnancy but are by far more frequent during the female's mid-cycle estrous periods.

Precise information on the improvement of reproductive capability with aging is being gathered in a captive colony at Berkeley, where ages are known. By 9 years a female is bearing regularly, and evidence from both field and colony suggests strongly that she continues to do so for many years. No doubt, if she survives long enough, she will eventually experience a diminution of reproductive ability, but very aged females are rarely observed to survive the rigors of feral existence. Estimating ages in the field is exceedingly difficult, and indicators such as worn teeth, wrinkles, bags under the eyes, or generally poor health and emaciation are extremely unreliable.

Adult female social ranks are seldom clearly apparent or linearly ordered, and neither age nor reproductive history is a good predictor of social rank. The older females are by no means usually the highest in social authority, nor do they automatically yield to younger females. The scales of power tip with the events of the reproductive cycle such as sexual receptivity, pregnancy, delivery, and lactation. Fluctuations occur within periods of months as well as over years, and a female may move up or down relative to other females, only to change her position in the hierarchy after some months. Body size does not determine status once females are mature.

Male rivalry

Juvenile males, in contrast to females of the same age, spend most of their time in rough play and rarely show interest in newborn infants. As a male grows he tends to become somewhat peripheralized to the troop because his boisterousness annoys adults. A young male may leave the group when he is sexually mature, and a few do even before that. An exit may be hastened or stimulated by tensions arising between him and the adult males. Such departures are not always permanent, and reentries as well as entries into new troops have been recorded. The adult males in a troop can usually be ranked in a linear hierarchy of status, but there is no clear correlation between size or age and position in the hierarchy.

Rivalry with other males is inevitably an extremely important factor in a male langur's life. Since it is one of the most significant variables in langur behavior, and since it is probably at the root of the infant disappearances reported from Abu and other sites, its expression within the multi-male troop demands close scrutiny.

One end of the male-rivalry spectrum is shown by the relaxed behavior and absence of fighting in the troop at Orcha. Further along the spectrum, rivalry was far more apparent among the Kaukori langurs, where a stable dominance hierarchy was easily discernible among the troop's six adult males. Prolonged harassment with only occasional outbreaks of fighting accompanied the gradual exchange of the two top male positions. Notably, this dominance shift took place *within* the troop; after the exchange was effected, the defeated male assumed a stable, albeit lower, rank within the hierarchy.

In contrast to Kaukori, male rivalry was dramatic among the Junbesi langurs, and its effects were the most striking so far reported within the multi-male troop pattern. Relations among the adult males were tense throughout the 16 months of observation. During the first winter of the study, one of the troop's four original males repeatedly left after skirmishes with the other three males. These departures were temporary and are best described as peripheralizations, since the male did not leave the troop's home range, which was so large that he could stay within it and rarely be in close contact with the troop. He rejoined the troop in late winter but was frequently driven from it during the peak mating season in summer. He was again able to remain within the home range, and as the mating season ended he was gradually able to shorten his enforced distance from the troop. He was one of the two adult males in the troop when the study ended.

The other two males had left the troop permanently after repeated aggressive exchanges and violent fights at the beginning of the summer. One of them was later discovered in a different troop some distance away. A similar gradual process had apparently led to his successful entry, and he was a well-assimilated member of the multi-male troop.

Relations among males within the multi-male troop can thus follow several patterns: males can coexist amicably, with little apparent aggression, as at Orcha. Rivalry can be resolved within a troop by threat, fighting, or temporary peripheralization, or it can lead more dramatically to a male's permanent exclusion from his troop. Males can also enter strange troops or reenter their own with very little fighting. This is a remarkable range of behavior, and at present it is difficult to identify or explain the critical elements that cause the variation.

We do not know, for instance, what causes some males to accept a lowered position in the dominance ranking and why others elect to leave the troop. We do not know how some males successfully manage to enter strange troops while others, as witnessed at Kaukori, are actively rebuffed. One generalization, however, is possible. The patterns of behavior that allow males to remain in or to enter troops in the face of male rivalry take time, and, perhaps most important, they take *space*. The dominance shift at Kaukori took $2\frac{1}{2}$ weeks, and the reentry of the Junbesi male into his troop took even longer. At first the male in the troop threatened the other when he was 1 km away, but as the mating season ended, his threats became less intense, and the distance gradually decreased. When troops are compressed into a restricted space that forces them to interact frequently and aggressively over limited food, water, and places of safety, or when they are harassed by humans, male rivalry may produce new and maladaptive results.

Behavior at crowded sites

The langurs at Abu (Hrdy 1974, 1977) and those watched by Mohnot (1971) at Jodhpur and by Sugiyama (1965, 1966, 1967) and his colleagues at Dharwar shared a pattern of behavior that falls outside the broad range so far discussed. Male behavior was much less variable. These investigators reported that the expression of male rivalry extended far beyond threat and fighting to include the routine killing of infants sired by competing males. This pattern differed further in being associated with only very specific ecological conditions. The habitats at all three sites were greatly disturbed by human activities, and langur population densities were either absolutely or effectively extremely high. The Jodhpur langurs lived on the edge of the desert and spent most of their time in very limited areas near water sources or gardens. Much of the 40-hectare home ranges of some of the Abu troops lay in the town itself. The forest near Dharwar had recently been cleared and the langurs concentrated

in what little remained, with the result that their population density reached 134 langurs per km².

Troops at these sites typically contained only one adult male, and strange males did not enter the troops by a gradual process similar to that observed at Junbesi. Nontroop males were therefore numerous, and these joined together in male groups of as many as 60 members. Both male groups and lone males attacked bisexual troops. In successful attacks—*male takeovers*—they drove out the male originally in the troop, and a *usurper* became the troop's new

Figure 3. A group of females huddle around an infant in a meadow above Junbesi. Their troop, on the flanks of 6,959-meter Numbur peak, lives at higher elevations than any other langurs in this Himalayan region.

leader. Several infants disappeared during some of these takeovers, and both Sugiyama and Hrdy assumed that the usurping males killed the missing infants. Both observers noted that some of the infants' mothers soon became pregnant, and Hrdy elaborated a model in which infant killing is an evolved behavioral trait of male langurs, selected for because it renders victims' mothers quickly available for insemination. We will review the events observed at Dharwar, Jodhpur, and Abu before turning to the model.

The circumstances surrounding the disappearances of some 40 infant langurs at these sites are known in varying degrees. Infant disappearance

following a male takeover was first noted at Dharwar in 1962. Eyewitness accounts of the takeover of two troops at Dharwar are available, and Sugiyama's (1965) description of the takeover of the 30th troop there is perhaps the most detailed. The downfall of Z, the original male leader of the 30th troop, began when three of his females, one in estrus and one accompanied by her infant, approached and actually entered an outside male group. Four days later, following repeated fighting in which Z suffered serious wounds, L, the largest member of the male group, was established as the lone male of the "new" 30th

troop. Fighting among L, Z, the rest of the male group, and the 30th troop females and juveniles continued sporadically for 15 days after the start of the takeover, and during this period two infants disappeared. Another infant survived the protracted battling with what appeared to be a canine bite wound. After six days, still alive, it was abandoned by its mother.

Despite the continued fighting, L was quickly able to consolidate his position, and ten days after the start of the takeover, he was leader of the new troop. As he gradually stopped attacking the females, they stopped running away screaming when he approached.

This reduction of tension was apparently somewhat illusory, because four more infants disappeared in the month and a half after the last observed contact between the 30th troop and either Z or other members of the male group. Like the earlier disappearances and wounding of infants, three of these new disappearances were discovered in the morning at the beginning of daily observation periods. The fourth death followed the only male attack on an infant that was actually observed at Dharwar: L attacked a mother-infant pair and bit the infant near the base of the tail.

Three other takeovers at Dharwar are mentioned by Sugiyama (1967), and four or five infants reportedly disappeared during one of these. A more complete account is available of the experimentally induced takeover of the 2nd troop at Dharwar (Sugiyama 1966). Investigators removed the male leader from the troop, and males from two adjacent troops then contested control over it. Events did not differ drastically from those already observed during the takeover of the 30th troop. Four infants disappeared, two of them after receiving wounds adjudged the result of langur canine bites. Unfortunately, neither the attacks of the two wounded infants nor the events leading to the disappearance of the other two were witnessed.

The next reports of infant disappearance followed three cases of infanticide at Jodhpur (Mohnot 1971). In another somewhat artificial removal of a troop leader, the male and 70 other members of an 82-member one-male troop died of disease or poisoning. An outside male group attacked the survivors, and one of this male group established himself as the new leader. Mohnot's report includes his eyewitness account of this male killing three infants within a month of his takeover. A fourth infant disappeared during this time, and a fifth vanished six months later.

The most recent reports of infant disappearance and male takeover come from Abu, where Hrdy (1974, 1977) observed langurs for five 2-to-3-month periods between June 1971 and June 1975. On the basis of her own observations, from reports of Abu townspeople, and from changes in troop membership, she inferred that at least nine takeovers of the Abu

B-6 and B-3 troops occurred during this 4-year period. At least 17 infants disappeared from the two troops, and infant mortality in the B-6 troop between 1971 and 1974 reached an astonishing 83%. The events surrounding these dramatic levels of infant death remain somewhat obscure. Hrdy witnessed male attacks on infants after three of the takeovers, but although one such attack led to serious injury, none inflicted fatal wounds. Only the reports of casual bystanders linked male langurs to any infant deaths.

When the evidence is reviewed carefully, the immediate causes of infant death remain perhaps the least understood aspect of the takeover/disappearance phenomenon. There is no question that males have definitely injured infants following some takeovers; however, the motivations behind their aggression are far from clear. Hrdy described the male assaults on mother-infant pairs that she saw as "highly goal-directed and organized" attacks specifically aimed at the infants, but Mohnot attributed the cases of actual infanticide that he witnessed to the male's "simultaneous sexual excitement and enragement." It is noteworthy that at least two of the three Jodhpur infanticides occurred during battles between the attacking male and members of his old male group.

It is quite probable that the instance Hrdy witnessed in which an infant was seriously injured may have occurred in a similar contest, because the male from a nearby troop chased the infant's attacker only moments after the assault. At Dharwar, as at Jodhpur, male attacks on females were frequent, particularly when the females approached male groups. This behavior raises serious doubts as to whether *infants* are the targets of the attacks; after the takeover of the 30th troop at Dharwar, the new leader continued to attack at least four females even after they had lost their infants.

The role of females in the takeover/disappearance complex is also far from understood. The central question is whether the deaths of the infants hastened the subsequent pregnancies of their mothers or whether the females might have given birth when they did even had they *not* lost their infants.

A great increase in sexual behavior typically followed takeovers, but data on subsequent pregnancies are available in only four instances. Langurs have a gestation period of approximately 6½ months. Following the takeovers of the 30th and 2nd troops at Dharwar, and the 1972 takeover of the B-3 troop at Abu, 70% (11 of 15) of the females who lost infants gave birth within 8 months. At Jodhpur, however, none of the 4 females who lost infants had given birth even a year after the takeover.

Even where births have followed infant disappearances, the significance of the timing is difficult to evaluate. Data available from the infanticide sites indicate that the age of the missing infants is a crucial factor. At the time of their death the infants at Jodhpur were, at most, 3 months old, and the mother of a 2-week-old infant experimentally removed from the 2nd Dharwar troop had not given birth even a year after the infant's loss. In marked contrast, the pregnancies begun just after the takeovers at Dharwar and Abu followed the disappearances of infants aged 6 months to a year. These relatively advanced ages lead inevitably to the question of how far apart langur births are when infants survive. Although we do not have a definite answer, estimates of langur birth intervals in the wild usually range from 1½ to 2 years. At the Berkeley colony, langur mothers with surviving infants give birth to subsequent offspring after an average interval of 16.3 months (Vonder Haar Laws, ms.), and data from Junbesi indicate that similar intervals occur in the wild.

Such short birth intervals mean that the infant disappearances at Dharwar and Abu in fact may not have drastically altered normal female reproductive cycles. Females who became pregnant after the loss of their infants may have done so simply because of the passage of time, their receptivity perhaps accelerating only slightly as a result of the sudden presence of new males. Data on the behavior of females obtained immediately prior to takeovers are too few to support any definitive statements, but it is interesting that the mother of an infant who subsequently disappeared approached the male group—with the infant—at the very start of the takeover of the 30th troop at Dharwar. Future research will reveal whether changes in female behavior—such as those induced by females' attraction to male groups—have an important role in precipitating takeovers.

Sugiyama (1965, 1966) reported that female sexual activity was noticeable within hours after the invasion of a troop by a male group. He surmised that the sudden appearance of the new males might be effective in inducing receptivity and explained the desertion of wounded infants by their mothers as a result of the mothers' already developing bonds with new males, and thus their loss of motivation to care for injured offspring.

After reviewing the evidence, we adjudge that it is at present not at all clear that female receptivity *follows* infant disappearance. If birth intervals are short, and strange males can indeed induce estrus in the mothers of older infants, the weaning infants' disappearance can more easily be considered to follow from the unfortunate combination of newly estrous females and sexually active and competitive males at a time when the infants are at a particularly vulnerable age.

Explaining infant death

The data from the three "infanticide" sites are too equivocal to guide interpreters toward any single conclusion. Hrdy's recent model of infanticide as a reproductive strategy carries the interpretation of the data further than any other hypothesis, but to do so it requires many assumptions.

According to the model, the history of a langur troop is cyclic. Adult males tolerate their own sons, and thus one-male troops may grow to include several adult males belonging to different generations. However, takeover by outside males is a normal process that checks development into a multi-male troop. Takeovers are most frequently successful in densely populated areas because of the high numbers of nontroop males. Most troops in such areas contain only one adult male, and the length of time that he remains in the troop is short.

Infanticide is explained as an evolutionary adaptation to this brief male tenure. When a male enters a troop after a successful takeover, the un-

weaned infants already in the troop are unrelated to him; because their presence postpones their mothers' fertility, they are obstacles to his reproductive success, and he systematically kills them. Natural selection is viewed as favoring male behavior that will remove the infants, and infanticide is seen as an easy solution.

Selective pressures favoring infanticide are especially powerful where there are many takeovers. First, situations in which males enter new troops—opportunities for "worthwhile" infanticide—are frequent. Second, and more important, a male can "expect" that he himself will soon be driven from his new troop, and his promptness in fathering infants is crucial to his evolutionary success. If killing their infants brings mothers into estrus, infanticide evolves essentially as an extension of male rivalry (and competition between male and female) into the past. An infanticidal male enjoys accelerated and lengthened access to receptive females at their expense, that of their infants, and that of the male who preceded him. The genes of the infanticidal male, rather than those of the more tolerant deposed male, are transmitted to the next generation and incorporated into the gene pool.

Under this model, infanticide can drastically affect the reproductive success of langurs of either sex, and many behaviors function to control or mitigate its effects. Not only must male langurs kill the progeny of their competitors, they must also *not* injure their own. Males moved back and forth quite frequently at Abu, and they sometimes usurped troops that probably contained their own offspring, and thus the model proposes a mechanism that enables males to avoid attacking their own children. Interestingly, the mechanism essentially concerns females, not infants: a male does not attack a female-infant pair if he was in consort with the female at the time of the infant's conception. The troubling questions raised by the gray langur's cooperative mothering and the possibility of adoption remain unanswered.

Since loss of her infant is always detrimental to a female, female counterstrategies against infanticide are thought to be even more elaborate. These strategies range from her in-

terference with a male's "paternity testing" to her ability to select the lesser of two evils. According to the model, a female who has already conceived will again become receptive after a takeover in order to deceive the usurping male into sparing her infant as though it were also his own. Females know when to cut their losses; a mother will desert her wounded infant in the face of a male attack after "practical evaluation of what *this* infant's chances are, weighed against the probability that her next infant will survive." A female finally mates with the killer of her infant because "a female who sexually boycotted infanticides would do so to the detriment of her male progeny. Her sons would suffer in competition with the offspring of nondiscriminating mothers." (Hrdy 1977:48–49).

Hrdy's model does not necessarily impute the langurs' decisions to conscious processes, but langurs are credited with remarkable abilities. Before committing infanticide, a usurping male in effect judges his potential victim's age and counts back one langur gestation period—about 6½ months from his birth date. Armed then with his target's date of conception—18 months in the past in the case of a yearling infant—he guards against eroding his own reproductive success by remembering whether he was in consort with the infant's caretaker at the crucial time. Females make similar calculations; before abandoning her infant, a mother evaluates its cost to her so far in terms of time and energy, the cost of its defense, the likelihood of its survival, and the cost and chances of a new infant.

This is an elegant and remarkably thorough model, but if we recall what is actually *known* of the events at Dharwar, Jodhpur, and Abu, it is clear that the model explains not the results of observation but the products of assumption. The primary assumption concerns the occurrence of infanticide itself. As we have seen, from all of India there has been clear-cut evidence of only four langur infants being killed by males. This is despite the fact that infanticide is a cause of infant death that would be likely to be seen. Visibility at Dharwar, Jodhpur, and Abu was apparently good; langurs are diurnal, and infanticide presumably takes place in the daylight; and, since habituated animals were studied, the presence of observers presumably would not deter the behavior. In spite of this, the model rests on the assumption that infanticide was the leading cause of death among the langurs at the three sites. The disappearance of 39 infants became evidence of the regular occurrence of infanticide.

The assumption does not stop at the cause of infant death; the identity and motivations of the males who committed the purported infanticides are also crucial to the model. Usurping males killed the infants in "highly goal-directed and organized attacks" which are largely the result of their particular genetic endowments. When an infant langur disappears in a situation marked by restricted space, violent fighting and chasing among males and females, and frequent harassment by men and dogs, we are asked to attribute his death to deliberate attack by a member of his own species and to name a specific male as his attacker.

The second major assumption concerns the effects of the purported infanticides on the reproductive behavior of the victims' mothers. Some females appear to have become receptive before takeovers; some have become so following their exposure to strange, usurping males but prior to the disappearance of their infants; and some, indeed, have come into estrus following the deaths of their infants. The available data seem to indicate that in some cases a mother's return to estrus can endanger her infant, but the model asks us to assume one direct relation: because the infant is killed, the mother comes into estrus and is impregnated.

These two assumptions are necessary to impart an easily explainable *order* to the chaotic events at Dharwar, Jodhpur, and Abu. A third assumption is needed to justify an *evolutionary* rationale for "infanticide as a reproductive strategy": the takeovers at these three sites represent normal langur behavior over the centuries. It is supposed that takeovers do occur at locations such as Orcha, Kaukori, and Junbesi, but are infrequent because of low numbers of nontroop males. This rarity explains why takeovers are never observed and why multi-male troops persist at those sites.

The evolutionary context

Our studies lead us to argue from the opposite direction. The pattern that, we believe, represents the typical and evolutionarily adaptive behavior of gray langurs includes male rivalry but not any egregious interference with female reproduction or the killing of infants, and does not have any systematic effect on troop structure. Males contest for status directly among themselves, by means of

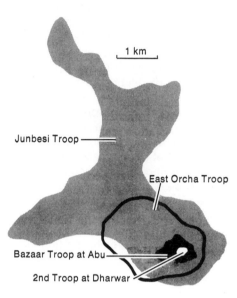

Figure 4. A comparison of the ranges of four langur troops illustrates variation in range size, which seems to influence troop structure. Troops at Junbesi and Orcha contained several adult males, but those at Abu and Dharwar contained only one, and new males could enter only after driving out the old.

threat or combat. The occasional expulsion of males from troops during conflict is balanced both by their frequent return and the ability of strange males sometimes to enter troops without major disruption. The latter counters any tendency toward a pattern of one-male troops and reduces the numbers of males living outside troops.

This pattern has been reported from almost every ecological condition in which the gray langur lives; it has been found wherever langurs have been studied in habitats that remained close to the aboriginal condition; and it has persisted in spite of greatly varying population densities. In sharp contrast, the one-male troop/male takeover pattern has been observed *only* where langur popula-

tion densities are high, and where human influence on the ecology is immense. This close correspondence of the specialized pattern with distinctive ecological conditions raises the possibility that the pattern arises directly as an effect of the conditions, and the conditions have not been prevalent for long.

The crowding associated with the pattern of male takeover has almost certainly arisen because of habitat destruction and predator removal in historical—not evolutionary—time. Agriculture and human handouts helped support the langurs at Abu and Jodhpur in their constricted, suburban home ranges, and at Dharwar recent logging has concentrated the langurs in what remains of the forest. Predator removal is another factor of undoubted importance that has been dramatic only since the development of agriculture and urbanization. Schaller's (1967) demonstration that the large Indian cats prey substantially on langurs, as well as the low langur densities at Orcha, Melemchi, and Junbesi, where predator populations remain effectively undisturbed, indicate that predation normally exercises an important restraint on langur population sizes. Habitat disruption and faunal impoverishment even approaching present levels began only after the advent of the Mughal Empire (Mukherjee 1974); both the introduction of firearms and an increase in poaching after Indian independence have since led to acceleration of the destruction of Indian wildlife.

Small home ranges stand out as a major characteristic of the "infanticide" sites, particularly when they are compared to similar sites elsewhere (Fig. 4). The 40-hectare home ranges at Abu should be compared to the 770- and 1,200-hectare ranges at Kaukori and Junbesi, respectively, and the 16.8-hectare ranges in the Dharwar forest should similarly be contrasted with the 380- and 220-hectare ranges at Orcha and Melemchi.

Since langur crowding is almost certainly a very recent phenomenon in terms of evolutionary time, it is appropriate to search for the origins of the male-takeover pattern in the effects of crowding on those patterns of behavior that probably *did* exist in the langur's evolutionary past. Although, as we have seen, there are powerful male rivalries within the widespread multi-male troop pattern, such rivalries are balanced by processes that allow outside males to enter troops. These balancing processes, however, require time and space. Crowding removes the space requisite for the resolution of male conflict within troops, or the peaceful entry of strange males into the troops. The levels of human harassment typical of crowded sites effectively remove the requisite time; the gradual reduction of tension between males is impossible if the animals are being chased frequently.

Clearly, it will be difficult for multi-male troops to exist in crowded habitats. Crowding increases male tension by enforcing close proximity between animals, and by increasing competition for food—particularly where humans feed the animals—or for receptive females. Crowding also escalates the consequences of male agonistic behavior. Any serious fighting will easily drive defeated males beyond home range or territorial boundaries and into direct confrontation with other troops. A neighboring troop is extremely unlikely to tolerate a precipitate entry into *its* range, particularly since it has probably had previous territorial encounters with the intruders. The defeated male will probably be unable to remain in the area; an interaction that might lead merely to temporary peripheralization in an uncrowded habitat will drive him permanently from his troop in a crowded area. High density increases the frequency of aggression and greatly magnifies its effect; losers become nontroop males.

The argument from here to the pattern of male takeover observed at Dharwar, Jodhpur, and Abu is short. The exiles form male groups, and, again because of crowding, contacts between these male groups and the troops are frequent. Since the male groups are numerically superior to the lone males in reproductive troops, takeover attempts are often successful.

A pattern of one-male troops and male takeovers does not represent normal langur behavior either today or in the evolutionary past. Neither does the one-male troop/male-take-over complex represent normal langur

social organization; it is instead a symptom of the failure of the multi-male troop structure to endure environmental conditions far different from those in which it evolved.

This argument leads to a simple explanation of infant mortality at Dharwar, Jodhpur, and Abu. Infanticide cannot be an evolutionary adaptation to a takeover pattern that is itself only an artifact of recent environmental pressures. Infant mortality at the three sites need have no evolutionary significance beyond the obviously detrimental. The observed events need not be interpreted as evidence of intricate but hidden patterns; they are, more simply, the indicators of chaos that they appear to be. Male takeovers involve fighting, chasing, and display among both males and females, approach of male groups by mothers carrying infants, and male attacks on mother-infant pairs. Any of these creates danger for infants. Removing the need to ascribe evolutionary function to infant death makes it unnecessary to assume that the 35 out of 39 deaths at the three sites which resulted from unknown causes were all products of infanticide by adult male langurs possessing the capacity to discern their own offspring.

The relationship between infant death and female reproductive cycles represents a similar situation. Pregnant females sometimes exhibit estrous cycles, just as they do in undisturbed habitats and in captivity, and females under attack sometimes abandon their infants. Again, there is no reason to extract from the observed variability of behavior either elegant "female counterstrategies" or the single rule that estrus is caused by infant loss.

The evolution of langur behavior took hundreds of thousands of years, and we can be confident that langurs were much as they are today long before *Homo sapiens'* rise to eminence. In the last few thousand years, at most, civilization has transformed India, and there now are langurs living under conditions that simply did not exist ten thousand years ago. To study the evolution of langur behavior, we must consider carefully the whole range of behavior that exists today. Behaviors that appear everywhere, or almost everywhere, and that exist in habitats little altered by man

are those most likely to have survived from the evolutionary past. Behaviors that occur only at selected sites and only under novel environmental conditions are quite likely to be artifacts of modern conditions. These new patterns of behavior are of great interest—not because they reveal the course of langur evolution, but because they demonstrate the disturbing results when even a remarkably adaptable species is pushed beyond the range of its flexibility.

References

Bishop, N. H. Social behavior of langur monkeys (*Presbytis entellus*) in a high altitude environment. 1975 diss., Univ. of California, Berkeley.

Curtin, R. A. Socioecology of the common langur, *Presbytis entellus,* in the Nepal Himalaya. 1975 diss., Univ. of California, Berkeley.

Dolhinow, P. 1972. The north Indian langur. In *Primate Patterns,* ed. P. Dolhinow. Holt, Rinehart and Winston.

Hrdy, S. B. 1974. Male-male competition and infanticide among the langurs (*Presbytis entellus*) of Abu, Rajasthan. *Folia Primat.* 22:19–58.

————. 1977. Infanticide as a primate reproductive strategy. *Am. Sci.* 65:40–49.

Mohnot, S. M. 1971. Some aspects of social change and infant-killing in the Hanuman langur, *Presbytis entellus* (Primates: Cercopithecidae) in Western India. *Mammalia* 35:175–98.

Mukherjee, A. K. 1974. Some examples of recent faunal impoverishment and regression. In *Ecology and Biogeography in India,* ed. M. Mani. The Hague: D. W. Junk.

Ripley, S. 1967. Intertroop encounters among Ceylon gray langurs. In *Social Communication among Primates,* ed. S. Altmann. Univ. of Chicago Press.

Schaller, G. B. 1967. *The Deer and the Tiger: A Study of Wildlife in India.* Univ. of Chicago Press.

Sugiyama, Y. 1965. On the social change of Hanuman langurs (*Presbytis entellus*) in their natural condition. *Primates* 6:381–417.

————. 1966. An artificial social change in a Hanuman langur troop. *Primates* 7:41–72.

————. 1967. Social organization of Hanuman langurs. In *Social Communication among Primates,* ed. S. Altmann. Univ. of Chicago Press.

Vonder Haar Laws, J. Female langur monkey (*Presbytis entellus*) reproduction and sexual behavior: Records and analyses from captive social groups. Manuscript.

PART II
The Mechanisms of Behavior

The study of animal behavior involves research on many different topics. On the one hand, we can explore how elements of an animal's internal machinery make behavior possible—that is, *how* the nerve cells, hormones and genetic mechanisms within an individual enable it to do something. On the other hand, we can try to figure out *why* these mechanisms (and the behaviors they control) evolved and spread through the species in the past. These are called the proximate (mechanistic) and ultimate (evolutionary) causes of behavior, respectively. The eight articles in this section show how these two types of causes can be subdivided and yet integrated to achieve a more thorough understanding of behavior.

Kay Holekamp and Paul Sherman use the dispersal behavior of Belding's ground squirrels to make the point that all the causes of this or any other behavior can be placed into one of four "levels of analysis," two that address proximate causes and two that address ultimate causes. The four levels are (1) the underlying developmental or (2) physiological causes of the behavior *or* (3) the evolutionary history or (4) fitness consequences of the behavior. These are complementary, not mutually exclusive, approaches: an answer at the developmental level, for example, can be correct without eliminating (or minimizing the importance of) hypotheses about the fitness consequences or evolution of the behavior.

Knowing what is meant by different levels of analysis should help eliminate certain academic disagreements. For example, how would Holekamp or Sherman respond to someone who claimed that the hypothesis that male dispersal is caused by male hormones cannot be correct because male dispersal is due to a long evolutionary history of greater male movement among mammals? Or someone who claimed that because young males leave home to avoid mating incestuously, the hypothesis that males are more exploratory and less fearful than juvenile females cannot be correct?

Not everyone, however, agrees that four fundamental questions underlie the entire field of behavioral research. Some persons have challenged the reasoning used by Holekamp and Sherman when they say that male Belding's ground squirrels disperse because this behavior reduces opportunities for deleterious father–daughter and sister–brother inbreeding. These critics claim that the fitness consequences of a behavior cannot be invoked as a *cause* of that behavior when they really are an *effect* of the behavior. Do Holekamp and Sherman provide the basis for a suitable response to this criticism?

In *From Society to Genes with the Honey Bee*, Gene Robinson focuses on the multiple kinds of proximate explanations for the ability of honey bee workers to adopt different roles in their marvelously complex societies. Most people seem to think that insects are highly programmed robotic automatons, when in reality they are not, as the honey bee demonstrates. These insects not only can switch from being hive cleaners to nurses to comb builders to hive defenders and foragers, but can also adjust the timing of transitions between these professions. Their capacity to do so can be due to their sensitivity to certain environmental cues, to hormonal changes that take place within their bodies, to the developmental changes that occur within their brains, and to the genes that reside inside the nuclei of their cells. Robinson explains how he and his colleagues have explored each of these different types of proximate causes of bee behavior and how they are interrelated.

A challenging assignment would be to diagram the series of events, beginning with the development of a honey bee egg, that underlie the behavioral transition made by a worker as she leaves the hive on her first foraging flight. Such a diagram could include the full spectrum of genetic-developmental, hormonal, and neurophysiological factors discussed by Robinson.

The honey bee story also offers us a special opportunity to think about the connection between genes and behavior. As the title of Robinson's article suggests, the honey bee's social behavior has genetic causes, and yet the bee's behavior does not consist of a set of rigid instincts of the sort commonly associated with genetic determination. How can flexible, adaptable behavior be dependent upon an animal's DNA? The honey bee provides an answer that is worth examining.

Donald Griffin also makes the case that flexibility of behavior is not limited to human beings. He uses evidence on this point to raise the possibility that the adaptability of our fellow creatures and their ability to adjust at times to novel conditions could essentially have the same kind of underlying proximate foundation as our own capacity to choose one behavioral option over another. That is, that they have conscious awareness and the ability to think through a problem. He encourages researchers to try to get inside the brains of other animals to determine what they are "aware" of. If we could, our understanding of animal behavior would be greatly advanced, according to Griffin.

Griffin's article raises the question of how one would decide whether an animal's actions are based on conscious, as opposed to unconscious, decision making. Of course, the difficulty lies in finding a situation in which conscious and unconscious thought yield different predictions about how an animal will behave. However, given that humans have consciousness, it seems reasonable to suppose that some other species, especially primates, also experience some form of consciousness. Indeed, to think otherwise would be to presume an evolutionary discontinuity between humans and all other creatures. Thus, it seems reasonable to seriously consider the possibility that some animals engage in conscious thinking.

How would it change your view of "lower animals" if you knew that their decisions were indeed conscious? If crickets (or fish or snakes) could think, would you feel any different about the need to treat them humanely in scientific experiments? Griffin points out that many scientists resist the notion that animals other than ourselves are consciously aware of certain things. He speaks out strongly against this attitude, but readers may wish to ask why so many of Griffin's fellow scientists still question the utility of trying to examine consciousness in other animals.

Griffin also discusses the possible fitness consequences of consciousness mechanisms, and so his article integrates several different levels of analysis of behavior. He proposes that animals might gain reproductive benefits by having conscious thoughts in certain situations. But are there any reproductive costs associated with this ability? If so, what would determine the spread of genes underlying the development of the brain systems required to produce consciousness in a killdeer, a sea otter, or any other animal, ourselves included? What sorts of human responses are conscious and which are unconscious, and how might knowing this help us understand the costs and benefits of consciousness in ourselves and other creatures?

Next, Marc Hauser shows us that some animals possess an ability that most people consider a special attribute of the human intellect: the ability to count. Indeed, Hauser makes the case that this mental capacity is far more widespread than is generally acknowledged. If other species have a number sense, we would understand more about the evolutionary origins (see Part III) and adaptive value (Part IV) of our own special numerical abilities. But the main focus of Hauser's research is on the proximate mechanisms that make a numerical sense possible. He, like Griffin, wants to get inside the heads of other animal species to figure out what numerical operations, if any, are occurring there. The way that he deduces what computations take place in the brains of monkeys (and those of nonverbal human infants) may be worth your analyzing formally in terms of causal question, hypothesis, prediction, test and conclusion. When you have completed this analysis, you can then return to Donald Griffin's article, which is similar in its goals. How do the approaches of the two researchers differ? How do the two articles compare in terms of the persuasiveness of their conclusions, and why?

Meredith West and Andrew King study the developmental level of analysis of a behavior—in this case,

singing by starlings. According to modern developmental theory, all behavior is the ontogenetic product of gene–environment interactions. Thus, the physiological systems that make it possible for a starling to sing are believed to have been produced through the interplay of genetic information and the bird's environment. Developmental theory, however, does not specify what elements of the inanimate or animate environment will affect the development of any given behavior.

West and King wondered how the social environment might shape the development of starling song, since these birds are skillful vocal mimics capable of imitating human speech. The authors describe how they determined whether or not and in what ways social cues have developmental effects on communication in the starling. In doing so, they clear up an entertaining mystery involving Mozart and his pet starling. If West and King are correct, what kinds of bird species should be immune to social influences on the development of their vocalizations? How do you suppose the brood-parasitic birds that are raised individually in the nests of other species acquire species-typical songs, or indeed, even come to recognize members of their own species? Does the evidence on how starling song develops suggest testable hypotheses about how humans acquire language?

When genes and environmental factors interact, they help build the physiological foundations of behavior, including hormonal and neuronal systems. The chapter by John Wingfield and his coworkers teaches us how a single hormone influences many elements of the behavior of breeding birds. The hormone in question is testosterone, a chemical long implicated in the regulation of sexual and aggressive behavior in vertebrates.

As Wingfield and his coauthors make clear, it has not been easy to determine precisely what behavior patterns are controlled by testosterone. (Why has the task been so difficult?) However, Wingfield's team believes that testosterone's main function in birds is to facilitate male aggression against rival males, something that is essential for males when they are establishing their territories and defending them against intruders during the breeding season. How do these authors use measurements of testosterone concentrations in the blood as a means of testing the "challenge hypothesis," which states that testosterone primes males to respond to aggressive challenges from others? How do they explain results that contradict this hypothesis? Are they striving to falsify the hypothesis using the "strong inference" approach or are they committed to the challenge hypothesis and eager to "explain away" any evidence to the contrary?

Stephan Schoech's article also illustrates how it is possible to study the physiological causes of behavior, including its hormonal bases, in a most interesting bird, the Florida scrub-jay. This jay is one of a minority of avian species in which young adults may remain with their parents for a year or two as nonbreeding family members who help their parents rear their younger brothers and sisters. Helping-at-the-nest raises all sorts of intriguing evolutionary issues because it is hard to imagine how not breeding can be selectively advantageous for a young scrub-jay. Some answers will be offered in Part V.

But helping behavior is also fascinating at the physiological level of analysis, since something about the internal mechanisms of young adult scrub-jays keeps them from even trying to reproduce. Instead, they put all their time and energy into feeding and protecting their siblings. Schoech is concerned primarily (but not exclusively) with the possible hormonal causes of delayed breeding and cooperative helping. As noted above, testosterone is widely believed to regulate male sex drive in many vertebrate species. Perhaps the failure of young male helpers to breed in the Florida scrub-jay is linked to low testosterone concentrations. Schoech collected data pertinent to this hypothesis. What were these data and what conclusion do they support?

In general, how different are helper and breeder jays in their hormonal state, and what do these findings have to say about the general theory that reproductive and parental behaviors of vertebrates are under hormonal control? Note also that Schoech mentions that an increase in, say, testosterone could actually be caused by presence of a sexual partner rather than a rise in testosterone causing the male to seek out a sexual partner. If hormone changes could be a *response to* changes in behavior, rather than *causing* the changes in behavior, how might we interpret the data presented in Figures 4 through 8 of Schoech's article? Do Schoech's results contradict those of Wingfield and his colleagues and, if so, can you see a way the two sets of findings can be reconciled? Finally, what hypotheses does Schoech reject in the course of his paper? Is research that leads to the rejection of a hypothesis as satisfying or as important as research that enables one to confirm a hypothesis? What might Woodward and Goodstein (Part I) have to say about this?

Hormonal factors also play a role in Andrew Bass's article on the mechanisms that control song in the midshipman fish. In this strange animal, two forms of sexually mature males compete for females: a large type that defends nesting territories under intertidal rocks where it produces a loud humming song that females find attractive, and a small, nonterritorial type that cannot sing, but only grunt, and "sneaks" matings instead of calling females to a nest site. The large singing males can produce so much noise that owners of houseboats floating above the midshipmen's habitat complain of insomnia in the spring when the fish are calling. Bass's article explores a large number of mechanisms, including certain hormonal ones, that are responsible for the differences between the two types of males. Thus, the discovery that testosterone concentrations differ dramatically in the blood of large and small males suggests that this hormone is involved in the development of the distinc-

tive song machinery that large males possess.

However, Bass devotes most of his article to describing how he and others have been able to discover the means by which nerve cells in the brain control the muscles involved in sound production by the large male midshipmen. In addition, he discusses the possible evolutionary reasons as to why two such different kinds of males coexist in this species. Readers of this article may wish to organize the various hypotheses reviewed by Bass according to the levels of analysis scheme developed by Holekamp and Sherman in their article on ground squirrels. Is their system of classification adequate to deal with the large number of causal explanations in this chapter? Can each one of Bass's hypotheses be unambiguously assigned to one of the categories outlined by Holekamp and Sherman? If not, should new levels of analysis be created, or should the existing four levels be subdivided?

In the final article of Part II, Mike May describes how he attempted to identify the physiological mechanisms that make it possible for certain night-flying insects to detect and escape from bats, a topic that has long fascinated biologists. A particularly appealing element of May's article is his reconstruction of his thoughts as he designed and conducted his doctoral research and how one experiment led to another, allowing readers to catch a glimpse of a young researcher's own ontogeny.

In the course of his article, May notes that "Not all useful observations come as a result of premeditated experimental design; sometimes it's useful just to play with the equipment." (The same point is raised by Winston and Slessor in Part V, where they describe their discovery of a method to test how certain chemicals affect the behavior of worker honey bees.) What do you think about this claim? Do you believe that May had no hypothesis at all in mind while he was playing around with a cricket and his experimental equipment? Was May being a Baconian inductivist (see Woodward and Goodstein's article in Part I) during this phase of his work, or was he actually testing specific predictions even though he had not written them down beforehand? Is it possible to make scientific progress without using the scientific method? Does the method always require premeditation?

The articles in Part II show that many different avenues exist for the study of behavioral mechanisms. Moreover, they show how studies of mechanism, ontogeny, function, and evolution are interrelated. An understanding of precisely how a physiological element works helps researchers understand what natural selection designed it to do. And by considering the ultimate, evolutionary aspects of behavior, scientists can better identify what reproduction-enhancing mechanisms to look for in a given species.

Why Male Ground Squirrels Disperse

Kay E. Holekamp
Paul W. Sherman

When they are about two months old, male Belding's ground squirrels *(Spermophilus beldingi)* leave the burrow where they were born, never to return. Their sisters behave quite differently, remaining near home throughout their lives. Why do juvenile males, and only males, disperse? This deceptively simple question, which has intrigued us for more than a decade *(1, 2)*, has led us to investigate evolutionary, ecological, ontogenetic, and mechanistic explanations. Only recently have answers begun to emerge.

Dispersal, defined as a complete and permanent emigration from an individual's home range, occurs sometime in the life cycle of nearly all organisms. There are two major types: breeding dispersal, the movement of adults between reproductive episodes, and natal dispersal, the emigration of young from their birthplace *(3, 4)*. Natal dispersal occurs in virtually all birds and mammals prior to first reproduction. In most mammals, young males emigrate while their sisters remain near home (the females are said to be philopatric); in birds, the reverse occurs *(4–6)*. Although naturalists have long been aware of these patterns, attempts to understand their causal bases have been hindered by both practical and theoretical problems. The former stem from difficulties of monitoring dispersal by free-living animals, and of quantifying the advantages and disadvantages of emigration *(6)*. The latter stem from failure to distinguish the two types of dispersal, and from confusion among immediate and long-term explanations for each type.

We begin with a discussion of the latter point and

A multilevel analysis helps us to understand why male and not female Belding's ground squirrels leave the area where they were born

develop the idea that natal dispersal, like other behaviors and phenotypic attributes, can be understood from multiple, complementary perspectives. Separating these levels of analysis helps organize hypotheses about cause and effect in biology *(7)*. In the case of natal dispersal, this approach can minimize misunderstandings in terminology and allow for clearer focus on the issues of interest.

Questions of the general form "Why does animal A exhibit trait X?" have always caused confusion among biologists. And even today, the literature is full of examples. The nature-nurture controversy, which arose over the question of whether behaviors are innate or acquired through experience, is a classic case *(8)*. After two decades of spirited but inconclusive argument in the nature-nurture debate, it became apparent to Mayr *(9)* and Tinbergen *(10)* that a lack of consensus was caused by the failure to realize that such questions could be analyzed from multiple perspectives.

In 1961, Mayr proposed that causal explanations in biology be grouped into proximate and ultimate categories. Proximate factors operate in the day-to-day lives of individuals, whereas ultimate factors encompass births and deaths of many generations or even entire taxa. Pursuing this theme in 1963, Tinbergen further subdivided each of Mayr's categories. He noted that complete proximate explanations of any behavior involve elucidating both its ontogeny in individuals and its underlying physiological mechanisms. Ultimate explanations require understanding both the evolutionary origins of the behavior and the behavior's effects on reproduction. The former involves inferring the phylogenetic history of the behavior, and the latter requires comparing the fitness consequences of present-day behavioral variants.

There are two key implications of the Mayr-Tinbergen framework. First, competition among alternative hypotheses occurs within and not between the four analytical levels. Second, at least four "correct" answers to any question about causality are possible, because explanations at one level of analysis complement rather than supersede those at another. Deciding which explanations are most interesting or satisfying is largely a matter of training and taste; debating the issue is usually fruitless *(7)*.

With the Mayr-Tinbergen framework in mind, let us turn to the question of natal dispersal in ground squir-

Kay Holekamp is a research scientist in the Department of Ornithology and Mammalogy at the California Academy of Sciences. She received a B. A. in psychology in 1973 from Smith College, and a Ph.D. in 1983 from the University of California, Berkeley. From 1983 to 1985 she studied reproductive endocrinology as a postdoctoral fellow at the University of California, Santa Cruz. She is currently observing mother-infant interactions and the development of social behaviors in hyenas in Kenya. Paul Sherman is an associate professor of animal behavior at Cornell. He received a B. A. in biology in 1971 from Stanford, and a Ph.D. in zoology in 1976 from the University of Michigan. Following a postdoctoral appointment at the University of California, Berkeley (1976–78), he joined the psychology faculty there. He moved to his present position at Cornell in 1980, and is currently studying the behaviors of naked mole-rats, Idaho ground squirrels, and wood ducks. Address for Dr. Sherman: Section of Neurobiology and Behavior, Seeley G. Mudd Hall, Cornell University, Ithaca, NY 14853.

Figure 1. A female Belding's ground squirrel *(Spermophilus beldingi)* sits with two of her pups in the central Sierra Nevada of California. The pups are about four weeks old, and have recently emerged above ground. At about six or seven weeks of age, male ground squirrels begin to disperse; young females always remain near home. The causes of male dispersal in ground squirrels and many other mammals are complex, but can be explained by using a multilevel analytical approach in which four categories of causal factors are considered separately. (Photo by Cynthia Kagarise Sherman.)

rels. Following analyses of why natal dispersal occurs from each of the four analytical perspectives, we attempt an integration and a synthesis. Our studies reaffirm the usefulness of levels of analysis in determining biological causality.

From 1974 through 1985 we studied three populations of *S. beldingi* near Yosemite National Park in the Sierra Nevada of California (Figs. 1 and 2). In each population, the animals were above ground for only four or five months during the spring and summer; during the rest of the year they hibernated *(1, 2)*. Females bore a single litter of five to seven young per season, and reared them without assistance from males. Most females began to breed as one-year-olds, but males did not mate until they were at least two. Females lived about twice as long as males, both on average (four versus two years) and at the maximum (thirteen versus seven years) *(11)*.

During each field season ground squirrels were trapped alive, weighed, and examined every two to three weeks. About 5,300 different ground squirrels were handled. The animals were marked individually and observed unobtrusively through binoculars for nearly 6,000 hours. Natal dispersal behavior was measured by a combination of direct observations, livetrapping, radio telemetry, and identification of animals killed on nearby roads *(12)*. The day on which each emigrant was last seen

within its mother's home range was defined as its date of dispersal. Only those juveniles that were actually seen after leaving their birthplace were classified as dispersers.

Observations of marked pups revealed that natal dispersal was a gradual process, visually resembling the fissioning of an amoeba (see Fig. 3). Young first emerged from their natal burrow and ceased nursing when they were about four weeks old. Two or three weeks later some youngsters began making daily excursions away from, and evening returns to, the natal burrow. Eventually these young stopped returning, restricting their activities entirely to the new home range; by definition, dispersal had occurred.

As shown in Figure 4, natal dispersal is clearly a sexually dimorphic behavior. In our studies, every one of over 300 surviving males dispersed by the end of its second summer; a large majority (92%) dispersed before their first hibernation, by the age of about 16 weeks. In contrast, only 5% of over 250 females recaptured as two-year-olds had dispersed from their mother's home range. The universality of natal dispersal by males suggested no plasticity in its occurrence; however, there was variation among individuals in the age at which dispersal occurred.

During the summer following their birth, males that

Figure 2. *S. beldingi* in the central Sierra Nevada are found above ground only four or five months of the year, during the spring and summer; they hibernate during the rest of the year. The group above is emerging from an underground burrow. Female adults bear litters of five to seven young each year and rear them in underground burrows without assistance from males. (Photo by George D. Lepp)

had dispersed as juveniles often moved again, always farther from their birthplace (Fig. 4). Yearling males were last found before hibernation an average of 170 m from their natal burrow, whereas yearling females moved on average only 25 m from home in the same time period. As two-year-olds, males mated at locations that were on average ten times farther from their natal burrows than the mating locations of females (13).

By the time they were two years old, male *S. beldingi* had attained adult body size. In the early spring they collected on low ridges beneath which females typically hibernated. As snow melted and females emerged, the males established small mating territories. Only the most physically dominant males—especially the old, heavy ones—retained territories throughout the three-week mating period. Although dominant males usually copulated with multiple females, the majority of males rarely mated. After mating, the most polygynous males again dispersed. They typically settled far from the places where they had mated; indeed, their new home ranges usually did not include their mating territories. Less successful males tended not to move, and they attempted to mate the following season in the same area where they were previously unsuccessful.

Females were all quite sedentary. After mating on a ridge top close to her hibernation burrow, each female dug a new nest burrow or refurbished an old one—sometimes her own natal burrow. There she reared her pups. As a result of philopatry, females spent their lives surrounded by and interacting with female relatives. Close kin cooperated to maintain and defend nesting territories and to warn each other when predators approached (13, 14). Natal philopatry has facilitated similar nepotism, or favoring of kin, among females in many other species of ground-dwelling sciurid rodents (15).

Physiological mechanisms

We began our analysis of natal dispersal in *S. beldingi* by considering physiological mechanisms. Of the two broad categories of such mechanisms, neuronal and hormonal, we were most interested in the latter. Gonadal steroids can influence the development of a specific behavior in two general ways: through organizational effects, which are the result of hormone action, in utero or immediately postpartum, on tissues destined to control the behavior, and through activational effects, which result from the direct actions of hormones on target tissues at the time the behavior is expressed (16). We suspected that gonadal steroids might mediate natal dispersal, and so we tested for organizational versus activational effects of androgens.

Under the activational hypothesis, levels of circulating androgens should be elevated in juvenile males at the time of natal dispersal. Conversely, in the absence of androgens, males should not disperse. To test this, we studied male pups born and reared in the laboratory. Blood samples were drawn every few weeks for four months (17). We also conducted a field experiment: soon after weaning but prior to natal dispersal, a number of juvenile males and females were gonadectomized; sham operations were performed on a smaller sample of each sex. After surgery, these juveniles were released into their natal burrow and subsequent dispersal behavior was monitored.

Castration was found to have little effect on natal dispersal. Although castrated males and those subjected to sham operations dispersed a few days later than untreated males, probably because of the trauma of surgery, castration did not significantly reduce the fraction that dispersed. Likewise, removal of ovaries did not increase the likelihood of dispersal by juvenile females. Finally, radioimmunoassays revealed only traces of testosterone in the blood of lab-reared juvenile males throughout their first four months, and no increase in circulating androgens was detected at the age when natal dispersal typically occurs (7–10 weeks).

Sex and body mass together were the most consistent predictors of dispersal status

Under the organizational hypothesis, exposing perinatal or neonatal females to androgens should masculinize subsequent behavior, including natal dispersal. We tested this idea by capturing pregnant females and housing them at a field camp until they gave birth. Soon after parturition, female pups were injected with a small amount of testosterone propionate dissolved in oil; a control group was given oil only. After treatment, the pups and their mothers were taken back to the field, where the mothers found suitable empty burrows and successfully reared their young.

Twelve of the female pups treated with androgens were located when they were at least 60 days old, and

75% of them had dispersed *(17)*. The distances they had traveled and their dispersal paths closely resembled those of juvenile males. By comparison, only 8% of untreated juvenile females in the same study area had dispersed by day 60, whereas 60% of juvenile males from the transplanted litters and 74% of males from unmanipulated litters born in the same area had dispersed by day 60.

It is possible that transplantation and not treatment with androgens caused the juvenile females in our experiment to disperse; unfortunately, we were unable to test this because none of the transplanted females treated with only oil were recovered. However, transplantation did not seem to affect the behavior of the juvenile males in the experiment. Also, other behavioral evidence linked natal dispersal in the females with androgen treatment. For example, treated juvenile females did not differ significantly from untreated juvenile males of the same age, but did differ from control females with respect to several indices of locomotor and social behavior. Androgen treatment masculinized much of the behavior of juvenile females, apparently including the propensity to disperse.

These results, which suggest an organizational role for steroids in sexual differentiation of *S. beldingi,* are consistent with those from studies of many other vertebrates *(18).* In mammals, females are homogametic (XX) and males are heterogametic (XY), whereas in birds the situation is reversed. In each taxon, natal dispersal occurs primarily in the heterogametic sex. In both birds and mammals, sex-typical adult behavior in the homogametic sex can often be reversed by perinatal exposure to the gonadal steroid normally secreted at a particular developmental stage by the heterogametic sex. These considerations suggest that natal dispersal in mammals and birds has a common underlying mechanism, namely the organizational effects of gonadal steroids on the heterogametic sex.

Ontogenetic processes

Natal dispersal might be triggered during development by changes in either the animal's internal or external environment. We tested two hypotheses about external factors. First, natal dispersal might be caused by aggression directed at juveniles by members of their own species. Under this hypothesis, prior to or at the time of dispersal, the frequency or severity of agonistic behavior between adults and juvenile males should increase. However, observations revealed that adults neither attacked nor chased juvenile males more frequently or vigorously than juvenile females *(19),* and there was no increase in aggression toward juvenile males at the time of dispersal. Moreover, there were no differences between juvenile males and females in the number and severity of wounds inflicted by other ground squirrels. Thus the data offered no support for the social aggression hypothesis.

A second hypothesis is that natal dispersal occurred because juvenile males attempted to avoid their littermates (current and future competitors) or their mother *(20).* For a large number of litters, we found no significant relationship between litter size or sex ratio and

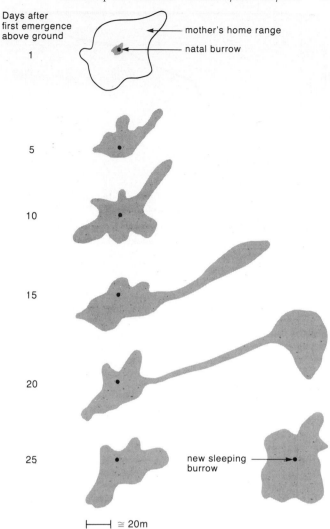

Figure 3. The process by which *S. beldingi* males disperse visually resembles the fissioning of an amoeba. When a male first emerges from the natal burrow, at an age of about four weeks, his daily range of movement is restricted to the immediate vicinity of the burrow. He soon enlarges that range into an amorphous shape, the boundaries of which are established by topographic features or the presence of other animals. By about the 15th day above ground, his range has surpassed the scope of his mother's home range. At this time he may spend long periods far from the natal burrow, yet he will return home at nightfall. Near the 25th day, when he is roughly seven weeks old, he will cease returning at dusk, thereby accomplishing dispersal. (After ref. *35.*)

dispersal behavior *(2, 19).* Males who dispersed during their natal summer were not from especially large or small litters, or predominantly male or female litters. Also, the timing of juvenile male dispersal depended neither on the mother's age nor on whether the mother was present or deceased. Thus the ontogeny of natal dispersal was apparently not linked to either of the exogenous (external) influences usually invoked to account for it.

In view of these results, we suspected that natal dispersal was triggered by endogenous (internal) factors. In particular, we hypothesized that males might stay home until they attained sufficient size or energy reserves to permit survival during the rigors of emigration. This ontogenetic-switch hypothesis predicts that juvenile

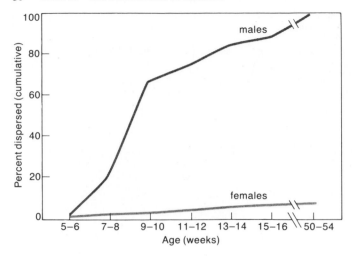

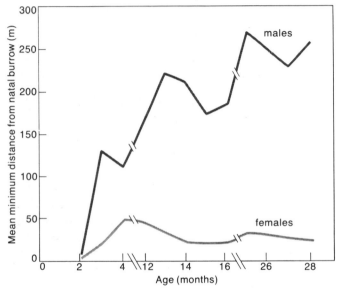

Figure 4. Although a small fraction of female *S. beldingi* disperse, the behavior is very evidently male-biased. The majority of male pups disperse by the 10th week; by about the 54th week, all males have dispersed *(above)*. Although many other mammals exhibit male-biased natal dispersal, *S. beldingi* is unusual in that all males eventually disperse. Males also move considerably farther from their natal burrows than do females, and they continue to move away from home throughout their first three years *(below)*. (After refs. 2 and 12.)

males will disperse when they attain a threshold body mass and that dispersers should be heavier, or exhibit different patterns of weight gain, than predispersal males of equivalent ages.

Our data were consistent with the ontogenetic-switch hypothesis. Emigration dates were correlated with the time at which males reached a minimum body weight of about 125 g, as shown in Figure 5. Emigrant juveniles were significantly heavier than male pups that had not yet dispersed. Most males attained the threshold weight during their natal summer, and dispersed then. Only the smallest males, who did not put on sufficient weight in the first summer, overwintered in their natal area. All these males dispersed the following season once they had become heavy enough.

Sex and body mass together were the most consistent predictors of dispersal status. Occasionally, how-

ever, predispersal and immigrant juvenile males with body weights exceeding the threshold were captured in the same area. This observation suggested that something closely associated with body weight, such as fat stores, may be the actual dispersal trigger.

Behavioral changes also accompanied natal dispersal. The frequencies of movement and distances moved per unit time by juvenile males were found to be greater than those of females, and these behaviors peaked at the time of dispersal. Relative to juvenile females, juvenile males also spent significantly more time climbing and digging and exploring nonfamilial burrows and novel objects—for example, a folding footstool; they also reemerged from a burrow into which they had been frightened much sooner than did females. These observations of spontaneous ontogenetic changes in the behavior of young males reinforced the hypothesis that endogenous factors triggered natal dispersal.

Effects on fitness

Natal dispersal might enable juvenile males to avoid fitness costs associated with life in the natal area and might allow them to obtain benefits elsewhere *(6)*. Possible disadvantages of remaining at home include shortages of food or burrows *(21)*, ectoparasite infestations or diseases, competition with older males for mates *(5, 22)*, and nuclear family incest *(4, 23, 24)*. We examined each of these hypotheses as functional explanations for natal dispersal in *S. beldingi*.

If natal dispersal occurs because of food shortages, then juveniles whose natal burrow is surrounded by abundant food should be more philopatric than those from food-poor areas; immigration to food-rich areas should exceed emigration from them; dispersing individuals should be in poorer condition (perhaps weigh less) than males of the same age residing at home; and, based on the strong sexual dimorphism in natal dispersal, food requirements of young males and females should differ.

Detailed observations revealed that juvenile males and females ate similar amounts of the same plants and at similar rates. Juvenile males spent only slightly more time foraging than did juvenile females. The diets and foraging behaviors of males that had not yet dispersed and males that had immigrated to that same area were indistinguishable. As discussed previously, dispersing males were significantly heavier than predispersal males, a result contrary to that predicted in the scenario of emigration because of lack of food. Finally, juvenile male immigration equaled emigration every year. This is important because preferred foods were unevenly distributed within and among populations *(1, 2)*. Evidence consistently suggested no link between immediate food shortages and natal dispersal.

A second reason for natal dispersal might be to locate a nest burrow. Ground squirrels depend on burrows for safety from predators, as places to spend the night, and as nests in which to hibernate *(25)*. Given the sexual dimorphism in natal dispersal, this hypothesis predicts differences between males and females in the type or location of habitable burrows and implies that dispersers should emigrate from areas of high population density or low burrow quality to areas where unoccupied holes of high quality are available. To test this idea we monitored population density each week and counted burrow entrances in the territories of lactating females. We found that neither the probability of juvenile male dispersal nor its timing was significantly related to population density or burrow availability near home, and that dispersers did not settle in areas of higher burrow density.

The only unusual aspect is that every male eventually leaves home

Another cause of natal dispersal might be ectoparasite infestation. If parasites build up in the natal nest and if juvenile males are more affected by them than are juvenile females, then males in particular might emigrate to avoid them. We examined this hypothesis indirectly, by counting the number of fleas and ticks on every captured juvenile. We found low levels of ectoparasitism throughout the animals' natal summer, and no consistent differences between infestations in males and females prior to or at the time of dispersal.

Do juvenile males disperse to avoid future competition with older males for sexual access to females? Because males always emigrated, it was not possible to determine if dispersers experienced less severe mate competition than hypothetical nondispersers. However, the mate competition hypothesis was examined indirectly by comparing, at sites where males were born and on ridge tops where those males mated two or more years later, three parameters: the density of breeding adult males, the mean number of fights adults engaged in for each successful copulation, and the mean daily ratio of breeding males to receptive females. We found no significant differences in any of these parameters, suggesting that dispersing males did not find better access to females than they would have if they had remained at home.

Do juvenile males disperse to avoid future nuclear family incest? A test of this hypothesis requires comparing the reproductive consequences of various degrees of inbreeding (26, 27). However, of more than 500 copulations observed, none occurred between close kin; therefore we could not directly test this hypothesis. Nonetheless, the nonrandom movements of males away from the natal area clearly resulted in complete avoidance of kin as mates (Fig. 4). Furthermore, during post-breeding dispersal, the highly polygynous males moved farthest. Under this hypothesis, the polygynous males who had sired many female pups in an area would have the most to gain by emigrating. Under the mate-competition hypothesis, successful males would be expected to stay put, while unsuccessful males might gain by dispersing. The observed pattern is thus most consistent with avoiding inbreeding.

Belding's ground squirrels are not unusual in the rarity of close inbreeding. Consanguineous mating is minimized in most mammals and birds (23, 24, 28, 29), often via the mechanism of sex-specific natal dispersal. But why are males the dispersive sex in mammals generally and ground squirrels particularly? The answer probably relates to a sexual asymmetry in the significance

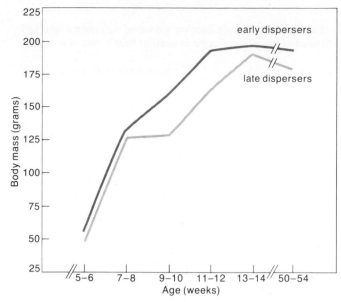

Figure 5. Weight gain among juvenile *S. beldingi* has been positively correlated with the onset of dispersal. Early dispersers (most males) left home at 7–10 weeks of age; late dispersers, in contrast, remained at home until they were 11–14 weeks old. Dispersal seems to occur when a threshold body mass of 125–150 g is attained. (After ref. *19*.)

of the location and quality of burrows for procreation (6, 30). The depth and dryness of nest burrows, their proximity to food, and their degree of protection from both inter- and intraspecific predators are vital to pup survival (31). The significance of the burrow, in turn, favors females who seek out and defend high-quality nest sites and who remain in them from year to year (25). The quality of a nursery burrow is of negligible significance to nonparental males. To avoid predators and inclement weather, and to forage, males can move frequently without jeopardizing the survival of their young. Thus the sexual bias in natal dispersal might occur because inbreeding is harmful to both sexes and males incur lower procreative costs by leaving home.

Sexual selection could reinforce a sex-bias in natal dispersal generated by incest avoidance. If consanguineous mating is indeed harmful, then the philopatric females should prefer to mate with unrelated (unfamiliar) males. A reproductive advantage should therefore accrue to males that seek and locate unfamiliar females (32).

Evolutionary origins

The fourth component of our investigation of natal dispersal was an attempt to infer evolutionary origins. A first hypothesis was that the male bias in natal dispersal arose in an evolutionary ancestor of *S. beldingi* as a developmental error (for example, in the timing of hormone secretion) or as a by-product of natural selection on males for the high levels of activity associated with finding mates and defending mating territories. Alternatively, perhaps natal dispersal was favored directly by selection, for example, as a mechanism to avoid inbreeding, throughout the evolutionary history of *S. beldingi*.

One way to evaluate these alternatives is to consider

Why do juvenile male Belding's ground squirrels disperse? Answers have been found at each of four levels of analysis.	
Level of analysis	**Summary of findings**
Physiological mechanisms	Dispersal by juvenile males is apparently caused by organizational effects of male gonadal steroid hormones. As a result, juvenile males are more curious, less fearful, and more active than juvenile females.
Ontogenetic processes	Dispersal is triggered by attainment of a particular body mass (or amount of stored fat). Attainment of this mass or composition apparently also initiates a suite of locomotory and investigative behaviors among males.
Effects on fitness	Juvenile males probably disperse to reduce chances of nuclear family incest.
Evolutionary origins	Strong male biases in natal dispersal characterize all ground squirrel species, other ground-dwelling sciurid rodents, and mammals in general. The consistency and ubiquity of the behavior suggest that it has been selected for directly across mammalian lineages.

the taxonomic distribution of male-biased natal dispersal. If selection has consistently and directly favored dispersal by juvenile males, then phylogenetic relatives of *S. beldingi* should share this trait to a greater degree than if it were a hormonally mediated side effect or developmental error. This is because any hormonal link between adult male sexual activities and dispersal by juveniles two years previously could presumably be broken by mutation in some species through evolutionary time. This, in turn, would lead to a spotty taxonomic distribution of the behavior if it were neutral for fitness.

Members of the squirrel family first appeared in the fossil record 35 to 40 million years ago; thus they are one of the most ancient of extant rodent families *(33)*. Belding's ground squirrel is one of 32 species in the genus *Spermophilus*; this genus is more closely related to marmots and prairie dogs than to tree squirrels *(34)*. Strongly male-biased natal dispersal occurs in all 12 *Spermophilus* species that have been studied in this regard *(5, 15, 35)*. Male-biased natal dispersal patterns are also the rule in marmots *(35)* and prairie dogs *(36)*. The dispersal behavior of *S. beldingi* is therefore probably a conservative rather than a derived trait; in other words, it is likely quite ancient.

As far as we know, the only unusual aspect of natal dispersal in *S. beldingi* is that every male eventually leaves home, whereas in a few other species a tiny fraction of males are philopatric. Male-biased natal dispersal is widespread among mammals *(4–6, 30, 32, 37)*, suggesting that this behavior may predate the appearance of the squirrel family. The ubiquity of natal dispersal seems more consistent with the hypothesis that it has been favored directly by natural selection in various

lineages than that it originated as a mistake or a correlated response to selection for some other male attribute and is maintained by phylogenetic inertia rather than adaptive value.

Synthesis

Our data reveal that there are at least four types of answers to the question of why juvenile male Belding's ground squirrels disperse (see the box). These answers complement rather than supersede each other. Clearly, however, the causal variables we have identified within each analytical level do not operate in isolation, and it seems appropriate to consider how they may interrelate.

During embryogenesis, sex chromosomes cause the formation of testes in male *S. beldingi*. The gonads secrete a pulse of androgens before birth, which, we hypothesize, sets up an ontogenetic switch, presumably by modifying the morphology or behavior of neurons or nuclei in the brain. When juvenile males have accumulated sufficient weight or fat stores, the switch turns on. The young males then boldly explore their environment, making increasingly longer forays away from home. The timing of dispersal by each individual may be influenced by any environmental factor that accelerates or delays arrival at the dispersal threshold (for example, food abundance or scarcity). The main cost of natal dispersal is probably mortality during emigration; the main benefits are likely related to reduced inbreeding and optimal outbreeding. Male biases in natal dispersal occur consistently across modern mammalian taxa *(37)*, suggesting an evolutionary history of natural selection favoring such behavior directly, and a taxon-wide consistency of function.

By employing the levels-of-analysis framework for developing and testing hypotheses, we have come to appreciate the complexity of what at first appeared to be a simple behavior. We suspect that our explanations for the proximate and ultimate causes of natal dispersal in *S. beldingi* will be applicable to other species. Perhaps equally important, our study illustrates that there can be multiple correct answers to questions of causality in behavioral biology *(38)*. The usefulness of the levels-of-analysis approach is thereby reemphasized.

References

1. P. W. Sherman. 1976. Natural selection among some group-living organisms. Ph.D. thesis, Univ. of Michigan.
2. K. E. Holekamp. 1983. Proximal mechanisms of natal dispersal in Belding's ground squirrels *(Spermophilus beldingi beldingi)*. Ph.D. thesis, Univ. of California.
3. W. Z. Lidicker, Jr. 1975. The role of dispersal in the demography of small mammals. In *Small Mammals: Their Productivity and Population Dynamics*, ed. F. B. Golley, K. Petruscewicz, and C. Ryszkowski, pp. 103–28. Cambridge Univ. Press.
4. P. J. Greenwood. 1980. Mating systems, philopatry and dispersal in birds and mammals. *Animal Behav.* 28:1140–62.
5. F. S. Dobson. 1982. Competition for mates and predominant juvenile male dispersal in mammals. *Animal Behav.* 30:1183–92.
6. A. E. Pusey. 1987. Sex-biased dispersal and inbreeding avoidance in birds and mammals. *Trends in Ecol. and Evol.* 2:295–99.
7. P. W. Sherman. 1988. The levels of analysis. *Animal Behav.* 36:616–19.
8. D. S. Lehrman. 1970. Semantic and conceptual issues in the

nature-nurture problem. In *Development and Evolution of Behavior*, ed. L. R. Aronson, E. Tobach, D. S. Lehrman, and J. S. Rosenblatt, pp. 17–52. W. H. Freeman.

9. E. Mayr. 1961. Cause and effect in biology. *Science* 134:1501–06.

10. N. Tinbergen. 1963. On aims and methods of ethology. *Zeitschrift für Tierpsychologie* 20:410–33.

11. P. W. Sherman and M. L. Morton. 1984. Demography of Belding's ground squirrels. *Ecology* 65:1617–28.

12. K. E. Holekamp. 1984a. Natal dispersal in Belding's ground squirrels (*Spermophilus beldingi*). *Behav. Ecol. Sociobiol.* 16:21–30.

13. P. W. Sherman. 1977. Nepotism and the evolution of alarm calls. *Science* 197:1246–53.

14. P. W. Sherman. 1981a. Kinship, demography, and Belding's ground squirrel nepotism. *Behav. Ecol. Sociobiol.* 8:251–59.

15. G. R. Michener. 1983. Kin identification, matriarchies, and the evolution of sociality in ground-dwelling sciurids. In *Recent Advances in the Study of Mammalian Behavior*, ed. J. F. Eisenberg and D. G. Kleiman, pp. 528–72. Am. Soc. Mammal.

16. C. H. Phoenix, R. W. Goy, A. A. Gerall, and W. C. Young. 1959. Organizing action of prenatally administered testosterone propionate on the tissues mediating mating behavior in the female guinea pig. *Endocrinology* 65: 369–82.

17. K. E. Holekamp, L. Smale, H. B. Simpson, and N. A. Holekamp. 1984. Hormonal influences on natal dispersal in free-living Belding's ground squirrels (*Spermophilus beldingi*). *Hormones and Behavior* 18:465–83.

18. E. Adkins-Regan. 1981. Early organizational effects of hormones: An evolutionary perspective. In *Neuroendocrinology of Reproduction*, ed. N. T. Adler, pp. 159–228. Plenum Press.

19. K. E. Holekamp. 1986. Proximal causes of natal dispersal in Belding's ground squirrels (*Spermophilus beldingi*). *Ecol. Monogr.* 56: 365–91.

20. S. Pfeifer. 1982. Disappearance and dispersal of *Spermophilus elegans* juveniles in relation to behavior. *Behav. Ecol. Sociobiol.* 10:237–43.

21. F. S. Dobson. 1979. An experimental study of dispersal in the California ground squirrel. *Ecology* 60:1103–09.

22. J. Moore and R. Ali. 1984. Are dispersal and inbreeding avoidance related? *Animal Behav.* 32:94–112.

23. A. E. Pusey and C. Packer. 1987. The evolution of sex-biased dispersal in lions. *Behaviour* 101:275–310.

24. A. Cockburn, M. P. Scott, and D. J. Scotts. 1985. Inbreeding avoidance and male-biased natal dispersal in *Antechinus* spp. (Marsupialia: Dasyuridae). *Animal Behav.* 33:908–15.

25. J. A. King. 1984. Historical ventilations on a prairie dog town. In *The Biology of Ground-dwelling Squirrels*, ed. J. O. Murie and G. R. Michener, pp. 447–56. Univ. of Nebraska Press.

26. W. M. Shields. 1982. *Philopatry, Inbreeding, and the Evolution of Sex*. State Univ. of New York Press.

27. P. J. Greenwood, P. H. Harvey, and C. M. Perrins. 1978. Inbreeding and dispersal in the great tit. *Nature* 271:52–54.

28. J. L. Hoogland. 1982. Prairie dogs avoid extreme inbreeding. *Science* 215:1639–41.

29. K. Ralls, P. H. Harvey, and A. M. Lyles. 1986. Inbreeding in natural populations of birds and mammals. In *Conservation Biology: The Science of Scarcity and Diversity*, ed. M. E. Soulé, pp. 35–56. Sinauer.

30. P. M. Waser and W. T. Jones. 1983. Natal philopatry among solitary mammals. *Q. Rev. Biol.* 58:355–90.

31. P. W. Sherman. 1981b. Reproductive competition and infanticide in Belding's ground squirrels and other animals. In *Natural Selection and Social Behavior*, ed. R. D. Alexander and D. W. Tinkle, pp. 311–31. Chiron Press.

32. A. E. Pusey and C. Packer. 1986. Dispersal and philopatry. In *Primate Societies*, ed. B. B. Smuts, D. L. Cheney, R. M. Seyfarth, R. W. Wrangham, and T. T. Struhsaker, pp. 250–66. Univ. of Chicago Press.

33. W. P. Luckett and L. J. Hartenberger, eds. 1985. *Evolutionary Relationships among Rodents*. Plenum Press.

34. D. J. Hafner. 1984. Evolutionary relationships of the nearctic Sciuridae. In *The Biology of Ground-dwelling Squirrels*, ed. J. O. Murie and G. R. Michener, pp. 3–23. Univ. of Nebraska Press.

35. K. E. Holekamp. 1984b. Dispersal in ground-dwelling sciurids. In *The Biology of Ground-dwelling Squirrels*, ed. J. O. Murie and G. R. Michener, pp. 297–320. Univ. of Nebraska Press.

36. M. G. Garrett and W. L. Franklin. 1988. Behavioral ecology of dispersal in the black-tailed prairie dog. *J. Mammal.* 69:236–50.

37. B. D. Chepko-Sade and Z. T. Halpin, eds. 1987. *Mammalian Dispersal Patterns*. Univ. of Chicago Press.

38. P. W. Sherman. 1989. The clitoris debate and the levels of analysis. *Animal Behav.* 37:697–98.

From Society to Genes with the Honey Bee

A combination of environmental, genetic, hormonal and neurobiological factors determine a bee's progression through a series of life stages

Gene E. Robinson

On September 9, 1997, an article in *The New York Times* announced the discovery of the "first gene for social behavior." Anthony Wynshaw-Boris, of the National Human Genome Research Institute, and his colleagues had discovered odd behavior in laboratory mice lacking a gene called *disheveled*-1. These mice interacted and huddled with others less than normal, and they failed to perform an important social duty, trimming the whiskers of fellow mice. Whether or not this is really the first gene discovered "for social behavior," no one should lean toward the notion that genes play an exclusive role in regulating behavior. Biologists long ago came to realize that behavior is influenced by genes, the environment and interactions between the two. To better understand this regulatory combination, scientists can turn to an organism, such as the honey bee, whose behavior can be studied in the field under natural conditions.

A discussion of "genes for behavior" might raise anxiety over the implications of attributing so much control to strings of nucleic acids, or DNA. In particular, some people fear that the concept of *biological determinism*—the notion that genes play a dominant, if not exclusive, role in regulating behavior—might creep in and diminish our appreciation for the role of the environment in shaping behavior. Nevertheless, genes never

Gene E. Robinson is professor in the Department of Entomology and the Neuroscience Program at the University of Illinois at Urbana-Champaign. He earned his B.S. in 1977 at Cornell University, worked in the bee industry until 1980, and then earned his M.S. and Ph.D. at Cornell in 1982 and 1986, respectively. His research employs honey bees for interdisciplinary studies of mechanisms of social behavior. Address: Department of Entomology, University of Illinois, Urbana, IL 61801.

act alone. They must operate in an environment, where they code for proteins that participate in many systems in an organism. In fact, genes themselves depend on many of those proteins for replicating DNA and linking together amino acids, which are the fundamental units of proteins. Consequently, biologists need to take a broad approach in assessing the impact of any gene.

To properly appreciate the influence of genes on behavior, we need behavioral studies that demonstrate—at the molecular level—the influences of genes, the environment and their interactions. Social behavior is ideally suited for this challenge because it is especially sensitive to environmental influence. Moreover, these influences are in many cases mediated by specific social signals communicated from individual to individual, which can make them easier to study experimentally. Molecular-genetic studies of social behavior will show how an animal's phenotype, which includes social behavior, arises from both its genotype and environment. Making that connection, however, requires identifying genes that influence social behavior, revealing how those genes regulate the neural and endocrine mechanisms through the production of proteins, and, finally, exploring how specific manipulations of an animal's social environment affect gene expression.

My research group uses the Western honey bee, *Apis mellifera*, to understand how genes and the environment govern social behavior. As I shall show, we study the development of naturally occurring social behavior, from society to gene. Honey bees are particularly useful for studying social behavior because, like humans, they experience *behavioral development*. In other words, honey bees pass through different life

stages as they age, and their genetically determined behavioral responses to environmental and social stimuli change in predictable ways. Often these responses increase in complexity and involve learning. We hope to explain the function and evolution of behavioral mechanisms that integrate the activity of individuals in a society, neural and neuroendocrine mechanisms that regulate behavior within the brain of an individual, and genes that influence behavior by encoding these mechanisms.

Basics of Bee Behavior

The so-called social insects, including honey bees, live in societies that rival our own in complexity and internal cohesion. For instance, honey bees always follow three rules: They live in colonies with overlapping generations, they care cooperatively for offspring other than their own and they maintain a reproductive division of labor. A colony arises from a *queen* that performs one task, laying lots of eggs, sometimes as many as 2,000 in one day. Her daughters, called *workers*, basically take care of the colony—doing everything from foraging for food to building the hive—but they generally do not reproduce. As one might expect, it takes many workers to run a hive, and some honey bee colonies consist of as many as 60,000 workers. Finally, a colony's males, called *drones*, can usually be found in the hive, where they do essentially nothing. The drones specialize in reproduction, which takes just a couple of hours on a sunny day when they fly to mating areas away from the hive. Once a drone mates, he dies.

A further division of labor exists among the workers. Although a worker's adult life span is just four to seven weeks, it undergoes a series of transitions.

Figure 1. Honey bees pass through different life stages as they age, and their behavioral responses to environmental and social stimuli change in predictable ways. For instance, older females forage for food, as shown by this bee collecting pollen from a passion flower. By studying a honey bee's naturally occurring social behavior, from society to gene, the author hopes to explain the function and evolution of behavioral mechanisms that integrate the activity of individuals in a society, of neural and neuroendocrine mechanisms that regulate behavior within the brain of an individual and of genes that influence this social behavior.

A worker usually spends its first few weeks tending to duties in the hive and its last few weeks foraging for food outside the hive. During the hive phase, a worker starts out with a couple of days of cell cleaning, literally removing debris from cells in the hive that are used to raise other bees or to store food. Next, a worker serves as a nurse, caring for and feeding larval bees. Toward the end of the hive phase, a worker spends its time processing and storing food and maintaining the nest, including building new sections of hive. Some workers also perform a few other tasks along the way, including guarding the hive or removing corpses. Finally, a worker switches to foraging, which is probably the most challenging task of all. To be a successful for-

ager, a bee must learn how to navigate in the environment and obtain nectar and pollen from flowers. Foragers also communicate the location of new food sources by means of the famous "dance language." These transitions in occupation typically do not arise abruptly. For example, a worker might slowly decrease its nursing duties and become gradually more involved in maintaining the hive.

Behavioral development in honey bees is a powerful system for integrated analysis. Although it occurs naturally in the field, some underlying mechanisms can be analyzed in the laboratory. Moreover, honey bees have been closely associated with humans for millennia because of their special status as prolific producers of honey and wax and as premier pollina-

tors for our food and fiber crops. As a result, we know more about honey bees than just about any other animal on earth. One consequence of this wealth of knowledge is that the natural social life of honey bees can be extensively manipulated with unparalleled precision.

Colony Adjustments

Although worker bees go through a rather consistent path of behavioral development, it is not rigidly determined. Bees can accelerate, retard or even reverse their behavioral development in response to changing environmental and colony conditions. For example, favorable environmental conditions in the late spring might cause a surge in worker birth rates, and that could re-

Figure 2. Worker bees, daughters of the queen, go through a series of transitions in their month-or-two-long lives. A worker bee starts as a hive cleaner and then advances to nursing larval bees (*top*). Next, a worker helps with maintaining a hive, including building new sections (*middle*). Finally, the oldest workers forage for food (*bottom*).

sult in a colony with a reduced percentage of foragers. Under these circumstances, young bees compress their period of hive work from three weeks to one week and become "precocious foragers." Conversely, a new colony founded by a swarm—a fragment of an old colony that leaves to establish a new colony—soon reaches a point at which it contains predominantly older individuals. In that case, some colony members retard their development and serve as overaged nurses. In those bees, *hypopharyngeal glands* that produce food for larvae continue with this function rather than producing other substances.

How does the behavior of thousands of individual bees generate a smoothly functioning colony? It seems unlikely that individual bees could monitor the state of their entire colony and then perform the tasks that are needed. Although some workers play special roles in organizing specific tasks, such as leading other bees to a new nest site during swarming, there is no evidence for real leaders or individuals—not even the queen—that perceive all or most of a colony's requirements and direct the activities of other colony members from one task to another. The challenge is understanding the mechanisms of integration that enable individual bees to respond to fragmentary information with actions that are appropriate to the state of the whole colony.

My colleagues and I and others have discovered that *juvenile hormone*—one of the most important hormones influencing insect development—helps to time the pace of behavioral maturation in honey bees. This hormone comes from the *corpora allata*, a gland that lies near a honey bee's brain. Indirect evidence for this hormone's role exists in the fact that young bees working in a hive have low levels of this hormone and older foragers have higher levels. Direct proof has also been obtained: Young bees given juvenile-hormone treatments become precocious foragers. Recently, my University of Illinois colleague Susan Fahrbach, graduate student Joseph Sullivan, undergraduate Omar Jassim and I found that removing the corpora allata does not prevent a bee from developing into a forager but does delay it for a few days on average. Juvenile-hormone treatments, however, eliminate that delay.

Manipulating hormone levels on a bee-by-bee basis is one thing, but demonstrating that bees alter hormone levels themselves in response to changing conditions is another. To show how the environment can modulate hormone levels, Robert Page of the University of California at Davis, Colette and Alain Strambi of the Centre National de la Recherche Scientifique in Marseille, France, and I induced precocious foraging by establishing colonies that consisted of only very young bees. Then we tested their blood levels of juvenile hormone and found that one-week-old precocious foragers had approximately 100 nanograms of juvenile hormone per milliliter of blood, which is about the same as that in three-week-old foragers and higher than the 5–20 nanograms usually found in one-week-old nurses. Two weeks later, we obtained overaged nurses from these colonies by preventing new adults from emerging, and these old nurses had levels of juvenile hormone that resembled young nurses, rather than foragers. From work with other experimental colonies, we found that bees that reverted from foraging to nursing were also "young" in terms of their levels of juvenile hormone.

Inhibitory Interactions

How do bees perceive changes in colony needs and adjust their behavioral development to perform the tasks most in demand? Postdoctoral associate Zachary Huang and I found that the rate of endocrine-mediated behavioral development is influenced by inhibitory social interactions. That is, older bees inhibit the behavioral development of younger bees. Bees reared in isolation in a laboratory for seven days have forager-like levels of juvenile hormone and forage precociously when placed in colonies. By carefully manipulating a colony's age demography but keeping other characteristics unchanged, we found that the rate of behavioral development is negatively correlated with the proportion of older bees in a colony. So depleting a colony's foragers stimulates younger bees to forage earlier than normal. Conversely, younger bees forage later than normal if a colony's foragers stay in the hive for several days because a sprinkler aimed at the hive entrance makes them think it's raining.

Someone might imagine that bees could learn about their colony's condition by monitoring the combs in their hive. For instance, a young bee might notice a food shortage in the combs, which might result in a neuorendocrine response that triggers precocious behavioral development. To explore the possibility that bees pay attention to the combs in this way, Huang, graduate student David Schulz and I recently tested the effects of starvation on the rate of behavioral development. Young bees from starved colonies do start foraging a few days earlier than bees from well-fed colonies. This starvation effect, however, is not mediated by perceiving a shortage of food in the honeycomb. We showed this by keeping a colony well fed from a sugar feeder while we constantly—but discretely—vacuumed any stored food out of their honeycomb. This was accomplished by drilling small holes at the base of each honeycomb cell. Well-fed bees in an empty hive started to forage at ages similar to bees in colonies with ample food stores and not nearly as early in life as did bees in truly starved colonies.

Inhibitory social interactions that influence the rate of behavioral development involve chemical communication between colony members. This is strikingly similar to pheromone regulation of sexual maturation in rodent societies. For example, a queen's mandibular glands produce a pheromone that inhibits behavioral development. (See "The Essence of Royalty: Honey Bee Queen Pheromone" by Mark Winston and Keith Slessor in the July–August 1992 issue of *American Scientist*.) *Queen mandibular pheromone* has been known for some time to exert long-lasting effects on worker physiology and behavior by inhibiting the rearing of new queens. More recently, Mark Winston and Tanya Pankiw of Simon Fraser University, Huang and I demonstrated that queen mandibular pheromone depresses blood levels of juvenile hormone and, more important, delays the onset of foraging.

The primary modulator of behavioral development, however, appears to come from the workers themselves. The mandibular glands of workers contain compounds similar to those found in queen mandibular glands. Huang, Erika Plettner, a graduate student at Simon Fraser University, and I recently found that there must be direct social contact between bees for older ones to inhibit the development of younger ones. Moreover, older bees with their mandibular glands removed do not inhibit behavioral development. The mandibular glands of workers contain compounds similar to those found in queen mandibular glands. The inhibition that

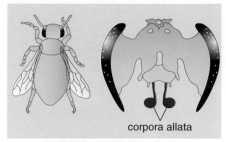

Figure 3. Hormones play a fundamental role in a worker's behavioral development. For example, the *corpora allata* (purple)—two tiny glands located just beneath a bee's brain *(shown in close-up on the right)*—produce juvenile hormone, which helps to time a bee's rate of maturation. Young workers in the hive have low levels of juvenile hormone in comparison with the older foragers. Moreover, juvenile-hormone treatments cause young workers to forage precociously.

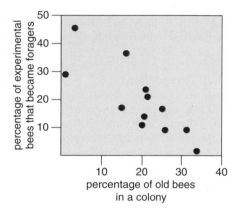

Figure 4. Older bees inhibit the development of younger ones. The author and his postdoctoral associate Zachary Huang varied a colony's age demography—producing a range of percentages of older bees in a hive—and then measured the percentage of experimental bees that became foragers by 14–32 days of age. The resulting data show that this percentage, a measure of the rate of behavioral development, is negatively correlated with the proportion of older bees in a colony. The primary modulator of behavioral development appears to come from the workers themselves, perhaps through chemical communication, because there must be direct social contact for older bees to inhibit the development of younger ones.

results from worker-worker interactions might come from exchanging a pheromone, which might be in the mandibular glands or somewhere else. When we removed the glands, that could have eliminated the inhibition because it removed the source of the pheromone or it simply blocked the pheromone's flow from another location. Clearly, more work must be done here.

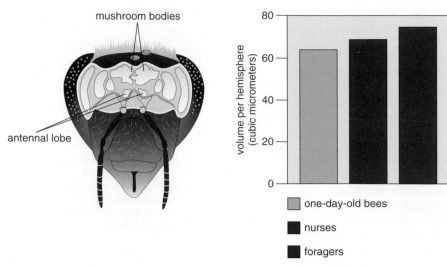

Figure 5. Changes in a bee's brain accompany its behavioral development. For example, portions of the *mushroom bodies (left)*, a region essentially in the middle of a bee's brain, increase in size by as much as 20 percent as a worker matures from a 1-day-old to a nurse to a forager *(right)*. This size increase apparently arises from expanding connections between the neurons, which probably affects how information gets processed in this region. Consequently, increases in the size of this brain structure—considered to be the center of learning and memory in insects—might allow bees to perform new behaviors.

Brain Remodeling

How does a bee's brain support the striking changes in behavior that take place during maturation? A small part of the answer lies in the *mushroom bodies*, a brain region thought to be the center of learning and memory in insects. Graduate student Ginger Withers, Fahrbach and I discovered about a 20 percent increase in the volume of a specific area of the mushroom bodies as worker honey bees mature. This volume increase occurs in a mushroom-body subregion where synapses, or connections, are made between neurons from other brain regions that are devoted to sensory input. This was the first report of such brain plasticity in an invertebrate, and it was particularly exciting because volume increases in brain regions in vertebrates reflect increases in certain cognitive abilities.

It seemed that the increase in the mushroom bodies might be learning-related. Young workers take orientation flights prior to the onset of foraging to learn their way around outside the hive, and the increase in volume in the mushroom bodies begins at that time. To test flying's effect on mushroom-body volume, Withers, Fahrbach and I made what we called "big-back bees." By attaching a large tag to each bee's back and placing a screen at the hive's entrance, we prevented some workers from flying out of the hive but allowed them to interact with other bees. Big-

back bees showed normal increases in mushroom-body volume despite their deprivation from orientation flights. So far, the volume increase is unstoppable. Fahrbach, Darrell Moore of East Tennessee State University, graduate student Sarah Farris, postdoctoral associate Elizabeth Capaldi and I showed that it takes place even in bees reared in social isolation and complete darkness in a laboratory.

Still, it might be premature to exclude the idea of a connection between the plasticity of the mushroom bodies and orientation flights in honey bees. Our results indicate that a bee's mushroom bodies need not increase *because* of taking orientation flights, but we have not ruled out a volume increase that *prepares* a bee for those flights. In other words, the mushroom bodies might need to increase in volume to provide the necessary brain space for a bee to learn how to get around outside its hive, and how to get back.

After learning to orient outside the hive, a bee learns to forage, and that might also involve an increase in the mushroom bodies. Withers, Fahrbach and I showed that the mushroom bodies increase in volume more rapidly in precocious foragers than in nurse bees of the same age. This result has been confirmed in the laboratory of Randolf Menzel in Berlin, using a somewhat different neuroanatomical analysis. These results suggest that the structure of the

mushroom bodies might be sensitive to changes in social context that are associated with the onset of foraging.

While we continue our efforts to unravel the significance of a volume increase in the mushroom bodies, we also wonder how the region gets bigger. The number of cells in the mushroom bodies is highly stable in adult life. The production of new neurons is not detectable, and there is no evidence for cell death, according to research with Fahrbach that was performed by undergraduates Jennifer Strande and Jennifer Mehren. Accordingly, the volume increase in the mushroom bodies probably represents an increased arborization of some subpopulation of brain cells that already exists. This increased proliferation of neuronal branches would likely result in an increase in the number of synapses per neuron, which would impact the processing of information in the mushroom bodies.

Beyond structural changes in a worker bee's brain, neurochemical analyses have revealed striking changes in levels of biogenic amines, which are well known as modulators of nervous-system function and organismal behavior in animals, including humans. Alison Mercer, her colleagues from the University of Otago in New Zealand and I found changes in brain levels of two biogenic amines—dopamine and serotonin—during behavioral development. Jeffrey Harris and Joseph Woodring at Louisiana State University reported similar findings. Recently, graduate students Christine Wagener-Hulme and David Schulz, research technician Jack Kuehn and I showed that another biogenic amine, octopamine, appears to be most important in honey bee behavioral development. When a bee receives treatments of juvenile hormone, levels of octopamine increase, but dopamine and serotonin do not. Looking specifically at the *antennal lobes*, a brain region that receives sensory information from a bee's antennae, we found high levels of octopamine in the antennal lobes of foragers as compared with nurse bees, regardless of worker age. In contrast, levels of all three amines in the mushroom bodies are intimately associated with worker age, but not behavioral status.

These results suggest that octopamine might influence behavioral development by modulating a bee's sensitivity to the stimuli that elicit the performance of age-specific tasks. We presume that these stimuli are mostly chemical, because bees live in a dark hive and possess modest auditory acuity, at least relative to their renowned chemosensory prowess. That is why we are so encouraged to find behaviorally related changes specifically in the antennal lobes. The hypothesis that octopamine is playing a causal role in behavioral development is currently being tested by chronic administration of octopamine to the brain, followed by behavioral assays. Studies of biogenic amines might also provide some of the missing links between endocrine regulation and behavioral development in honey bees.

Sociogenomics

Molecular-genetic research in my laboratory has only recently begun, and it currently involves selecting candidate genes and exploring their possible involvement in social behavior. This is done by studying whether differences in social behavior—within and between individuals—are correlated with variation in gene-transcription regulation, gene structure or both. Graduate student Daniel Toma and I are exploring the role of the *period (per)* gene in honey bee behavioral development. In the fruit fly *Drosophila melanogaster, per* is a principal component of the fly's circadian clock. The protein that *per* encodes is thought to help create circadian rhythms of activity by orchestrating the transcription of other genes according to a precisely timed schedule. We chose *per* because we have found intriguing links between division of labor and circadian behavioral rhythms in honey bees.

Moore, Fahrbach, undergraduates Iain Cheeseman and Jennifer Angel and I discovered that foragers have pronounced circadian rhythms of activity—including being more active during the day than the night—but workers in the early part of their hive phase do not show such rhythms. This difference is obvious both in beehives and in assays of individually isolated bees in the laboratory. For example, consider the behavior of nurse bees that need to feed bee larvae around the clock. How is this accomplished? If nurse bees have a circadian rhythm of brood care, one might expect to find evidence of "shift work," or groups of nurse bees on different schedules. Alternatively, if nurse bees perform brood care with no circadian rhythm, one might expect to find them performing it randomly with respect to time. Monitoring individually tagged bees every three hours around the clock in glass-walled observation hives, we

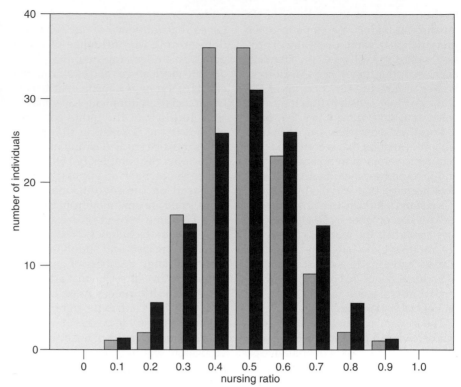

Figure 6. Younger bees show no overt behavioral circadian rhythms, but older bees do. The author and his associates monitored patterns of task performance in glass-walled observation hives at regular intervals around the clock. The results for nursing behavior are shown here. The nursing ratio was calculated for each bee by dividing the number of observations of nursing during the day by those in both the day and night. A ratio of 0.5 indicates arrhythmicty, or no difference in the number of times nursing was performed in the day versus the night; a ratio of 0.0 indicates that a bee had only nocturnal nursing activity; and a ratio of 1.0 indicates that a bee had only diurnal nursing activity. A statistical analysis demonstrated that nursing is not performed according to a circadian rhythm: There was no difference (P > 0.5) in the observed distribution *(orange)* of bees compared to an expected binomial distribution *(black)* with a probability of 0.5. By exploring the changes in gene expression that underlie these and other developmental changes in behavior, the author expects to find that two-way interactions between the nervous system and the genome contribute fundamentally to the control of social behavior. (Adapted from Moore *et al.*)

found no evidence of rhythmicity or shift work for nursing. Nurse bees were arrhythmic in the performance of this task.

Moreover, genetic factors appear to influence both the plasticity in a honey bee's behavioral development and its circadian rhythm. Using colonies composed of workers with identifiable genotypes, graduate student Tugrul Giray and I learned that workers of some genotypes are more likely to consistently mature rapidly and forage precociously, even in different social environments, and workers of other genotypes are more apt to develop into overaged nurses. Working with such "fast" and "slow" genotypes, Moore, Giray and I found that fast-genotype bees developed a circadian rhythm of locomotor behavior in the laboratory at younger ages than did slow-genotype bees. Fast-genotype bees also had a periodicity to their rhythm of loco-

motor activity that was faster than that of slow-genotype bees. These results are reminiscent of the strikingly diverse effects of the *per* gene in fruit flies. This gene governs not only circadian rhythms in flies but some elements of their mating song as well.

Toma and I have cloned a putative bee homologue of the *per* gene. We hope to determine whether variation in gene expression and gene structure is correlated with variation in the ontogeny of behavioral circadian rhythmicity and the rate of behavioral development. The next step is to manipulate a hive's age demography to determine the association between gene expression and behavioral status.

Another approach to discovering genes involved in a honey bee's behavioral development is to build on our neuroanatomical work by identifying molecular mechanisms that contribute to the increase in the volume of the mushroom bodies. We hope to determine whether variation in the expression or structure of genes preferentially expressed in a honey bee's mushroom bodies affects the functioning of this brain structure, which in turn may affect behavioral development in honey bees.

Luckily, it is relatively easy to find behavioral variation that is correlated with genotypic variation in honey bees. Several laboratories, including mine, have demonstrated that variation in a worker's genotype influences many aspects of the division of labor in honey bee colonies, including the tendency to specialize in rare tasks such as removing corpses and guarding a nest's entrance. In addition, controlled mating via instrumental insemination facilitates research on honey bee behavioral genetics.

The molecular-genetic analysis of social behavior in general is a fertile new field. Results of classical quantitative genetic studies have indicated that there are strong correlations between genetic variation and variation in social behavior among individuals, but none of the genes has been identified. (Much more work must be done on *disheveled*-1 before it could meet these criteria.) The idea that gene expression in the brain is sensitive to social context is supported by recent findings from the laboratories of Fernando Nottebohm of Rockefeller University and David Clayton of the University of Illinois on bird song, Donald Pfaff of Rockefeller University and Thomas Insel of Emory University on rodents, Russell Fernald of Stanford University on cichlid fish and Edward Kravitz of Harvard University and Donald Edwards of Georgia State University on lobsters. I propose that two-way interactions between the nervous system and the genome contribute fundamentally to the control of social behavior. Information about social conditions that is acquired by the nervous system is likely to induce changes in genomic function that in turn adaptively modify the structure and function of the nervous system.

With the presence of abundant genetic variation in behavior and a growing selection of tools needed to exploit it, the prospects are good that honey bees can be used as a new model for molecular genetic analyses of social behavior. Nevertheless, I believe that the difficulty in studying the genetic basis of social behavior demands a bold, new initiative, which I call *sociogenomics*. In essence, this means taking a wide-ranging approach to identify genes that influence social behavior, determining the influence of these genes on underlying neural and endocrine mechanisms and exploring the effects of the environment—particularly the social environment—on gene action. Implicit in the name sociogenomics is the realization that many genes must be studied simultaneously to decipher the complexity behind social behavior. Such an approach could be based on the revolutionary advances that are emerging from the Human Genome Project. For example, there are new techniques for sequence-variation analysis and simultaneously screening large numbers of genes for differences in expression that are correlated with differences in behavioral state that can contribute significantly to gene discovery in bees.

In continued studies of honey bees, investigators will probably find common mechanisms that govern life in both invertebrate and vertebrate societies. If so, the identification of genes influencing social behavior in honey bees—guided by our emerging understanding of the underlying neural and endocrine mechanisms—will likely yield insights that go well beyond a beehive.

Acknowledgments

Work in the author's laboratory has been supported by grants from the National Institute of Mental Health, National Institutes of Health, National Science Foundation and U.S. Department of Agriculture. The author thanks his colleagues, postdoctoral associates, technician and graduate and undergraduate students for their contributions, and in particular Professors Susan Fahrbach and Robert Page for outstanding collaboration and stimulating camaraderie over the years.

Bibliography

Capaldi, E. A., S. E. Fahrbach and G. E. Robinson. In press. Neuroethology of spatial learning: The birds and the bees. *Annual Review of Psychology.*

Fahrbach, S. E., D. Moore, E. A. Capaldi, S. M. Farris and G. E. Robinson. In press. Experience-expectant plasticity in the mushroom bodies of the honey bee. *Learning and Memory.*

Giray, T., and G. E. Robinson. 1994. Effects of intracolony variability in behavioral development on plasticity of divison of labor in honey bee colonies. *Behavioral Ecology and Sociobiology* 35:13–20.

Huang, Z.-Y., E. Plettner and G. E. Robinson. In press. Effects of social environment and worker mandibular glands on endocrine-mediated behavioral development in honey bees. *Journal of Comparative Physiology A.*

Huang, Z.-Y., and G. E. Robinson. 1992. Colony behavioral integration in honey bees: worker-worker interactions mediate plasticity in hormonally regulated division of labor. *Proceedings of the National Academy of Sciences* 89:11726–11729.

Huang, Z.-Y., and G. E. Robinson. 1996. Regulation of division of labor in honey bees via colony age demography. *Behavioral Ecology and Sociobiology* 39:147–158.

Moore, D., I. M. Cheeseman, J. E. Angel, S. E. Fahrbach and G. E. Robinson. In press. Integration of circadian rhythms and division of labor in the honey bee colony. *Behavioral Ecology and Sociobiology.*

Pankiw, T., Z.-Y. Huang, G. E. Robinson and M. L. Winston. In press. Effects of queen mandibular pheromone on behavioural ontogeny and juvenile hormone titres in honey bees. *Journal of Insect Physiology.*

Robinson, G. E. In press. Integrative animal behaviour and sociogenomics. *Trends in Ecology and Evolution.*

Robinson, G. E. 1992. The regulation of division of labor in insect societies. *Annual Review of Entomology* 37:637–665.

Robinson, G. E., S. E. Fahrbach and M. L. Winston. 1997. Insect societies and the molecular biology of social behavior. *BioEssays* 19:1099–1108.

Robinson, G. E., and R. E. Page. 1988. Genetic determination of guarding and undertaking in honey-bee colonies. *Nature* 333:356–358.

Robinson, G. E., R. E. Page, C. Strambi and A. Strambi. 1989. Hormonal and genetic control of behavioral integration in honey bee colonies. *Science* 246:109–112.

Schulz, D. J., Z.-Y. Huang and G. E. Robinson. 1998. Effects of colony food shortage on behavioral development in honey bees. *Behavioral Ecology and Sociobiology* 42:295–303.

Sullivan, J. P., O. Jassim, G. E. Robinson and S. E. Fahrbach. 1996. Foraging behavior and mushroom bodies in allatectomized honey bees. *Society for Neuroscience Abstract* 22:1144.

Taylor, D. J., G. E. Robinson, B. J. Logan, R. Laverty and A. R. Mercer. 1992. Changes in brain amine levels associated with the morphological and behavioural development of the worker honey bee. *Journal of Comparative Physiology* A 170:715–721.

Withers, G. S., S. E. Fahrbach and G. E. Robinson. 1993. Selective neuroanatomical plasticity and division of labour in the honey bee (*Apis mellifera*). *Nature* 364:238–240.

Withers, G. S., S. E. Fahrbach and G.E. Robinson. 1995. Effects of experience on the organization of the mushroom bodies of the honey bee brain. *Journal of Neurobiology* 26:130–144.

Links to Internet resources for further exploration of "From Society to Genes with the Honey Bee" are available on the *American Scientist* Web site:

http://www.amsci.org/amsci/ articles/98articles/robinson.html

Animal Thinking

Donald R. Griffin

Ethologists are once again investigating the possibility that animals have conscious awareness

What is it like to be an animal? What do monkeys, dolphins, crows, sunfishes, bees, and ants think about? Or do nonhuman animals experience any thoughts and subjective feelings at all? Aside from Lorenz (1963) and Hediger (1947, 1968, 1980) very few ethologists have discussed animal thoughts and feelings. While seldom denying their existence dogmatically, they emphasize that it is extremely difficult, perhaps impossible, to learn anything at all about the subjective experiences of another species. But the difficulties do not justify a refusal to face up to the issue. As Savory (1959) put the matter, "Of course to interpret the thoughts, or their equivalent, which determine an animal's behaviour is difficult, but this is no reason for not making the attempt to do so. If it were not difficult, there would be very little interest in the study of animal behaviour, and very few books about it" (p. 78).

Most biologists and psychologists tend, explicitly or implicitly, to treat most of the world's animals as mechanisms, complex mechanisms to be sure, but unthinking robots nonetheless. Mechanical devices are usually considered to be incapable of conscious thought or subjective feeling, although it is currently popular to ascribe mental experiences to computer systems. John (in Thatcher and John 1977), among others, has equated consciousness with a sort of internal feedback whereby information about one part of a pattern of information flow acts on another part. This may be a necessary condition for conscious thinking, but it is also an aspect of many physiological processes that operate without any conscious awareness on our part.

Many comparative psychologists seem petrified by the notion of animal consciousness. Historically, the science of psychology has been reacting for fifty years or more against earlier attempts to understand the workings of the human mind by introspective self-examination—trying to learn how we think by thinking about our thoughts. This effort led to confusing and

Donald R. Griffin, a professor at The Rockefeller University, is an authority on animal physiology and behavior, best known for his work on echolocation in bats and other animals. He is the author of numerous articles and books on ethology and comparative physiology, including Listening in the Dark (*1958*), Echoes of Bats and Men (*1959*), Bird Migration (*1964*), and The Question of Animal Awareness (*1976*). *The present article is adapted by permission of the publisher from* Animal Thinking, *published in April 1984 by Harvard University Press. Address: The Rockefeller University, 1230 York Avenue, New York, NY 10021.*

contradictory results, so in frustration experimental psychologists largely abandoned the effort to understand human consciousness, replacing introspection with objective experiments. While experiments have been very helpful in analyzing learning and other human abilities, the rejection of any concern with consciousness and subjective feelings has gone so far that many psychologists virtually deny their existence or at least their accessibility to scientific analysis.

In one rather extreme form of this denial, Harnad (1982) has argued that only after the functioning of our brains has determined what we will do does an illusion of conscious awareness arise, along with the mistaken belief that we have made a choice or had control over our behavior. The psychologists who thus belittle and ignore human consciousness can scarcely be expected to tell us much about subjective thoughts and feelings of animals. If we cannot gather any verifiable data about our own thoughts and feelings, the argument has run, how can we hope to learn anything about those of other species?

A long-overdue corrective reaction to this extreme antimentalism is well under way. To a wide range of scholars, and indeed to virtually the whole world outside of narrow scientific circles, it has always been self-evident that human thoughts and feelings are real and important (see, for example, MacKenzie 1977 and Whiteley 1973). This is not to underestimate the difficulties that arise when one attempts to gather objective evidence about other people's feelings and thoughts, even those one knows best. But it really is absurd to deny the existence and importance of mental experiences just because they are difficult to study.

Why do so many psychologists appear to ignore a central area of their subject matter when most other branches of science refrain from such self-inflicted paralysis? The usual contemporary answer to such a question is that a relatively new sort of cognitive psychology has developed during the past twenty or thirty years, based in large part on the analysis of human and animal behavior in terms of information-processing (reviewed in Norman 1981). Analogies to computer programs play a large part in this approach, and many cognitive psychologists draw their inspiration from the success of computer systems, feeling that certain types of programs can serve as instructive models of human thinking. Words that used to be reserved for conscious human beings are now commonly used to describe the impressive accomplishments of computers. Despite the

Figure 1. A chimpanzee, having selected a suitably shaped small branch, strips it of twigs and leaves, transforming it into a serviceable tool for capturing termites. The chimpanzee will then walk to a termite nest, often at a considerable distance, and probe it with the stick for termites. Such behavior, which differs radically from other activities of chimpanzees, seems difficult to understand unless the animal is consciously thinking about gathering termites while preparing the probe. (Photograph by Linda Koebner, Bruce Coleman Inc.)

an empirical test, but the extent and the complexity of information-processing in our brains is so great that available procedures can detect only a tiny fraction of it, and even if it could be monitored in full detail, we do not know whether any computer system could duplicate it.

The difference between conscious and nonconscious states is a significant one, yet most scientists concerned with animal behavior have felt that looking for consciousness in animals would be a futile anachronism. This defeatist attitude is based in part on convincing evidence that we do a great deal of problem-solving, decision-making, and other kinds of information-processing without any consciousness of what is going on. Harnad (1982) bases his belief that human consciousness is merely an illusion on the fact that we are conscious of only the tip of the iceberg of information-processing in our brains. Indeed the ratio of conscious to unconscious brain activity is probably even smaller than the density ratio of ice to water. The intellectual excitement of this discovery has obscured the obvious fact that we are conscious some of the time, and we certainly do experience many sorts of thoughts and feelings that are very important to us and our companions. If the choice were open, would anyone prefer a lifelong state of sleepwalking?

optimism of computer enthusiasts, however, it is highly unlikely that any computer system can spontaneously generate subjective mental experience (Boden 1977; Dreyfus 1979; Baker 1981).

Conspicuously absent from most of contemporary cognitive psychology is any serious attention to conscious thoughts or subjective feelings. For example, Wasserman (1983) defends cognitive psychology to his fellow behaviorists by arguing that it is not subjective and mentalistic. Analyzing people as though they were computers may be useful as an initial, limited approach, just as physiologists began their analysis of the functioning of hearts by drawing analogies to mechanical pumps. But it is important to recognize the limitations inherent in this approach; it suffers from the danger of leading us into what Savory (1959) called by the apt but unfortunately tongue-twisting name of "the synechdochaic fallacy." This means the confusion of a part of something with the whole, or as Savory put it, "the error of nothing but." Information-processing is doubtless a necessary condition for mental experience, but is it sufficient? Human minds do more than process information; they think and feel. We experience beliefs, desires, fears, expectations, and many other subjective mental states.

Many cognitive psychologists imply that a computer system that could process information exactly as the human brain does would duplicate all essential elements of thinking and feeling; others simply feel that subjective experience is beyond the reach of scientific investigation. Perhaps the issue will someday be put to

What behavior suggests conscious thinking?

Just what is it about some kinds of behavior that leads us to feel that it is accompanied by conscious thinking? Comparative psychologists and biologists worried about this question extensively around the turn of this century. No clear and generally accepted answers emerged from their thoughtful efforts, and this is one reason why the behavioristic movement came to dominate psychology.

Complexity is often taken as evidence that some behavior is guided by conscious thinking. But complexity is a slippery attribute. One might think that simply running away from a frightening stimulus was a rather simple response, yet if we make a detailed description of every muscle contraction during turning and running away, the behavior becomes extremely complex. But, one might object, this complexity involves the physiology of locomotion; what is simple is the direction in which the animal moves. If we then ask what sensory and central nervous mechanisms cause the animal to move in this direction, the matter again becomes complex. Does the animal continuously listen to the danger signal and push more or less hard with its right or left legs in order to keep the signal directly behind it? Or does it head directly toward some landmark? If the latter,

Figure 2. The assassin bug at the top has camouflaged itself chemically and tactilely by gluing bits of a termite nest all over its body. In this way it is able to capture a termite at the opening of the nest without alarming the soldier termites. After sucking out the termite's semifluid organs, the assassin bug jiggles the empty exoskeleton in front of the nest opening in order to attract another termite worker, which will normally attempt to consume or dispose of the corpse. When a second termite worker seizes the first, it is then captured and consumed itself, as shown in the photograph below, and the process may be repeated continuously many times by the same assassin bug. The extraordinary complexity and coordination of these actions strongly suggest conscious thought, even though the assassin bug's central nervous system is very small. (Photographs by Raymond A. Mendez.)

how does it coordinate vision and locomotion? Again one might say that the direction of motion is simple, and it is irrelevant to worry about the complexities of the physiological mechanisms involved.

But how is this simple direction "away from the danger" represented within the animal's central nervous system? Does the animal employ the concepts of *away from* and *danger*? If so, how are such concepts established? Even though we cannot answer the question in neurophysiological terms, it is clear that running away from something is a far simpler behavior than, say, the

construction of a bird's nest. Conversely, even the locomotor motions of a caterpillar that will move toward a light with a machine-like consistency hour after hour are not simple when examined in detail. What is simple is the abstract notion of *toward* or *away*, but the mechanistic interpretation of animal behavior tends to deny that the animal could think in terms of even such a simple abstraction.

One very important attribute of animal behavior that seems intuitively to suggest conscious thinking is its adaptability to changing circumstances. If an animal repeats some action in the same way regardless of the results, we assume that a rigid physiological mechanism is at work, especially if the behavior is ineffective or harmful to the animal. When a moth flies again and again at a bright light or burns itself in an open flame, it is difficult to imagine that the moth is thinking, although one can suppose that it is acting on some thoughtful but misguided scheme. When members of our own species do things that are self-damaging or even suicidal, we do not conclude that their behavior is the result of a mechanical reflex. But to explain the moth flying into the flame as thoughtful but misguided seems far less plausible than the usual interpretation that such insects automatically fly toward a bright light, which leads them to their death in the special situation where the brightest light is an open flame.

Conversely, if an animal manages to obtain food by a complex series of actions that it has never performed before, intentional thinking seems more plausible than rigid automatism. For example, Japanese macaques learned a new way to separate grain from inedible material by throwing the mixture into the water; the kernels of grain would float while the inorganic sand and other particles tended to sink (Kawai 1965). These new types of food handling were first devised by a few monkeys, then were gradually acquired by other members of their social group through observational learning.

Behavioral versatility came into play in a spectacular fashion in the 1930s when two species of tits discovered that milk bottles delivered to British doorsteps could be a source of food (Fisher and Hinde 1949; Hinde and Fisher 1951). At that time milk-bottle tops were made of soft metal foil, and the milk was not homogenized, so the cream rose to the top of the bottle. One or more birds discovered that the same type of behavior used to get at insects hidden under tree bark could also be used to get cream from milk bottles. The people whose milk was disturbed immediately noticed it, and careful studies were made of the gradual spread of this behavior throughout much of England. A change in the technology of covering milk bottles eventually ended the whole business, but meanwhile thousands of birds had learned, almost certainly through observation, to exploit a newly available food source.

Connected patterns of behavior

Another criterion upon which we tend to rely in inferring conscious thinking is the element of interactive steps in a relatively long sequence of appropriate behavior patterns. Effective and versatile behavior often entails many steps, each one modified according to the results of the previous actions. In such a complex sequence the animal must pay attention not only to the immediate stimuli, but also to information obtained in the past. Psychologists once postulated that complex behavior can be understood as a chain of rigid reflexes, the outcome of one serving as stimulus for the next. Students of insect behavior have generally accepted this explanation for such complex activities as the construction of elaborate shelters or prey-catching devices, ranging from the underwater nets spun by certain caddis-fly larvae to the magnificent webs of spiders. But the steps an animal takes often vary, depending on the results of the previous behavior and on many influences from the near or distant past. The choice of *which* past events to attend to may be facilitated by conscious selection from a broad spectrum of memories.

An outstanding example of such sequences of interactive behaviors is the use of probes by chimpanzees to gather termites from their mounds (Goodall 1968, 1971). The chimpanzee prepares a probe by selecting a suitable branch, pulling off its leaves and side branches, breaking the stick to the right length, carrying it—often for several minutes—to a termite mound, and then probing into the openings used by the termites (see Fig. 1). If the hole yields nothing, the chimpanzee moves to another one. Even after the tool has been prepared, its use is far from stereotyped. When curious scientists try to imitate the chimpanzees' techniques, they find it rather difficult and seldom gather as many termites. It is especially interesting that the young chimpanzees seem to learn this use of tools by watching their mothers or other members of their social group. Youngsters have been observed making crude and relatively ineffectual attempts to prepare and use their own termite probes; the termite "fishing" of chimpanzees gives every evidence of being learned.

Examples can be found among insects of long, complex sequences of behaviors that are as suggestive of thought as the chimpanzee's use of a termite probe. McMahan (1982) has discovered that in tropical rain forests one species of assassin bug, a predatory insect, uses two effective tricks, illustrated in Figure 2, to capture the workers of termite colonies. The bug glues small bits of the outer layers of a termite nest to its head, back, and sides. Then it stands near an opening to the termite colony. The bits of termite nest on the assassin bug apparently smell and perhaps feel familiar to the termites, so no alarm signals are emitted, which otherwise would attract the well-armed members of the soldier caste that attack intruders. Although the assassin bug's actions often attract soldier termites, its camouflage seems to prevent them from recognizing it as an intruder, and they return to the nest. This chemical and tactile camouflage allows the assassin bug to reach into the opening and capture a termite worker, which it kills and con-

Figure 3. With an apparent deliberateness that suggests intentional thinking, nesting killdeer will conspicuously lead a potential predator away from their nest or young, adjusting the speed of their abnormally awkward movements away from the nest so as to remain within sight but out of reach of the intruder. Quite often the killdeer acts as if it is injured, as shown here, a behavior that would make it more attractive to a predator. The killdeer will sometimes employ a very different tactic in response to approaching cattle, which may trample on eggs or nestlings but will not eat them; the birds will then stand close to the nest and spread their wings in a conspicuous display that usually causes the cattle to step aside. (Photograph by Noble S. Proctor.)

sumes by sucking out all the semifluid internal organs, leaving only the exoskeleton.

Such camouflage-assisted prey capture is remarkable enough, but the next step is even more thought-provoking. The assassin bug pushes the empty exoskeleton of its victim into the nest opening and jiggles it gently. Another termite worker seizes the corpse as part of a normal behavior pattern of devouring the body of a dead sibling or carrying the corpse away for disposal. The assassin bug pulls the exoskeleton of the first victim out with the second worker attached. This one is eaten and its empty exoskeleton used in another "fishing" effort. In one case an assassin bug was observed to thus devour thirty-one termites before moving away with a fully distended abdomen.

When chimpanzees fashion sticks to probe for termites, their actions are considered among the most convincing cases of intentional behavior yet described for nonhuman animals. When McMahan discovers assassin bugs carrying out an almost equally elaborate feeding behavior, must we assume that the insect is only a genetically programmed robot incapable of understanding what it does? Perhaps we should be ready to infer conscious thinking whenever any animal shows such ingenious behavior, regardless of its taxonomic group and our preconceived notions about limitations of animal consciousness.

An example even more suggestive of thoughtful behavior among insects is the so-called "dance language" of honeybees, which was discovered by the remarkably brilliant and original experiments of von Frisch (reviewed by von Frisch 1967, 1972). One significant reaction to von Frisch's discovery was that of Jung (1973). Late in his life he wrote that although he had believed insects were merely reflex automata:

this view has recently been challenged by the researches of Karl von Frisch. . . . Bees not only tell their comrades, by means of a peculiar sort of dance, that they have found a feeding-place, but they also indicate its direction and distance, thus enabling beginners to fly to it directly. This kind of message is no different in principle from information conveyed by a human being. In the latter case we would certainly regard such behavior as a conscious and intentional act and can hardly imagine how anyone could prove in a court of law that it had taken place unconsciously. . . . We are . . . faced with the fact that the ganglionic system apparently achieves exactly the same result as our cerebral cortex. Nor is there any proof that bees are unconscious.

In many cases the networks of informative events that suggest animal thinking are sufficiently complicated that we are not sure what the animal is doing even when we know most of the relevant facts. Consider, for example, how certain ground-nesting birds, such as the killdeer or piping plovers, lead predators away from their nests or young. At a considerable distance, long before an approaching large intruder, such as a person or other mammal, can see the cryptically colored bird or its eggs, the plover may stand up and walk slowly to a point a few meters from the nest. Then the bird may flutter slowly but conspicuously away from the nest, staying relatively close to the intruder. It almost always makes loud piping or peeping sounds similar to those a bird makes when disturbed or mildly irritated.

It is common for the bird to hold its tail or wing in an abnormal position as it moves. Often the tail almost drags on the ground, and the wings are slightly extended, sometimes one more than the other, strongly suggesting some weakness or injury. After running a few meters, the bird may flop about on the ground, extending one or both wings, as if injured. This behavior, shown in Figure 3, is often called the "broken-wing display," and it requires considerable effort for an observer to believe that the bird is really quite healthy. Predators are extremely sensitive to minor differences in the gait and demeanor of potential prey and are much more likely to attack animals that are behaving abnormally.

Throughout most of this predator-distraction behavior, the bird watches the intruder. Typically it does not move in a straight line and stops from time to time. If the intruder approaches, the bird moves farther ahead. If not, the bird usually flies back closer to the intruder and repeats the behavior. The bird will allow the intruder to approach quite close, sometimes within two meters, but it always moves just fast enough and far enough to avoid capture. Typically the bird continues the injury simulation while leading the intruder some distance away from the nest or young. Finally, however, it flies away rapidly, usually in the same direction, then circling back to the general vicinity, though seldom to the exact spot, where the eggs or young are located. One Wilson's plover, a close relative of the killdeer and piping plover, led me more than 300 meters along a sandy beach before flying off.

Killdeer have been observed to use very different tactics when their nests are approached by cattle, which may trample on eggs or nestlings but will not eat them. Rather than moving away from the nest and fluttering as though injured, the birds stand close to the nest and spread their wings in a conspicuous display that usually causes the cattle to step aside (Skutch 1976).

Adaptations to novelty

One further consideration can help refine the criteria for determining the presence of conscious thought. We can easily change back and forth between thinking consciously about our own behavior and not doing so. When we are learning some new task such as swimming, riding a bicycle, driving an automobile, flying an airplane, operating a vacuum cleaner, caring for our teeth by some new technique recommended by a dentist, or any of the large number of actions we did not formerly know how to do, we think about it in considerable detail. But once the behavior is thoroughly mastered, we give no conscious thought to the details that once required close attention.

This change can also be reversed, as when we make the effort to think consciously about some commonplace and customary activity we have been carrying out for some time. For example, suppose you are asked about the pattern of your breathing, to which you normally give no thought whatsoever. But you can easily take the trouble to keep track of how often you inhale and exhale, how deeply, and what other activities accompany different patterns of breathing. You can find out that it is extremely difficult to speak while inhaling, so talking continuously requires rapid inhalation and slower exhalation. This and other examples that will readily come to mind if one asks the appropriate questions show that we can bring into conscious focus activities that usually go on quite unconsciously.

The fact that our own consciousness can be turned on and off with respect to particular activities tells us that in at least one species it is not true that certain behavior patterns are always carried out consciously while others never are. It is reasonable to guess that this is true also for other species. Well-learned behavior patterns may not require the same degree of conscious attention as those the animal is learning how to perform. This in turn means that conscious awareness is more likely when the activity is novel and challenging; striking and unexpected events are more likely to produce conscious awareness.

Thus it seems likely that a widely applicable, if not all-inclusive, criterion of conscious awareness in animals is *versatile adaptability of behavior to changing circumstances and challenges.* If the animal does much the same things regardless of the state of its environment or the behavior of other animals nearby, we are less inclined to judge that it is thinking about its circumstances or what it is doing. Consciously motivated behavior is more plausibly inferred when an animal behaves appropriately in a novel and perhaps surprising situation that requires specific actions not called for under ordinary circumstances. This is a special case of versatility, of course, but the rarity of the challenge combined with the appropriateness and effectiveness of the response are important indicators of thoughtful actions.

For example, Janes (1976) observed nesting ravens make an enterprising use of rocks. He had been closely observing ten raven nests in Oregon, eight of which were near the top of rocky cliffs. At one of these nests two ravens flew in and out of a vertical crack that extended from top to bottom of a twenty-meter cliff. Janes and a companion climbed up the crevice and inspected the six nearly fledged nestlings. As they started down,

Figure 4. When sea otters cannot open shellfish with their claws or teeth, they often employ a stone for the purpose, smashing it against the shells. The apparently conscious, intentional nature of the behavior is further indicated by the fact that an otter carefully selects a stone of suitable size and weight and often carries the stone under its armpit for considerable periods. (Photograph by William F. Bryan.)

both parents flew at them repeatedly, calling loudly, then landed at the top of the cliff, still calling. One of the ravens then picked up small rocks in its bill and dropped them at the human intruders. Several of the rocks showed markings where they had been partly buried in the soil, so the birds presumably had pried them loose. Only seven rocks were dropped, but the raven seemed to be seeking other loose ones and apparently stopped only because no more suitable rocks were available.

While many birds make vigorous efforts to defend their nests and young from intruders, often flying at people who come too close, regurgitating or defecating on them, and occasionally striking them with their bills, rock throwing is most unusual. Nor do ravens pry out rocks and drop them in other situations. It is difficult to avoid the inference that this quite intelligent and adaptable bird was anxious to chase the human intruders away from its nest and decided that dropping rocks might be effective.

There are limits to the amount of novelty with which a species can cope successfully, and this range of versatility is one of the most significant measures of mental adaptability. This discussion of adaptable versatility as a criterion of consciousness implies that conscious thinking occurs only during learned behavior, but we should be cautious in accepting this belief as a rigid doctrine.

Another aspect of conscious thinking is anticipation and intentional planning of an action with conscious awareness of its likely results. An impressive example is the use of small stones by sea otters to detach and open shellfish (Kenyon 1969). These intelligent aquatic carnivores feed mostly on sea urchins and mollusks. The sea otter must dive to the bottom and pry the mollusk loose with claws or teeth, but some shells, especially abalones, are tightly attached to the rocks and have shells that are too tough to be loosened in this fashion. The otter will search for a suitable stone, which it carries while diving, then uses the stone to hammer the shellfish loose, holding its breath all the while.

The otter usually eats while floating on its back, as shown in Figure 4. If it cannot get at the fleshy animal inside the shell, it will hold the shell against its chest with one paw and pound it with the stone. The otter often tucks a good stone under an armpit as it swims or dives. Although otters do not alter the shapes of the stones, they do select ones of suitable size and weight and often keep them for considerable periods. The otters use tools only in areas where sufficient food cannot be obtained by other methods. In some areas only the young and very old sea otters use stones; vigorous adults can dislodge the shellfish with their unaided claws or teeth. Thus it is far from a simple stereotyped behavior pattern, but one that is used only when it is helpful. Sea otters sometimes use floating beer bottles to hammer open shells. Since the bottles float, they need not be stored under the otter's armpit.

Anticipation and planning are of course impossible to observe directly in another person or animal, but indications of their likelihood are often observable. As early as the 1930s Lorenz studied the intention movements of birds (Lorenz 1971), and other ethologists have noted that these movements, small-scale preliminaries to major actions such as flying, often serve as signals to others of the same species. Although Lorenz interpreted the movements as indications that the bird was planning and preparing to fly, the term *intention movement* has been quietly dropped from ethology in recent years. I suspect this is because the behavioristic ethologists fear that the term has mentalistic implications. Earlier ethologists such as Daanje (1951) described a wide variety of intention movements in many kinds of animals, but their interest was in whether the movements had gradually become specialized communication signals in the course of behavioral evolution. The possibility that intention movements indicate the animal's conscious intention has been totally neglected by ethologists during their behavioristic phase, but we may hope that the revival of scientific interest in animal thinking will lead cognitive ethologists to study whether such movements are accompanied by conscious intentions.

Animal communication

The very fact that intention movements so often evolve into communicative signals may reflect a close linkage between thinking and the intentional communication of thoughts from one conscious animal to another. These considerations lead us directly to a recognition that because communicative behavior, especially among social animals, often seems to convey thoughts and feelings from one animal to another, it can tell us something about animal thinking: it can be an important "window" on the minds of animals.

Human communication is hardly limited to formal language; nonverbal communication of mood or intentions also plays a large and increasingly recognized role in human affairs. We make inferences about people's feelings and thoughts, especially those of very young children, from many kinds of communication, verbal and nonverbal; we should similarly use all available evidence in exploring the possibility of thoughts or feelings in other species. When animals live in a group and depend on each other for food, shelter, warning of dangers, or help in raising the young, they need to be able to judge correctly the moods and intentions of their companions. This extends to animals of other species as well, especially predators or prey. It is important for the animal to know whether a predator is likely to attack or whether the prey is so alert and likely to escape that a chase is not worth the effort. Communication may either inform or misinform, but in either case it can reveal something about the conscious thinking of the communicator.

Vervet monkeys, for example, have at least three different categories of alarm calls, which were described by Struhsaker (1967) after extensive periods of observation. He found that when a leopard or other large carnivorous mammal approached, the monkeys gave one type of alarm call; quite a different call was used at the sight of a martial eagle, one of the few flying predators that captures vervet monkeys. A third type of alarm call was given when a large snake approached the group. This degree of differentiation of alarm calls is not unique, although it has been described in only a few kinds of animals. For example, ground squirrels of western North America use different types of calls when frightened by a ground predator or by a predatory bird such as a hawk (Owings and Leger 1980).

The question is whether the vervet monkey's three types of alarm calls convey to other monkeys information about the type of predator. Such information is important because the animal's defensive tactics are different in the three cases. When a leopard or other large carnivore approaches, the monkeys climb into trees. But leopards are good climbers, so the monkeys can escape them only by climbing out onto the smallest branches, which are too weak to support a leopard. When the monkeys see a martial eagle, they move into thick vegetation close to a tree trunk or at ground level. Thus the tactics that help them escape from a leopard make them highly vulnerable to a martial eagle, and vice versa. In response to the threat of a large snake they stand on their hind legs and look around to locate the snake, then simply move away from it, either along the ground or by climbing into a tree.

To answer this question, Seyfarth, Cheney, and Marler (1980a, b) conducted some carefully controlled playback experiments under natural conditions in East Africa. From a concealed loudspeaker, they played tape recordings of vervet alarm calls and found that the playbacks of the three calls did indeed elicit the appropriate responses. The monkeys responded to the leopard alarm call by climbing into the nearest tree; the martial eagle alarm caused them to dive into thick vegetation; and the python alarm produced the typical behavior of standing on the hind legs and looking all around for the nonexistent snake.

Inclusive behaviorists—that is, psychologists interested only in contingencies of reinforcement during an individual's lifetime, and ethologists or behavioral ecologists solely concerned with the effects of natural selection on behavior—insist on limiting themselves to stating that an animal benefits from accurate information about what the other animal will probably do. But within a mutually interdependent social group, an individual can often anticipate a companion's behavior most easily by empathic appreciation of his mental state. The inclusive behaviorists will object that all we need postulate is behavior appropriately matched to the probabilities of the companions *behaving* in this way or that—all based on contingencies of reinforcement learned from previous situations or transmitted genetically.

But empathy may well be a more efficient way to gauge a companion's disposition than elaborate formulas describing the contingencies of reinforcement. All the animal may need to know is that another is aggressive, affectionate, desirous of companionship, or in some other common emotional state. Judging that he is aggressive may suffice to predict, economically and parsimoniously, a wide range of behavior patterns depending on the circumstances. Neo-Skinnerian inclusive behaviorists may be correct in saying that this empathy came about by learning, for example, the signals that mean a companion is aggressive. But our focus is on the animal's possible thoughts and feelings, and for this purpose the immediate situation is just as important as the history of its origin.

Humphrey (1976) has extended an earlier suggestion by Jolly (1966) that consciousness arose in primate evolution when societies developed to the stage where it became crucially important for each member of the group to understand the feelings, intentions, and thoughts of others. When animals live in complex social groupings, where each one is crucially dependent on cooperative interactions with the others, they need to be "natural psychologists," as Humphrey puts it. They need to have internal models of the behavior of their companions, to feel with them, and thus to think consciously about what the other one must be thinking or feeling.

Following this line of thought, we might distinguish between the animals' interactions with some feature of the physical environment or with plants, and their interactions with other reacting animals, usually their own species, but also predators and prey. Although Humphrey has so far restricted his criterion of consciousness to our own ancestors within the past few million years, it could apply with equal or even greater force to other animals that live in mutually interdependent social groups.

All this adds up to the simple idea that when animals communicate to one another they may be conveying something about their thoughts or feelings. If so, eavesdropping on the communicative signals they exchange may provide us with a practicable source of data about their mental experiences. When animals devote elaborate and specifically adjusted activities to communication, each animal responding to messages from its companion, it seems rather likely that both sender and receiver are consciously aware of the content of these messages.

The adaptive economy of conscious thinking

The natural world often presents animals with complex challenges best met by behavior that can be rapidly adapted to changing circumstances. Environmental conditions vary so much that for an animal's brain to have programmed specifications for optimal behavior in all situations would require an impossibly lengthy instruction book. Whether such instructions stem from the animal's DNA or from learning and environmental influences within its own lifetime, providing for all likely contingencies would require a wasteful volume of specific directions. Concepts and generalizations, on the other hand, are compact and efficient. An instructive analogy is provided by the hundreds of pages of official rules for a familiar game such as baseball. Once the general principles of the game are understood, however, quite simple thinking suffices to tell even a small boy approximately what each player should do in most game situations.

Of course, simply thinking about various alternative actions is not enough; successful coping with the challenges of life requires that thinking be relatively rapid and that it lead both to reasonably accurate decisions and to their effective execution. Thinking may be economical without being easy or simple, but consideration of the likely results of doing this or that is far more efficient than blindly trying every alternative. If an animal thinks about what it might do, even in very simple terms, it can choose the actions that promise to have desirable consequences. If it can anticipate probable events, even if only a little way into the future, it can avoid wasted effort. More important still is being able to avoid dangerous mistakes. To paraphrase Popper (1972), a foolish impulse can die in the animal's mind rather than lead it to needless suicide.

I have suggested that conscious thinking is economical, but many contemporary scientists counter that the problems mentioned above can be solved equally well by unconscious information-processing. It is quite true that skilled motor behavior often involves complex, rapid, and efficient reactions. Walking over rough ground or through thick vegetation entails numerous adjustments of the balanced contraction and relaxation of several sets of opposed muscles. Our brains and spinal cords modulate the action of our muscles according to whether the ground is high or low or whether the vegetation resists bending as we clamber over it. Little, if any, of this process involves conscious thought, and yet it is far more complex than a direct reaction to any single stimulus.

We perform innumerable complex actions rapidly, skillfully, and efficiently without conscious thought. From this evidence many have argued that an animal does not need to think consciously to weigh the costs and benefits of various activities. Yet when we acquire a new skill, we have to pay careful conscious attention to details not yet mastered. Insofar as this analogy to our own situation is valid, it seems plausible that when an animal faces new and difficult challenges, and when the stakes are high—often literally a matter of life and death—conscious evaluation may have real advantages.

Inclusive behaviorists often find it more plausible to suppose that an animal's behavior is more efficient if it is automatic and uncomplicated by conscious thinking. It has been argued that the vacillation and uncertainty involved in conscious comparison of alternatives would slow an animal's reactions in a maladaptive fashion. But when the spectrum of possible challenges is broad, with a large number of environmental or social factors to be considered, conscious mental imagery, explicit anticipation of likely outcomes, and simple thoughts about them are likely to achieve better results than thoughtless reaction. Of course, this is one of the many areas where we have no certain guides on which to rely. And yet, as a working hypothesis, it is attractive to suppose that if an animal can consciously anticipate and choose the most promising of various alternatives, it is likely to succeed more often than an animal that cannot or does not think about what it is doing.

References

Baker, L. R. 1981. Why computers can't act. *Am. Philos. Q.* 18:157–63.

Boden, M. A., ed. 1977. *Artificial Intelligence and Natural Man.* Basic Books.

Daanje, A. 1951. On the locomotory movements in birds and the intention movements derived from them. *Behaviour* 3:48–98.

Dreyfus, H. L. 1979. *What Computers Can't Do: The Limits of Artificial Intelligence*, rev. ed. Harper and Row.

Fisher, J., and R. A. Hinde. 1949. The opening of milk-bottles by birds. *Brit. Birds* 42:347–57.

Frisch, K. von. 1967. *The Dance Language and Orientation of Bees*. Harvard Univ. Press.

———. 1972. *Bees, Their Vision, Chemical Senses and Language*, 2nd ed. Cornell Univ. Press.

Goodall, J. van Lawick. 1968. Behaviour of free-living chimpanzees of the Gombe Stream area. *Anim. Beh. Monogr.* 1:165–311.

———. 1971. *In the Shadow of Man*. Houghton Mifflin.

Harnad, S. 1982. Consciousness: An afterthought. *Cog. Brain Theory* 5:29–47.

Hediger, H. 1947. Ist das tierliche Bewusstsein unerforschbar? *Behaviour* 1:130–37.

———. 1968. *The Psychology of Animals in Zoos and Circuses*. Dover.

———. 1980. *Tiere verstehen, Erkenntnisse eines Tierpsychologien*. Munich: Kindler.

Hinde, R. A., and J. Fisher. 1951. Further observations on the opening of milk bottles by birds. *Brit. Birds.* 44:393–96.

Humphrey, N. K. 1976. The social function of intellect. In *Growing Points in Ethology*, ed. P. P. G. Bateson and R. A. Hinde. Cambridge Univ. Press.

Janes, S. W. 1976. The apparent use of rocks by a raven in nest defense. *Condor* 78:409.

Jolly, A. 1966. Lemur social behavior and primate intelligence. *Science* 153:501–06.

Jung, C. G. 1973. *Synchronicity, a Causal Connecting Principle*. Princeton Univ. Press.

Kawai, M. 1965. Newly acquired pre-cultural behavior of the natural troop of Japanese monkeys on Koshima Islet. *Primates* 6:1–30.

Kenyon, K. W. 1969. *The Sea Otter in the Eastern Pacific Ocean*. North American Fauna, no. 68. US Bureau of Sport Fisheries and Wildlife.

Lorenz, K. 1963. Haben Tiere ein subjectives Erleben? *Jahr. Techn. Hochs. München.* Eng. trans., Do animals undergo subjective experience? In *Studies in Animal and Human Behavior*, vol. 2. Harvard Univ. Press.

———. 1971. *Studies in Animal and Human Behavior*, vol. 2. Harvard Univ. Press.

MacKenzie, B. D. 1977. *Behaviorism and the Limits of Scientific Method*. London: Routledge and Kegan Paul.

McMahan, E. A. 1982. Bait-and-capture strategy of termite-eating assassin bug. *Insectes Sociaux* 29:346–51.

Norman, D. A., ed. 1981. *Perspectives on Cognitive Science*. Hillsdale, N. J.: Erlbaum.

Owings, D. H., and D. W. Leger. 1980. Chatter vocalizations of California ground squirrels: Predator- and social-role specificity. *Z. Tierpsychol.* 54:163–84.

Popper, K. R. 1972. *Objective Knowledge*. Oxford Univ. Press.

Savory, T. H. 1959. *Instinctive Living, a Study of Invertebrate Behaviour*. London: Pergamon.

Seyfarth, R. M., D. L. Cheney, and P. Marler. 1980a. Monkey responses to three different alarm calls: Evidence for predator classification and semantic communication. *Science* 210:801–03.

———. 1980b. Vervet monkey alarm calls: Semantic communication in a free-ranging primate. *Anim. Beh.* 28:1070–94.

Skutch, A. F. 1976. *Parent Birds and Their Young*. Univ. of Texas Press.

Struhsaker, T. T. 1967. *The Red Colobus Monkey*. Univ. of Chicago Press.

Thatcher, R. W., and E. R. John. 1977. *Foundations of Cognitive Processes*. Hillsdale, N. J.: Erlbaum.

Wasserman, E. A. 1983. Is cognitive psychology behavioral? *Psychol. Record.* 33:6–11.

Whiteley, C. H. 1973. *Mind in Action, an Essay in Philosophical Psychology*. Oxford Univ. Press.

What Do Animals Think About Numbers?

Many animals have basic numerical abilities, but some experiences can transform their minds and ultimately change how they think about numbers

Marc D. Hauser

The British philosopher Bertrand Russell once said that "it must have required many ages to discover that a brace of pheasants and a couple of days were both instances of the number two." That discovery, however, was made not by the brace of pheasants, but the philosopher himself, presumably as an adult human being. And what of the pheasants? Are they capable of understanding that as a pair they represent the number two?

Birding wisdom holds that to watch most birds without disturbing them, it is best to hide behind a blind. If the bird sees you enter, however, you're not much better off because it is now aware of the blind. One way around this problem is for two people to enter the blind together. Some time later, one person leaves and the bird, apparently assuming the coast is clear, goes back to business as usual. Why? Because most birds observed in this situation are incapable of computing a simple subtraction: $2 - 1 = 1$!

It would seem that, if birds are any indication, animals are far from the most astute of mathematicians. But we do know that animals must have at least limited numerical capabilities. Species with widely different life cycles, ecologies and mating systems have been known to engage in varied forms of mental calculus designed to maximize energetic intake: They calculate average rates of return in food patches, use information about search costs and search speed to assess optimal rates of return, obey Bayes's theo-

rem (a calculation of the probability of future returns based on prior experience) and hide seeds over a broad swath of turf, returning months later to retrieve their stash. These calculations show that animals are indeed equipped with some form of powerful "number-crunching" device.

When facing competition within and between groups, social animals often adhere to the dictum "there is strength in numbers." In chimpanzees, for example, attacking and killing a member of another community occurs only if the intruder is alone and there are at least three adult males in the attacking party. Within social groups, individuals may form coalitions of two or more members to increase their relative dominance over a third individual. In bottlenose dolphins, such coalitions can reach exceptional levels of sophistication. Two to three males in one coalition join up with a second coalition to defeat a third. Occasionally, a large group of 14 male dolphins forms a team that readily overpowers smaller groups. And for what? A single, sexually receptive female. It remains to be seen, however, whether a group's superiority arises from overall numerical superiority or from some other, as yet undiscovered factor. For example, what if the total number of individuals in the two united coalitions is four and the number of individuals in the single coalition is five? Here, two coalitions are greater than one, but five is greater than four. If dolphins are truly counting the number of individuals, then such differences matter.

It is clear that animals have need for certain basic arithmetic calculations. We know little, however, about how they represent those calculations, and the corresponding numbers, in their heads. There are two popular experimental

designs from which we hope to gain insight into the conceptual representation of numbers in both animals and people: those that explore spontaneous representations of number and those that involve training.

Magic with Numbers

One recent approach to understanding animal cognition takes advantage of a technique originally developed for human infants. Called the *expectancy-violation technique*, it uses the twisted logic of a magic show to explore spontaneous representation in animals. Here's the basic principle: Imagine a magic show in which the magician walks on stage, saws a human body in half, separates the pieces and, with a wave of the wand, brings them together again. The victim sits up, perfectly aligned and good as new. The audience stares in amazement. Why? Because adult humans know that the magician has violated a fundamental physical principle: Human bodies can't be sawed in half and then brought back together again. But would human infants watching the same show be equally amazed? Do they respond differently to "magical" deviations of physical principles than to events consistent with those principles? If so, the logic goes, then they have detected a violation and have expressed some level of understanding of physical principles.

Developmental psychologist Karen Wynn of Yale University used the expectancy-violation technique to explore whether five-month-old human infants can compute simple math problems, such as $1 + 1 = 2$. To remove the effects of novelty, the experimenter must first familiarize the infant with the key objects and non-magic events. In this particular study, the experimenter showed an infant either one, two or three Mick-

Marc D. Hauser is a professor in the Department of Psychology and Program of Neuroscience at Harvard University. Address: Department of Psychology, 33 Kirkland Street, Harvard University, Cambridge, MA 02138. Internet: hauser@wjh.harvard.edu

Figure 1. Counting termites? It seems unlikely this chimpanzee is actually counting the calories in his next meal, but research suggests that many animals have a rudimentary appreciation of small numbers. The challenge for scientists is to devise experiments that reveal what the animal does understand about numerical concepts. Such research may ultimately lead to a better understanding of how numerical abilities evolved in *Homo sapiens*.

ey Mouse dolls on a stage, as well as a screen that moved up and down. Test trials started once the infant was bored, looking away from the stage. In the "expected" test (1 + 1 = 2), an infant watched as an experimenter lowered one Mickey Mouse doll onto an empty stage. A screen was then placed in front of the doll. The experimenter then produced a second Mickey Mouse doll and placed it behind the screen. When the screen was removed, the infant saw the expected outcome: two Mickeys on the stage. No magic.

In the "unexpected" test, the infant watched the same sequence of actions, involving the same two Mickey Mouse dolls but with one crucial change—a bit of backstage magic. When the experimenter removed the screen, the infant

saw either one doll (1 + 1 = 1) or three (1 + 1 = 3). Wynn found that five-month olds consistently look longer (about two seconds more) when the outcome is one or three Mickeys than when the outcome is two. And precisely the same kind of result emerges from an experiment involving subtraction (2 – 1 = 1) instead of addition.

Wynn concluded that infants have an innate capacity to do simple arithmetic. By simple, she meant addition and subtraction with a small number of objects. By innate, she meant that the general capacity to track objects and perform arithmetical operations on them comes standard as part of our genetic equipment.

Wynn's results raise fundamental questions about the development and

evolution of nonlinguistic representations. For example, what kind of representation does an infant have while watching Mickey Mouse dolls come and go? Do infants have access to nonlinguistic mental symbols—what cognitive scientists Randy Gallistel and Rochel Gelman at the University of California, Los Angeles call "numerons"—that allow them to tag individual objects as they appear and then disappear behind the screen? When an infant sees one doll, does some kind of nonlinguistic symbol light up in her brain? When a second doll is introduced, does a different symbol, one representing the number two, light up? Or maybe infants lack such symbols altogether. Perhaps they assess number by storing information about each ob-

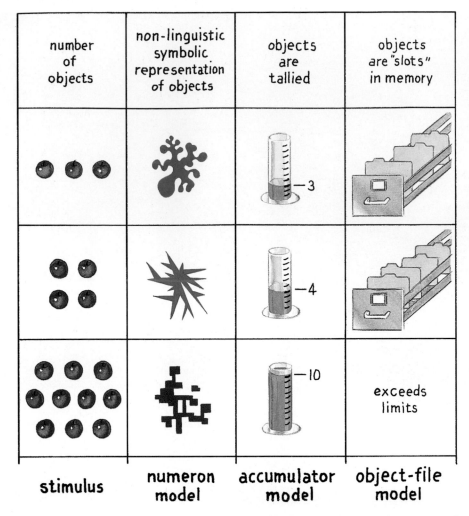

number of objects	non-linguistic symbolic representation of objects	objects are tallied	objects are "slots" in memory
		—3	
		—4	
		—10	exceeds limits
stimulus	**numeron model**	**accumulator model**	**object-file model**

Figure 2. Three psychological models attempt to explain how numbers may be represented in the human mind and nonhuman-animal mind. The *numeron* model proposes that "non-linguistic symbols" (here represented abstractly) in the mind code for integers so that each number is represented by a unique symbol. Thus, the concept of three (apples) would be represented by a symbol different from symbols representing four or 10 (apples). The *accumulator* model suggests that individual objects are tallied by the mind in a manner analogous to the way water rises in a graduated cylinder. The relative analog amount is then converted into a digital value. The *object-file* model suggests that "objects" are stored in memory "slots," perhaps analogous to an office filing system. Thus, when two objects are placed behind a screen, two slots or files in memory are opened. In this model, the limitations of short-term memory explain the inability of animals and human infants to track numbers exceeding approximately four.

ject in their memory banks, folders within the brain's "filing cabinets" that can later be retrieved and processed. Alternatively, they might compute number by means of an internal metronome that tallies the number of objects perceived. As each object is registered, a record might accumulate, much the way mercury registers changes in temperature or sand fills a graduated cylinder. These are all possible mechanisms by which number representation might occur in human infants. And, as several experiments now show, they are possible mechanisms for non-human animals as well.

One, Two, Three, Many?

Working with some of my students, I set up a magic show for the rhesus monkeys living on the Puerto Rican island of Cayo Santiago. More than 1,000 monkeys live wild on the island; on encountering a solitary curious individual, we set up to begin a trial. We started with a virtually identical version of Wynn's 1 + 1 = 2 task for human infants. Rather than Mickey Mouse dolls, however, we used bright purple eggplants. After familiarization to the eggplants and display box, the rhesus were presented with a series of test trials. In each trial, the subject watched as an ex-

perimenter placed two eggplants behind a screen and then removed the screen. Just as the human infants had done, the rhesus tended to look longer when the test outcome was one or three eggplants than when it was the expected two. Rhesus monkeys appear to understand that 1 + 1 = 2. They also seem to understand that 2 + 1 = 3, 2 − 1 = 1, and 3 − 1 = 2—but fail, however, to understand that 2 + 2 = 4.

Based on studies using the expectancy-violation procedure, the limit for spontaneous number discrimination in human infants and rhesus monkeys is approximately three. Given their ability to visually discriminate small numerosities, I then asked the question, along with developmental psychologist Susan Carey and my veterinarian wife, Lilan Basse Hauser, whether infants and monkeys are capable of acting on this knowledge. We have recently completed a set of experiments on rhesus monkeys that builds on the logic of the earlier eggplant tests but requires active searching rather than mere looking.

Attempting to simulate a natural foraging experiment, we offered the rhesus a choice between two food quantities. In the experiment, the monkey sat and watched as an experimenter put several pieces of apple into one box, followed by several pieces of apple into a second box. Initially, one box always received one more piece of apple than the other. As soon as both boxes were loaded with apple slices, the experimenter walked away and allowed the subject to approach. The rhesus monkeys consistently selected the box with the larger quantity of apple slices, discriminating quantities up to four versus three pieces. At five versus four pieces, however, the monkeys failed to show any systematic preference, sometimes selecting the box with five apple slices and sometimes the one with four. When the interval was increased to five versus three, the monkeys consistently selected five.

These results suggest that rhesus monkeys have a system of spontaneous quantification that translates to something like *one, two, three* and *many*. It must be remembered, however, that these experiments may be confounded by time. Simply, it takes longer to place four slices of apple into the box than it does to place three. Thus, the rhesus monkey could pick the box with more food based on the time it takes to place apple slices into the box rather than the actual number of slices. To eliminate this complicating factor, we reran the experiments controlling for time: We placed three apple slices in one box versus two apple slices and a rock in the second box. Although both the number of objects and the time required to place objects into each box were equated in the two boxes, the subjects continued consistently to select the box with more apple slices. And, as in the first condition, their ability to select the box with the larger amount of food was limited to a discrimination between four and three slices.

Studies of rhesus-monkey foraging decisions indicate that animals spontaneously, and without training, exhibit rudimentary numerical abilities. From a comparative perspective, the limit on spontaneous number estimation in rhesus monkeys based on looking and foraging is interesting because it parallels the limits of human infants. It also corresponds to the range in the syntax of natural language: Human languages tend to have words for *one, two* and *three* and use an expression such as *many* to denote entities greater than three. Such convergence suggests that the brain mechanism underlying spontaneous numerical estimation is shared across a diversity of animals. This hypothesis raises several crucial questions: First, are there conditions under which more sophisticated numerical capacities can be elicited in animals? And what social or ecological pressures would favor a mind capable of more precise numerical quantification? The answer to the first question is "yes," but in order to fully comprehend how and why, we must consider the second category of experimental approaches, those that involve extensive training.

In an example of a typical animal-learning experiment, rats or pigeons are placed in a Skinner box—a cage with a few lights, a food dish, several buttons and a system of wire relays—and set on a schedule of reinforcement in which every press or peck on a button yields one food pellet. With proper training, rats and pigeons easily learn this task. Now change the reinforcement schedule: For every three presses, one food pellet is delivered. The animals quickly adapt. They are even capable of grasping that to retrieve one pellet, they must press 24 times, no more, no less. Even more astonishing is this: Place a pigeon in a Skinner box

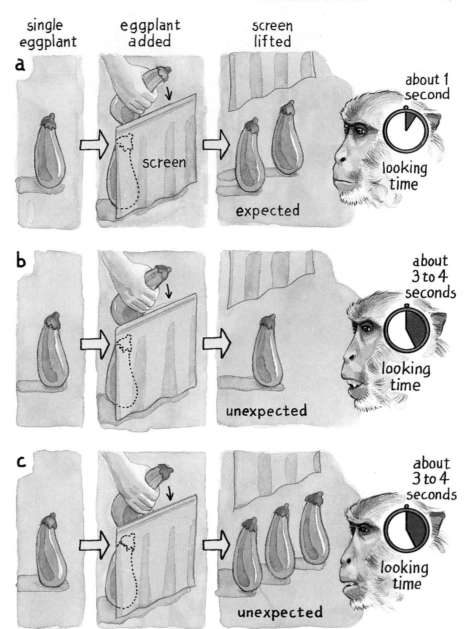

single eggplant — eggplant added — screen lifted

a screen expected about 1 second looking time

b unexpected about 3 to 4 seconds looking time

c unexpected about 3 to 4 seconds looking time

Figure 3. Creative experiments—involving "back-stage magic"—must be employed by scientists to determine whether rhesus monkeys can compute simple arithmetic problems. A single eggplant—a favorite food item that is sure to attract the monkey's attention—is first presented on "stage" in front of the animal. A screen is lowered in front of this eggplant, and a second eggplant is added behind the screen in plain view of the monkey. In control situations (*a*), the experimenters then simply raise the screen, revealing two eggplants, as would be expected (1 + 1 = 2). In such cases the monkey will look at the two eggplants for about one second. In two test conditions, an experimenter secretly removes an eggplant (*b*) or adds a third eggplant (*c*) behind the screen so that the monkey is unaware. In such instances, the unexpected outcomes (*b*, 1 + 1 = 1) and (*c*, 1 + 1 = 3) resulted in significantly longer "looking times" (about three to four seconds), suggesting that the monkeys detected a violation of an arithmetic operation. Scientists believe that such experiments reveal a capacity in rhesus monkeys to carry out simple arithmetic (with low numbers). Human infants respond similarly to experiments involving dolls rather than eggplants.

with three buttons, the left of which provides food if pecked 45 times, the right if pressed 50 times. When the center button lights up, the pigeon pecks away until the experimenter turns the light off, after either 45 or 50 pecks.

Next, both side buttons are illuminated. The pigeon must recall how many times it pecked before the experimenter turned the light off, and then peck the side button associated with this number. Thus, if the pigeon

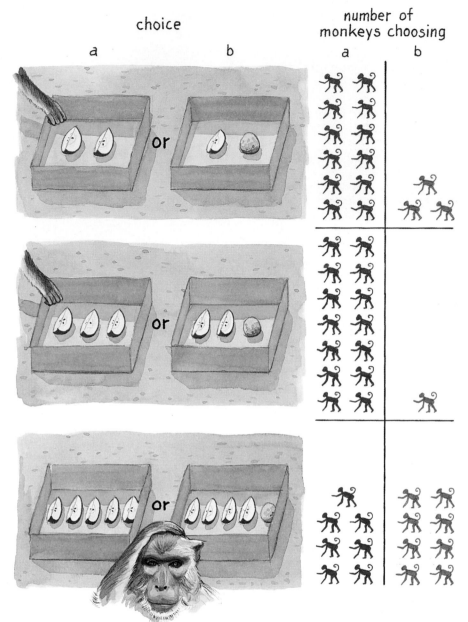

number of
monkeys choosing

Figure 4. Spontaneous counting abilities in rhesus monkeys are tested in an experiment involving apple slices and a rock. The monkeys are presented with a choice of two boxes in three experimental protocols. In each instance, only apple slices are added to one box, whereas apple slices and a rock are placed in the other box, so that the total number of objects is the same in both boxes. The rock serves as a control for the amount of time it takes to place an equal number of objects in both boxes. Hence the monkey's choice is based solely on the number of apple slices. In the first two situations (top and middle), when there are never more than three apple slices in a box, most monkeys are able to distinguish which box has the greater number of apple slices. However, when four or five apple slices are presented (bottom), the monkeys are no longer able to choose which box has the greater number of food items. Such experiments suggest that the monkey's ability to count may be limited to three; any number above three may be construed as many. These limitations are similar to those of human infants and may be indicative of a shared mechanism for counting across species.

pecked 45 times on the center button, it must turn and peck the left button. Incredibly, pigeons solve this problem.

There's more. Teach rats to press one button when they see two light flashes and another button for four light flashes. No problem. Now, present either two or four sound beeps in succession. The rats immediately transfer their knowledge from the visual to the auditory modality and do so even when the duration of the flashes or tones varies. All that seems to matter is the number of cues. In this study and many others, researchers have systematically varied the length of time a cue is presented, how hard it is to depress the lever, the amount of time elapsed between presentations and the subject's hunger level. All of these factors could serve as cues for figuring out when to press and how often. Rats and pigeons nevertheless ignore these cues, using only the number of presses to maximize the amount of food obtained.

Numbers: Category vs. Concept

It is clear that animals have a *number category*, which, like the categories *solid object* and *verb*, is a category by virtue of the fact that it refers to specific things on the basis of their properties. In the case of number, the essential property is the countable item, action or event, independent of its physical attributes. Thus, the category *seven* can refer to seven dolphins, seven pecks, seven sins, seven wonders of the world, seven days or seven licks on an ice-cream cone.

But do animals have a *number concept*? Could the pigeons understand that to peck one less than three times for a pellet requires them to peck twice? A number concept represents a symbol that has a particular relation to other symbols within the number domain. Like nouns and verbs that hold a particular relation to each other in the structure of a sentence (the domain of grammar), number concepts have unique roles by virtue of the arithmetical operations performed upon them. Thus, the concept of *seven* is unique because it is a prime number, the sum of one plus six, the number of pigeons left when two leave a flock of nine, and the only integer that is less than eight and greater than six. The category/concept distinction is important because a human child pointing to a flock of pigeons and saying *seven* may know that there are seven pigeons, but she may not know that this represents one less than a flock of eight pigeons, six more than one pigeon, and so forth. The same argument applies to animals pressing a button seven times or touching a symbol for seven on a screen.

Recent experiments by psychologists Elizabeth Brannon and Herb Terrace of Columbia University show that captive rhesus monkeys can understand the ordinal relations among the numbers one to nine. In the first phase of their experiment, experimenters presented the subject with a touch-sensitive monitor showing four different images. Each image displayed a different number of objects, varying from one to four. If the subject touched the images according to their ordinal relations—1, 2, 3, 4—it was rewarded. To ensure that subjects were attending to number, rather than some other feature, the experimenters varied size, color, and shape of the objects within each image. For example, in one trial there might be one large blue square, two small red triangles, three horses and four tiny circles. The rhesus monkeys' performance was excellent—but only after receiving hundreds of training trials.

Next, in the critical-generalization condition, the team introduced novel numbers and number combinations to the trials. Thus, in addition to seeing images with one to four objects, the rhesus monkeys now saw images with five to nine objects. Once again, Brannon and Terrace forced the monkeys to attend to number by varying size, shape and color of the objects. Surprisingly, the rhesus maintained their high level of performance, correctly pressing the images in ordinal fashion. Whether they see 1, 4, 7, 8 or 5, 6, 7, 8, the monkeys respond correctly. Clearly, rhesus monkeys are capable of understanding the principle of ordinality for the numbers one to nine.

Given that animals have limited spontaneous numerical abilities. which with training improve dramatically, we must return to a question first posed earlier: What kind of evolutionary or ecological pressures would have favored the numerical competence found in *Homo sapiens?* I suggest that in nature animals confront situations where relative, rather than absolute, quantification is sufficient. In many amphibian and fish species, for example, extremely large numbers of eggs are laid and guarded, and parents are aware of neither the initial number of eggs laid nor the number that have died or been removed by predators. Many avian species have their nests parasitized by a member of another species, one example being the cuckoo. Rather than rear its own young, the cuckoo deposits its eggs in a foreign nest, allowing the host parents to incubate and feed its young. Sometimes the parasite does a one-for-one swap, knocking out the host egg and replacing it with one

of its own. In other cases, eggs are added without removing a host egg, or the number of eggs added exceeds the number removed. Despite the number differences in a parasitized nest, there is no evidence that the host shifts allocation of parental care. The host parents don't seem to be counting at all! For most species, parental investment appears to be guided by approximations rather than absolute numbers.

While competing for resources, animals track only a small number of individuals: a few competitors of higher or lower rank, a coalition of two or three allies and a small number of potential mates. In a troop of 50 baboons, individuals might notice the disappearance of a troop member but are unlikely to think "Geez, we're down to 49. That puts us at a disadvantage against our neighbors who have 50." It seems highly unlikely, therefore, that animals living under natural conditions would confront ecological problems that would select for greater numerical competence. But, as has been shown, it is quite possible to elicit exceptional numerical abilities in animals through training in a laboratory environment.

Countdown to a Combinatorial System

What mechanisms evolved in our most immediate ancestors that enabled them to represent and conceptualize numbers with greater competence than their animal neighbors? At this point, we can only offer a highly speculative answer to this question.

People are endowed with two cognitive talents that animals naturally lack: first, the capacity to spontaneously assign arbitrary symbols to objects and events in the world, and second, the ability to manipulate the sequence and order of a string of those symbols to alter their meaning—a combinatorial system. The explosion in a child's numerical competence, lacking in monkeys and apes, arises from her capacity to formally manipulate symbols. As she amasses a growing lexicon and learns to manipulate words, the child acquires the ability to juggle number symbols. Some of the basic elements of the number system are in place before the elements of language have been fully mastered. For example, a child understands that any solid object or discrete action, such as a star in the sky or the bounce of a basketball, can be counted—the principle of property indifference—but non-solid objects like sand, water and

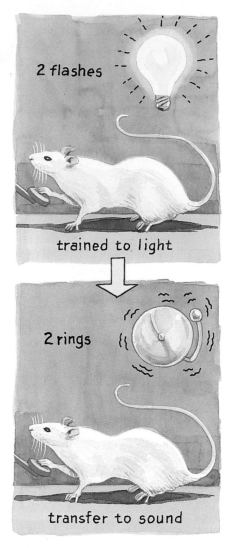

Figure 5. Rat in a Skinner box can transfer its understanding of number across different modalities, such as from a visual stimulus (light flashes) to an auditory stimulus (ringing sounds). In an experiment a rat was trained *(top)* to press one lever if the light flashed twice and to press another lever if the light flashed four times. When placed in another Skinner box where the stimulus involved either two rings or four rings of a bell *(bottom),* the rat immediately transferred its knowledge of number, correctly pressing the levers associated with the number of stimuli presented.

pudding cannot. It is for this reason that a child knows to ask for two bowls of pudding but not for two puddings. In contrast, more abstract elements of number emerge after the child has acquired a reasonable command of words. A child may not, for instance, have developed the concepts of count sequence and cardinality, in which the last label applied to the sequence represents the total number of items counted. Consequently, after counting a plate

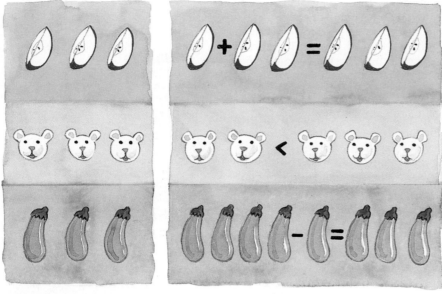

number category   number concept

Figure 6. Two types of number comprehension can be distinguished in adult human beings and, to some extent, in children and monkeys. A number category refers to a specific thing on the basis of its properties as a countable item, action or event. Thus, the category "three" can refer to three apple slices, three mice or three eggplants. A number concept represents a symbol that may be manipulated arithmetically along with other symbols in the number domain. The number concept "three" is understood to be one more than two (*top*), a number greater than two (*middle*) and one less than four (*bottom*). Adult human beings have a firm grasp of the number concept and are able to manipulate numbers far greater and more abstract than a couple of eggplants. Human infants and monkeys, however, may have a limited understanding of the number concept—one that extends only up to three or four. It is not yet known why the number concept continues to develop in the human infant as she matures, whereas a monkey's spontaneous development is arrested at numerical representations of approximately three or four.

Margaret Miller (Photo Researchers, Inc.)

Figure 7. Human children have the ability to assign meaning to symbols and to manipulate the order of those symbols, thereby increasing the number and complexity of possible variations in a "combinatorial system." Such qualities are central to the effectiveness of human language and may prove to be the foundation of the adult human's facility with numbers. In contrast, the number systems of animals lack any spontaneous demonstration of combinatorial power. By studying the development of numerical competence in children, scientists hope to gain some understanding of the remarkable evolution of numerical abilities in *Homo sapiens*.

full of cookies and saying that there are "five," she will start counting again from one when asked the total number on the plate. In this sense, a child's numerical abilities are less mature than her linguistic abilities. The pattern of development proceeds with some aspects of numerical competence emerging before linguistic competence, and others emerging afterwards.

The combinatorial engine underlying our number and language systems allows for a finite number of elements to be recombined into an infinite variety of expressions. The evolutionary origin of this capacity remains unclear. Did it evolve for number, language or both? Clearly, the number system of animals shows no sign of combinatorial power, nor do their natural communication systems show any sign of combinatorial organization. At present, therefore, research on animals does little to further our understanding of this evolutionary mystery, but developmental data on children help a bit. Because children are capable of producing sentences long before they grasp the idea of counting, it would appear that recombination occurs first in the language system and then, somewhat later, in the number system. Studies of brain-damaged patients show that some individuals may suffer linguistic deficits without significant loss of numerical competence. Conversely, other individuals might be inflicted with severe numerical deficits while maintaining functioning linguistic abilities. This suggests that separate computational systems are responsible for language and number.

What I propose is that the selective pressure responsible for the emergence of a numerical combinatorial system, one that allowed ancestral humans to enumerate at a more precise level than other animals, is the emergence of exchange systems—trading, to be precise. Whether trading spears, mongongo nuts, goats for a dowry or coins, it is essential to know how much you are getting and that it is a fair exchange. Approximations are doomed to failure in this kind of system. And although some animals do engage in reciprocal exchanges, they are not based on any kind of quantitative precision. Vampire bats regurgitate blood to those that have regurgitated to them in the past, but they don't count milliliters. Bonobo males trade access to food for sex, but they don't count the amount of food dispensed nor tally the number of resulting copulations. In all of these interactions, the system functions on the basis of approximate returns. When social exchange of material goods came onto the scene, selection favored those individuals capable of enumeration and combinatorial computation with symbols. Early humans evolved to demand precise reciprocal exchange, providing the groundwork for a multitude of extraordinary mathematical systems.

Today, while sitting in mathematics classes or perusing library bookshelves, we can study trigonometry, algebra, calculus and set theory. These systems showcase the endless creativity of the human mind and its invention of symbolic notation. We must not forget, however, that such systems stand on a foundation left behind by our animal ancestors. At present we do not understand how these two domains of knowledge affect each other during the course of evolution or that of development. Some day we will.

Bibliography

Boysen, S. T. 1997. Representation of quantities by apes. *Advances in the Study of Behavior* 26:435–462.

Brannon, E. M., and H. S. Terrace. 1998. Ordering of the numerosities 1 to 9 by monkeys. *Science* 282:746–749.

Davis, H., and R. Perusse. 1988. Numerical competence in animals: definitional issues, current evidence and a new research agenda. *Behavioral and Brain Sciences* 11:561–579.

Gallistel, C. R. 1990. *The Organization of Learning.* Cambridge: MIT Press.

Hauser, M. D., P. MacNeilage and M. Ware. 1996. Numerical representations in primates. *Proceedings of the National Academy of Sciences* 93:1514–1517.

Hauser, M. D. 1997. Tinkering with minds from the past. In *Characterizing Human Psychological Adaptations*, ed. M. Daly. New York: John Wiley & Sons. pp. 95–131.

Hauser, M. D., and S. Carey. 1998. Building a cognitive creature from a set of primitives: Evolutionary and developmental insights. In *The Evolution of Mind*, ed. D. Cummins and C. Allen. Oxford: Oxford University Press. pp. 51–106.

Hauser, M. D. 2000. *Wild Minds: What Animals Really Think.* New York: Holt.

Hauser, M. D., S. Carey and L. B. Hauser. In press. Spontaneous number presentation in semi-free-ranging rhesus monkeys. *Proceedings of the Royal Society, London: Biological Sciences.*

Matsuzawa, T. 1996. Chimpanzee intelligence in nature and in captivity: isomorphism of symbol use and tool use. In *Great Ape Societies*, eds. W. C. McGrew, L. F. Nishida and T. Nishida. Cambridge: Cambridge University Press. pp. 196–209.

Shettleworth, S. 1998. *Cognition, Evolution and Behavior.* New York: Oxford University Press.

Internet links for further exploration of "What Do Animals Think About Numbers?" are available on the *American Scientist* Web site:

http://www.amsci.org/amsci/articles/00articles/hauser.html

Mozart's Starling

Meredith J. West
Andrew P. King

On 27 May 1784, Wolfgang Amadeus Mozart purchased a starling. Three years later, he buried it with much ceremony. Heavily veiled mourners marched in a procession, sang hymns, and listened to a graveside recitation of a poem Mozart had composed for the occasion (1). Mozart's performance has received mixed reviews. Although some see his gestures as those of a sincere animal lover, others have found it hard to believe that the object of Mozart's grief was a dead bird. Another event in the same week has been put forth as a more likely cause for Mozart's funereal gestures: the death of his father Leopold (2).

The scholars who have reported and interpreted this historical incident knew much about Mozart but little, if anything, about starlings. To put the incident into better perspective, we will provide here a profile of the vocal capacities of captive starlings. Mozart's skills as a musician and composer would have rendered him especially susceptible to the starling's vocal charms, and thus we will also propose that the funeral and the poem are not the end of the story. Mozart may have left another memorial to his starling, an offbeat requiem for rebels.

Mozart's starling was a European starling, *Sturnus vulgaris*. The species was later introduced to North America on an artistic note. The birds were imported from England in the 1890s in an effort to represent the avian cast of Shakespeare's plays in this country (3). Fewer than 200 birds were released in New York's Central Park. Population estimates in the 1980s hovered around 200,000,000 birds, a millionfold increase, making starlings one of the most successful road shows in history.

The vocal talents of starlings have been known since antiquity (4). The species possesses a rich repertoire of calls and songs composed of whistles, clicks, rattles, snarls, and screeches. In addition, starlings copy the sounds of other birds and animals, weaving these mimicked themes into long soliloquies that, in captive birds,

Meredith J. West and Andrew P. King received their Ph.D.s from the Department of Psychology at Cornell University. Meredith West is a professor of psychology at Indiana University, and Andrew King is a research associate professor at Duke University. Their research interests include learning, development, and communication. Address: Laboratory of Avian Behavior, Route #2 Box 315, Mebane, NC 27302.

Like echo-locating bats or dolphins, some birds may bounce sounds off the animate environment, using behavioral reverberations to perceive the consequences of their vocal efforts

can contain fragments of human speech. Pliny reported individual birds, mimicking Greek and Latin, that "practiced diligently and spoke new phrases every day, in still longer sentences." Shakespeare knew enough about their abilities to have Hotspur propose teaching a starling to say the name "Mortimer," an earl distrusted by Henry IV, to disturb the king's sleep (*Henry IV, Part I*, act 1, scene 3). In the song cycle *Die schöne Müllerin*, Schubert set to music a poem in which a starling is given a romantic mission: "I'd teach a starling how to speak and sing, / Till every word and note with truth should ring, / With all the skill my lips and tongue impart, / With all the warmth and passion of my heart" (5).

Despite this wealth of anecdotal information, few scientists have studied the vocal behavior of starlings under the conditions necessary to separate fact from fiction. The problem with starlings is that they vocalize too much, too often, and in too great numbers, sometimes in choruses numbering in the thousands (a flock of starlings is labeled a murmuration). Even the seemingly elementary step of creating an accurate catalogue of the vocal repertoires of wild starlings is an intimidating task because of the variety of their sounds. Other well-known avian mimics, such as the mockingbird (*Mimus polyglottos*), have proved as challenging, leaving unanswered key questions about the development and functions of mimetic behavior.

Some of the problems involved in the study of nonmimetic songbirds arise with mimics as well. Researchers must be able to find and raise songbirds from a young age or ideally from the egg under conditions in which their exposure to social and acoustic stimulation can be controlled. The birds must be observed for many months or sometimes years to capture fully the processes of cultural evolution and transmission of vocal motifs from generation to generation. And for all species, researchers must acquire expertise in the acoustic analysis of sounds to overcome their inability to hear much of the fine detail in avian vocalizations.

Because of these difficulties, many "definitive" pieces of work have been based on small sample sizes, often fewer than ten individuals, sometimes fewer than five. Larger samples are possible only with avicultural favorites, such as canaries (*Serinus canaria*) or zebra finches (*Poephila guttata*). Even with these subjects, research schedules must be accommodated to seasonal cycles. The kinds of vocalizations produced by a species can differ considerably throughout the year, with the most "interesting" sounds in the form of territorial or mating signals occurring for only a few months each year. In sum, songbirds are a handful.

Mimetic species add another layer of difficulty by including sounds made by other birds, other animals, and even machines. Thus, in addition to exploring how members of a mimetic species develop species-typical calls and songs—that is, vocalizations with many shared acoustic properties within a population—investigators routinely encounter individual idiosyncrasies. Why does one starling mimic a goat and another a cat? Given the abundance of sounds in the world, what processes account for the selection of models?

Baylis *(6)* advocated studying just part of the mimic's repertoire as a first step, suggesting the example of mockingbirds frequently mimicking cardinals *(Cardinalis cardinalis)*. Although mockingbirds mimic many species, cardinals are a favorite. Why? What consequences accrue for mimic or model? By focusing on one model-mimic system, scientists might answer a number of questions surrounding the nature and function of mimicry. Further control of the model-mimic system can be gained by exposing birds to human speech, a vocal code with a more favorable "signal-to-noise" ratio. This heightens the probability that investigators can detect mimicry and makes it easier to identify the origin of mimicked sounds and the environmental conditions facilitating or inhibiting interspecific mimicry *(7)*. Here, the use of human language is not comparable to efforts with apes or dolphins aimed at uncovering possible analogues to human language. Rather, the use of speech sounds is more properly compared to the use of a radioactive isotope to trace physiological pathways. Thus, when a captive starling utters, "Does Hammacher Schlemmer have a toll-free number?" it is easier to trace the phrase's origin and how often it has been said than to trace the history of the bird's production of "breep, beezus, breep, beeten, beesix."

Over the past decade, we have studied nine starlings, each hand-reared from a few days of age *(8)*. We have also collected information on the behavior of five other starlings (Fig. 1), raised under similar conditions by individuals unaware of our work and unaware of starlings' mimicking abilities when their relationship with the birds began *(9)*. Although many questions remain about the species's vocal capacities, the findings shed light on Mozart's response to his starling's death.

The 14 starlings experienced different social relationships with humans. Eight birds lived individually in what is called interactive contact with the humans who

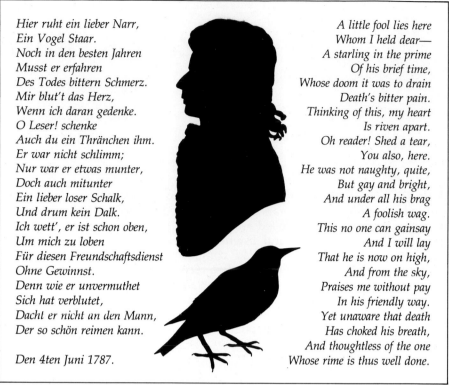

Hier ruht ein lieber Narr,	A little fool lies here
Ein Vogel Staar.	Whom I held dear—
Noch in den besten Jahren	A starling in the prime
Musst er erfahren	Of his brief time,
Des Todes bittern Schmerz.	Whose doom it was to drain
Mir blut't das Herz,	Death's bitter pain.
Wenn ich daran gedenke.	Thinking of this, my heart
O Leser! schenke	Is riven apart.
Auch du ein Thränchen ihm.	Oh reader! Shed a tear,
Er war nicht schlimm;	You also, here.
Nur war er etwas munter,	He was not naughty, quite,
Doch auch mitunter	But gay and bright,
Ein lieber loser Schalk,	And under all his brag
Und drum kein Dalk.	A foolish wag.
Ich wett', er ist schon oben,	This no one can gainsay
Um mich zu loben	And I will lay
Für diesen Freundschaftsdienst	That he is now on high,
Ohne Gewinnst.	And from the sky,
Denn wie er unvermuthet	Praises me without pay
Sich hat verblutet,	In his friendly way.
Dacht er nicht an den Mann,	Yet unaware that death
Der so schön reimen kann.	Has choked his breath,
	And thoughtless of the one
Den 4ten Juni 1787.	Whose rime is thus well done.

English translation reprinted with permission of Charles Scribner's Sons, an imprint of Macmillan Publishing Co., from *Mozart*, by Marcia Davenport. Copyright 1932, and renewed 1960, by Marcia Davenport.

hand-reared them. Their cages were placed in busy parts of the home, and the birds had considerable freedom to associate with their caregivers in diverse ways: feeding from hands; perching on fingers, shoulders, or heads; exploring caregivers' possessions; and inserting themselves into activities such as meal preparation, piano lessons, baths, showers, and telephone conversations (Fig. 2). The humans spontaneously talked to the birds, whistled to them, and gestured by kissing, snapping fingers, and waving good-bye.

Explicit procedures to teach human words using methods prescribed for other mimicking species were not used. Six of the eight caregivers did not know that such training would have an effect until the birds themselves demonstrated their mimicking ability, and two refrained because they were instructed by us to do so. The birds could obtain food and water (and avian companionship in five of eight cases) without interacting with humans.

Three other starlings lived under conditions of limited contact with humans. After 30 days of hand rearing by us, they were individually placed in new homes, along with a cowbird *(Molothrus ater)*. They lived in cages, rarely flew free, and were passively exposed to humans. They heard speech but were not "spoken to" because they did not engage in the kinds of social interactions described for the first group. The final three starlings lived together in auditory contact with humans. They were housed in an aviary on a screened porch of the caregivers raising one of the freely interacting birds. As a result, their auditory environment was loosely yoked to that of the other bird.

The information gathered on the starling's mimicry

differed by setting and caregiver. Extensive audio taping was carried out for the nine subjects studied under our supervision. For three of the remaining birds involved in interactive contact, we used repertoires available in published works, supplemented by personal inquiries. For the last two we obtained verbal reports from caregivers.

Social transmission of the spoken word

The starlings' mimetic repertoires varied consistently by social context: only the birds in interactive contact mimicked sounds with a clearly human origin. None of the other subjects imitated such sounds, although all mimicked their cowbird companions, each other, wild birds, and mechanical noises. For the purposes of this article, we have elected to focus solely on the actions of the birds in interactive contact.

All of these birds mimicked human sounds—including clear words, sounds immediately recognizable as speech but largely unintelligible, and whistled versions of songs identified as originating from a human source—and mechanical sounds whose source could be identified within the households. For the three audiotaped birds, roughly two-thirds of their vocalizations were related to the words or actions of caregivers. The same categories applied to the remaining five birds, who mimicked speech, whistles, and human-derived or mechanical sounds (Table 1).

Many of the more impressive properties of the starlings' vocal capacities defy simple categorization. The most striking feature was their tendency to mimic con-

Fewer than 200 starlings were released in Central Park in the 1890s; population estimates in the 1980s hovered around 200,000,000 birds, a millionfold increase

nected discourse, imitating phrases rather than single words. Words most often mimicked alone included the birds' names and words associated with humans' arrivals and departures, such as "hi" or "good-bye." All phrases were frequently recombined, sometimes giving the illusion of a different meaning. One bird, for example, frequently repeated, "We'll see you later," and "I'll see you soon." The phrase was often shortened to "We'll see," sounding more like a parental ploy than an abbreviated farewell. Another bird often mimicked the phrase "basic research" but mixed it with other phrases, as in "Basic research, it's true, I guess that's right."

The audiotapes and caregivers' reports made clear, however, that nonsensical combinations (from a human speaker's point of view) were as frequent as seemingly sensible ones: the only difference was that the latter were more memorable and more often repeated to the birds. Sometimes, the speech utterances occurred in highly incongruous settings: the bird mentioned above blasted his owners with "Basic research!" as he struggled frantically with his head caught in string; another screeched, "I have a question!" as she squirmed while being held to have her feet treated for an infection. The tendency for the birds to produce comical or endearing combinations did much to facilitate attention from humans. It was difficult to ignore a bird landing on your shoulder announcing, "Hello," "Give me a kiss," or "I think you're right."

The birds devoted most of their singing time to rambling tunes composed of songs originally sung or whistled to them intermingled with whistles of unknown origin and starling sounds. Rarely did they preserve a melody as it had been presented, even if caregivers repeatedly whistled the "correct" tune. The tendency to sing off-key and to fracture the phrasing of the music at unexpected points (from a human perspective) was reported for seven birds (no information on the eighth). Thus, one bird whistled the notes associated with the words "Way down upon the Swa-," never adding "-nee River," even after thousands of promptings. The phrase was often followed by a whistle of his own creation, then a fragment of "The Star-spangled Banner," with frequent interpositions of squeaking noises. Another bird whistled the first line of "I've Been Working on the Railroad" quite accurately but then placed unexpectedly large accents on the notes associated with the second line, as if shouting, "All the livelong day!" Yet another routinely linked the energetically paced *William Tell* Overture to "Rockaby Baby."

One category of whistles escaped improvisation. Seven of the eight caregivers used a so-called contact whistle to call the birds, typically a short theme (e.g., "da da da dum" from Beethoven's Fifth Symphony). This fragment of melody escaped acoustic improvisation in all cases, although the whistles were inserted into other melodies as well. One bird, however, often mimicked her contact whistle several times in succession, with each version louder than the preceding one (perhaps a quite accurate representation of the sound becoming louder as her caregiver approached her).

All the birds in interactive contact showed an interest in whistling and music when it was performed. They often assumed an "attentive" stance, as shown in Figures 1 and 2: they stood very quietly, arching their necks and moving their heads back and forth. The birds did not vocalize while in this orientation. Records for all eight subjects contained verbal or pictorial reports of the posture.

Clear mimicry of speech was relatively infrequent, due in large part to the birds' tendency to improvise on the sounds, making them less intelligible although definitely still speechlike. Other aspects of their speech imitations were also significant. First, the birds would mimic the same phrase, such as "see you soon" or "come here," but with different intonation patterns. At times, the mimetic version sounded like a human speaking in a pleasant tone of voice, and at other times in an irritated tone. Second, when the birds repeated speech sounds, they frequently mimicked the sounds that accompany speaking, including air being inhaled, lips smacking, and throats being cleared. One bird routinely preceded his rendition of "hi" with the sound of a human sniffing, a combination easily traced to his caregiver being allergic to birds. Finally, the quality of the mimicry of the human voice was surprisingly high. Many visitors who heard the mimicry "live" looked for an unseen human. Those listening to tapes asked which sounds were the starlings' and which the humans', when the only voices were the birds'.

The particular phrases that were mimicked varied, although a majority fell into the broad semantic category of socially expressive speech used by humans as greetings or farewells, compliments, or playful responses to children and pets (see Table 1). Several of the starlings used phrases of greeting or farewell when they heard the sound of keys or saw someone putting on a coat or approaching a door. Several mimicked household events such as doors opening and closing, keys rattling, and dishes clinking together. One bird acquired the word "mizu" (Japanese for water), which she routinely used after flying to the kitchen faucet. Another chanted "Defense!" when the television was on, a sound that she apparently had acquired as she observed humans responding to basketball games.

Figure 1. Kuro is a starling who was hand-reared in captivity. Living in daily close contact with the Iizuka family, she has spontaneously developed, like other starlings in similar circumstances studied by the authors, a rich repertoire of imitations of human speech, songs, and household sounds. Here Kuro listens to whistling. (Photo by Birgitte Nielsen; reprinted by permission of Nelson Canada from *Kuro the Starling,* by Keigo Iizuka and family.)

Caregivers reported that it took anywhere from a few days to a few months for new items to appear in the birds' repertoires. Acquisition time may have depended on the kind of material: one of the birds in limited contact, housed with a new cowbird, learned its companion's vocalization in three days, while one bird in interactive contact took 21 days to mimic his cowbird companion. The latter bird, however, repeated verbatim the question, "Does Hammacher Schlemmer have a toll-free number?" a day after hearing it said only once.

Starlings copy the sounds of other birds and animals, weaving these mimicked themes into long soliloquies that, in captive birds, can contain fragments of human speech

---

Some whistled renditions of human songs also appeared after intervals of only one or two days. An important variable in explaining rate of acquisition and amount of human mimicry may be the birds' differential exposure to other birds. The three birds without avian cage mates appeared to have more extensive repertoires, but they were also older than the other subjects.

The birds did not engage much in mutual vocal exchanges with their caregivers—that is, a vocalization directed to a bird did not bring about an immediate vocal response, although it often elicited bodily orientation

and attention. Thus, the mimicry lacked the "conversational" qualities that have been sought after in work with other animals (10). As no systematic attempt had been made to elicit immediate responding by means of food or social rewards, reciprocal exchanges may nevertheless be possible. Ongoing human conversation not involving the starlings, however, was a potent stimulus for simultaneous vocalizing. The birds chattered frequently and excitedly while humans were talking to each other in person or on the telephone.

The starlings' lively interest and ability to participate in the activities of their caregivers created an atmosphere of mutual companionship, a condition that may be essential in motivating birds to mimic particular models, as indicated by the findings with the birds in limited and auditory contact. The capacity of starlings to learn the sounds of their neighbors fits with what is known about their learning of starling calls, especially whistles, in nature. They learn new whistles as adults by means of social interactions, an ability that is quite important when they move into new colonies or flocks (11). Analyses of social interactions between wild starling parents and their young also indicate the use, early in ontogeny, of vocal exchanges between parent and young and between siblings (12). Thus, the capacities identified in the mimicry of human speech and their dependence on social context seem relevant to the starling's ecology.

Other mimics and songsters

Studies of another mimic, the African gray parrot (*Psittacus erithacus*), also indicate linkages between mimicry and social interaction (13). This species mimics human speech when stimulated to do so by an "interactive modeling technique" in which a parrot must compete for the attention of two humans engaged in conversation. Extrinsic rewards such as food are avoided. The reinforcement is physical acquisition of the object being talked about and responses from human caregivers. Such procedures lead to articulate imitation and often highly appropriate use of speech sounds. Pepperberg reports that one bird's earliest "words" referred to objects he could use: "paper," "wood," "hide" (from rawhide chips), "peg wood," "corn," "nut," and "pasta" *(14)*. The parrot also employed these mimicked sounds during exchanges with caregivers in which he answered questions about the names of objects and used labels identifying shape and color in appropriate ways. The parrot's use of "no" and "want" also suggested the ability to form functional relationships between speech and context, a capacity perhaps facilitated by the trainer's explicit attempts to arrange training sessions meaningful for the student.

Explanations of mimicry of human sounds in this and other species originate in the idea that hand-reared birds perceive their human companions in terms of the social roles that naturally exist among wild birds. Lorenz and von Uexküll elaborated on the kinds of relationships between and among avian parents, offspring, siblings, mates, and rivals *(15)*. In the case of captive birds, humans become the companion for all seasons, with the nature of the relationship shifting with the changing developmental and hormonal cycles in a bird's life.

Mimics are not the only birds to show clear evidence of the effects of companions on vocal capacities. Two examples from nonmimetic species are relevant. In the white-crowned sparrow (*Zonotrichia leucophrys*), the capacity to learn the songs of other males differs according to the tutoring procedure used. For example, young males learn songs from tape recordings until they are 50 days of age but not afterward. They do acquire songs well after 50 days from live avian tutors with whom they can interact, copying the song of another species, even if

---

Explanations of mimicry of human sounds originate in the idea that hand-reared birds perceive their human companions in terms of the social roles that naturally exist among wild birds

---

they can hear conspecifics in the background. The potency of social tutors has led to a comprehensive reinterpretation of the nature of vocal ontogeny in this species *(16)*. We tried tutoring nine of the starlings using tapes of the caregiver's voice singing songs and reciting prose. There was no evidence of mimicry, except that one bird learned the sound of tape hiss. And thus, if we had relied on tape tutoring, as has been done with many species to assess vocal capacity, we would have vastly underestimated the starlings' skills.

What are the characteristics of live tutors that make them so effective? The studies of white-crowned sparrows suggest that it is not the quality of the tutor's voice, but the opportunity for interaction. Indeed, we have studied a case where voice could not be a cue at all because the "tutor" could not sing. In cowbirds, as in many songbirds, only males sing. Females are frequently the recipients of songs and display a finely tuned perceptual sensitivity to conspecific songs *(17)*. We have documented that acoustically naive males produce distinct themes when housed with female cowbirds possessing different song preferences. We have also identified one important element in the interaction. When males sang certain themes, females responded with distinctive wing movements. The males responded in turn to such behavior by repeating the songs that elicited the females' wing movements. Such data show that singers attend to visual, as well as acoustic, cues and that tutors can be salient influences even when silent. In this species, the social, as distinct from the vocal, conduct of a male's audience is of consequence.

Figure 2. Kuro adopts a listening posture during a music lesson, with neck arched and head moving back and forth. (Photo by Birgitte Nielsen; reprinted by permission of Nelson Canada from *Kuro the Starling*, by Keigo Iizuka and family.)

Studies of another avian group, domestic fowl (*Gallus gallus*), also direct attention to the importance of a signaler's audience *(18)*. In this species, male cockerels produce different calls in the presence of different social companions. Emitting a food call in the presence of food is not an obligatory response but one modulated by the signaler's observations of his audience. Similar findings with cockerel alarm calls indicate the need to consider the multiple determinants of vocal production. Taken as a whole, the findings reveal that, for many birds, acoustic communication is as much visual as vocal experience.

Mozart as birdcatcher

Mozart knew how to look at, as well as listen to, audiences, especially when one of his compositions was the object of their attention. After observing several audiences watching *The Magic Flute*, he wrote to his wife, "I have at this moment returned from the opera, which was as full as ever. . . . But what always gives me most pleasure is the *silent approval!* You can see how this opera is becoming more and more esteemed" *(19)*. Mozart's enjoyment of the less obvious reactions of his audience suggests that, like a bird, he too was motivated not only by auditory but by visual stimuli. The German word he used can be translated "applause" as well as "approval," suggesting his search for rewards more meaningful than the expected clapping of hands. We now turn to the case of Mozart's starling and to the kinds of social and vocal rewards offered to him by his choice of an avian audience.

Mozart recorded the purchase of his starling in a diary of expenses, along with a transcription of a melody whistled by the bird and a compliment (Fig. 3). He had begun the diary at about the same time that he began a catalogue of his musical compositions. The latter effort was more successful, with entries from 1784 to 1791, the year of his death. His book of expenditures, however, lapsed within a year, with later entries devoted to practice writing in English *(20)*. The theme whistled by the starling must have fascinated Mozart for several reasons. The tune was certainly familiar, as it closely resembles a theme that occurs in the final movement of the Piano Concerto in G Major, K. 453 (see Fig. 3). Mozart recorded the completion of this work in his catalogue on 12 April in the same year. As far as we know, just a few people had heard the concerto by 27 May, perhaps only the pupil for whom it was written, who performed it in public for the first time at a concert on 13 June. Mozart had expressed deep concern that the score of this and three other concertos might be stolen by unscrupulous copyists in Vienna. Thus, he sent the music to his father in Salzburg, emphasizing that the only way it could "fall into other hands is by that kind of cheating" *(21)*. The letter to his father is dated 26 May 1784, one day before the entry in his diary about the starling.

Mozart's relationship with the starling thus begins on a tantalizing note. How did the bird acquire Mozart's music? Our research suggests that the melody was certainly within the bird's capabilities, but how had it been transmitted? Given our observation that whistled tunes are altered and incorporated into mixed themes, we assume that the melody was new to the bird because

Table 1. Sounds mimicked by starlings

Greetings and farewells

hi	hey there	I'll (we'll) *see you* soon
good morning	*c'mon, c'mere*	breakfast
hello	go to your cage	it's time
hey buddy	night night	

Attributions

you're a crazy bird	nutty bird	you're gorgeous
good girl	rascal	see you soon baboon
pretty bird	you're kidding	baby
silly bird		

Conversational fragments

it's true	OK	have the kids called
I suggest	I have a question	*whatcha doing*
that's right	defense	what's going on
basic research	thank you	all right you guys
because	*right*	this is Mrs. Suthers
I guess	who is coming	calling

Human sounds

sighing	*sniffing*	kissing
coughing	*lip smacking*	wolf whistle
throat clearing	*laughing*	

Household sounds

door squeaking	alarm clock	dishes clinking
cat meowing	telephone beep	gun shots
dog barking	keys rattling	

Categories refer to social contexts in which humans produced the sounds, not necessarily the ones in which starlings repeated them. Italicized entries were imitated by four or more birds.

it was so close a copy of the original. Thus, we entertain the possibility that Mozart, like other animal lovers, had already visited the shop and interacted with the starling before 27 May. Mozart was known to hum and whistle a good deal. Why should he refrain in the presence of a bird that seems to elicit such behavior so easily?

A starling in May would be either quite young, given typical spring hatching times, or at most a year old, still young enough to acquire new material but already an accomplished whistler. Because it seems unlikely to us that a very young bird could imitate a melody so precisely, we envision the older bird. The theme in question from K. 453 has often been likened to a German folk tune and may have been similar to other popular tunes already known to the starling, analogous to the highly familiar tunes our caregivers used. But to be whistled to by Mozart! Surely the bird would have adopted its listening posture, thereby rewarding the potential buyer with "silent applause."

Given that whistles were learned quite rapidly by the starlings we studied, it is not implausible that the Vienna starling could have performed the melody shortly after hearing it for the first time. Of course, we cannot rule out a role for a shopkeeper, who could have repeated Mozart's tune from its creator or from the starling. In any case, we imagine that Mozart returned to the shop and purchased the bird, recording the expense

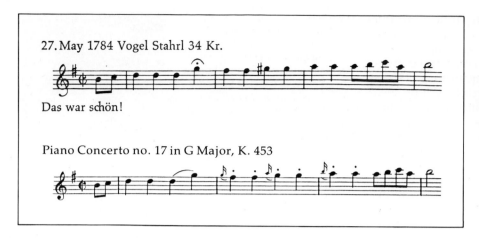

27. May 1784 Vogel Stahrl 34 Kr.

Das war schön!

Piano Concerto no. 17 in G Major, K. 453

Figure 3. Wolfgang Amadeus Mozart was also the delighted owner of a pet starling. He recorded the purchase of the starling in an expense book, noting the date, price, and a musical fragment the bird was whistling. The pleasure he expressed at hearing the starling's song—"Das war schön!" (that was beautiful!)—is all the more understandable when one compares the beginning of the last movement of his Piano Concerto in G Major, K. 453, which was written about the same time. Somehow the bird had learned the theme from Mozart's concerto. It did however sing G sharp where Mozart had written G natural, giving its rendition a characteristically off-key sound.

out of appreciation for the bird's mimicry. Some biographers suggest an opposite course of transmission—from the starling to Mozart to the concerto—but the completion date of K. 453 on 12 April makes this an unlikely, although not impossible, sequence of events.

Given the sociable nature of the captive starlings we studied, we can imagine that some of the experiences that followed Mozart's purchase must have been quite agreeable. Mozart had at least one canary as a child and another after the death of the starling, suggesting that it would not be hard for him to become attached to so inventive a housemate. Moreover, he shared several behavioral characteristics with captive starlings. He was fond of mocking the music of others, often in quite irreverent ways. He also kept late hours, composing well into the night *(22)*. The caregivers of the starlings we

---

The mimicry of vocal acts such as lip noises, sniffs, and throat clearing brought to the attention of caregivers routine dimensions of their own behavior that they rarely took notice of

---

studied uniformly reported—and sometimes complained about—the tendency of their birds to indulge in more than a little night music.

The text of Mozart's poem on the bird's death suggests other perceptions shared with the caregivers. Mozart dubbed his pet a "fool"—the German word could also be translated as "clown" or "jester"—an attribution in keeping with the modern starlings' vocal productions of "crazy bird," "rascal," "silly bird," and "nutty bird" and the even more frequent use of such terms in the written description of life with starlings. Mozart gets to the heart of the starling's character when he states that the bird was "not naughty quite, / But gay and bright, / And under all his brag, / A foolish wag." And thus, when we contemplate Mozart's emotions at the bird's death, we see no reason to invoke attributions of displaced grief. We regard Mozart's sense of loss as genuine, his epitaph as an apt gesture.

No other written records of Mozart's relationship with his pet are known. He may have said more, given his prolific letter writing, but much of his correspondence during this period has been lost. The lack of other accounts, however, cannot be considered to indicate a lack of interest in his starling. We are inclined to believe that other observations by Mozart on the starling do exist but have not been recognized as such. Our case rests in part on recent technical analyses of the original (autograph) scores of Mozart's compositions, investigations describing changes in handwriting, inks, and paper. Employing new techniques to date paper by analyzing the watermarks pressed into it at the time of its manufacture, Tyson *(23)* has established that the dates and places assigned to some of Mozart's compositions can be questioned, reaching the general conclusion that many pieces were written over an extended period of time and not recorded in his catalogue until the time of completion. The establishment of an accurate chronology of Mozart's compositions is obviously essential to those attempting to understand the development of his musical genius. It also serves our purposes in reconstructing events after the starling's funeral.

One composition examined by Tyson is a score entered in Mozart's catalogue on 12 June 1787, the first to appear after the deaths of his father and the starling. The piece is entitled *A Musical Joke* (K. 522). Consider the following description of it from a record jacket: "In the first movement we hear the awkward, unproportioned, and illogical piecing together of uninspired material . . . [later] the andante cantabile contains a grotesque cadenza which goes on far too long and pretentiously and ends with a comical deep pizzicato note . . . and by the concluding presto, our 'amateur composer' has lost all control of his incongruous mixture" *(24)*. Is the piece a musical joke? Perhaps. Does it bear the vocal autograph of a starling? To our ears, yes. The "illogical piecing together" is in keeping with the starlings' intertwining of whistled tunes. The "awkwardness" could be due to the starlings' tendencies to whistle off-key or to fracture musical phrases at unexpected points. The presence of drawn-out, wandering phrases of uncertain structure also is characteristic of starling soliloquies. Finally, the abrupt end, as if the instruments had simply ceased to work, has the signature of starlings written all over it.

Tyson's analysis of the original score of K. 522

indicates that it was not written during June 1787, but composed in fragments between 1784 and 1787, including an excerpt from K. 453. This period coincides with Mozart's relationship with the starling. A common interpretation is that *A Musical Joke* was meant to caricature the kinds of music popular in Mozart's day. Writing such music, a course of action urged on him by his father, might have earned Mozart more money. And thus, the composition has also been interpreted in regard to the father/son relationship *(25)*. Tyson disputes this view on the basis of the physical nature of the autograph score, as much of it was written before Leopold's death, and the lack of solid evidence that Mozart's relationship with his father was bitter enough to cause him to commemorate his first and foremost teacher with a parody.

Although we do not presume to explain all the layers of compositional complexity contained in K. 522, we propose that some of its starling-like qualities are pertinent to understanding Mozart's intentions in writing it. Given the propensities of the starlings we studied and the character and habits of Mozart, it is hard to avoid the conclusion that some of the fragments of K. 522 originated in Mozart's interactions with the starling during its three-year tenure. The completion of the work eight days after the bird's death might then have been motivated by Mozart's desire to fashion an appropriate musical farewell, a requiem of sorts for his avian friend.

Last words

We have offered these observations on starlings and on Mozart for two reasons. First, to give music scholars new insights with which to evaluate one of the world's most studied composers. The analyses of the autograph scores and recent reinterpretations of Mozart's illnesses and death demonstrate the power of present-day knowledge to inform our understanding of the past. We have provided the profile of captive starlings as another way to gain perspective on Mozart's genius.

Second, we hope to spark further interest in the analysis of the social stimulation of vocal learning. Although the role of social companions in motivating avian vocal learning is now well established, the mechanisms by which social influence exerts its effects have only begun to be articulated *(26)*. Part of the problem is defining the nature of social contexts. To say birds interact is to say something quite vague. Interact how? By fighting? By feeding? By flocking? By sitting next to one another? Measuring sound waves is easy compared to calibrating degrees of social influence. Moreover, social signals are multi-modal. The species described here make much use of visual, as well as vocal, stimulation. By what means do they link sights and sounds? Why are only certain linkages made? Answering these questions is the next challenge for students of communication.

One of the founders of the study of bird song, W. H. Thorpe, speculated that birds' imitation of sounds represents a quite simple cognitive process: "The essence of the point may be summed up by saying that while it is very difficult for a human being (and perhaps impossible for an animal) to see himself as others see him, it is much less difficult for him to hear himself as others hear him"

Figure 4. Relationships between starlings and human beings appear to reflect the behavior of birds in the wild. Hand-reared starlings interact with their human companions in terms of the social roles of wild birds. In particular, they learn by observing vocal and other responses to their own expressive efforts. (Photos by Birgitte Nielsen.)

(27). Although we recognize the law of parsimony in Thorpe's remark, we are led by the evidence to seek a phylogenetic middle ground between self-awareness and vocal matching. We propose that some birds use acoustic probes to test the contingent properties of their environment, an interpretation largely in keeping with concepts of communication as processes of social negotiation and manipulation *(28)*. An analogy with the capacities of echo-locating animals may be appropriate. Like bats or dolphins emitting sounds to estimate distance, some birds may bounce sounds off the animate environment, using behavioral reverberations to gauge the effects of their vocal efforts. They are not using Thorpe's behavioral mirror, necessary for self-reflection, but instead a social sounding board with which to shape functional repertoires.

In the case of our starlings, we also conclude that social sonar works two ways: human caregivers cast many sounds in the direction of their starlings and were often educated by the messages returned. The mimicry of vocal acts such as lip noises, sniffs, and throat clearing brought to the attention of caregivers routine dimensions of their own behavior that they rarely took notice of. The birds' echoing of greetings, farewells, and words of affection conveyed a sense of shared environment with another species, a sensation hard to forget (Fig. 4). The caregivers' sadness in response to the illnesses, absence, or death of their avian companions also suggests that they had been beguiled by the chance to glimpse a bird's-eye view of the world. Most found themselves at a loss for words. And thus we turn to Mozart for fitting emotional expressions—his poem, his *Musical Joke,* and his appropriately grand burial for a "starling bird."

References

1. G. Nottebohm. 1880. *Mozartiana*. Breitkopf and Härtel.
   O. E. Deutsch. 1965. *Mozart: A Documentary Biography*. Stanford Univ. Press.

2. O. Jahn. 1970. *Life of Mozart*, trans. P. D. Townsend. Cooper Square.
   B. Brophy. 1971. In *W. A. Mozart. Die Zauberflöte*. Universe Opera Guides.
   W. Hildesheimer. 1983. *Mozart*, trans. M. Faber. Vintage.
   P. J. Davies. 1989. *Mozart in Person: His Character and Health*. Greenwood.

3. F. M. Chapman. 1934. *Handbook of Birds of Eastern North America*. Appleton.
   E. W. Teale. 1948. *Days without Time*. Dodd, Mead.

4. E. A. Armstrong. 1963. *A Study of Bird Song*. Oxford Univ. Press.
   C. Feare. 1984. *The Starling*. Oxford Univ. Press.

5. R. Dyer-Bennet, trans. 1967. Impatience. In *The Lovely Milleress (Die schöne Müllerin)*. Schirmer.

6. J. R. Baylis. 1982. Avian vocal mimicry: Its function and evolution. In *Acoustic Communication in Birds*, vol. 2, ed. D. E. Kroodsma and E. H. Miller, pp. 51–84. Academic Press.

7. D. Todt. 1975. Social learning of vocal patterns and models of their application in grey parrots. *Zeitschrift für Tierpsychologie* 39:178–88.
   I. M. Pepperberg. 1981. Functional vocalizations by an African Grey Parrot *(Psittacus erithacus)*. *Zeitschrift für Tierpsychologie* 55:139–60.

8. M. J. West, A. N. Stroud, and A. P. King. 1983. Mimicry of the human voice by European starlings: The role of social interactions. *Wilson Bull.* 95:635–40.

9. H. B. Suthers. 1982. Starling mimics human speech. *Birdwatcher's Digest* 2:37–39.
   M. S. Corbo and D. M. Barras. 1983. *Arnie the Darling Starling*. Houghton Mifflin.
   K. Iizuka. 1988. *Kuro the Starling*. Nelson.
   M. S. Corbo and D. M. Barras. 1989. *Arnie and a House Full of Company*. Fawcett Crest.
   A. DeMotos, pers. com.
   W. R. Fox, unpubl. data.
   A. Peterson and T. Peterson, pers. com.

10. I. M. Pepperberg. 1986. Acquisition of anomalous communicatory systems: Implication for studies on interspecies communication. In *Dolphin Behavior and Cognition: Comparative and Ethological Aspects*, ed. R. J. T. Schusterman and F. Wood, pp. 289–302. Erlbaum.

11. M. Adret-Hausberger. 1982. Temporal dynamics of dialects in the whistled songs of sedentary starlings. *Ethology* 71:140–52.
    ———. 1986. Species specificity and dialects in starlings' whistles. In *Acta 19th Congr. Intl. Ornithol.*, vol. 2, pp. 1585–97.

12. M. Chaiken. 1986. Vocal communication among starlings at the nest: Function, individual distinctiveness, and development of calls. Ph.D. diss., Rutgers Univ.

13. I. M. Pepperberg. 1988. An interactive modeling technique for acquisition of communication skills: Separation of "labeling" and "requesting" in a psittacine subject. *App. Psycholing.* 9:59–76.

14. Pepperberg. Ref. *7*.

15. K. Lorenz. 1957. Companionship in bird life. In *Instinctive Behavior: The Development of a Modern Concept*, ed. C. H. Schiller, pp. 83–128. International Universities Press.
    J. von Uexküll. 1957. A stroll through the world of animals and men. In *Instinctive Behavior: The Development of a Modern Concept*, ed. C. H. Schiller, pp. 5–82. International Universities Press.

16. L. F. Baptista and L. Petrinovich. 1984. Social interaction, sensitive periods, and the song template hypothesis in the white-crowned sparrow. *Animal Behav.* 36:1753–64.

L. Petrinovich. 1989. Avian song development: Methodological and conceptual issues. In *Contemporary Issues in Comparative Psychology*, ed. D. A. Dewsbury, pp. 340–59. Sinauer.

17. A. P. King and M. J. West. 1988. Searching for the functional origins of song in eastern brown-headed cowbirds, *Molthrus ater ater. Animal Behav.* 36:1575–88.
    M. J. West and A. P. King. 1988. Female visual displays affect the development of male song in the cowbird. *Nature* 334:244–46.

18. P. Marler, A. Dufty, and R. Pickert. 1986. Vocal communication in the domestic chicken. II. Is a sender sensitive to the presence of a receiver? *Animal Behav.* 34:194–98.
    S. J. Karakashian, M. Gyger, and P. Marler. 1988. Audience effects on alarm calling in chickens *(Gallus gallus)*. *J. Comp. Psychol.* 102:129–35.

19. E. Anderson, ed. 1989. *The Letters of Mozart and His Family*, p. 907. Norton.

20. Jahn. Ref. *2*.

21. Anderson. Ref. *19*, p. 877.

22. F. Niemtschek. 1956. *Life of Mozart*, trans. H. Mautner. Leonard Hyman.
    Jahn. Ref. *2*.
    Davies. Ref. *2*.

23. A. Tyson. 1987. *Mozart: Studies of the Autograph Scores*. Harvard Univ. Press.

24. W. A. Mozart. *A Musical Joke*. Liner notes by P. Cohen. Deutsche Grammophon. 400 065–2.

25. Ref. *2*.

26. Ref. *16*.

27. W. H. Thorpe. 1961. *Bird-Song*, p. 79. Cambridge Univ. Press.

28. D. W. Owings and D. F. Hennessy. 1984. The importance of variation in sciurid visual and vocal communication. In *The Biology of Ground-dwelling Squirrels: Annual Cycles, Behavioral Ecology, and Sociality*, ed. J. O. Murie and G. R. Michener, pp. 167–200. Univ. Nebraska Press.

Testosterone and Aggression in Birds

John C. Wingfield, Gregory F. Ball, Alfred M. Dufty, Jr.,
Robert E. Hegner, Marilyn Ramenofsky

The familiar spring sound of birdsongs heralds the onset of territory formation and a complex sequence of interrelated events that make up the breeding period. Such songs are an integral part of the repertoire of aggressive behaviors that males use to advertise and defend territorial boundaries and to attract mates (Fig. 1). It is well established that hormones, particularly testosterone, have stimulatory effects on aggression in reproductive contexts. The prevailing "challenge" hypothesis asserts that testosterone and aggression correlate only during periods of heightened interactions between males. Under more stable social conditions, according to the hypothesis, relationships among males are maintained by other factors such as social inertia, individual recognition of status, and territorial boundaries, and testosterone levels remain low. Recent research has suggested ways in which the hypothesis should be modified or extended. In this article we will consider the complexities of aggressive behaviors and their regulation, focusing specifically on species differences in territorial behavior of male birds as models for the multiple interactions of hormones, environment, and behavior.

The secretion of testosterone by interstitial cells in the testis is controlled primarily by a glycoprotein, luteinizing hormone, secreted from the anterior pituitary gland (Fig. 2). Testosterone stimulates the development of secondary sex characteristics such as wattles, combs, spurs, the cloacal protuberance (a copulatory organ), and

Testosterone may not trigger aggressive behavior but may facilitate responses to it

in some species bright-colored skin and nuptial plumage. These characteristics are used extensively in sexual and aggressive displays (Witschi 1961).

Testosterone is also transported in the blood to the brain, where it influences the expression of reproductive behaviors. Classical experiments conducted on a variety of vertebrates, including birds, showed that if the testes are removed, there is a decline in the frequency and intensity of aggressive and sexual behaviors such as singing (or equivalent vocalizations), threat postures, and actual fights. If exogenous testosterone is given to these castrates, the frequency of aggressive behaviors increases again (for reviews on birds see Harding 1981; Balthazart 1983).

The extent to which aggressive behaviors decline after castration or increase after administration of exogenous testosterone varies greatly from species to species, in part because of the different ways in which testosterone can influence behavior. Two mechanisms have been proposed involving organizational and activational effects. Organizational effects of testosterone occur early in development, often immediately after hatching, and once adulthood is reached the neurons involved can operate independently. Activational effects require the immediate presence of testosterone for the sensitive neurons to function normally. Whether organizational or activational effects predominate depends on context and stage in the breeding period. However, in birds it appears that testosterone may have important activational effects regulating short-term changes in territorial aggression within the breeding season.

Over the past 15 years, radioimmunoassay has been used to determine circulating levels of testosterone. If testosterone does activate aggressive behavior, plasma levels should correlate with the behavior in reproductive contexts. Recent work on rodents (Schuurman 1980; Brain 1983; Sachser and Pröve 1984) and primates (Eaton and Resko 1974; Dixson 1980; Phoenix 1980; Bernstein et al. 1983; Sapolsky 1984) suggests that there are such correlations, but that they depend to a great extent on taxonomic class, age, experience, social context, and other environmental influences. The mechanisms underlying such variation are still largely unknown.

In birds, the evidence for correlations of testosterone and aggression is more convincing, although not completely so. Once again, social context must be taken

John C. Wingfield is an associate professor at the University of Washington. He has combined laboratory techniques in comparative endocrinology with field investigations to study the responses of birds to their social and physical environment. He obtained his Ph.D. from the University College of North Wales in 1973 and was on the faculty of The Rockefeller University before moving to the University of Washington. Gregory F. Ball obtained his Ph.D. at the Institute of Animal Behavior, Rutgers University, and is now assistant professor at The Rockefeller University; Alfred M. Dufty performed his doctoral work at SUNY Binghamton and is a post-doctoral fellow at The Rockefeller University; Robert E. Hegner graduated with a Ph.D. from Cornell and completed postdoctoral work at Oxford, The Rockefeller University, and the University of Washington; and Marilyn Ramenofsky was awarded a Ph.D. from the University of Washington, was visiting assistant professor at Vassar College, and is now a research associate at the University of Washington. Address for Professor Wingfield: Department of Zoology, NJ-15, University of Washington, Seattle, WA 98195.

Figure 1. As part of the annual ritual of establishing territories and attracting mates, male birds engage in a variety of aggressive behaviors. In the photograph on the left, a male song sparrow (*Melospiza melodia*) assumes the posture that heralds an attack on an intruder, in this case a decoy in a cage. Recent research has shown how the steroid hormone testosterone stimulates aggression in response to such perceived threats. Mist nets stretched between aluminum poles are used to catch birds in the field. After removing a bird from the net (*right*), the scientist collects a blood sample from a wing vein. The bird is then released unharmed. (Photographs by J. C. Wingfield.)

into account, as well as environmental influences such as length of day, presence of a mate, and nest sites (Wingfield and Ramenofsky 1985). At least some of this confusion can be eliminated by bringing a comparative approach to bear on a variety of avian species. Birds are ideal for this kind of research because there is much diversity in social systems across species. They are also relatively easy to study under free-living conditions, enabling us to conduct parallel field and laboratory investigations.

Seasonal changes

If testosterone is as intimately involved with territorial aggression in birds as is usually presumed, testosterone levels in the blood should parallel the expression of seasonal territoriality. This relationship has been investigated in several species of birds under free-living conditions, thus reducing possible artifacts of captivity (see Wingfield and Farner 1976).

It is crucial when analyzing these kinds of data to determine the precise stage in the reproductive period at which each individual is sampled. This point is illustrated in Figure 3, which depicts plasma levels of luteinizing hormone and testosterone in free-living house sparrows (*Passer domesticus*). If plasma levels of a number of individuals are organized by calendar date, several stages of reproductive activity (prelaying, laying, incubating, renesting) are averaged out on any given date, and the result is a pair of curves, with luteinizing hormone and testosterone rising in spring, remaining relatively high during the breeding season, and then declining to basal as reproduction ends in August and September. If the data are reorganized according to the phase of the breeding cycle, the true pattern of hormone variation is revealed, making allowance for the average

time it takes a pair to progress through each stage (about 4 to 6 days to lay the first egg, 5 days to produce a clutch, and 11 to 14 days to incubate).

Figure 4 compares levels of testosterone in several monogamous species sampled in free-living conditions. Typically, testosterone is highest when territories are first established and aggressive interactions among males are most frequent. For the song sparrow (*Melospiza melodia*), there are two peaks of testosterone, the first associated with the establishment of territory and the second with the egg-laying period for the first clutch, when the male guards his sexually receptive mate. Plasma levels of testosterone decline markedly just prior to or during the parental phase (incubation) and gradually diminish to basal concentrations by the end of the breeding season.

There is no increase in plasma levels of testosterone during the egg-laying period of the second brood for many of the species with open-cup nests, such as the song sparrow and the European blackbird (*Turdus merula*), because there are virtually unlimited sites for these nests, and competition focuses on maintaining territorial boundaries and guarding mates. However, species such as the house sparrow and the European starling (*Sturnus vulgaris*) that nest in holes, a limited resource for which there often is intense competition (in addition to guarding mates), do show an increase in testosterone level with each egg-laying period (see Figs. 3 and 4).

An interesting contrast is provided by the western gull (*Larus occidentalis wymani*). Individuals of this species are long-lived, may breed for 20 years or more, usually mate for life, and return to the same breeding territory year after year. Furthermore, there is an excess of females and no shortage of nest sites at one of the breeding colonies, Santa Barbara Island (Hunt et al. 1980). As a result, competition between males is minimal, and it is not surprising, given the low level of aggression, that the cycle of plasma testosterone in male western gulls is of very low amplitude (Wingfield et al. 1982).

As Figure 5 shows, the same relationship of testosterone levels and aggression can be found in polygamous and promiscuous species, but males of these species have high levels of testosterone for longer periods than do monogamous species. For example, male red-winged blackbirds (*Agelaius phoeniceus*) generally do not feed young but rather display at one another throughout the breeding season in an attempt to maintain territorial boundaries and retain females.

Both monogamous and polygynous males are found within populations of the pied flycatcher (*Ficedula hypoleuca*). Monogamous males have testosterone levels similar to those of monogamous males in other species, but polygynous males maintain high levels of testosterone until the second female has begun incubating. Only then do levels decline rapidly, followed by a return of the male to his first mate, whose young he helps to feed (see Silverin and Wingfield 1982).

Male brown-headed cowbirds (*Molothrus ater*) are unusual in that they do not defend a territory but form dominance hierarchies for access to females. They are brood parasites, showing no parental care. Males spend the entire breeding season guarding females from the attentions of other males. Accordingly, we see prolonged high levels of testosterone that decline only gradually during the season (Dufty and Wingfield 1986a).

Laboratory tests of the challenge hypothesis

As we have seen, field investigations of free-living birds suggest that testosterone is elevated during periods of elevated competition between males, and that parental behavior in males is preceded by a decline in testosterone. Only in species in which males do not feed young or are exposed to intense competition do plasma levels of testosterone remain elevated. These results have led to the challenge hypothesis.

What is the experimental evidence in support of the hypothesis? A positive correlation of aggressive displays with plasma testosterone was found when Japanese quail (*Coturnix coturnix*) were paired in a tournament lasting several days, but the correlation was apparent only immediately prior to the first fighting day and during the following three days (Ramenofsky 1984). From the fifth day onward, levels of testosterone in quail that won fights were indistinguishable from levels in those that lost. By that time, dominance relationships had been established. This may explain why Balthazart and his colleagues (1979) and Tsutsui and Ishii (1981) could find no correlation of plasma testosterone level

and dominance status in groups of male quail with well-established social relationships.

Other experiments confirmed these findings. Captive flocks of house sparrows formed social hierarchies in which dominant individuals had higher plasma levels of testosterone than subordinates only during the first week after the birds were grouped. Before grouping, and more than one week after, there were no correlations of

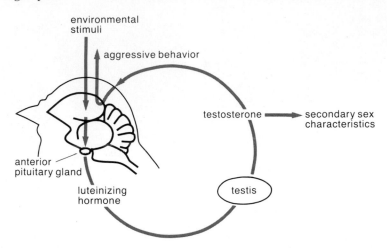

Figure 2. The system through which testosterone influences aggressive behavior begins with the secretion, in response to environmental stimuli, of the glycoprotein luteinizing hormone from the anterior pituitary gland at the base of the brain. Luteinizing hormone in turn stimulates secretion of testosterone by interstitial cells in the testis, where testosterone is produced. In addition to arousing aggressive behavior, testosterone contributes to the development of secondary sex characteristics, such as combs, spurs, and bright plumage.

testosterone level and social status (Hegner and Wingfield 1986). This is consistent with the challenge hypothesis, since testosterone levels were elevated only for a short period as relationships were established. Similarly in the brown-headed cowbird, three males grouped with a single female formed social relationships, and the dominant male gained access to the female. Plasma levels of testosterone in dominant males were elevated one day after grouping, but not before or one week after (Dufty and Wingfield 1986b).

What happens if exogenous testosterone is given to individuals? Do they rise in status, gain a territory, or enlarge an existing one? If a testosterone implant was given to an identified subordinate of a regularly matched pair of Japanese quail, he became more aggressive and fought more persistently with other males. Nevertheless, these subordinates did not win a sufficient number of fights to be considered dominant (Ramenofsky 1982). This suggests that testosterone is not sufficient in itself to heighten aggressive displays to the point of overthrowing previously established relationships. Similar results have been obtained in dominance hierarchies of California quail (*Lophortyx californica*), free-living Harris's sparrows (*Zonotrichia querula*), and sharp-tailed grouse (*Tympanuchus phasianellus*) (Emlen and Lorenz 1942; Trobec and Oring 1972; Rohwer and Rohwer 1978).

Another laboratory experiment sheds more light on the challenge hypothesis. Castrated male white-crowned sparrows (*Zonotrichia leucophrys gambelii*) were given implants of testosterone that maintained circulating levels very similar to those observed during the spring (see Wingfield and Farner 1978a, 1978b). Castrated controls were given empty implants. Songs and threat displays often seen during the establishment of territories in the field increased in both groups after treatment, but there was no significant difference in the frequency of these actions between the two groups despite the wide difference in testosterone level (Wingfield 1985a).

This apparent contradiction of the challenge hypothesis can perhaps be attributed to the fact that the birds had been housed together for over six months. It was thus likely that social relationships among individuals had been established for some time. When a new male was introduced in an adjacent cage, there was an immediate increase in aggression in both groups, and the males with higher levels of testosterone showed significantly more aggressive displays than did the controls. By the next day, the frequency of aggression had dropped dramatically, illustrating how quickly social relationships can be established and emphasizing the ephemeral nature of the correlation between testosterone and aggressive behavior.

It is of little surprise that some investigations have identified hormone-behavior relationships and some have not, particularly since social contexts vary across the studies. Experiential factors such as the development of dominance relationships among individuals can exert a strong influence on the degree to which the circulating levels of testosterone affect frequency and intensity of aggressive behavior. Nevertheless, there is little doubt that testosterone is requisite for increased frequency of aggressive behavior when an individual is challenged in a territorial or other reproductive context.

Environmental cues and testosterone

What controls the timing and amplitude of changes in plasma levels of testosterone so that they occur at appropriate stages in the reproductive period? Clearly, environmental cues play a major role, and one obvious candidate is the annual change in the length of day. It is well known that the vernal increase in length of day promotes secretion of luteinizing hormone and steroid hormones such as testosterone (e.g., Farner and Follett 1979; Wingfield and Farner 1980). Experiments with male white-crowned and song sparrows demonstrated that spermatogenesis is completed, secondary sex characteristics are developed, and the full repertoire of reproductive behaviors (both territorial and sexual) are expressed when birds are transferred from short to long days (see Wingfield and Moore 1986). However, the seasonal changes in testosterone in free-living males are dramatically different from those generated solely by exposing captive males to long days in the laboratory, and the absolute levels can reach an order of magnitude higher than those of males maintained in captivity. Since it has also been shown that high circulating levels of testosterone are not required for the expression of sexual behavior (Moore and Kranz 1983), it is possible that elevated levels in free-living males are involved solely in the regulation of aggression.

What other environmental cues influence the secretion of testosterone and aggressive behavior? Two possibilities spring to mind: stimuli from the territory itself or signals emanating from a challenging male. To evaluate these possibilities, male song sparrows were captured

and removed from their territories, thus creating a vacant spot within the local population. Usually another male claimed the spot within 12 to 72 hours. The result was an increase in conflict between the replacement male and the neighbors, who reestablished territorial boundaries with the newcomer. During this period of social instability, blood samples were collected from replacement males and neighbors. Samples were also

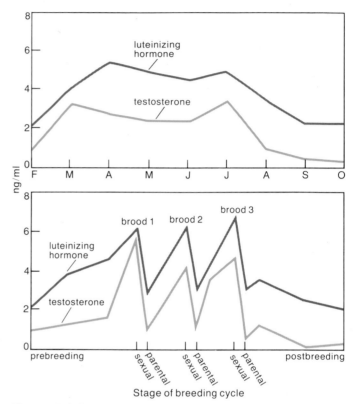

Figure 3. It is important when analyzing seasonal changes in luteinizing hormone and testosterone to distinguish between organization of data by calendar date and by stage in the breeding cycle. If plasma levels of a number of individuals are organized by calendar date (*top*), the various stages in the breeding cycle average out, and the result for both luteinizing hormone and testosterone is a curve with two peaks. If on the other hand the data are organized by stages in the breeding cycle (*bottom*), a much more complicated pattern of hormone variation appears. The data displayed here are from free-living male house sparrows (*Passer domesticus*). (After Hegner and Wingfield 1986.)

collected from control males in a separate area in which boundaries had been stable for some time.

The results were quite clear: plasma levels of testosterone were higher in the replacement males and in their otherwise untreated neighbors than in the controls. Both the neighbors of the replacement and the controls had territories, yet the latter had much lower levels of testosterone. These two groups differed only in that the neighbors were reestablishing territorial boundaries whereas the controls were not. This suggests that the stimulus for increased secretion of testosterone may be not the territory per se (although the data do not disprove a possible effect) but the challenging behavior of the replacement male as he attempts to establish new territorial boundaries (Wingfield 1985b).

To test this further, intrusions were simulated with a decoy male song sparrow in a cage placed in the center

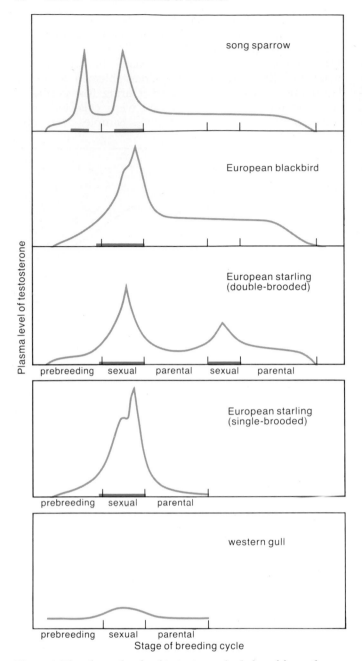

Figure 4. The plasma levels of testosterone in males of four other monogamous species are quite different from those of the house sparrow shown in Figure 3, although in all cases a relationship between testosterone and aggressive behavior is discernible. (Periods when confrontations between males are most frequent are indicated by bars.) There are two peaks for the song sparrow (*Melospiza melodia*), the first associated with the establishment of territory and the second with the egg-laying period of the first brood. The European blackbird (*Turdus merula*) has a single peak during the first brood. Neither the song sparrow nor the European blackbird has a peak during subsequent broods, because these species nest in open cups, for which there are unlimited sites and thus little competition. In these species, competition between males is most intense early in the season. The European starling (*Sturnus vulgaris*), on the other hand, nests in holes, for which competition is keen, and so starlings with double broods have a second peak during their second brood; those with single broods have the expected single peak. The western gull (*Larus occidentalis wymani*) has a distinctively different pattern because of the relative lack of conflict in its breeding colonies. (After Schwabl et al. 1980; Wingfield et al. 1982; Dawson 1983; Wingfield 1984a; Ball and Wingfield 1986.)

of a territory (see Fig. 1). Tape-recorded songs also were broadcast through a speaker placed alongside the decoy. The territorial male almost invariably attacked and attempted to drive the simulated intruder away. He was captured after skirmishing with the intruder for 5 to 60 minutes, and a blood sample was drawn. Controls were captured at the same time of day as the simulated intrusions. Males exposed to a challenge from a simulated intruder showed an increase in testosterone compared with controls. Essentially the same result was obtained in early April and in May through June, indicating that this effect could occur at any time during the breeding period.

It is important to note that the response required about ten minutes before the increase in testosterone was significant. We know that males tend to trespass on other territories regularly and are quickly chased out when seen by the owner (Wingfield 1984b). The confrontations usually last only a few seconds, and thus an increase in testosterone level is unnecessary. However, if an intruder persists and attempts to take over the territory, prolonged fights lasting several hours or even days may result. In such cases an increase in plasma testosterone is appropriate.

There is a third line of evidence suggesting that encounters between males can result in an increase in plasma levels of testosterone. Implants of testosterone in free-living male song sparrows resulted in heightened aggression for longer periods than in control males. In turn, plasma levels of testosterone were elevated in neighbors of testosterone-implanted males compared with neighbors of controls. This effect was most apparent early and late in the season. At other times no effect was noted, because factors such as the presence of young possibly overrode the effect of the aggressive male neighbor.

It was also found that males who had a territory at least one removed from a testosterone-implanted male did not have elevated levels, even though they could hear and see encounters between their immediate neighbors and the testosterone-implanted males (Wingfield 1984b). It appears that an individual male must be involved directly in an agonistic encounter for a hormonal response to be initiated. This blocking of a ripple effect may be adaptive; otherwise, a wave of responses would pass indiscriminately through the local population, affecting males that were not involved in the original skirmish. Moreover, functionally irrelevant surges of testosterone could interfere severely with other reproductive activities such as the feeding of young.

The environmental stimuli generated in the course of an agonistic encounter could enter the central nervous system by several routes: visual, auditory (songs and other vocalizations), tactile (fights), or chemical (pheromones). We can rule out tactile stimuli, because several of the experiments outlined above show that testosterone levels increase in response to a caged male with whom contact is precluded. Also we can probably rule out pheromonal cues, since these are largely regarded as being absent in birds (although it is important to note that this point has not been rigorously investigated). Thus we are left with visual and auditory information influencing secretion of testosterone.

Are both components required for the response?

Recent field experiments showed that if male song sparrows are exposed to a playback of tape-recorded songs (auditory but no visual stimulus), a devocalized male (visual but no auditory stimulus), or a playback plus a devocalized male (visual and auditory stimuli), only those males exposed to both visual and auditory cues have elevated levels of testosterone. Auditory or visual cues alone do not result in a significant increase in testosterone. It was also found that the response is specific: captive male song sparrows showed an increase in plasma levels of testosterone following a challenge from another song sparrow but not following a challenge from a house sparrow (a heterospecific).

Now that the external receptors for environmental cues have been identified and the endocrine response and the specificity of that response determined, we can investigate the neural pathways by which environmental information controls reproductive function.

What is testosterone doing?

This may appear to be an odd question, since it is well established that testosterone has direct effects on aggressive territorial behavior. There is no doubt that testosterone has organizational effects insofar as it influences the formation of song control nuclei in the brain during development and seasonal breeding (e.g., Nottebohm 1981). It is also clear that high levels of testosterone during establishment of a territory are playing some activational role, at least early in the breeding season. However, many of our observations do not fit neatly into these categories (see also Arnold and Breedlove 1985).

Responses to challenges outside the normal seasonal pattern suggest another role for testosterone in the arousal of aggressive behavior. The initial response to a challenging male is to attack vigorously even though the circulating level of testosterone may be much lower than in early spring. Only *after* the attack does testosterone increase, and this appears to take at least ten minutes. Clearly testosterone cannot be playing an activational role in the literal sense of the word, since it increases after the fact. Is it possible that testosterone is playing a facilitative role for the neurons involved during extended periods of intense aggression?

This role would require a very rapid action of a steroid hormone on a target cell. The classical mode of action is through the genome, a process that can take many hours (typically 16 to 30). But recently compiled evidence from mammals suggests that steroid hormones can also have very rapid effects. For example, steroids have been shown to influence rates of gene transcription in rats within 15 minutes, and estradiol can have morphological effects on neuronal cell nuclei within two hours (Jones et al. 1985; McEwen and Pfaff 1985). Even more striking is the demonstration in vitro that estradiol injected directly onto the membrane of an excitable cell induces action potentials within one minute (Dufy et al. 1979). Furthermore, Towle and Sze (1983) found that several steroid hormones, including testosterone, bind to synaptic membranes in the rat brain with high specificity and affinity. Thus the potential exists for very rapid actions of testosterone on the central nervous system through membrane receptors, although more research is required to confirm this in avian systems.

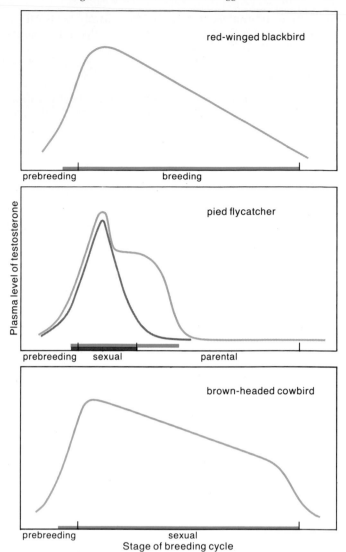

Figure 5. Males of polygynous and promiscuous species have their own characteristic patterns of circulating testosterone: levels remain high for longer periods than they do in monogamous species (see Fig. 4). These correlate with the greater amount of time that such males must spend defending territories and females. (Periods of frequent conflicts between males are again indicated by bars.) In red-winged blackbirds (*Agelaius phoeniceus*), the breeding period includes both sexual and parental stages, since each male may have several females on his territory, some of which may be in the sexual or parental stage at any one time. The pied flycatcher (*Ficedula hypoleuca*) includes both monogamous (*gray line and bar*) and polygynous males (*colored line and bar*). The brown-headed cowbird (*Moluthrus ater*) is a brood parasite that has no territory and performs no parental duties but rather spends the breeding season guarding females from competing males. (After Silverin and Wingfield 1982; Dufty and Wingfield 1986a. Additional data supplied by W. A. Searcy.)

Such a concept is speculative, but the possibility arises that in addition to the two classical modes of genomic action of steroid hormones, involving organizational and activational effects, a third mode of action—supporting or facilitative—could arise during periods of heightened agonistic encounters. Mediated either through rapid-acting membrane receptors or genomically, the third mode would influence the function of brain nuclei involved in the control of aggression. Whether this may ultimately prove to be simply a form of activational effects of testosterone, or indeed a separate mode of action, remains to be seen.

References

Arnold, A. P., and S. M. Breedlove. 1985. Organizational and activational effects of sex steroids on brain and behavior: A reanalysis. *Horm. Beh.* 19:469–98.

Ball, G. F., and J. C. Wingfield. 1986. Changes in plasma levels of sex steroids in relation to multiple broodedness and nest site density in male starlings. *Physiol. Zool.* 60:191–99.

Balthazart, J. 1983. Hormonal correlates of behavior. In *Avian Biology*, ed. D. S. Farner, J. R. King, and K. C. Parkes, vol. 7, pp. 221–366. Academic Press.

Balthazart, J., R. Massa, and P. Negri-Cesi. 1979. Photoperiodic control of testosterone metabolism, plasma gonadotropins, cloacal gland growth, and reproductive behavior in the Japanese quail. *Gen. Comp. Endocrinol.* 39:222–35.

Bernstein, I. S., T. P. Gordon, and R. M. Rose. 1983. The interaction of hormones, behavior, and social context in non-human primates. In *Hormones and Aggressive Behavior*, ed. B. Svare, pp. 535–62. Plenum.

Brain, P. F. 1983. Pituitary-gonadal influences on social aggression. In *Hormones and Aggressive Behavior*, ed. B. Svare, pp. 3–26. Plenum.

Dawson, A. 1983. Plasma gonadal steroid levels in wild starlings (*Sturnus vulgaris*) during the annual cycle and in relation to the stages of breeding. *Gen. Comp. Endocrinol.* 49:286–94.

Dixson, A. F. 1980. Androgens and aggressive behavior in primates: A review. *Aggressive Beh.* 6:37–67.

Dufty, A. M., and J. C. Wingfield. 1986a. Temporal patterns of circulating LH and steroid hormones in a brood parasite, the brown-headed cowbird, *Molothrus ater*. I. Males. *J. Zool. London (A)* 208:191–203.

———. 1986b. Endocrine changes in breeding brown-headed cowbirds and their implications for the evolution of brood parasitism. In *Behavioural Rhythms*, ed. Y. Quéinnec and N. Delvolvé, pp. 93–108. Toulouse: Université Paul Sabatier.

Dufy, B., et al. 1979. Membrane effects of thyrotropin-releasing hormone and estrogen shown by intracellular recording from pituitary cells. *Science* 204:509–11.

Eaton, G. G., and J. A. Resko. 1974. Plasma testosterone and male dominance in Japanese macaque troops with repeated measures of testosterone in laboratory males. *Horm. Beh.* 5:251–59.

Emlen, J. T., and F. W. Lorenz. 1942. Pairing responses of free-living valley quail to sex-hormone pellets. *Auk* 59:369–78.

Farner, D. S., and B. K. Follett. 1979. Reproductive periodicity in birds. In *Hormones and Evolution*, ed. E. J. W. Barrington, pp. 829–72. Academic Press.

Harding, C. F. 1981. Social modulation of circulating hormone levels in the male. *Am. Zool.* 21:223–32.

Hegner, R. E., and J. C. Wingfield. 1986. Behavioral and endocrine correlates of multiple brooding in the semi-colonial house sparrow *Passer domesticus*. I. Males. *Horm. Beh.* 20:294–312.

Hunt, G. L., Jr., J. C. Wingfield, A. Newman, and D. S. Farner. 1980. Sex ratio of western gulls on Santa Barbara Island, California. *Auk* 97:473–79.

Jones, K. J., D. W. Pfaff, and B. S. McEwen. 1985. Early estrogen-induced nuclear changes in rat hypothalamic ventromedial neurons: An ultrastructural and morphometric analysis. *J. Comp. Neurol.* 239:255–66.

McEwen, B. S., and D. W. Pfaff. 1985. Hormone effects on hypothalamic neurons: Analysing gene expression and neuromodulator action. *Trends Neurosci.*, March, pp. 105–10.

Moore, M. C., and R. Kranz. 1983. Evidence for androgen independence of male mounting behavior in white-crowned sparrows (*Zonotrichia leucophrys gambelii*). *Horm. Beh.* 17:414–23.

Nottebohm, F. 1981. A brain for all seasons: Cyclical anatomical changes in song control nuclei of the canary brain. *Science* 214:1368–70.

Phoenix, C. H. 1980. Copulation, dominance, and plasma androgen levels in adult rhesus males born and reared in the laboratory. *Archives Sexual Beh.* 9:149–68.

Ramenofsky, M. 1982. Endogenous plasma hormones and agonistic behavior in male Japanese quail, *Coturnix coturnix*. Ph.D. diss., Univ. of Washington.

———. 1984. Agonistic behavior and endogenous plasma hormones in male Japanese quail. *Animal Beh.* 32:698–708.

Rohwer, S., and F. C. Rohwer. 1978. Status signalling in Harris' sparrows: Experimental deceptions achieved. *Animal Beh.* 26:1012–22.

Sachser, N., and E. Pröve. 1984. Short-term effects of residence on the testosterone responses to fighting in alpha male guinea pigs. *Aggressive Beh.* 10:285–92.

Sapolsky, R. M. 1984. The endocrine stress-response and social status in the wild baboon. *Horm. Beh.* 16:279–92.

Schuurman, T. 1980. Hormonal correlates of agonistic behavior in adult male rats. *Prog. Brain Res.* 53:415–520.

Schwabl, H., J. C. Wingfield, and D. S. Farner. 1980. Seasonal variation in plasma levels of luteinizing hormone and steroid hormones in the European blackbird, *Turdus merula*. *Vogelwarte* 30:283–94.

Silverin, B., and J. C. Wingfield. 1982. Patterns of breeding behaviour and plasma levels of hormones in a free-living population of pied flycatchers, *Ficedula hypoleuca*. *J. Zool. London (A)* 198:117–29.

Towle, A. C., and P. Y. Sze. 1983. Steroid binding to synaptic plasma membrane: Differential binding of glucocorticoids and gonadal steroids. *J. Steroid Biochem.* 18:135–43.

Trobec, R. J., and L. W. Oring. 1972. Effects of testosterone propionate implantation on lek behavior of sharp-tailed grouse. *Am. Midland Nat.* 87:531–36.

Tsutsui, K., and S. Ishii. 1981. Effects of sex steroids on aggressive behavior of adult male Japanese quail. *Gen. Comp. Endocrinol.* 44:480–86.

Wingfield, J. C. 1984a. Environmental and endocrine control of reproduction in the song sparrow, *Melospiza melodia*. I. Temporal organization of the breeding cycle. *Gen. Comp. Endocrinol.* 56:406–16.

———. 1984b. Environmental and endocrine control of reproduction in the song sparrow, *Melospiza melodia*. II. Agonistic interactions as environmental information stimulating secretion of testosterone. *Gen. Comp. Endocrinol.* 56:417–24.

———. 1985a. Environmental and endocrine control of territorial behavior in birds. In *Hormones and the Environment*, ed. B. K. Follett, S. Ishii, and A. Chandola, pp. 265–77. Springer-Verlag.

———. 1985b. Short-term changes in plasma levels of hormones during establishment and defense of a breeding territory in male song sparrows, *Melospiza melodia*. *Horm. Beh.* 19:174–87.

Wingfield, J. C., and D. S. Farner. 1976. Avian endocrinology—field investigations and methods. *Condor* 78:570–73.

———. 1978a. The endocrinology of a naturally breeding population of the white-crowned sparrow (*Zonotrichia leucophrys pugetensis*). *Physiol. Zool.* 51:188–205.

———. 1978b. The annual cycle in plasma irLH and steroid hormones in feral populations of the white-crowned sparrow, *Zonotrichia leucophrys gambelii*. *Biol. Reprod.* 19:1046–56.

———. 1980. Environmental and endocrine control of seasonal reproduction in temperate-zone birds. *Prog. Reprod. Biol.* 5:62–101.

Wingfield, J. C., and M. C. Moore. 1986. Hormonal, social, and environmental factors in the reproductive biology of free-living male birds. In *Psychobiology of Reproductive Behavior: An Evolutionary Perspective*, ed. D. Crews, pp. 149–75. Prentice-Hall.

Wingfield, J. C., A. Newman, G. L. Hunt, and D. S. Farner. 1982. Endocrine aspects of female-female pairing in the western gull (*Larus occidentalis wymani*). *Animal Beh.* 30:9–22.

Wingfield, J. C., and M. Ramenofsky. 1985. Testosterone and aggressive behavior during the reproductive cycle of male birds. In *Neurobiology*, ed. R. Gilles and J. Balthazart, pp. 92–104. Springer-Verlag.

Witschi, E. 1961. Sex and secondary sexual characters. In *Biology and Comparative Physiology of Birds*, ed. A. J. Marshall, pp. 115–68. Academic Press.

Physiology of Helping in Florida Scrub-Jays

When these birds are young, they delay reproduction and help others raise their offspring. The hormone prolactin may influence that cooperation

Stephan J. Schoech

If you are in the right habitat in south-central Florida, you can frequently watch groups of Florida scrub-jays. If you make a hissing sound—a technique that birders call *pishing*—you might attract half a dozen of these birds, which resemble the common blue jay in size and shape, minus the crest. The blue wings and tail combined with a mostly blue head, a gray-brown triangular "cape" on a blue-gray back and a streaked blue-gray "necklace," or breast band, identify the scrub-jay. The pishing calls in the birds from the local territory, where they work together to raise offspring, an approach called *cooperative breeding*.

In cooperatively breeding species, one or more helpers assist a breeding pair in rearing young. Alexander Skutch, an American who has observed birds in Costa Rica for more than 50 years, first described cooperative breeding in 1935. Since then, the overwhelming majority of research in this field has examined the *ultimate*, or evolutionary, factors behind this approach to reproduction. In brief, many of those studies suggest that young birds of species that have relatively strict habitat requirements often remain at home when all of the available nesting territories are occupied. Remaining at home sets the stage

Stephan J. Schoech is a research associate at the department of biology and the Center for the Integrative Study of Animal Behavior at Indiana University. He earned a B.S. and an M.S. in zoology from Arizona State University. He earned his Ph.D. at the University of Washington, where he worked under John C. Wingfield. Schoech's research interests include the hormonal mechanisms that facilitate parental and helping behaviors, as well as other aspects of reproduction, especially in birds. Address: Department of Biology, Indiana University, Bloomington, IN 47405. Internet: sschoech@indiana.edu.

for helping to rear closely related nestlings (most are full siblings). If the helping effort results in survival for more young that share genes with the helper, the nonbreeding helper may gain some measure of evolutionary success by helping to get some of its genes into the population, where they have a chance of surviving and reproducing. This is termed *indirect fitness*, and it might be thought of as making the best of a bad situation.

Few studies have addressed the *proximate*, or causal, mechanisms that facilitate cooperative behavior in birds, although many investigators have suggested that physiological mechanisms could help explain either delayed breeding or helping behavior. Biologists are just beginning to explore that area, largely because many of the techniques required to examine the reproductive or energetic physiology of free-living animals in the field have only recently been developed.

The Florida scrub-jay may be the most studied cooperatively breeding species in the world. In 1969, Glen Woolfenden of the University of South Florida, John Fitzpatrick of Cornell University and subsequently their colleagues and students began studying a population of these birds at Archbold Biological Station. In 1987, my colleague, Ron Mumme of Allegheny College, started working with a population immediately adjacent to Woolfenden and Fitzpatrick's. Fortunately, Mumme chose me as his field assistant in 1989, and I have been wintering in southern Florida ever since.

The Florida scrub-jay lives only on the state's peninsula. On average, a group consists of three birds, but it can range from two to eight. They work together to defend a territory that aver-

ages about nine hectares. Although the helpers (nonbreeding birds) do not assist in building nests, incubating eggs or brooding nestlings, they perform many tasks, including defending the territory, acting as sentinels and defending against predators, and the vast majority also provide food for the nestlings. More than 50 percent of the helpers are one-year-old birds, but a few of them may be as old as seven, and male and female birds perform helping duties with equal likelihood. In general, helpers are the offspring of the breeding pair from previous years. Florida scrub-jays are monogamous and—with the exception of an occasional "divorce"—pairs remain together until one member dies. To switch from helper to breeder, males generally stay on their home territory and either "bud" a small section on its periphery, inherit it when the breeding male dies or find an opportunity when a neighboring males dies. Females usually disperse more widely in search of breeding opportunities.

Although some populations have adapted to suburban living, the majority live in dry oak scrub dominated by vegetation that is usually less than two meters tall. Suitable scrub habitat is maintained by fire. Fire suppression leads to taller oaks, which seem to favor blue jays that apparently exclude scrub-jays. The combination of fire suppression and increased human habitation of the peninsula has made the Florida scrub-jay a threatened species.

Readers may be most familiar with cooperative breeders—including social ants, bees, wasps and naked-mole rats—in which only one female, the queen, can breed. Other females are not capable of breeding, so they help instead. Consequently, my first question was: Are helper scrub-jays physiologically capable of reproducing? As I shall show, a combination of hormonal measurements together with observations of these birds indicates that at least some of the male and female helpers could reproduce, given the proper circumstances. Second, I wondered if a hormonal mechanism facilitates helping behavior. The hormone prolactin influences parental behaviors, including nest building and other aspects of caregiving, in many animals. Prolactin, therefore, was a likely candidate for mediating helping behavior in scrub-jays.

Dissecting the Delay

If helpers could not reproduce, differences might exist in the *reproductive axes* of breeders and helpers. The reproductive axis consists of the hypothalamus (a region of the brain), the pituitary (an endocrine organ immediately below the brain) and the gonads (testes or ovaries). The components of the hypothalamo-pituitary-gonadal axis communicate with one another via endocrine or neuroendocrine secretions. For example, the hypothalamus produces a neurohormone called gonadotropin-releasing hormone, which travels to the anterior pituitary. In response, the pituitary produces and releases luteinizing hormone and follicle stimulating hormone into the blood system. During puberty or at the onset of a breeding season, luteinizing hormone and follicle stimulating hormone initiate gonadal growth and maturation. Later, these hormones maintain gonadal function, including the production of sperm and the development of ovarian follicles. The gonads are the primary producers of the sex steroid hormones, including the best-known ones, testosterone and estradiol, an estrogen. These hormones affect gamete development and maturation, as well as secondary sex characteristics, such as coloration of bare skin (a chicken's comb or a turkey's wattles) and specialized courtship plumage in herons and egrets. Testosterone and estradiol can also induce sexual behav-

Figure 1. Florida scrub-jay *(top)* raises its young through cooperative breeding. In this social system, one or more young birds serve as helpers that assist a breeding pair in caring for offspring. The helpers—usually one-year-old offspring of the breeding pair—perform a variety of tasks, including defending the group's territory and bringing food to nestlings *(bottom)*. The author searches for the physiological mechanisms that cause helping.

A. & E. Morris/VIREO

Reed Bowman

Figure 2. Dry scrub-oak habitat on the Florida peninsula is where scrub-jays live. This limited habitat constrains the number of breeding pairs, so these birds reproduce cooperatively. (Except where noted, photographs courtesy of the author.)

iors, including courtship behavior, through their effects on the brain.

During three years of fieldwork, I collected more than 400 blood samples to compare hormone levels of breeders and nonbreeding helpers in search of a physiological basis for reproductive inactivity among helpers. To examine the performance of a helper's hypothalamus and pituitary, I measured levels of luteinizing hormone. My results show that the luteinizing-hormone levels in both male and female helpers are statistically equivalent to those of breeders (*Figure 4*). The relatively high levels seen in female breeders during the time when they are nest building, copulating and laying eggs reflects their participation in these activities. Overall, the similarity in luteinizing-hormone levels indicates that a helper's hypothalamus and pituitary are fully functional.

The condition of a helper's gonads can be inferred from the levels of sex steroids (*Figure 5*). Male breeders had higher levels of testosterone than male nonbreeding helpers, but the seasonal pattern of testosterone secretion in helpers mirrors that of breeders. A male breeder's high level of testosterone during nest building and egg laying might be attributed to participating in those activities, but male helpers show similar increases even though they do not participate at that time. The increased levels of testosterone in helpers may reflect heightened interactions with jays of either sex or solely be a response to the suite of environmental cues that herald the onset of spring and summer. Nevertheless, a male helper's lack of a mate or its young age might explain the failure

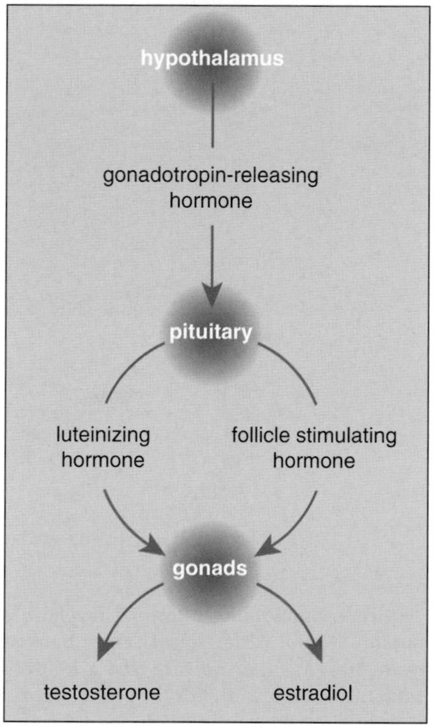

Figure 3. Reproductive axis can be studied to detect physiological differences between birds that breed and those that help. This axis consists of the hypothalamus (in the brain), the pituitary gland and the gonads. In one example of communication between these components, the hypothalamus secretes gonadotropin-releasing hormone, which goes to the pituitary. That hormone causes the pituitary to release luteinizing hormone and follicle-stimulating hormone. These hormones affect the seasonal development of a bird's gonads and the later development of sperm and eggs in adults. The gonads produce sex-steroid hormones—primarily, testosterone in males and estradiol in females—that also affect the development of sperm and egg cells, as well as secondary sex characteristics.

of helpers to achieve the same testosterone levels as breeders.

Beyond secreting testosterone, can a male helper's testes also produce sperm? A few observations suggest that they can. In my population during 1993 and 1994, 10 one-year-old males bypassed the helper stage and ascended directly to breeder—meaning that they occupied and defended a territory with a female. In the end, eight of those pairs built nests and seven of the females laid eggs. Of the seven clutches, three hatched, three were lost to predators and two were infertile. Although the rate of infertile clutches is higher than that of experienced breeders, whether the infertility of the eggs resulted from the absence of viable sperm as opposed to a problem with the female's ovarian follicle or was the result of unsuccessful copulations owing to the inexperience of the young male cannot be determined.

In contrast to the males, the females' levels of their primary sex-steroid hormone, estradiol, did not differ statistically between breeders and helpers. That suggests that a female helper's ovaries function, at least in terms of hormone production and secretion. Also in contrast to the males, the seasonal changes in a female helper's levels of estradiol do not parallel those of breeders. Female breeders secrete the highest levels of estradiol during the prenesting period, whereas helpers' highest levels are between the nest-building and incubating periods, which may reflect their dispersal in search of a breeding territory. In 1993, I captured 17 female helpers during the nest-building, egg-laying and incubation periods. Of those, the four that were captured away from their home territories had estradiol levels nearly 10 times higher than the 13 helpers that were captured on their home territories. The high levels of estradiol might have prompted dispersal, or the dispersing females may have interacted with potential mates, which could also induce the high levels of estradiol. Paradoxically, the few dispersing females that I sampled in 1994—the only other year that I captured female helpers away from home—did not have elevated estradiol levels.

Additional evidence indicates that helpers are reproductively capable and that they delay breeding only because they lack the opportunity. In each of three years—1989, 1994 and 1997—my colleagues and I found two females sharing incubation duties at a nest with seven eggs. Because the maximum clutch size for a scrub-jay is five, both females had clearly produced eggs with the resident male. Moreover, Woolfenden and Fitzpatrick reported a few cases of breeding by one-year-old females. In addition, some males and females remain helpers until they are two, three or older, long after they can breed. Finally, if a female or male breeder's mate dies, they occasionally revert to helping, which provides further evidence that helping is not solely a function of being too young to breed. As mentioned earlier, some one-year-old males can also breed.

Body Condition and Stress

Based on their studies of austral blackbirds, brown-and-yellow marshbirds and bay-winged cowbirds in Argentina, Gordon Orians of the University of Washington and his colleagues hypothesized that insufficient food resources might explain delayed breeding in those species. This hypothesis predicts that helpers will be in poorer condition and weigh less than breeders and that supplemental food should increase the number of helpers that become breeders. Considerable evidence shows that young birds—and helpers of most species are young birds—are generally less adept at foraging than their elders.

To test the first prediction, I weighed jays over the three years of my study, and the helpers weighed less than same-sex breeders. Before drawing any conclusions, I factored in the effect of a general size difference. Using wing-chord dimensions—essentially the length of a bird's wing, a measurement that is often used as an indicator of size—I found that the size of helpers is 97.3 percent of breeders. That figure corresponds closely with the body-mass data that show that male helpers weigh 98.0 percent as much as male breeders. Statistical analysis confirmed that the size difference explains the weight differences. So, rather than being in poorer condition, the male helpers are just relatively small, young birds that are not yet fully grown.

The differences between the body masses of female breeders and helpers cannot be explained solely by a size difference. The female helpers' wing-chord measurements show that an average helper is 97.5 percent as large as a breeder, but the average helper's body mass is

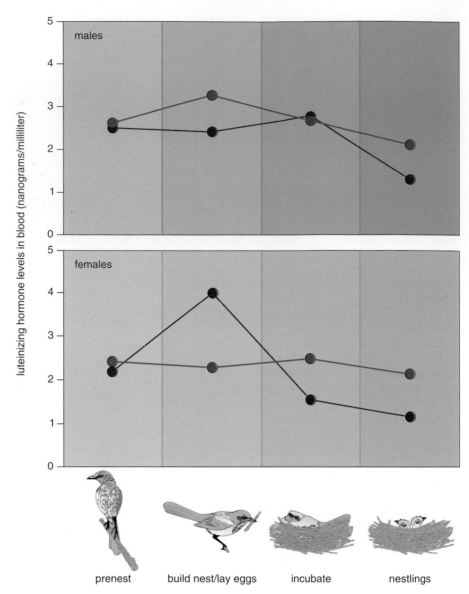

Figure 4. Luteinizing hormone levels in scrub-jays, measured throughout the breeding season, do not differ significantly between helpers *(purple)* and breeders *(red)* of either sex. These results indicate that a helper has a functional hypothalamus and pituitary, which are responsible for the secretion of this hormone.

only 94.1 percent that of an average breeder *(Figure 6)*. A female breeder's higher body mass may be explained by seasonal variation. At the onset of the breeding season, female breeders and helpers weigh about the same amount, suggesting that they are in equally good condition and that helpers have foraged as efficiently as breeders during the winter months. As the breeding season progresses, the difference in weight develops when female breeders increase their body mass by 11 percent in a matter of weeks as they undergo physiological and anatomical changes in preparation for laying.

In 1993, I tested the second prediction of Orians's hypothesis by providing

nine groups of jays with twice-a-day supplements of dried dog food, peanuts and meal worms *(Figure 7)*. These intelligent birds soon learned that I came bearing snacks, and they would usually be waiting for me at the food station. Nevertheless, the supplemental food failed to increase the proportion of helpers that switched to breeders. That was the case for both males and females—suggesting, again, that opportunity, not body condition, determines whether a helper becomes a breeder.

Another potential physiological explanation for helping comes from Jerram Brown of the State University of New York at Albany, who proposed the psychological-castration hypothesis,

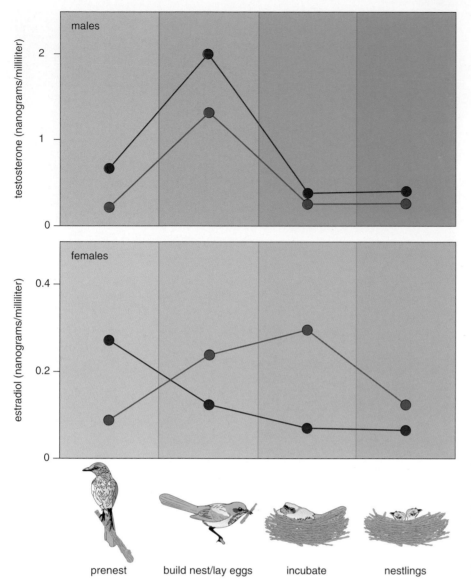

Figure 5. Sex steroid levels vary throughout the breeding season. Among male scrub-jays, breeders *(red)* secrete more testosterone than helpers *(purple)*, but both types of males have similar changes in testosterone levels across the season. In females, the overall levels of estradiol do not differ significantly between breeders and helpers. However, female breeders secrete peak levels of estradiol during the prenesting period, and helpers secrete the most during the nest-building and egg-incubating periods, when this hormone may stimulate helper females to disperse in search of breeding grounds.

which suggests that the presence of a breeding pair "forces" helpers to remain reproductively quiescent. Although Brown did not invoke an endocrine mechanism, John Wingfield of the University of Washington and I—along with our colleagues—postulated independently that dominance interactions could activate the stress, or adrenal, axis, which could suppress breeding. Abundant evidence from many animals shows that numerous stressors, including social conflict, can increase adrenal secretion of the stress hormone corticosterone. (Some groups of animals produce the slightly different steroid molecule, cortisol, and yet others produce both.) Elevated corticosterone levels can inhibit the reproductive axis at many levels, which could force helpers to remain reproductively inactive.

As is true for most cooperatively breeding species, Florida scrub-jay helpers are subordinate to the same-sex breeder on the territory. Nevertheless, I found that helpers and breeders have statistically equivalent levels of corticosterone throughout the breeding season. These findings agree with Wingfield and his colleagues' data from white-browed sparrow weavers, the only other cooperatively breeding

bird species in which basal levels of corticosterone have been assessed. Although Nora Mays and her colleagues at the University of Arizona measured corticosterone and found no differences between breeders and helpers in the cooperatively breeding Harris' hawk, the difficulty of trapping this raptor made measuring basal levels impossible.

I would argue that there are no fundamental physiological differences between breeders' and helpers' reproductive axes that provide a mechanism that explains the lack of reproduction in helpers. Even though younger jays differ to a small degree in their reproductive axes, ontogeny by itself does not explain delayed reproduction. Similarly, despite differences in body mass and body condition, these are as likely an effect of a helper's reproductive quiescence rather than its cause. Florida scrub-jay helpers are merely breeders in waiting. Given the opportunity, they are fully capable of breeding, but given the constraints under which they live, they are forced to bide their time.

Parenting, Helping and Prolactin
Although I have not found a clear-cut physiological mechanism that causes delayed breeding in Florida scrub-jays, a hormonal mechanism could still be responsible for scrub-jays performing parental and helping behavior. In many animals, the pituitary hormone prolactin has been associated with parental behavior, which made it a natural candidate for examination in scrub-jays. From scrub-jays in my study population, I found the following seasonal profile of prolactin secretion: low levels early in the breeding season and the highest levels when the birds are caring for eggs and nestlings. In addition, females have higher levels of prolactin than males, and breeders have higher levels than helpers *(Figure 8)*.

Before trying to interpret those data, I shall explain a fundamental truism about the relation between hormones and behavior. When asked whether a hormone's presence causes a behavior or the behavior causes the hormone's presence, I often answer somewhat flippantly, "Yes!" In other words, a hormone may increase the likelihood that a behavior will occur, and engaging in the behavior often induces increased secretion of the same hormone.

How does this relate to the prolactin profiles of Florida scrub-jays? As is true of other species for which there are prolactin data, the member of the breeding pair that provides the greater portion of care of the eggs and young invariably has higher levels of prolactin. This is thought to be partially attributable to the greater levels of exposure to the stimuli of the nest, eggs and young that a primary caregiver experiences. Given that scrub-jay female breeders are the sole incubators of eggs and brooders of nestlings, it is not surprising that they have the highest prolactin levels. Similarly, helpers probably have lower levels of prolactin because they are excluded from the nest area by the breeders until after the young have hatched, thereby eliminating their exposure to the potential stimuli of a nest and eggs.

Perhaps more interesting, the seasonal profile of prolactin secretion provides temporal evidence for a link between the hormone and parental and helping behaviors. To better examine this relationship, during 1993 and 1994 I watched 27 nests to quantify the number of nest visits and how much food each group member contributed to nestling care. Not surprisingly, given their greater degree of investment in the nestlings, breeders contributed more than helpers. There were, however, no differences by sex.

To correlate an individual's parental or helping behavior with its prolactin level, I captured as many of the animals as possible within one day of the nest watch and measured their levels of the hormone. Then, I compared that level with a *feeding score*, a quantification of how much a bird fed the nestlings. Scrub-jays carry food to the nest in their bill, throat sac or both. The amount of food an individual brought to the nest was scored numerically based on the degree of throat-sac distension as follows: a 1 if there was no visible distention but the bird did transfer food to at least one nestling, a 2 for a noticeable bulge in the throat sac and a 3 if the throat sac was extremely full. When my colleagues and I could not clearly see the throat sac but the individual fed, it was scored as a 1 to err on the side of caution. The number of nest visits and feeding scores were totaled and averaged over the two nest watches and then expressed as feeding score per hour.

The results show a positive correlation between feeding score and prolactin levels when all jays are consid-

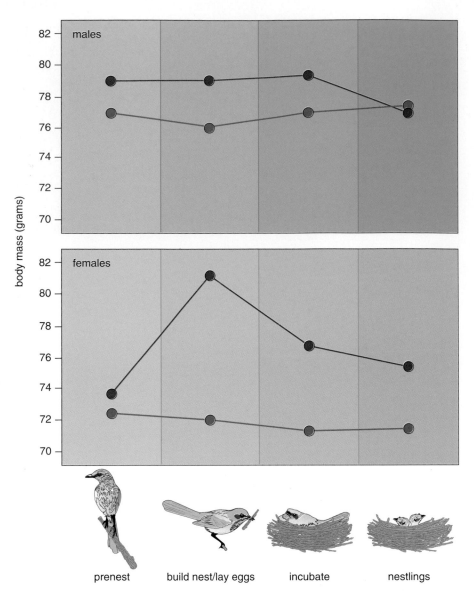

Figure 6. Body weight indicates a bird's overall condition. In males and females, breeders *(red)* weigh more than helpers *(purple)*. In males, however, the difference in body weight comes entirely from a difference in overall size—the helpers are younger and smaller, not necessarily in poor condition. The weight difference among females, however, cannot be explained through a size difference alone. Female breeders experience an 11 percent increase in body weight before laying eggs, but female helpers keep a consistent weight throughout a breeding season.

Figure 7. Supplemental feeding might encourage helpers to become breeders, if insufficient food contributes to delayed breeding. In an experiment, the author provided some scrub-jays with additional food, but it did not cause more helpers to become breeders.

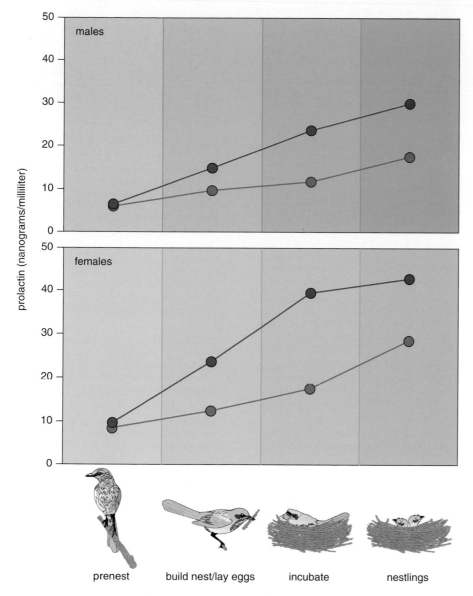

Figure 8. Pituitary hormone prolactin stimulates parental behavior in many animals. In scrub-jays, breeding *(red)* and helping *(purple)* males and females secrete more prolactin as the breeding season progresses. On average, females secrete more prolactin than males, and breeders secrete more prolactin than helpers. This indicates that the birds that perform the most nestling care also secrete the most prolactin.

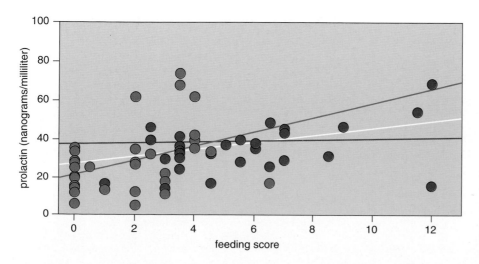

ered together *(Figure 9)*. Moreover, helpers show the strongest positive correlation between feeding score and prolactin. A closer look at those data reveal that a number of helpers did not feed nestlings, thereby receiving feeding scores of zero. Comparing the prolactin levels of these "nonhelpers" with those of helpers that fed nestlings shows that the latter birds—the *helping* helpers—had significantly higher levels of prolactin. Such correlations between helping and prolactin are exciting and provide some support for the hypothesis that helping behavior is mediated by prolactin.

Paradoxically, when the relation between prolactin and the amount of food delivered by breeders is considered, no correlation was found. Although disconcerting, this is not surprising, because I did not examine all of the parental behaviors that prolactin has been shown to promote in other animals. For example, prolactin has been implicated in distraction displays, defense of young and maternal aggression in birds and mammals. Given the evidence that prolactin may mediate a diverse array of parental behaviors, I would like to study a broader range of behavioral measures in the future.

In fact, many more experiments remain to confirm prolactin's role in helping. Historically, attributing a function to a hormone requires three steps. First, a correlation between the hormone and a behavior is noted. Second, the hormone is removed—often by excision of the source tissue or through more modern pharmacological methods that obviate the need for surgical excision—and then investigators check to see if the potentially hormone-mediated behavioral or physiological response has also been removed. Third, replacement of the hormone should reinstate the behavioral or

Figure 9. Prolactin levels correlate with how much a bird feeds nestlings. The feeding score provides a measurement of how much food a bird brings to nestlings each hour. The linear regression of all of these data points *(white line)* shows a positive correlation between prolactin levels and feeding score. The regression line for helpers alone *(blue line)* shows an even stronger positive correlation—suggesting that prolactin stimulates helping behavior. Among the breeders *(red line)*, the data show no correlation between prolactin levels and feeding score, which may reflect the fact that the study did not examine all of the parental behaviors that prolactin affects.

physiological response. Additionally, if a hormone is thought to cause a behavior, increasing levels of the hormone should also increase the behavior. Although most of these experiments remain to be completed on scrub-jays, my colleagues and I are excited about the potential that this research has in illuminating one of the mechanisms underlying cooperative behavior in Florida scrub-jays.

Acknowledgments

I would like to thank my collaborators, Ronald L. Mumme and John C. Wingfield, without whom this research would not have been possible. Field assistants Alison Banks and Rob McMonigle helped immeasurably with data collection, and Artie Fleischer has helped keep the population going in our absence. My wife, Sarah Kistler, held down the fort and tolerated my long absences, not to mention making invaluable editorial comments on this manuscript. My gratitude goes out to Ellen Ketterson, who has provided me with space and guidance at Indiana University. I would also like to thank the staff at American Scientist for their editorial assistance. Finally, I thank Peter Kareiva.

Bibliography

Brown, J. L. 1987. *Helping and Communal Breeding in Birds: Ecology and Evolution*. Princeton: Princeton University Press.

Mumme, R. L. 1992. Do helpers increase reproductive success? An experimental analysis in the Florida scrub jay. *Behavioral Ecology and Sociobiology* 31:319–328.

Mumme, R. L., and W. D. Koenig. 1991. Explanations for avian helping behavior. *Trends in Ecology and Evolution* 6:343–344.

Orians, G. H., C. S. Orians and K. J. Orians. 1977. Helpers at the nest in some Argentine blackbirds. In *Evolutionary Ecology*, ed. B. Stonehouse and C. Perrins. New York: University Park Press.

Schoech, S. J. 1996. The effect of supplemental food on body condition and the timing of reproduction in a cooperative breeder, the Florida scrub-jay. *Condor* 98:234–244.

Schoech, S. J., R. L. Mumme and M. C. Moore. 1991. Reproductive endocrinology and mechanisms of breeding inhibition in cooperatively breeding Florida Scrub Jays (*Aphelocoma c. coerulescens*). *Condor* 93:354–364.

Schoech, S. J., R. L. Mumme and J. C. Wingfield. 1996. Prolactin and helping behaviour in the cooperatively breeding Florida scrub-jay (*Aphelocoma coerulescens*). *Animal Behaviour* 52:445–456.

Schoech, S. J., R. L. Mumme and J. C. Wingfield. 1996. Delayed breeding in the cooperatively breeding Florida scrub-jay (*Aphelocoma coerulescens*): Inhibition or the absence of stimulation. *Behavioral Ecology and Sociobiology* 39:77–90.

Schoech, S. J., R. L. Mumme and J. C. Wingfield. 1997. Corticosterone, reproductive status, and body mass in a cooperative breeder, the Florida scrub-jay (*Aphelocoma coerulescens*). *Physiological Zoology* 70:68–73.

Skutch, A. F. 1935. Helpers at the nest. *Auk* 52:257–273.

Wingfield, J. C., R. E. Hegner and D. M. Lewis. 1991. Circulating levels of luteinizing hormone and steroid hormones in relation to social status in the cooperatively breeding white-browed sparrow weaver, *Plocepasser mahali*. *Journal of Zoology, London* 225:43–58.

Woolfenden, G. E., and J. W. Fitzpatrick. 1996. Florida scrub-jay (*Aphelocoma coerulescens*). In *The Birds of North America, No. 228*, ed. A. Poole and F. Gill. Philadelphia and Washington, D.C.: The Academy of Natural Sciences and The American Ornithologists' Union.

Woolfenden, G. E., and J. W. Fitzpatrick. 1990. Florida scrub jays: a synopsis after 18 years of study. In *Cooperative Breeding in Birds: Long-term Studies of Ecology and Behavior*, ed. P. B. Stacey and W. D. Koenig. Cambridge: Cambridge University Press.

Woolfenden, G. E., and J. W. Fitzpatrick. 1984. *The Florida Scrub Jay: Demography of a Cooperative-breeding Bird*. Princeton: Princeton University Press.

Shaping Brain Sexuality

Male plainfin midshipman fish exercise alternative reproductive tactics.
The developmental trade-offs involved shape two brain phenotypes

Andrew H. Bass

Viewed from the perspective of an evolutionary biologist, life is a game whose object is to maximize the number of individuals carrying your genes into subsequent generations. Reproductive strategies vary with the species, but they always represent some kind of trade-off. Some animals try to beat the odds by producing as many offspring as possible, leaving little time or metabolic energy to care for them. Other animals try to maximize the survival rates of those they do produce by having fewer young, but tending to them until they go off on their own. Particular strategies adopted by animals to win this game have evolved as a response to the specific selection pressures on particular species.

Biologists have come to appreciate that selection pressures may act on individuals within a species, so that different individuals of the same sex may employ very different reproductive tactics. This view comes in part from studies of teleost fishes, where males may engage in one of several alternative reproductive tactics. The sequentially hermaphroditic reef fishes, such as wrasses, sea bass, gobies, parrotfish and anemonefish include individuals that can permanently change their sex. Behavioral sex change begins within minutes of a social cue, and complete

sex change can be achieved within days. Other teleosts, such as sunfish, swordtails, platyfish, salmon and the plainfin midshipman—the fish I study in my laboratory—cannot change sex. Rather, in these species, individual males develop into one of two types, or morphs.

For the midshipman, type I males are the larger of the two, and the only morph capable of attracting females.

Type I males build the nests in which females deposit their eggs and attract the females with their almost indefatigable humming, which has earned them the nicknames "California singing fish" and "canary bird fish." To acquire these abilities, type I males take longer to reach sexual maturity than do the second male morph, the type II males.

Type II males may become sexually mature earlier than type I's, but they

Figure 1. Underneath rocks along the intertidal and subtidal zones of the western coast of North America from Canada down to northern California, type I male plainfin midshipman fish build their nests. These fish are teleosts, many species of which manifest unusual reproductive strategies. Some are hermaphrodites, able to alter their sex in a matter of minutes. Others, such as the midshipman, have two forms of males—the nest-building type I, and the "sneaker males," or type II. Type I males also attract the females to the nest, coax them to lay their eggs and guard the nests, as they can be seen doing above and on the following page. Type II males, on the other hand, do not build nests or attract females on their own; they merely sneak into the type I's nests and deposit their sperm. The differing behavior of these two distinct reproductive morphs provides neuroscientists with a rare opportunity to study whether and how behavioral differences translate into differences in the brains of these fish. (All photographs courtesy of Margaret Ann Marchaterre.)

Andrew H. Bass is a professor in and chair of the Department of Neurobiology and Behavior at Cornell University and a research associate at the Bodega Marine Laboratory of the University of California at Davis. He is interested in the evolution of vertebrate brain and behavior with a focus on mechanisms of acoustic communication and reproductive plasticity in teleost fishes. He received his Ph. D. in zoology from the University of Michigan in 1979. Address: Section of Neurobiology and Behavior, Cornell University, Mudd Hall, Ithaca, NY 14853. Internet: ahb3@cornell.edu.

lose something in the bargain. They are smaller and have never been found to build nests or attract females. Their reproductive strategy is to sneak into the Type I's nest or lie perched outside the nest's entrance and deposit their sperm there, earning them the nickname of "sneaker males" or "satellite males."

Having two distinct male forms—each exhibiting distinct behaviors—presents neurobiologists such as myself with a unique opportunity to study a brain-behavior relationship. At the root of this issue is whether behavioral differences translate into differences in the structure and function of the nervous system. I have found that in fact they do.

The behavioral trade-offs exhibited by the two male morphs essentially reflect the sexual phenotype of the nervous system, which in turn directs the expression of an adult individual's sexual behavior. Armed with this knowledge, my colleagues and I have been exploring the factors that shape the sexual differentiation of the type I and type II brains throughout the animal's sexual development. We have focused our efforts on understanding how the structure and function of the brain might be shaped by early events in development that involve trade-offs between individual characters, such as growth rate and age or size at sexual maturity. An understanding of how and why early developmental events lead to alternative phenotypes for individuals within a species provides fertile ground for examining the linkages between neurobiology and behavioral ecology within a modern evolutionary framework.

Nests and Songfests

Along the western coast of North America, from southern Canada into northern California, from late spring

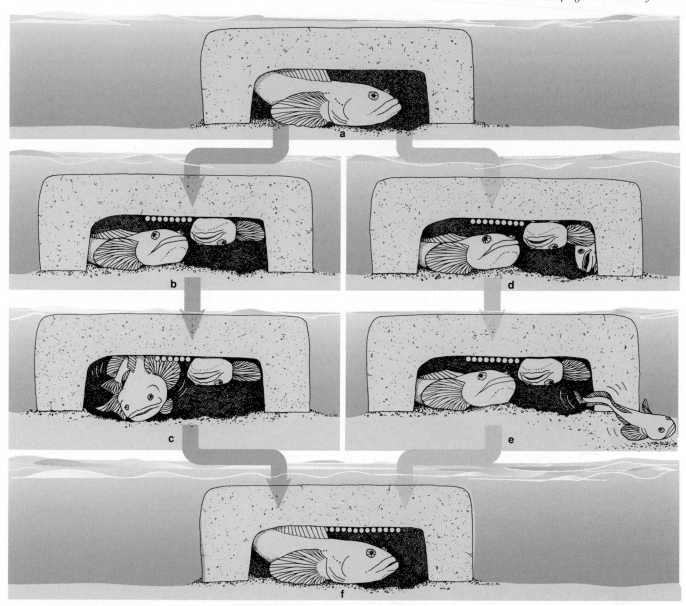

Figure 2. Spawning behavior of type I male plainfin midshipman differs dramatically from that of type II. A type I male generates advertisement calls in the form of low-frequency hums from inside his nest after nightfall (*a*). The male stops his humming after a gravid female enters the nest and they spawn. The female deposits eggs on the roof of the nest (*b*). The type I male rolls and quivers and releases sperm near the eggs (*c*). A variation on this sequence takes place when a type II male enters the picture. If he can get inside the nest, the type II male (*far right*) sneak spawns (*d*). Otherwise, the type II male releases sperm while fanning water towards the nests's opening (*e*). After egg laying is completed, the female leaves the nest, and the type I male remains to guard the eggs (*f*). He hums again the next evening to attract another female, and the sequence begins all over again. This sequence is based on a series of photographs of captive, reproductively active specimens that have taken up residence in artificial nests in aquaria. (Illustration adapted from Brantley and Bass 1994; courtesy of Margaret Nelson.)

through summer, the "song" of the male plainfin midshipman can be heard after nightfall. This song, really a low-frequency hum, may drone on to human ears, but it is highly attractive to female midshipman fish, who seek out the singers and mate with them. The males await the arrival of the females in nests that they build under rocks in the intertidal and subtidal zones. Females deposit their eggs and leave the male to guard the nest soon after spawning.

Not all male midshipman are vocal lotharios or nest builders. As mentioned before, these activities are the sole province of type I males. Richard Brantley, a former graduate student in my lab who is now at Vanderbilt University discovered that type II males exploit the type I's reproductive tactics. As mentioned earlier, they lie perched outside of or sneak into a type I male's nest and shed sperm in a competition with the type I male for the eggs. It is unknown whether type II males remain affiliated with one or several type I nests.

In addition to behavioral differences, each reproductive morph also has a characteristic suite of morphological traits. On average, type I males are about two times longer and eight times heavier than are type II males at the time of sexual maturity. Surprisingly, although they are smaller overall, type II males have the advantage in gonad size. The average ratio of gonad to body weight in type II males is nine times greater than in type I males. Type II males may therefore invest up to 15 percent of their weight in testes, compared with only one percent in type I males.

Gravid females resemble type II males in having a large gonad-to-body weight ratio. Both gravid females and type II males have a distended and firm belly, reflecting the large size of their gonads. In fact, it is easy for the untrained observer to mistake a type II male for a small, gravid female. Adding to the confusion, type II males and females are similarly colored. Although the backs of all three reproductive morphs are olive-gray, the bellies of the fish differ during breeding season. Type I males are typically light to dark gray on the under-side, whereas type II males are mottled yellow. Gravid females have a bronze or golden color on their bellies, and spent females are more like type II males in their appearance.

The females, for their part, apparently select only one type I male to mate with each season. The number of eggs per female increases with body size and may approach 200. Each female leaves her entire clutch in the nest of the cho-

sen male. Midshipman have large eggs, approximately 5 millimeters in diameter, which are attached by an adhesive disk to the roof of the nest. Embryos develop upside-down while attached to the yolk sac. After the fry hatch, they remain in the nest. But during their early pre-reproductive months, juveniles aged 5–12 months are found only in eel-grass beds, where adult morphs are also occasionally found. Whereas females apparently mate with only one male per season, type I males, and presumably type II's, mate with several different females. Each nest typically contains several thousand eggs—obviously origi-nating with many different females.

Nesting type I males generate two major classes of vocalizations. They make short grunts, which, along with their large body mass, form an effective threat to any potential intruder males. But the sounds that have been of partic-ular interest to members of my labora-tory are the mating sounds, the monot-onic hums that can last from minutes to over an hour at a time. Those of us who

study this phenomenon know that these are mating calls by observing the fish in experimental situations. For ex-ample, Jessica McKibben, a graduate student currently in my lab, played computer-synthesized acoustic signals that approximate these hums to fe-males, who are then attracted to under-water speakers in outdoor aquariums. Simulated grunts, on the other hand, seem to do nothing for the females. The hum may help females select the best male to mate with, or it may just serve as a beacon for females looking for nest sites. It may also attract the type II males to these sites. Type II males, like fe-males, do not produce hums at all, and only produce the occasional grunt in non-spawning situations.

Just as the differences between the two male types in their acoustic reper-toire are quite pronounced, so too are the differences in the level of develop-ment of the organs that produce those sounds. The vocal organ of a midship-man consists of a pair of sonic muscles attached to the walls of the swimblad-

Figure 3. Seen from above, the three reproductive morphs—type I males (large fish, *lower right*), type II males (four smaller fish, *left and center*) and females *(topmost fish)*—appear the same olive-gray. The undersides of the fish differ during breeding season. Type I males are typically light to dark gray on the underside, whereas type II males are mottled yellow. Gravid females have a bronze or golden color on their bellies, and spent females are more like type II males in their appearance.

sexually polymorphic traits	type I male	type II male	female
nest building	yes	no	no
egg-guarding	yes	no	no
body size	large	small	intermediate
gonad-size/body-size ratio	small	large	large(gravid), small(spent)
ventral coloration	olive-gray	mottled yellow	bronze (gravid), mottled (spent)
circulating steroids	testosterone, 11-ketotestosterone	testosterone	testosterone, estradiol
vocal behavior	hums, grunt trains	isolated grunts	isolated grunts
vocal muscle	large	small	small
vocal neurons	large	small	small
vocal discharge frequency	high	low	low

Figure 4. Traits of type I and type II males differ markedly. In many respects, type II males more closely resemble females than they do type I males.

der. Contracting the muscles causes the swimbladder to act like a drum—which produces the type I male's low-frequency hums and grunts. One would expect the huge disparity in the vocal capabilities of the two male morphs to be reflected in differences in the sonic musculature. Indeed this is the case.

The ratio of vocal muscle to body weight is six times greater in type I males than in type II males or in females. Furthermore, type I males have four times as many muscle fibers and these are five times larger in diameter than those of type II males and females. Margaret Marchaterre, an electron microscopist in my lab, and I discovered that the disparities are evident even at the subcellular level.

For example, the Z-lines of muscles, the points at which the actin filaments of muscle overlap, which show up under the electron microscope as a dark band, are much wider in the sonic muscles of type I than in type II males or females. The reservoir called the sarcoplasmic reticulum, which contains calcium ions required to mediate muscle-cell activities, is more highly branched in type I males. In addition, the muscles of these males contain vastly higher numbers of the subcellular energy-producing mitochondria than do the muscles of type II males and females. All of this—both the gross appearance of the muscle and the subcellular features—suggests that the sonic muscles of type I males are in every way better equipped than are the muscles of type II males and females

to sustain the continuous singing. (Imagine trying to sing without stopping for over an hour.)

The relative differences in muscle development in the two male morphs are impressive, but there is nothing especially surprising about it given the vocal prowess of the type I males. My interest as a neuroscientist is in determining the neuronal input into this differential behavior. To do that, my colleagues and I looked at the brains of these animals to see whether the difference in muscle size and use in the two male morphs is in any discernible way reflected in their control by nerve cells in the brain.

Brainwork

Before my colleagues and I could start to parse differences in the brains of type I and type II males, we had first to identify the neuronal pathway controlling vocalizations. In this task we were aided by the recent discovery in my lab that a tracer compound called biocytin, which consists of the amino acid lysine and biotin, a naturally occurring protein in neurons, turns out to be perfect for delineating entire circuits in the midshipman's brain. We learned that we could apply biocytin crystals to the cut ends of the motoneurons, the cells that innervate and stimulate the sonic muscles, and the biocytin would be carried backwards from the nerve ending along an axon to its parent cell body. Biocytin also completely filled the arbor of dendrites that extends from the cell body and receives inputs from other neurons. Furthermore, the biocytin did not stop in the first cell it encoun-

tered. Rather, it crossed into the synaptic space between that cell and the end-terminal of the one before it in the circuit. The biocytin travels backwards this way, all the way up to the first cells in the brain that we have so far identified as the sites that initiate vocalizations. Since the biocytin stains the cells brown, my colleagues and I could actually see the entire vocal neuronal circuit required for stimulating the sonic muscle. Biocytin staining confirmed what we had previously found using standard neurophysiological techniques and helped us find new components.

The cell bodies of the sonic motoneurons that stimulate the sonic muscle lie in two sausage-shaped clusters on both sides of the midline of the midshipman's brain, close to the junction of the brain and spinal cord. In adults, about 2,000 cells are found in each cluster, or nucleus. Axons exiting from the sonic motor nucleus bundle together and leave the brain to form the sonic nerve, which stimulates the activity of the sonic muscle.

Robert Baker, at the New York University Medical Center, and I demonstrated that the sonic motoneurons in midshipman receive direct input from a set of pacemaker cells that lie just adjacent to the motoneurons. Each pacemaker neuron connects to motoneurons on both sides of the brain and fire in a constant rhythm, setting the pace at which the sonic motor cells fire. The rhythm set by pacemaker cells corresponds exactly with the rhythm at which sonic motoneurons stimulate the sonic muscle. This in turn determines the frequency at which the muscles contract, which ultimately determines the pitch of the sound.

Our mapping studies with biocytin allowed us to discover another set of neurons in this circuit that had previously been unknown. We found a cluster of cells just in front of the sonic motor nucleus, which we called the ventral medullary neurons. These neurons form the major route connecting the two sides of the pacemaker–motoneuron circuit and so likely make a major contribution to coordinating the activities of both sides of the brain. This eventually leads to the simultaneous contraction of both sonic muscles.

We are just beginning to investigate the sensory stimuli that might activate the vocal motor system. One obvious candidate would be activation of this system by a neighboring midshipman's vocal signal. The inner ear of all vertebrates has a number of divisions. In the midshipman, the largest one is known as the sacculus, which is considered to be the main organ of hearing. It includes a palette of sensory cells linked by the eighth nerve to neurons in the hindbrain, which are the first in a chain of neurons forming a central auditory pathway that extends through all levels of the brain. Deana Bodnar, a research associate in my lab, has recently identified neurons in a midbrain auditory nucleus that encode information that could be used to recognize differences between the hums of neighboring males. This discovery, together with other anatomical data collected from our biocytin studies, suggest that midbrain auditory neurons along with neurons of the paraventricular and tegmental nuclei may form a vocal-acoustic network, which provides a circuit for vocalizations to be elicited by the sounds of neighboring midshipman.

Having worked out the entire circuit, members of my lab group were in a good position to make comparisons of the brains of the different morphs to determine whether there were any obvious differences between them. The first thing we discovered was that male morphs and females possess identical circuitry. We also found that they have the same ratio of nerve cells to body weight, so any differences in behavior could not be due to the number of cells.

We did find, however, that the pacemaker-motoneuron circuit in type I males fires at a frequency that is about 15 to 20 percent higher compared with type II males and females. This parallels sex differences in the frequency of natural vocalizations. We also found that the cell bodies, dendrites and axons are one to three times larger in type I males than in females and type II males. The junction between the nerve and muscle is also larger in type I males. Therefore differences within and between sexes in the organization of the vocal motor system depend upon a divergence in the morphological and physiological properties of individual nerve cells. It seems likely that the larger cells of the type I male are specifically adapted to fire more frequently and without attenuation for a longer period to support the activity of their much enlarged sonic muscle during prespawning periods of singing.

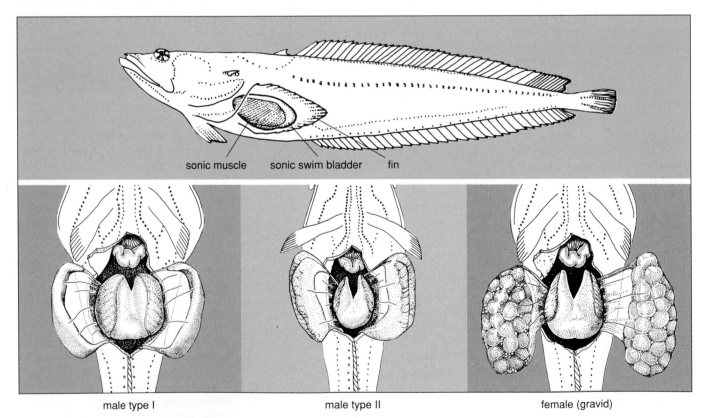

sonic muscle   sonic swim bladder   fin

male type I   male type II   female (gravid)

Figure 5. To attract females to their nests, type I males perform low-frequency hums throughout the night during breeding season. The vocal organ of the midshipman fish is a pair of sonic muscles attached to the walls of the swimbladder (*top*)**. Contracting the sonic muscles makes the swimbladder act as a kind of drum. Sonic muscles of type I males** (*bottom left*) **are extremely well developed in comparison with muscles from type II males** (*bottom center*) **or females** (*bottom right*)**. The ratio of sonic muscle to body weight is six times greater in type I males than in the other reproductive morphs. In contrast, the gonad to body weight ratio is nine times larger in type II males and 20 times larger in gravid females than in type I males and juveniles. (Illustration courtesy of Margaret Nelson.)**

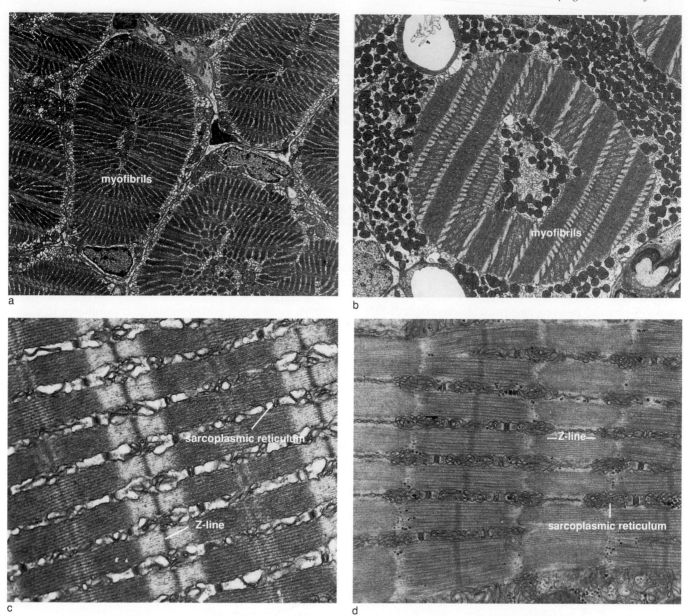

Figure 6. Morph-related differences in sonic muscles can be seen clearly in the electron microscope. In cross section, sonic muscles from type II males (*a*) and type I males (*b*) appear tubular. The myofibrils of juveniles, females and adult type II males are densely packed, surrounded by a thin rind of sarcoplasm, which contains the muscle-cell nuclei and a few mitochondria, which mediate muscle activity. Type I males have fibers with an inner doughnut-shaped core of myofibrils bordered by a large volume of sarcoplasm densely filled with mitochondria. When viewed in longitudinal section, additional differences between muscle fibers of type II (*c*) and type I males (*d*) become apparent. The width of Z-lines, the point at which the actin filaments of muscle overlap, and the degree of branching of sarcoplasmic reticulum are greater in magnitude in type I males. Females and juveniles resemble type II males. The bar scale is 6 micrometers in panels a and b, 0.8 micrometers in panels c and d. (Electron micrographs courtesy of the author; adapted from Bass and Marchaterre 1989.)

Getting Big

The studies performed by my colleagues and I thus suggest that alternative mating tactics among sexually mature males are paralleled by alternative phenotypes for the neurons in the relevant circuit. My colleagues and I wanted to learn the origins of these differences. We could envision at least two distinct scenarios.

In the first, juvenile males may transform into either type I or type II males. That is, both types of males follow mutually exclusive, nonoverlapping growth patterns. The other possibility is that the smaller type II males eventually change into type I males. To determine which of these was happening, my coworkers and I followed the development of these fish in the early juvenile stages just prior to and on into sexual maturity.

We exploited the properties of biocytin to map and trace the development of the neuronal vocal circuit in juvenile midshipman. These studies revealed that for type I males sexual maturation is preceded by growth of the mate-calling circuit and the sonic muscle. The size of motoneurons and the volume of the entire sonic motor nucleus, which likely reflects the growth of motoneuron dendrites as well, increase twofold just prior to the type I's sexual maturation. At the same time, the number of sonic muscle fibers increases by four.

With the onset of sexual maturity, the type I male experiences an additional, albeit more modest, growth in the size of motoneurons. This, however, is coupled with a large increase in the size of pacemaker neurons although the size of this increase does not compare to that of motoneurons. At this time, the sonic muscle undergoes its major expansion—a remarkable fivefold increase in the size of the muscle fibers, which accounts mostly for the large increase in sonic-muscle weight in type I males. It was surpising to see that the largest increase in muscle growth followed, rather than coincided with, the largest growth in the neural circuit. The neurons in the ventral medulla show similar growth increments during both stages.

In contrast, the transformation from juvenile to type II male or to adult female is not accompanied by the dramatic changes seen for type I males in the sizes of vocal neurons and muscle cells. In fact, these cells change little or not at all as type II males and females mature. The sum of these findings suggests that type I males and type II males and females grow along alternative growth trajectories, at least as it con-

cerns the neurons and muscles that determine morph-specific vocal behaviors.

As a next step in our research, my colleagues and I were interested in learning the rate at which each reproductive morph achieved sexual maturity. To do this, we had to be able to determine the precise age of the individual fish we studied, which turned out to be surprisingly easy.

The sacculus division of the inner ear of teleost fishes contains a structure called an otolith, which is mainly composed of calcium carbonate. As a fish grows, new layers of calcium carbonate are added to the otolith. These layers of calcium carbonate can be read, like the rings of a tree, to determine the individual's age. By reading the growth increments in midshipman otoliths, Ed Brothers of EFS Consultants in Ithaca, New York, and I have shown that type I and type II males not only overlap in age, but type II's become sexually mature earlier than do type I males. This finding, we believe, adds support for our hypothesis that alternative male

morphs in midshipman fish adopt nonsequential, mutually exclusive growth patterns during their first year of life.

An Early Start
At this point in our research, we were reasonably convinced that the path leading to one or another male morph was set very early in the animal's development. We know that for fish, as for all vertebrates, hormones are necessary for the development of secondary sexual characteristics. We were interested to see whether differences between type I and type II males relating to their reproductive strategies might be reflected in hormonal differences acting during maturation.

A number of studies have indicated that the hormonal cascade leading to sexual maturation is initiated in a part of the forebrain known as the preoptic area. Neurons in this region release a neurochemical called gonadotropin-releasing hormone (GnRH). GnRH is a 10 amino-acid-long peptide that has been identified in a wide range of vertebrate

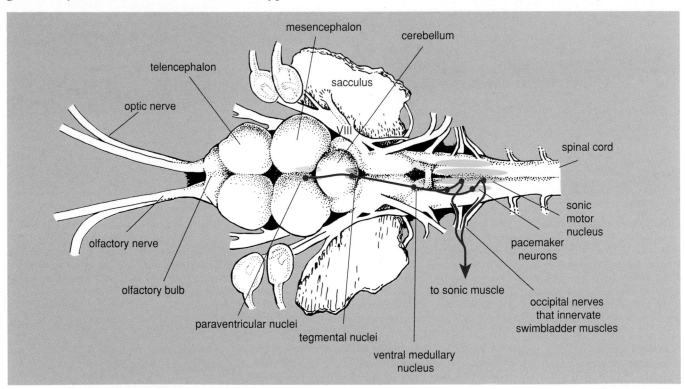

Figure 7. Differences in the vocal-motor circuit in the brain controlling vocalization parallel behavioral differences between the morphs. Here a schematic illustrates the entire circuit beginning in the brain. The initial stimulation comes from sites in the mesencephalon and hindbrain called paraventricular and tegmental nuclei. These stimulate ventral medullary neurons, which in turn stimulate pacemaker neurons. Pacemaker neurons fire at a set frequency equal to the one at which the sonic motoneurons fire and the sonic muscle contracts. In effect, the pacemaker neurons determine the pitch of sounds made by midshipman fish. The axons of the sonic motoneurons bundle together and eventually form the sonic nerve, after exiting the brain via paired occipital nerve roots. The sonic nerve stimulates the sonic muscle on that side of the body. The activity of the two sonic nerves emerging from each side of the brain are coordinated by ventral medullary and pacemaker neurons. In this way, both sonic muscles contract synchronously. The circuit in the three reproductive morphs is the same as is the ratio of neurons relative to body weight. The neurons in the circuit of type I males have larger cell bodies, dendrites and axons than do those in the other two morphs. These features likely help the neurons of type I males fire frequently without fatigue or attenuation. (Adapted from an illustration by Margaret Nelson.)

species. Working with Dean and Tami Myers, Cornell colleagues now at the University of Oklahoma, our lab showed that the gene sequence encoding the GnRH decapeptide in midshipman is remarkably similar to the GnRH coding sequence in other fishes, amphibians, birds and mammals, including people. This finding argues for a highly conserved function for the GnRH peptide among all vertebrates.

GnRH stimulates another set of cells in the pituitary gland, a structure found at the base of the brain. Once stimulated, the pituitary releases a family of hormones called gonadotropins. Gonadotropins act directly on the developing gonads, be they female or male. The gonads, in turn, release steroid hormones—androgens, such as testosterones and estrogens—that stimulate the development of secondary sexual characteristics. In midshipman, one of the effects of the steroid hormones is to mediate the development of the vocal motor system.

One of our first projects was to compare the steroid hormones produced by the three reproductive morphs. Working with Brantley and John Wingfield from the University of Washington, we found that the different morphs did indeed produce different levels of the various hormones.

Testosterone is detectable in all three morphs, although at progressively lower levels, with type II males producing the highest levels followed by females. Type I males produce the lowest amounts. Estrogen (in the form of 17β-estradiol) is detectable only in females, but at much lower concentrations than testosterone.

In addition to these common steroids, teleosts produce a unique form of testosterone known as 11-ketotestosterone. On average, type I males have five times as much 11-ketotestosterone as they do testosterone; 11-ketotestosterone is undetectable in type II males and females. This hormonal distribution is similar in all teleosts with two distinct male morphs. It seems likely, then, that 11-ketotestosterone is more potent than testosterone in supporting courtship behaviors, such as humming.

We were also interested in knowing whether hormonal differences could explain why type II males reach sexual maturity earlier than type I males. Martin Schreibman and his colleagues at Brooklyn College first showed in platyfish, which also have two male morphs that differ in age at sexual maturity, that the fish start to become sexually mature when the GnRH cascade is initiated. Our studies of midshipman sexual development have yielded similar results. Matthew Grober, a former postdoctoral associate in my lab and now at Arizona State University, led a study that found that the number and size of neurons re-

leasing GnRH increase as the animal is making the transition from juvenile to adult. At the point of sexual maturity, this region is equally as developed in all three morphs. That is, it is not the case that the more sexually precocious morphs have better-developed hormonal circuits. The difference, as indicated by otolith-aging studies, is that the cascade is turned on at least three to four months earlier in type II males and females than in type I males.

The question that remains for us to answer is whether these differences are genetic or environmental. When I began these studies, I believed firmly that the difference between type I and type II males was programmed into their genes. But recent studies with Christy Foran, a current graduate student in my laboratory, are casting some doubt on that assertion. We have some preliminary experimental data to suggest that the number of type II males produced is a function of population density. Our work suggests that under sparsely populated conditions, more type I males are produced. As the population density increases, so too does the percentage of type II males.

Alternative Male Morphs: Trade-Offs
Our work suggests that type II males reach sexual maturity earlier than type I males, but they remain physically and behaviorally immature with regard to their ability to vocalize. On the other

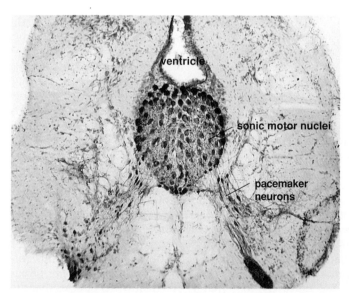
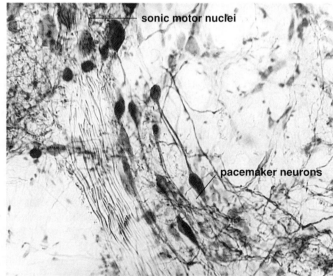

Figure 8. Neuronal circuitry of the vocal motor system can be traced using biocytin, a compound made of the amino acid lysine and the protein biotin. Biocytin crystals are applied to the cut ends of the axons innervating the sonic muscle. The compound is carried backwards to the cell bodies of these neurons, and further backwards still into the nerve cell innervating that one and on back to the first nerve cells in the circuit. In a low-power photomicrograph of the circuit in a type I male *(left)*, sonic motor nuclei, which contain the cell bodies and dendrites of the sonic motoneurons, appear brown when stained with biocytin. Also shown are the pacemaker neurons, which lie along the sonic motor nuclei. Biocytin transport results in extensive filling of the cell bodies, dendrites and axons of vocal pacemaker neurons, as shown in the right panel. (Photomicrographs courtesy of the author.)

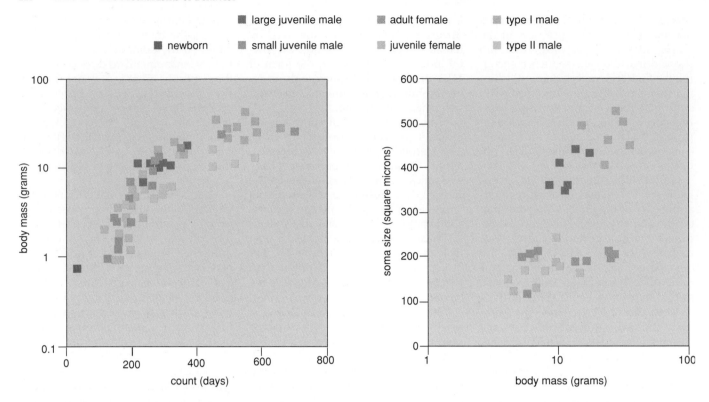

Figure 9. Alternative male morphs represent nonsequential, mutually exclusive life-history tactics, according to studies of their developmental trajectories. The age of fish can be determined by counting the growth increments on otoliths, calcium carbonate structures in the ear. The left graph includes data from 72 individuals representative of all juvenile and adult morphs. It is clear that type II males and females can reach sexual maturity earlier than do type I males and that one type does not change into the other. The right graph is a scatter plot of average values for the size of cell bodies of sonic motoneurons of juvenile and adult morphs. The data indicate alternate growth trajectories for the neurons that determine morph-specific vocal behaviors. This further supports the notion that type I and II males adopt mutually exclusive patterns of growth during their first year of life. (Adapted from Bass *et al.*, in press.)

hand, type I males delay maturation, but have a fully developed sexual behavioral repertoire.

In his book *Ontogeny and Phylogeny*, Stephen Jay Gould reviewed the extensive literature suggesting that the dissociation in time between sexual and physical maturity, referred to as heterochrony, is often characteristic of speciation events. But from our work, we see that heterochrony may also lead to behavioral innovation within a species. In a 1986 review article in the *Proceedings of the National Academy of Sciences*, Mary Jane West-Eberhard proposed a switch mechanism giving rise to alternative phenotypes within a species.

The evolutionary significance of a switch is that it determines which of an array of potential phenotypes will be expressed and, therefore, exposed to selection in a particular timespan and context. Insofar as one set of characters is independently expressed relative to another, it is independently molded by selection. Therefore, different covariant character sets evolve semi-

independently, taking on different forms in accord with their different functions. Like juvenile and adult forms, different alternative phenotypes of the same species may show dramatic differences in morphology, behavior and ecological niche. This is possible because once a switch mechanism is established, contrasting phenotypes can evolve simultaneously within the same genome—without reproductive isolation between forms.

It seems likely that the type II male morph developed under conditions of intense sexual selection, namely competition between males for access to females and nest sites. The switch mechanism is associated with a trade-off among midshipman males in the age and size at sexual maturity, as well as a multidimensional suite of secondary sexual characteristics. Thus whereas type I and II males share gonadal sex, they are highly divergent in behavioral, cellular, hormonal and vocal-motor traits. The convergence or monomorphism in behavioral and

physical traits between type II males and females reflects a common developmental pattern of trade-offs.

This implies that the type I male morph represents the ancestral behavioral state for male plainfin midshipman. This is supported by the available comparative data that show that other species of midshipman and their closest phylogenetic relative, the toadfish, have a single male reproductive phenotype resembling type I males. Therefore the most parsimonious conclusion is that the type II male morph represents a derived character state for this group of teleosts.

The close temporal onset of sexual maturity in type II male and female midshipman resembles many other teleosts. Graham Bell at McGill University shows in a review of approximately 100 freshwater species that the age at sexual maturity in males is close to or less than that of females. The exceptions are in species where males, like the type I midshipman morph, guard a territory or suitable substrate for a female's eggs. By adopting early maturity and thus an

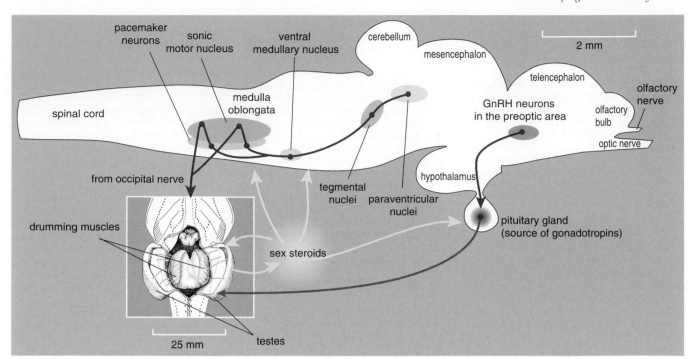

Figure 10. Proposed cascade of hormonal events leading to the expression of type I male, type II male and female reproductive morphs is shown in this schematic. Gonadotropin-releasing hormone (GnRH) is released from a region in the forebrain. GnRH stimulates the pituitary gland, at the underside of the midshipman brain, to release gonadotropins. Gonadotropins stimulate the gonads to release steroid hormones, such as testosterone. These steroids stimulate the differential development of secondary sexual characteristics, such as the ability to vocalize and hence the neuronal circuitry of the vocal motor system. Type I males have a type of testosterone, called 11-ketotestosterone, not found in the other two morphs. This hormone might be especially potent in the development of the sonic musculature and the neural circuitry. The hormonal cascade leading to sexual maturity is initiated three to four months earlier in sexually precocious type II males and females than it is in type I males. The contributions of ecological, behavioral and genetic factors to the activation of a GnRH-gate remain to be defined.

increased chance of surviving to become reproductively active, type II male midshipman essentially forgo direct competition with type I males to establish nest sites. Their investment in gonad development and in an expanded hormonal system is likely to be among the principal costs associated with early maturation.

Type I male midshipman derive at least two major growth-related benefits from delayed maturity. The first is increased body size, which would provide an advantage in combat and nest competition. The second is hypertrophy of a vocal motor system that generates vocalizations important, at the very least, for female attraction to nest sites and probably useful for intimidating other males. The principal costs for type I males in delaying maturity are the considerable structural and metabolic investment in physical and behavioral development and an increased chance of *not* surviving to sexual maturity.

Michael Taborsky at the Konrad Lorenz Institute in Vienna, Austria, points to external fertilization as a trait favoring the evolution of alterna-

tive reproductive tactics and morphs among teleosts because it "makes it difficult for male fishes to monopolize access to fertilizable eggs." The latter leads to males trying to "control preferential access to fertilizable eggs," as exemplified by the nest-building and vocal courtship tactic of type I males who alone guard substrate-attached eggs. (Egg-guarding would also incur additional costs to a type I male tactic.) Intense competition among males for nest sites, and thus access to a female's eggs, would then be considered a major selective force favoring the adoption of alternative type II-like male reproductive tactics.

Usually, when theorists consider the effects of evolution on an animal's physical traits, they consider body size or color, which in fact do differ between the reproductive morphs of the midshipman. My colleagues and I have considered how evolution also affects the neural substrates of display behaviors, which are also part of an animal's life history. Our studies of alternative phenotypes in midshipman fish show that sexuality for each reproductive morph or for that matter,

each individual, can be defined by developmental, sexual maturity-dependent trade-offs between suites of species-typical traits. As we have seen, these trade-offs shape the brain and its behavioral sex, and provide an important link between neurobiology and behavioral ecology.

Acknowledgments
The author is indebted to the collaboration of many colleagues at Cornell University and other institutions, as well as the generous financial support of Cornell University, the University of California's Bodega Marine Laboratory and the National Science Foundation, all of whom helped make this research possible. Special thanks to Peter Klimley for encouragement to write this essay, to Margaret Ann Marchaterre for the field photography and to Deana Bodnar, Matthew Grober, Rosemary Knapp and Margaret Ann Marchaterre for help with the manuscript.

Bibliography
Bass, A. H. 1990. Sounds from the intertidal zone: Vocalizing fish. *Bioscience* 40:249–258.

Bass, A. H. 1992. Dimorphic male brains and alternative reproductive tactics in a vocalizing fish. *Trends in Neurosciences* 15:139–145.

Bass, A. H. 1995. Alternative life history strategies and dimorphic males in an acoustic communication system. *Fifth International Symposium on Reproductive Physiology in Fish*, pp. 258–260.

Bass, A. H., B. J. Horvath and E. B. Brothers. In press. Non-sequential developmental trajectories lead to dimorphic vocal circuitry for males with alternative reproductive tactics. *Journal of Neurobiology*.

Bass, A. H., and M. A. Marchaterre. 1989. Sound-generating (sonic) motor system in a teleost fish (*Porichthys notatus*): Sexual polymorphism in the ultrastructure of myofibrils. *Journal of Comparative Neurology* 286:141–153.

Bass, A. H., M. A. Marchaterre and R. Baker. 1994. Vocal-acoustic pathways in a teleost fish. *Journal of Neuroscience* 14:4025–4039.

Bell, G. 1980. The costs of reproduction and their consequences. *American Naturalist* 116:45–76.

Bodnar, D. and A. H. Bass. 1996. The coding of concurrent signals (beats) within the auditory midbrain of a sound producing fish, the plainfin midshipman. *Association for Research in Otolaryngology*.

Brantley, R. K., and A. H. Bass. 1994. Alternative male spawning tactics and acoustic signalling in the plainfin midshipman fish, *Porichthys notatus*. *Ethology* 96:213–232.

Brantley, R. K., J. Tseng and A. H. Bass. 1993. The ontogeny of inter- and intrasexual vocal muscle dimorphisms in a sound-producing fish. *Brain, Behavior and Evolution* 42:336–349.

Brantley, R. K., M. A. Marchaterre and A. H. Bass. 1993. Androgen effects on vocal muscle structure in a teleost fish with inter and intrasexual dimorphisms. *Journal of Morphology* 216:305–318.

Brantley, R. K., J. Wingfield and A. H. Bass. 1993. Hormonal bases for male teleost dimorphisms: Sex steroid levels in *Porichthys notatus*, a fish with alternative reproductive tactics. *Hormones and Behavior* 27:332–347.

Caro, T. M. and P. Bateson. 1986. Organization and ontogeny of alternative tactics. *Animal Behaviour* 34:1483–1499.

Crews, D. 1987. Animal sexuality. *Scientific American* (January) 106–114.

DeMartini, E. E. 1990. Annual variations in fecundity, egg size and condition of the plainfin midshipman (*Porichthys notatus*). *Copeia* 3:850–855.

Emlen, S. T., and L. W. Oring. 1977. Ecology, sexual selection and the evoluton of mating systems. *Science* 197:215–223.

Fine, M. L., H. Winn and B. L. Olla 1977. Communication in fishes. In *How Animals Communicate*, ed. T. Seboek. Bloomington, Indiana: Indiana University Press, pp. 472–518.

Gould, S. J. 1977. *Ontogeny and Phylogeny*. Cambridge, Mass.: Belknap Press.

Grober, M. S., S. Fox, C. Laughlin and A. H. Bass. 1994. GnRh cell size and number in a teleost fish with two male reproductive morphs: Sexual maturation, final sexual status and body size allometry. *Brain, Behavior and Evolution* 43:61–78.

Grober, M. S., T. R. Myers, M. A. Marchaterre, A. H. Bass and D. A. Myers. 1995. Structure, localization and molecular phylogeny of a GnRH cDNA from a paracanthopterygian fish, the plainfin midshipman (*Porchthys no-*

tatus). *General and Comparative Endocrinology* 99: 85–99.

Gross, M. R. 1996. Alternative reproductive strategies and tactics: Diversity within sexes. *Trends in Ecology and Evolution* 11:92–98.

Gross, M. R., and R. C. Sargent. 1985. The evolution of male and female parental care in fishes. *American Zoologist* 25:807–822.

Halpern–Sebold, L. R., M. P. Schreibman and H. Margolis-Nunno. 1986. Differences between early- and late-maturing genotpyes of the platyfish (*Xiphophorus maculatus*) in the morphometry of their immunoreactive luteinizing hormone releasing hormone-containing cells. A developmental study. *Journal of Experimental Zoology* 240:245–257.

Ibara, R. M., L. T. Penny, A. W. Ebeling, G. van Dykhuizen and G. Cailliet. 1983. The mating call of the plainfin midshipman fish, *Porichthys notatus*. In *Predators and Prey in Fishes*, eds. D. L. G. Noakes et al. The Hague, The Netherlands: Dr. W. Junk Publishers, pp. 205–212.

Kelly, D. B. 1988. Sexually dimorphic behaviors. *Annual Review of Neuroscience* 11:225–251.

McKibben J., D. Bodnar and A. H. Bass. 1995. Everybody's humming but is anybody listening: Acoustic communication in a marine teleost fish. *Fourth International Congress of Neuroethology*, p. 351.

Moore, M. C. 1991. Application of organization-activation theory to alternative male reproductive strategies: a review. *Hormones and Behavior* 25:154–179.

Shapiro, D. Y. 1992. Plasticity of gonadal development and protandry in fishes. *Journal of Experimental Zoology* 261:194–203.

Thresher, R. E. 1984. *Reproduction in Reef Fishes*. Neptune, New Jersey: T. F. H. Publications, Ltd.

Stearns, S. C. 1992. *The Evolution of Life Histories*. New York:Oxford University Press.

Taborsky, M. 1994. Sneakers, satellites, and helpers: Parasitic and cooperative behavior in fish reproduction. In *Advances in the Study of Behavior*, volume 23, ed. P. J. B. Slater, J. S. Rosenblatt, C. T. Snowdon and M. Milinski. New York: Academic Press, pp. 1–100.

Walsh, P. J., T. P. Mommsen and A. H. Bass. 1995. Biochemical and molecular aspects of singing in batrachoidid fishes. In *Biochemistry and Molecular Biology of Fishes*, volume 4, ed. P. W. Hochachka and T. P. Mommsen. Amsterdam: Elsevier, pp. 279–289.

Warner, R. R. 1984. Mating behavior and hermaphroditism in coral reef fishes. *American Scientist* 72:128–136.

West-Eberhard, M. J. 1986. Alternative adaptations, speciation, and phylogeny. *Proceedings of the National Academy of Sciences* 83:1388–1392.

Aerial Defense Tactics of Flying Insects

Preyed upon by echolocating bats, some night-flying insects have developed acrobatic countermeasures to evade capture

Mike May

Walking home late one summer night, I glimpsed a small mass slip through the air, past the halo of a street lamp, and into the darkness. Although I was fatigued by a long day, my curiosity was piqued; I crouched down in the darkness and waited for another sign of movement. Within minutes the elusive flyer returned, swerving momentarily in the light, and then shooting back into the night. It was a bat—apparently foraging for its nightly meal of insects. Soon there were others, darting and weaving by the lamp as they attempted to scoop up the insects attracted to the light. As I watched the aerial display, I was impressed by the remarkable speed at which a bat could change its flight path. I tossed a few pebbles into the air and watched as the bats easily pursued the decoys, but turned away when the deception became apparent. Surely, I thought, there was little hope for an insect once a bat had homed in on it.

I gave the matter no more thought until several years later, when I began my doctoral research—perhaps not coincidentally concerned with the flying abilities of insects. As a graduate stu-

Mike May is a free-lance science writer. He acquired a taste for the breadth of biology as an undergraduate at Earlham College in Richmond, Indiana. While completing an M.S. in biological engineering at the University of Connecticut at Storrs, he discovered some electronic answers to biological questions. After pursuing bicycle mechanics for a year, he returned to biology and earned a Ph.D. as a biomechanic at Cornell University. Address: P.O. Box 141, Etna, New York 13062.

dent I learned there is a considerable history to the study of the aerial encounters between bat and insect. It proves to be a story with a number of surprising turns, and it begins almost 200 years ago with the discovery that bats use their ears, and not their eyes, to navigate.

Lazaro Spallanzani, an 18th-century pioneer of experimental biology, showed that blinded bats are not only able to avoid obstacles in their flight path—such as fine silk threads—but are also able to snag insects in midflight. After hearing of Spallanzani's research, Charles Jurine, a surgeon and entomologist, demonstrated that when the bats' ears are plugged, the animals collide with even relatively large objects in their path, and they are incapable of catching insects.

For over a century the observations of Spallanzani and Jurine were not widely accepted, primarily because no one could imagine how it was that a bat could hear the precise location of such small, essentially silent objects. No advance was made in understanding "Spallanzani's bat problem" until 1920, when the English physiologist H. Hartridge suggested that bats might somehow use sounds of very high frequency to detect the objects. Perhaps the frequencies might even extend beyond the upper limit of human hearing—about 20 kilohertz—to the part of the spectrum called ultrasound.

The mystery of bat navigation was ultimately solved by a Harvard undergraduate, Donald Griffin, in collaboration with the Harvard physicist G. W. Pierce—who invented a device that could detect ultrasound—and the Harvard physiologist Robert Galambos. In 1938 Pierce and Griffin pointed a "sonic detector" at bats flying in a room and found that the animals were, in fact, emitting signals at ultrasonic frequencies. Griffin and Galambos later showed that bats emit ultrasonic cries from their mouths and use their ears to detect the echoes of the sounds reflected from objects in their flight paths. Griffin called this process of navigation *echolocation*.

Echolocation turns out to be an extremely precise and effective method by which bats navigate and identify objects in the dark. In the early 1980s Hans-Ulrich Schnitzler and his colleagues at the Institute for Biology in Tübingen, and Nobuo Suga of Washington University, found that bats are able to analyze the ultrasonic echoes reflected from the bodies and wings of flying insects in such a way as to determine not only the location but also the speed and, perhaps, the type of insect that produces the echoes. All the evidence suggests that the echolocating bat is a very sophisticated hunter; not only is it an adept flyer, but it is equipped with a sensitive auditory system designed to locate and identify potential targets.

However, the bat's ability to find and capture a flying insect is just one side of the story. Some flying insects are able to detect the ultrasonic cries of a bat and take evasive action. Flying insects pursued by a bat do not follow simple ballistic trajectories; they are not such easy targets. To the contrary, the encounter between a bat and an insect is one that might rival the tactics of modern air-to-air combat, involving an efficient early-warning system, some clever aerodynamic engineering and

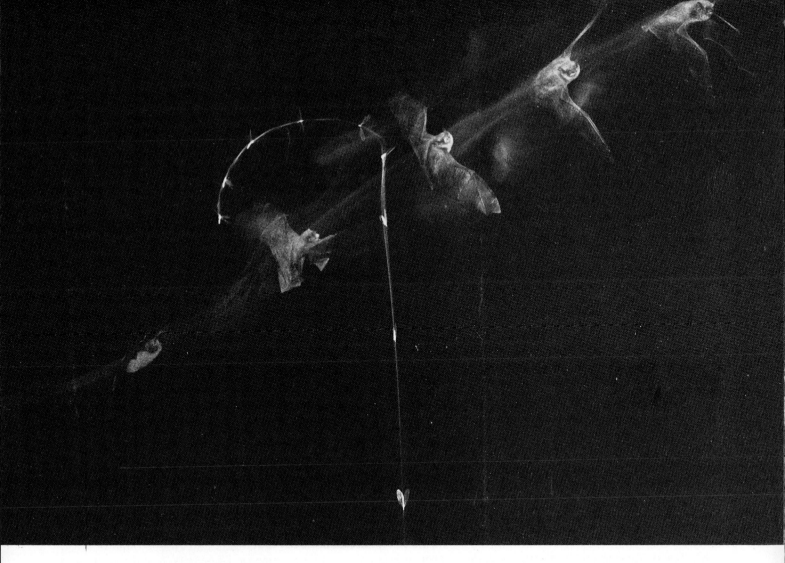

Figure 1. Aerial encounter between an insect-eating bat and a green lacewing reveals one of the evasive maneuvers—a passive dive—that an insect will use to escape a bat. Hunting bats locate their prey by emitting high-frequency (ultrasonic) cries and detecting the echoes of the sounds reflected from the insect's wings and body—a system of navigation called echolocation. Some insects are able to hear the bat's high-frequency sounds and can respond with rapid changes in their flight trajectories. In this stop-motion photograph, a stroboscopic flash reveals the relative positions of the insect and the bat—at intervals of less than one-tenth of a second—as they move from left to right in the scene. (Photograph courtesy of Lee Miller, Odense University, Denmark.)

the simple economics of making do with what is available.

Dodging a Speeding Bat

Almost 70 years before Griffin and Galambos demonstrated that bats can locate objects with ultrasound, F. Buchanan White of Perth, Scotland, proposed that moths can detect bats through the sense of hearing. Although White had no evidence for this conjecture, his idea was ultimately confirmed by behavioral studies in the 1950s, and especially by the work of Kenneth Roeder of Tufts University in the early 1960s. Roeder made hundreds of long-exposure photographs of

free-flying moths and recorded their aerial maneuvers in response to a stationary source of artificial ultrasound. He found that if the moths were more than 10 feet from the source of the ultrasound, they simply turned away. But if they were closer to the sound, the moths performed a variety of acrobatic maneuvers, including rapid turns, power dives, looping dives and spirals. For a more natural touch, Roeder photographed wild bats attacking the flying moths; clearly visible in the photographs is the track of the bat zipping across the scene and the evasive path of the moth as it escapes the attack—sometimes.

Since Roeder's studies, a number of nocturnal flying insects have been found to perform evasive aerial maneuvers in response to the ultrasonic cries of bats. In 1979 Lee Miller and Jens Olesen of Odense University found that hunting bats or artificial ultrasonic pulses induce erratic flight in free-flying green lacewings. More recently, David Yager of Cornell University, Brock Fenton of York University in Toronto, and I have shown that some species of praying mantis will perform different types of escape procedures depending on the loudness of the ultrasound emission—not unlike Roeder's moths. In these experiments we

Figure 2. Failure to escape results in death by devourment for a green lacewing performing a diving arc into the embracing wings and tail membrane of an approaching bat. The wings of the captured insect can be seen in the mouth of the bat as it descends to the right. (Photograph courtesy of Lee Miller, Odense University, Denmark.)

used an artificial source of ultrasound—sportively called a "batgun"—with which we "shot" free-flying mantises. At distances greater than 10 meters, most of the mantises did not respond to the ultrasound. Within seven to nine meters of the batgun, however, the mantises would make a slight turn or a shallow dive. At still closer range—within five meters—the mantises would perform steep dives ranging from 45 degrees to nearly a vertical drop, occasionally even in a spiral. Just as Roeder had found in his studies of the moths, we found that the praying mantis will make its most drastic evasive maneuvers when the ultrasound is loudest.

Artificial ultrasound has also been shown to induce changes in the flying patterns of other insects. Daniel Robert, now at Cornell University, found that flying locusts respond to ultrasonic pulses by steering away from the source of the sound and by increasing the rate at which they beat their wings. Similarly, Hayward Spangler of the Carl Hayden Bee Research Center in Tucson showed that, immediately after hearing artificial ultrasonic pulses, tiger beetles fly toward the ground and land. Frederic Libersat, now at the Hebrew University in Jerusalem, and Ronald Hoy of Cornell University discovered that a tethered, flying katydid will stop flying immediately after hearing an ultrasonic stimulus, suggesting that it would perform a dive. Although no one has reported interactions between these insects and bats, it seems likely that such bat-avoidance responses will be found in many night-flying insects that hear ultrasound.

Certainly the value of a rapid escape mechanism for the survival of a flying insect is no longer in doubt. Roeder's studies showed that moths that dive in response to an ultrasonic stimulus were 40 percent less likely to be captured by a bat. The green lacewings, studied by Miller and Olesen, were even more successful at escaping bats—being captured only 30 percent of the time. On the other hand, deafened green lacewings were captured about 90 percent of the time.

More recently, Yager and members of Fenton's research group performed a series of field experiments in which they exposed two species of praying mantis to wild, hunting bats. One species, *Parasphendale agrionina*, makes rapid changes in its flight path in response to artificial pulses of ultrasound or in response to hunting bats. The other species, *Miomantis paykullii*, is an excellent flyer, but does not change its flying pattern in response to artificial ultrasound or in response to hunting bats. Of five attacks on *P. agrionina* in which the mantises performed evasive maneuvers the insects successfully escaped the hunting bats in every case. In contrast, during three attacks on *M. paykullii* and three attacks on *P. agrionina* in which neither species performed evasive maneuvers the insects were captured in five of the six cases. These experiments provide strong evidence that ultrasonic hearing and rapid changes in trajectory can help a flying insect evade an attack by a predatory bat.

Mating, Death and Phonotaxis

It seems clear that some insects are able to detect the ultrasonic pulses emitted by bats, and then use this information as an early warning system—much the same way a combat pilot in a fighter plane might detect the radar of an enemy plane. And like the combat pilot, the targeted insect must perform evasive maneuvers or suffer the consequences of being captured. In the case of the fighter pilot, however, we know the physics and the engineering behind the detection of radar, and the aerodynamics of flight maneuvers is the stuff of textbooks in flight school. But how does an insect do it? How does it detect the ultrasound, convert this signal into a message that says "take evasive action," and then perform its spectacular acrobatic maneuvers? We don't as yet know all the answers, but bits and pieces of the story are coming to light.

Part of the answer lies in the behavior known as phonotaxis, the movement of an animal in a direction determined by the location of a sound source. Phonotactic behavior of insects has been especially well studied in certain species of crickets. Phonotaxis takes two forms in these animals, based on the direction the cricket moves with respect to the sound source. When a female cricket moves toward the source of the calling song of a courting male cricket—the familiar chirp we hear on summer nights—the locomotory behavior of the female is described as positive phonotaxis. On the other hand, the same female will respond to another sound, of a higher frequency, by flying away from the source—a display of negative phonotaxis. Positive and negative phonotaxis in these instances suggest that the cricket is able to discern at least two distinct aspects of the sound: its fre-

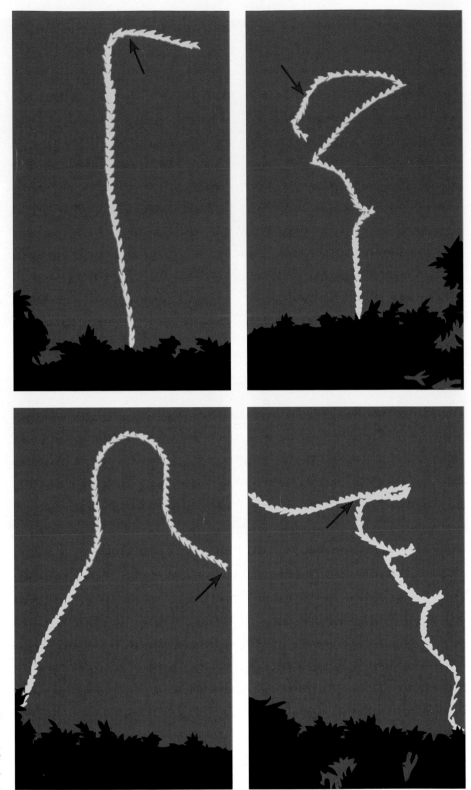

Figure 3. Flight paths of various insects in response to the onset of artificial pulses of ultrasound *(red arrows)* **demonstrate some of the tactics used to escape a hunting bat. A single species of insect will often have several different evasive maneuvers in its behavioral repertoire—preventing bats from anticipating any single response. A passive dive** *(upper left)*, **resulting from the absence of any wing motion, is the simplest type of response. Erratic flight movements** *(upper right)*, **consisting of a looping turn and ending with a passive dive, is one of the evasive maneuvers performed by a small geometer moth. A powered dive (assisted by wingbeats) may be preceded by a rapid ascent** *(lower left)*. **A series of tight turns may also make the insect's descent to the ground somewhat more gradual** *(lower right)*. **(Adapted from Roeder 1962.)**

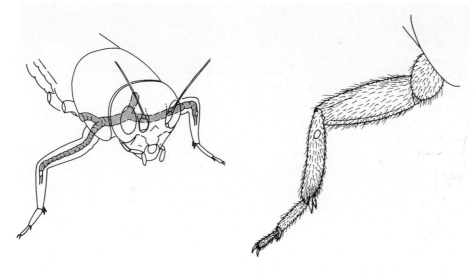

Figure 4. Crickets detect sounds—such as the cries of hunting bats—with a pair of ears located on the forelegs, just below the "knees." The ears are connected via air-filled tubes that meet at the insect's midline. Each of the tubes also has a branch leading to an opening, called a spiracle, behind each of the forelegs. The connections between the ears and the spiracles suggest that sound may reach an ear through separate channels. The presence of these different sound paths is thought to produce a differential response in the left and right ears that varies according to the location of the sound source. (Left illustration adapted from Hill and Boyan 1976.)

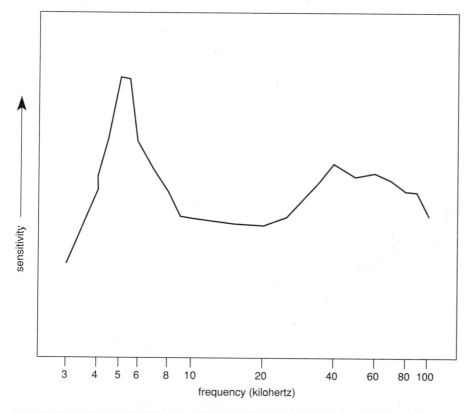

Figure 5. Frequency-sensitivity curve of a cricket's hearing exhibits two distinct peaks. A female cricket is most sensitive to sounds of about 5 kilohertz, which corresponds to the calling song of a courting male cricket. A second, broadly tuned peak—between 20 kilohertz and 100 kilohertz—lies within the frequency range of the ultrasonic cries emitted by hunting bats. The curve represents the average sensitivity of several animals as determined by observations of steering behavior toward or away from the sound. (Adapted from Moiseff, Pollack and Hoy 1978.)

quency and the location of its source.

An elegant demonstration of the frequency dependence of phonotactic behavior was devised in 1978 by Andrew Moiseff and his colleagues at Cornell. They investigated the behavior of flying female crickets (of the species *Teleogryllus oceanicus*) in response to electronically synthesized sounds ranging in frequency from 3 kilohertz to 100 kilohertz. The crickets were attached to a tether so that they were able to fly in place while suspended in an airstream—the aerial equivalent of a treadmill. Moiseff and his colleagues noticed that the sound stimulation caused the crickets to move their abdomens to one side or the other, an indication that they were attempting to steer in a particular direction. The female crickets seemed particularly responsive to two frequency ranges. They were sensitive to sounds with a frequency of about 5 kilohertz, but they also had a second sensitivity band that ranged from about 30 kilohertz to 90 kilohertz. The females steered toward the source of the 5-kilohertz sound—which corresponds to the frequency of the natural calling song of the male cricket—but steered away from the high-frequency sounds—which lie within the frequency range of the ultrasonic cries of echolocating bats. The phonotactic responses of the female cricket suggest that the auditory system of these animals is specialized not only for communication with other members of the same species but also for the detection and avoidance of the predatory bat.

The structure of the cricket's auditory system may help us to understand how it is that the cricket is able to determine the location of a sound source. The cricket's ears—consisting essentially of membranous eardrums—are not located on either side of its head but just below the "knee" on its foremost pair of legs. The ears on the left and right legs are connected via air-filled tubes that meet at the animal's midline. Each of the tubes also has a branch leading to an external opening, called a spiracle, on the cricket's body behind each of the forelegs.

Because the cricket's ears are connected to each other and to the spiracles, sound may reach the ear through any of three channels. First, of course, is the direct path, in which sound pressure waves strike the outside of the eardrum. But there are also two indirect paths, through which sound waves may strike the inside of the eardrum: through the air tube from the opposite ear and through the tube from the spiracles. When the sound originates from a source on one side of the cricket, the pressure wave takes a little longer to reach the eardrum on that side via the indirect routes than it does directly. The delay between the direct and the indirect routes is such that the sound pressure is at a maximum on the external part of the ear at the same time that it is at a minimum on the inside of the ear. Thus, when a sound wave coming from the cricket's left strikes the left ear, the eardrum is maximally excited. In contrast, a sound wave coming from the right strikes the left eardrum on the outside at about the same time that it reaches the inside of the left eardrum via the indirect route. Consequently, the internal and external pressures are the same, and there is little net movement of the eardrum. Such differential responses of the left and right ears to sound sources at various points in the cricket's acoustic space may allow the animal to determine the location of the sound source.

Having the means to discern the location of the sound source, the cricket must now translate this information into a movement toward or away from the sound. In 1983 Moiseff and Hoy investigated one part of the cricket's nervous system that may mediate this phonotactic response. They inserted a glass microelectrode into a cricket's prothoracic ganglion—a cluster of nerve cells that sends and receives information from the forelegs—while delivering sounds of various frequencies (from 3 kilohertz to 100 kilohertz) to the cricket's ear. They recorded from a nerve cell, which they called interneuron-1, that was excited by sounds over a wide range of frequencies (from 8 kilohertz to 100 kilohertz)—a range that covers the frequencies at which bats search during echolocation. Interneuron-1 was strongly excited when the sound stimuli mimicked not only the frequency but also the temporal patterns of the search stimulus used by the echolocating bat. Furthermore, sounds with frequencies of about 5 kilohertz—the frequency of the cricket's calling song—maximally inhibited the response of interneuron-1. Since interneuron-1 receives input from the auditory nerve that connects

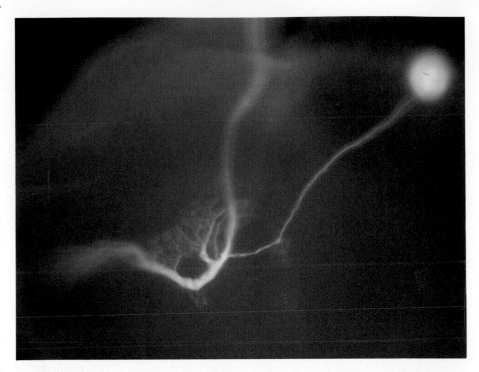

Figure 6. High-frequency-sensitive auditory neuron in a cricket—here labeled with a fluorescent dye called Lucifer Yellow—serves as an alarm that signals the presence of an echolocating bat. The neuron, called interneuron-1, is part of a neural circuit that elicits the insect's aerial escape maneuvers; excitation of the cell is both necessary and sufficient to elicit a bat-avoidance response. Fine branches called dendrites (*center*) receive signals from the auditory nerve (*not visible*); a single axon (*exiting at top*) relays the information to the brain. The spherical cell body is at the upper right. (Photomicrograph courtesy of Andrew Moiseff, University of Connecticut at Storrs.)

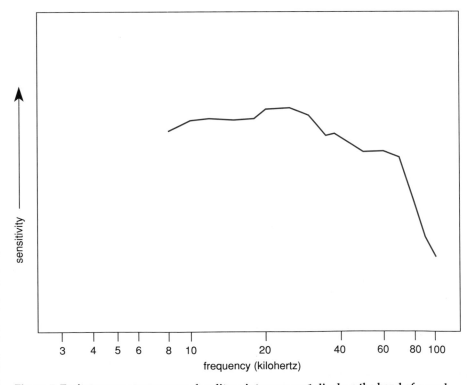

Figure 7. Excitatory response curve of auditory interneuron-1 displays the band of sound frequencies—ranging from 8 kilohertz to 100 kilohertz—that activate the nerve cell. The range of frequencies includes those sounds corresponding to the ultrasonic cries of hunting bats. The curve represents the average sensitivity of the neuron as determined by electrophysiological recordings in several animals. (Adapted from Moiseff and Hoy 1983.)

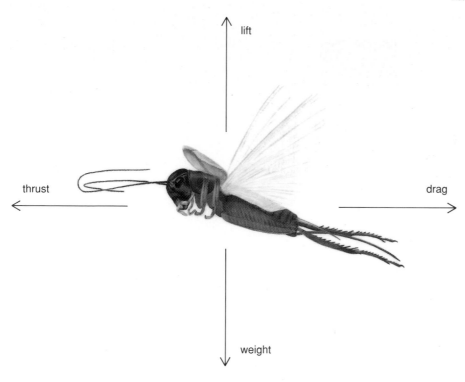

Figure 8. Balance of forces acting on a cricket in flight determines the insect's speed and altitude. Thrust propels the cricket forward and overcomes the drag imposed by the resistance of the air, whereas lift raises the cricket upward by overcoming the cricket's weight. A cricket modifies the balance of these forces by altering several factors: the frequency of its wingbeats, the extent of its wing strokes, the angle of its wings and the relative position of its body and its legs.

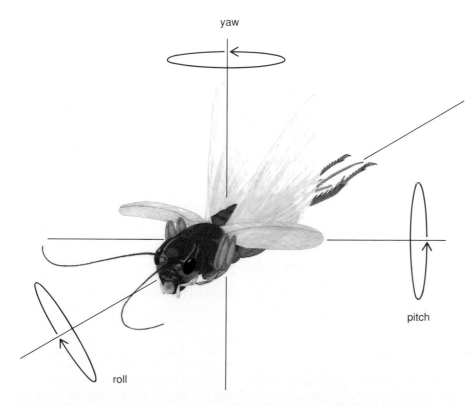

Figure 9. Body rotations in the three dimensions of space are employed by the cricket during evasive flight maneuvers. In response to ultrasonic stimulation from one side, a cricket will pitch downward, while yawing and rolling away from the sound source.

to the ear, and sends its signal to the brain, Moiseff and Hoy suggested that it might serve as an alarm that signals the presence of an echolocating bat.

Tom Nolen, now at the University of Miami, and Hoy later showed that interneuron-1 is both necessary and sufficient for eliciting negative phonotaxis in crickets. In other words, if interneuron-1 is inactivated, no level of ultrasound will induce a cricket to steer away. And, even without ultrasound, the activation of interneuron-1 by electrical stimulation will cause the cricket to steer away from the apparent source of the sound. This work seems to suggest a simple neuronal system: a single neuron that controls negative phonotaxis, perhaps by activating other neurons in the brain. Indeed, Peter Brodfuehrer, now at Bryn Mawr College, and Hoy showed that this bat-avoidance system diverges in the brain. At least 20 of the cricket's brain cells respond to ultrasound; they, in turn, must activate other neurons to produce the phonotactic response. At this time the complete circuit is still unknown.

In Two Strokes of a Cricket's Wing

My foray into the study of the cricket's response to ultrasound began in the fall of 1985, when I started my doctoral research in Hoy's laboratory at Cornell. The question I came to ask concerned not so much how a cricket knew to steer away from a predatory bat, but the aerodynamics behind the cricket's flying stunts. There were a number of ways the cricket might veer away from a bat; assessing the possibilities did not demand cleverness so much as a willingness to compromise. Although the creation of a natural environment plays a critical role—allowing the cricket to fly freely across the sky, for example—such a scenario allows few opportunities to measure subtle changes in the beating of the wings or the posture of the body. So, like the investigators before me, I accepted the now classical constraint of placing the insect on a tether.

The tether, a small wire attached to the cricket's back, holds the little acrobat in a flowing airstream where it can fly in place. In this situation the cricket beats its hindwings through large arcs, covering nearly 120 degrees, while the forewings essentially vibrate up and down. Both pairs of wings move rapidly, completing 32 wingbeats per second. In this situation a short pulse

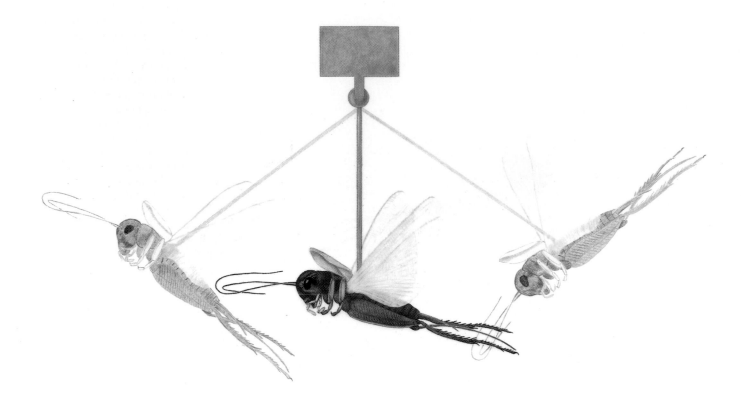

Figure 10. Crickets suspended on a pendulum tether may fly for over an hour, providing a useful model system in which to examine the aerodynamics of insect flight. The cricket swings forward by generating thrust with its wingbeats; it swings backward when the drag created by the airstream overpowers the cricket. When the thrust and the drag are equal, the cricket hangs directly below the pendulum, flying at a speed equal to that of the airstream. This technique reveals that crickets—and many other insects—fly at a speed of about two meters per second. Some hunting bats, on the other hand, fly at about nine meters per second.

of ultrasound (lasting 10 milliseconds), causes the cricket to respond immediately. The response is so strong that even a naive observer will interpret the cricket's movements as a turn, though the rigid tether prevents it.

One day in the fall of 1986, Brodfuehrer was watching the movements of a tethered, flying cricket. He noticed that whenever he jiggled a ring of keys—a makeshift source of ultrasound—the cricket tilted its forewings away from the sound. (This was the singular observation that launched me into the study of cricket aerodynamics.) Brodfuehrer and I made a few quick observations before I began a series of photographs to measure the angle of the forewing tilt. We noted that the cricket moved much like an airplane, banking its forewings into the turn. When I measured the angle of the tilt over a range of ultrasonic intensities, I found that the relationship was linear. That is, as the ultrasound gets louder, the cricket tilts its forewings farther away from the sound. This suggests that the degree of the turn is graded with the intensity of the ultrasonic stimulus.

Next, I decided to examine the frequency of the cricket's wingbeats. Others had found that a change in the frequency of the wingbeats accompanies steering in a number of insect species, including locusts, dragonflies, moths and several types of flies. Did the cricket also increase its pace in response to ultrasound? It did, by about three or four beats per second in response to a single pulse of ultrasound. Incredibly, the cricket will make this change of pace within the time it takes to beat its wings twice. This means that a cricket can detect the ultrasound, alter the rhythmic signals in its flight system and move its wings faster in as little as 60 milliseconds.

At this point I had a few valuable pieces of evidence. First, the linearity of the forewing tilt suggests that the cricket turns more sharply when the ultrasound is louder. That is important because an increase in the intensity of the ultrasound in the natural world corresponds to a closer bat; it behooves a cricket to turn a little more sharply. Moreover, by increasing the frequency of its wingbeats in response to ultrasound, the cricket probably flies faster and makes quicker changes in direction—useful attributes for any bat-avoidance system.

Flows Here, Forces There

I had learned a few things about the way a cricket flies, but I still had not answered the original question—how does a cricket evade ultrasound? I really wasn't sure that the forewing tilt made a cricket turn or that increasing the frequency of the wing stroke made it fly faster. The inference seemed reasonable, but I didn't want to be too hasty; more than any other endeavor I can think of, the field of aerodynamics punishes scientists for making too-quick assumptions. This happens largely because aerodynamics is extremely complicated; with flows here and forces there, how does one know which factors are going to be important?

To get an appreciation of the scientist's dilemma, consider an example from baseball: the curve ball. A well-thrown curve ball flies from the pitcher's hand, streaks toward the plate, and curves at the last instant. What does this say about cause and effect in aerodynamics? The cause (the manner in which the pitcher throws the ball) and the effect (the ball's curving trajectory) are substantially separated in time. It poses a challenge for a scientist intent on correlating cause and effect. Similar perplexities haunt all of aerodynamics.

In an attempt to avoid such problems, I decided to start by looking at the effect, and then to search for the cause. I began by measuring the aerodynamic movements of the cricket—in pursuit of the forces involved.

A flying insect is balanced by two pairs of forces. First, in the vertical plane gravity produces a downward force on the cricket's mass, namely the force measured as weight; a flying insect offsets its weight with the upward force called lift. As long as the lift and the weight are equal, the insect maintains a constant altitude. Along the horizontal dimension, the force driving the insect forward is thrust; it is resisted by friction with the air, or drag, which pulls the insect backward. If the thrust and the drag are equal, then the insect moves at a constant speed. If a cricket is to increase its airspeed, there must be an imbalance: The thrust must exceed the drag on the cricket's body. But how does a cricket increase its airspeed? Does a change in the frequency of the wingbeats cause an imbalance between thrust and drag? It was a hypothesis waiting to be tested.

I used a tether designed much like a pendulum; it allowed the cricket to swing freely, forward and backward. When a cricket attached to such a pendulum starts to fly, thrust pushes the cricket forward, and so the cricket swings up, like a child in a swing. By turning on a fan, I created an airstream that pushes the cricket backward by increasing the drag. I adjusted the speed of the airstream until the pendulum hung straight down; this meant that the cricket's flight speed equaled the speed of the airstream—about two meters per second. As soon as I turned on the ultrasound, the flying cricket swung forward on the tether—because it flew faster. Now I could say that the frequency of the wingbeat and the speed of the cricket both increase in response to ultrasound.

The literature on insect aerodynamics also suggests that rotations of the animal's body are important for steering. Like an airplane, a flying insect can rotate about three axes, called the

Figure 11. Tethered praying mantis is seen in flight before *(top)* and after *(bottom)* the onset of an ultrasonic stimulus. During normal flight, the mantis has a streamlined posture—tucking its legs neatly into its straight body. On hearing the ultrasound, the mantis extends its forelegs, rolls its head back and bends its abdomen—all within about a tenth of a second. During free flight, this suite of behaviors results in a turning or spiralling dive that helps the insect to escape from a bat. (Photographs courtesy of David Yager, University of Maryland at College Park. From Yager and May 1990. Reproduced by permission of the Company of Biologists Limited.)

pitch, roll and yaw axes, that all pass through the insect's center of gravity. A pitching motion is a rotation around the transverse axis (parallel to an airplane's wings); roll is rotation around the longitudinal axis (parallel to the fuselage), and yaw describes rotation around the vertical axis.

I designed a tether that allowed a flying cricket the freedom to rotate about the pitch axis, and watched the insect's response to ultrasonic stimulation. When I delivered a pulse of ultrasound to the cricket, it pitched downward. As with the tilting of the forewings, the louder the ultrasonic stimulus, the greater the amount of downward pitch. Here was another linear response, one which probably translates into a dive.

But does the cricket use its forewings or its hindwings to adjust its pitch attitude? A few simple experiments—removing one or the other pair of wings—showed that crickets use both their hindwings and forewings for the control of pitch. But there was more. I set up an experiment to test the relative contributions of the hindwings and the forewings to the degree of pitch motion. I set the ultrasound to a constant intensity and then measured the angle of pitch in three circumstances: a cricket with both pairs of wings, a cricket with only its forewings and a cricket with only its hindwings. One might assume that the relative contributions of the hindwings and the forewings would be additive. That is, by adding the angle of pitch for a cricket that only had its forewings to the angle of pitch for a cricket that only had its hindwings, it might be expected that the sum would equal the angle of pitch for a cricket with both pairs of wings. But this isn't the case; the sum of the individual pitch angles is always smaller than the angle of pitch produced by a cricket with both pairs of wings. It was another pitfall in aerodynamics—the whole can be more than the sum of the parts.

Next, with a minor modification to the tether, I measured the angle of roll. Again, I found that the crickets tended to roll away from the source of the ultrasound. And, much like the pitch measurements, the crickets could roll with either their hindwings or their forewings. But in this case the sum of the parts (rolling with forewings or hindwings alone) did equal the whole (rolling with both pairs of wings). The aerodynamic corollary to be found here was that the whole *might*

Figure 12. Tethered cricket is seen in flight before *(top)* and after *(bottom)* the onset of an ultrasonic stimulus. During normal flight, the cricket tucks its forelegs and middle legs just behind its head—which it holds slightly above horizontal—while extending its hindlegs. After hearing the ultrasound, the cricket pitches downward, its body now nearly level. A hinge just above the cricket's back allows the cricket to pitch freely. (Photographs courtesy of the author.)

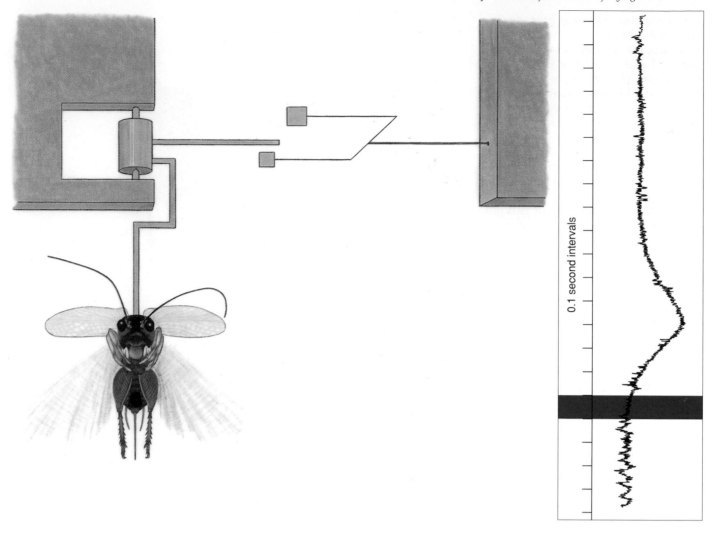

Figure 13. Yaw movements of a flying cricket are measured with a tether *(left)* **that allows lateral motions by means of a rotating cylinder. Rotations of the cylinder swing a wand that is located between two sensors. The sensors record the size, the direction and the duration of the yaw. The recording** *(right)* **shows that a cricket will yaw in less than a tenth of a second after the onset of an ultrasonic stimulus** *(red band)*.

equal the sum of the parts— sometimes.

One axis of rotation remained to be investigated, the yaw axis. When I had measured the pitch and the roll movements, my experimental apparatus determined only the change in the angle. For the measurement of yaw, I wanted both the angle itself and the change in the angle. This required the construction of a device that could keep track of the angle of yaw over time. The device also provided the latency of the response—the time from the onset of the stimulus to the onset of the yawing motion—and the angular velocity about the yaw axis.

I found that, like the pitching and rolling rotations, the magnitude of the yaw is linearly related to the intensity of the ultrasound. And both the forewings and the hindwings participate in the yaw movement. But I also found that the intensity of the ultrasound affects the time course of the yaw. As the ultra-sound gets louder, the latency of the yaw decreases, whereas the angular velocity increases. At the highest intensities of ultrasound— around 90 decibels, which corresponds to a shrieking bat within a few meters of a cricket's ear—the latency is down to about a tenth of a second, and the angular velocity is about 100 degrees per second.

There was one surprise in my studies of the cricket's ability to yaw. Not only is the cricket able to yaw with either pair of wings, but it can also yaw without any wings at all. The movement is small, but significant. This "body yaw" is reminiscent of an abdominal swing that crickets are known to perform in response to ultrasound. Previous studies had suggested that the abdomen might act like a rudder; by swinging the abdomen to one side, the cricket increased the drag on that side. It seems reasonable to think that increasing the drag on one side would produce yaw. Might that be the cause of the yaw in a wingless cricket? It could be another aerodynamic trap between cause and effect. I put the hypothesis to the test by removing the cricket's wings and fixing the abdomen so that it could not swing. Surprisingly, this treatment did not stop the crickets; they were still able to yaw. In a second approach I used wingless crickets again, but this time I freed the abdomen and fixed the front portion of the cricket's body, the thorax. This time the crickets were unable to yaw. The swinging abdomen did not produce the yaw; it was the twisting thorax that did it. The rudder is in the front.

Some Leg Action

Not all useful observations come as a result of premeditated experimental design; sometimes it's useful just to play with the equipment. That is what happened one day while I was toying with the ultrasound stimulus and watching the responses of a tethered cricket. I was looking at the cricket's wings—which appear to be a transparent blur, beating about 30 times a second—when I noticed that the hindwing farther from the source of the ultrasound didn't go down as far. With every burst of ultrasound, the blurred wing seemed to stop short. Was this another mechanism by which the cricket was trying to turn? If the wing farther from the ultrasound did not make a full stroke, it should produce less thrust—inducing a turn toward that side.

On closer inspection, more than the hindwing was changing. For every pulse of ultrasound, one of the cricket's hindlegs would swing up, and appear to collide with the hindwing. Was the hindleg impeding the movement of the hindwing? Jeffrey Camhi of the Hebrew University in Jerusalem had noticed a similar phenomenon in tethered locusts, but had not shown that there were any aerodynamic consequences.

To establish that there was an interaction between the hindwing and the hindleg I needed photographic evidence. I posed the crickets in three situations: an intact cricket flying without ultrasound, an intact cricket flying with ultrasound, and a cricket that had a hindleg removed and that was flying with ultrasound. The photographs confirmed my suspicions. When there is no ultrasonic stimulus, the hindleg does not interfere with the hindwing, but with the ultrasound the hindleg farther from the sound does impede the hindwing's downstroke. In a cricket without the hindleg, the hindwing completes a full stroke even during an ultrasound stimulus.

Is there an aerodynamic effect to sticking a leg into a beating wing? To test this possibility I tethered some crickets in the device I had used to measure yaw motions. The ultrasonic stimulus was given from the left and the right sides of the cricket, both before and after removing the cricket's left hindleg. Removing the left hindleg means that all of the ultrasonic pulses from the left serve as controls (the hindleg only hits the hindwing when the ultrasound comes from the *opposite* side). And indeed, the yaw induced by an ultrasound stimulus from the left looked the same before and after removing the left hindleg. Removing the

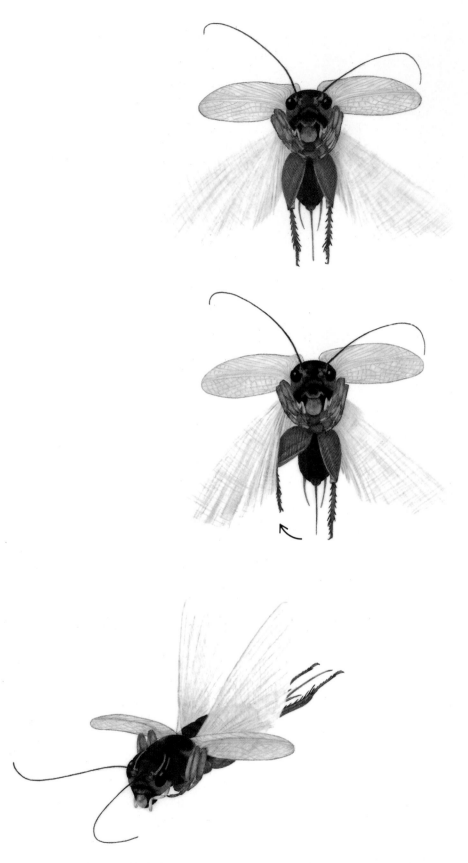

Figure 14. Hypothetical flight path shows how a free-flying cricket might respond to the ultrasonic cries of an echolocating bat. Normal flight *(top)* **is interrupted after the cricket detects the ultrasound coming from the insect's left. Within 40 milliseconds** *(middle)* **the cricket swings its right hindleg into its right wing, thereby reducing the thrust on that side. At about the same time, the cricket tilts its forewings, increases its speed, and pitches downward while rolling and yawing to its right. All factors should contribute to a powered dive away from the hunting bat** *(bottom).*

hindleg that was on the same side as the ultrasound stimulus did not change the flight behavior. But when the ultrasound came from the right, there was a large change in the cricket's responses. When the cricket had both hindlegs, ultrasound from the right would cause the cricket to yaw after 100 milliseconds; after removing the left hindleg, the yaw did not begin until after 140 milliseconds. The latency of the yaw response increased by 40 percent. The magnitude of the yaw also changed: it was 17 degrees before removing the leg and only 9 degrees after removing the leg. The hindleg does affect steering—it makes the yaw bigger and faster.

The hindleg might affect steering in two different ways. It might work alone to cause some drag, or its interaction with the hindwing could be the crucial factor. I repeated the yaw experiments with just one change—I removed the hindwings. This time there was no change in the yaw; it looked the same whether the ultrasound came from the left or the right, or whether the hindleg was present or absent. The result was clear: The hindleg steers the cricket by impeding the hindwing's downstroke.

Although the mechanism may appear to be crude—somewhat like stopping a bicycle by sticking a tire pump into the spokes of the rear wheel—it is functional and economical engineering design. To appreciate this, consider another example from baseball. During a fast-ball pitch, the pitcher's arm is pulled through a high-speed arc to throw the ball toward the plate. Attempting to stop this motion halfway through the pitch is humanly impossible. A cricket winging its way from an echolocating bat faces the same difficulty. Its wing muscles are not designed to halt the downstroke in the brief time needed to steer out of the bat's way. But the hindleg is able to block the downstroke of the hindwing in such a way as to produce a nearly instantaneous change in the length of the stroke—which causes the cricket to veer away. It is a clever yet simple solution.

All told, however, the swing of the hindleg is merely one small part of the cricket's acrobatic maneuvers. We now know that the cascade of responses consists of swinging legs, tilting forewings, twisting thoraxes and rapidly beating wings. It all adds up—in a fraction of a second—to an elegant ballet that whisks the cricket beyond the grasp of a hungry bat.

Bibliography

Brodfuehrer, P. D. and R. Hoy. 1989. Integration of ultrasound and flight inputs on descending neurons in the cricket brain. *Journal of Experimental Biology* 146:157-171.

Brodfuehrer, P. D. and R. Hoy. 1990. Ultrasound-sensitive neurons in the cricket brain. *Journal of Comparative Physiology, A* 166:651-662.

Camhi, J. M. 1970. Yaw-correcting postural changes in locusts. *Journal of Experimental Biology* 52:519-531.

Cooter, R. J. 1979. Visually induced yaw movements in the flying locust, *Schistocerca gregaria* (Forsk). *Journal of Comparative Physiology* 99:1-66.

Cranbrook, T. E. O., and H. G. Barrett. 1965. Observations of nocturnal bats (*Nyctalus noctula*) captured while feeding. *Proceedings of the Zoological Society of London* 144:1-24.

Easteria, D. A. and J. O. Whitaker, Jr. 1972. Food habits of some bats from Big Bend National Park, Texas. *Journal of Mammalogy* 53(4):887-890.

Griffin, D. R. and R. Galambos. 1941. The sensory basis of obstacle avoidance by flying bats. *Journal of Experimental Zoology* 86:481-506.

Griffin, Donald R. 1984. *Listening in the Dark.* Dover Publications.

Hill, K. G. and G. S. Boyan. 1976. Directional hearing in crickets. *Nature* 262:390-391.

May, M. L., Brodfuehrer, P. D., and R. R. Hoy. 1988, Kinematic and aerodynamic aspects of ultrasound-induced negative phonotaxis in flying Australian field crickets (*Teleogryllus oceanicus*). *Journal of Comparative Physiology* 164:243-249.

May, M. L. and R. R. Hoy. 1990a. Ultrasound-induced yaw movements in the flying Australian field cricket (*Teleogryllus oceanicus*). *Journal of Experimental Biology* 149:177-189.

May, M. L. and R. R. Hoy. 1990b. Leg-induced steering in flying crickets. *Journal of Experimental Biology* 151:485-488.

Miller, L. A. and J. Olesen. 1979. Avoidance behavior in green lacewings. I. Behavior of free flying green lacewings to hunting bats and ultrasound. *Journal of Comparative Physiology* 131:113-120.

Moiseff, A., Pollack, G. S. and R. R. Hoy. 1978. Steering responses of flying crickets to sound and ultrasound: mate attraction and predator avoidance. *Proceedings of the National Academy of Sciences of the U.S.A. Biological Sciences.* 75:4052-4056.

Moiseff, A. and R. R. Hoy. 1983. Sensitivity to ultrasound in an identified auditory interneuron in the cricket: a possible neural link to phonotactic behavior. *Journal of Comparative Physiology* 152:155-167.

Nachtigall, W. and D. M. Wilson. 1967. Neuromuscular control of dipteran flight. *Journal of Experimental Biology* 47:77-97.

Nolen, T. G. and R. Hoy. 1984. Initiation of behavior by single neurons: The role of behavioral context. *Science* 226:992-994.

Pollack, G. S. and N. Plourde. 1982. Phonotaxis in flying crickets: Neural correlates. *Journal of Insect Physiology* 146:207-215.

Popov, A. V. and V. F. Shuvalov. 1977. Phonotactic behavior of crickets. *Journal of Comparative Physiology* 119:111-126.

Robert, D. 1989. The auditory behavior of flying locusts. *Journal of Experimental Biology* 147:279-301.

Roeder, K. D. 1962. The behavior of free flying moths in the presence of artificial ultrasonic pulses. *Animal Behavior* 10:300-304.

Roeder, K. D. 1967. Turning tendency of moths exposed to ultrasound while in stationary flight. *Journal of Insect Physiology* 13:873-888.

Roeder, K. D. and A. E. Treat. 1961. The detection and evasion of bats by moths. *American Scientist* 49:135-148.

Rüppell, G. 1989. Kinematic analysis of symmetrical flight manoeuvres of Odonata. *Journal of Experimental Biology* 144:13-42.

Schnitzler, H.-U., D. Menne, R. Kober and K. Heblich. 1983. The acoustical image of fluttering insects in echolocating bats. pp. 235-250, in *Neuroethology and Behavioral Physiology*, F. Huber and H. Markl (eds.). Springer: Heidelberg.

Spangler, H. G. 1988. Hearing in tiger beetles (Cicindelidae). *Physiological Entomology* 13:447-452.

Suga, N. 1984. Neural mechanisms of complex-sound processing for echolocation. *Trends in Neurosciences* 7:20-27.

Whitaker, J. O., Jr. and H. Black. 1976. Food habits of cave bats from Zambia, Africa. *Journal of Mammalogy* 57(1):199-205.

Yager, D. D., May, M. L., and B. M. Fenton. 1990. Ultrasound-triggered, flight-gated evasive maneuvers in the flying praying mantis, *Parasphendale agrionina*. I: Free flight. *Journal of Experimental Biology* 152:17-39.

Yager, D. D. and M. L. May. 1990. Ultrasound-triggered, flight-gated evasive maneuvers in the flying praying mantis, *Parasphendale agrionina*. II: Tethered flight. *Journal of Experimental Biology* 152:41-58.

PART III

The Evolutionary History of Behavior

Biologists have always been fascinated with the question of where the traits that exist today came from, i.e., their evolutionary history. But how can the history of behavioral mechanisms and the behaviors they control be explored? Unfortunately the fossil record is largely uninformative. Only under rare circumstances (e.g., the discovery of fossilized dinosaur nests) is enough information captured in fossils to enable us to draw inferences about the origins of complex social or reproductive behaviors and their subsequent evolution.

Recently, however, historical ecologists have suggested that one can reconstruct the history of behaviors using methods as rigorous and scientific as those used by any other kind of biologist. This approach relies on Darwin's "comparative method." This method takes advantage of the thousands of "natural experiments" that have occurred over evolutionary time as new species have formed and their special characteristics have evolved. The first step is to construct a phylogenetic tree for the group of species at issue by, for example, using morphological traits or DNA sequences. The second step is to "map" the character states of extant taxa onto ancestral nodes of the tree. Character mapping is accomplished by minimizing the number of character-state transitions (changes), thereby allowing inference of the shortest ("most parsimonious") evolutionary path taken by a character. Assuming that (i) the traits of extant species have remained the same since the last speciation event, and (ii) the shortest evolutionary path (fewest changes in character states) is most likely, an hypothesis can be made about the relative timing of trait origins.

The five articles in Part III introduce readers to this and other historical approaches that have been used to infer the evolutionary pathways of behaviors. These historical analyses complement studies of how internal mechanisms control behavior (Part II), as well as studies of the development (Part II) and adaptive value of behavior (Parts IV and V), as explained by Holekamp and Sherman in their article on the four levels of analysis that kicks off Part II.

Bert Hölldobler and Edward O. Wilson use the comparative method to get at the history of the amazing behaviors of modern weaver ants, several of which build elaborate silken nests in tropical trees, rather like the constructions of the more familiar North American tent caterpillars. These nest builders belong to a group of four living ant genera whose members weave with silk to varying degrees. The authors examine the shared behavioral characteristics of these genera as they try to de-termine what series of changes led to the complex nest-weaving of certain species. Detailed comparisons of adult and larval behaviors allow the authors to array the genera along a "phylogenetic grade" from least to most complex. For example, in the "advanced" genus *Oecophylla*, workers use middle-aged larvae that produce silk as living shuttles, carrying the larvae to and fro to bind leaves together into a nest; by contrast, in the "primitive" genus *Dendromyrmex*, larvae produce silk only when they are about to pupate, and workers simply plop the larvae down and let them create small patches of silk as they thrash about.

Hölldobler and Wilson suggest that the primitive *Dendromyrmex* weaver ants may be on an evolutionary trajectory that will carry them through stages exemplified by the genera of intermediate complexity until they can attain the advanced nest-weaving accomplishments of *Oecophylla*. What are the merits of arraying species according to phylogenetic grade? Is it reasonable to infer "progress" from simpler to more complex forms of behavior? If so, why are there so many "intermediate" and "primitive" species that appear to possess "imperfect" adaptations? Finally, why do you suppose only a handful of the estimated 20,000 species of ants have taken to building silken nests in treetops? Is it because most ants have never had the capacity to take the first steps toward silk weaving, or is treetop nesting actually disadvantageous for almost all ant species?

James Nieh does not use the concept of phylogenetic grade in his article on how stingless bees communicate information about new food sources. However, Nieh does use the comparative method in his attempt to understand how the complex communication systems of social bees has evolved. He was drawn to stingless bees in part by a conjecture of Karl von Frisch, the Austrian biologist who decoded the "dance language" of honey bees (*Apis mellifera*). Von Frisch reportedly said "we cannot believe that the bee dance of the European bees has come from heaven as it is and, since the Indian honeybees and the stingless bees there live in a more primitive social organization, we should expect some phylogenetically primitive stages of the bee dance." In fact, before Nieh's work began, it was thought that stingless bees might not possess a dance-language, but rather lay scent-trails to food.

Nieh studied a species of stingless bee called *Melipona panamica* that lives in Central American rainforests. He captured colonies, ensconced them in transparent observation nests, and "trained" them to visit feeders containing a scented sugar solution. By placing "control" feeders at different locations and counting the number of recruits that visited them versus the training feeder, Nieh could determine what information experienced foragers were transmitting to naive colony mates. He found that these apparently "primitive" bees indicate distance, direction, and smell of food sources. In other words, they can do everything the more "advanced" honey bees do—and more, including indicating the height of their find (something European honey bees cannot do).

Neih also found subtle differences between *M. panamica* and honey bees in the way foraging information is transmitted. Stingless bee foragers encode information on height and distance to food in pulsed sounds emitted during dances inside the nest; how they indicate direction is still unknown (although Nieh suspects that recruits follow foragers that are starting back to the site). *M. panamica* thus has a representational communication system that is different from, but not more primitive than, the communication system of honey bees. This finding forces us to reexamine the issue of phylogenetic grade. Does natural selection tend to promote complexity, or does this view derive from our admiration of the technological advances that have occurred in human societies? Are "advanced" species with complex behaviors or social structures really better adapted than "primitive" species? A useful exercise would be to compare and contrast the communication systems of *A. mellifera* and *M. panamica* and then relate the differences to each species' ecology. For example, why do you suppose stingless bees indicate the height of food sources whereas honey bees do not? Why don't stingless bees indicate the direction to the food in their dance, as do honey bees? And why do successful foragers in honey bees and stingless bees dance in the first place, instead of personally leading recruits back to food or laying scent trails?

The next article reveals that bowerbirds are an avian equivalent of weaver ants in that some species build highly complex structures using behavior patterns whose origin and subsequent modification are far from obvious. Thus, they also provide an intriguing historical puzzle for evolutionary biologists, a puzzle that Gerald Borgia attempts to solve. Although the article does not present an explicit phylogenetic hypothesis of how bowerbirds are related, readers should be able to draw one using the information that Borgia provides. You can then check your diagram against the one that Borgia and his colleagues constructed based on molecular data (see the *Proceedings of the Royal Society of London*, Series B, *264:* 307-314, 1997). Your diagram should reveal that among the living bowerbirds there are species whose ancestors probably engaged in complex activities no longer exhibited by any living species. What significance does this fact have for persons who assume that living species with simpler behavior patterns have retained the simpler traits of their ancestors? Why is it inappropriate to look at any one living species and infer a linear progression with one ancient species giving rise to another, which in turn produced yet another, in an inexorable march toward the current living species of interest?

Borgia does more than reconstruct the pattern of changes in bower building. He also addresses another level of analysis, namely the adaptive value of the characteristics that have evolved in one lineage or another. Borgia hypothesizes that male bowerbirds build bowers to protect females against forced copulation. What does this hypothesis imply about the importance of female choice of mates in bower-building species? How could males have evolved a trait that made it harder for them to reproduce compared to their male ancestors that sometimes engaged in forced matings? How could a trait that reduced a male's chances of siring offspring have spread through an ancestral population? And what about other bird species? Doesn't the "protection against forced copulation" hypothesis predict that (1) forced copulation will be a feature of the reproductive lives of other birds, at least those species closely related to bowerbirds, and that if so, (2) building bowers, or their analogs, should have evolved in other species whose ancestors engaged in occasional forced copulations. Can you acquire the information needed to test these predictions?

Next, Lyudmila Trut illustrates a completely different way to reconstruct the history of behavior. Forty years ago her colleague, Dmitry Belyaev, became intrigued by the question of how animals, in particular dogs, acquired their domesticated characteristics. Belyaev set out to examine this issue by capturing some wild foxes (*Vulpes vulpes*) and breeding them for tameness. Once a month, starting when each pup was a month old, an experimenter offered it food and tried to approach and handle it. When the juveniles were 7-8 months old, those that were most enthusiastic about human contact were selected as breeding stock.

Over time, the foxes became tamer and tamer due to this strong artificial selection regime. Today about three-fourths of the current generation are "eager to establish human contact, whimpering to attract attention and sniffing and licking experimenters like dogs." More remarkable still are the changes in morphology that have accompanied domestication as unselected side effects, including floppy ears, shortened legs and tails, tails curved upwards, underbites and overbites, and novel coat patterns (piebald) and colors (white blazes, gray hairs). The ontogeny of the foxes' social behavior also has changed. Their eyes open earlier and their fear response is intitiated later, which widens the time "window" for social bonding. At present the foxes are so docile and engaging that the experimenters plan to market them as house pets!

Trut believes that as the foxes' behavior evolved, changes took place in the mechanisms that regulate development, leading to shifts in the rates and timing of ontogenetic processes such as socialization. Although this explanation is persuasive for some novelties, such as earlier eye opening, it is less obvious how it can account for floppy ears, recurved tails, and bizarre colors. Trut suggests that these are genetically correlated traits—meaning traits whose development is affected by the same genes that result in tameness. But there are alternative possibilities. For example, perhaps the unusual behavioral and morphological phenotypes represent deleterious recessive traits that are expressed due to inbreeding; or to the breaking up of groups of genes that had once evolved into integrated, cooperative complexes; or perhaps they are developmental anomalies triggered by stresses of captivity. How would you evaluate these alternatives? Do the many similarities of the features of domesticated foxes to those of other domesticated species (e.g., pigs, horses, and cows) affect your thinking? Do you agree that the farm-fox experiment recreated the process by which wolves became domesticated house dogs 10,000-15,000 years ago? If so, could the dog breeds that we see today and the diversity of their behaviors be due to selection by early humans for differences in body shape and temperament among canid ancestors?

The final article in this section takes yet another approach to reconstructing the history of behaviors. Paul Sherman and Samuel Flaxman discuss the probable modern-day functions of two familiar phenomena, namely cooking with spices, and nausea and vomiting during early pregnancy. They propose that this analysis enables them to understand the usefulness of these behaviors in antecedent environments. The approach presumes that present advantages associated with certain traits are similar to those that occurred in the past, despite the many technological changes that have taken place. If a behavior currently provides higher fitness than its alternatives, perhaps natural selection acting in similar environments in the past caused the initial spread of the trait.

Spice use and morning sickness are two seemingly unrelated behaviors, one shaped by cultural traditions and the other by physiological processes. Sherman and Flaxman argue that they serve a common purpose today and probably in ancestral populations as well: protecting individuals from dangers lurking in foods, especially foodborne pathogens. At this point you may be thinking "Come on, guys! I use spices just because they taste good." But is this explanation really an alternative to the antimicrobial hypothesis (i.e., at the same level of analysis)? What about the possibility that spices are used to disguise the taste and smell of spoiled foods? Or to increase perspiration and thus evaporative cooling? Or perhaps people use whatever spices are available to make bland food taste more "interesting?" After decid-

ing which of these hypotheses are true alternatives and which are complementary, design ways they might be tested. Note also that Sherman and Flaxman found that vegetable dishes are less heavily spiced than meat dishes, which they interpret as further support for the antimicrobial hypothesis. Can you think of alternative explanations for that result?

Use of spices can involve costs as well as benefits because many plant chemicals, when ingested in sufficient quantities, can be mutagens, carcinogens, and abortifacients. Sherman and Flaxman argue that nausea and vomiting during the first trimester of pregnancy (commonly known as "morning sickness") protects the embryo by causing pregnant women to physically expel and subsequently learn to avoid foods containing potentially dangerous chemicals and microorganisms.

Until recently, morning sickness was generally thought of as a "disease" or as a side-effect of hormones associated with viable pregnancies. What are the medical implications of considering it to be an adaptation instead? If nausea and vomiting serve a useful purpose, should women attempt to alleviate these symptoms? And since refrigeration is widespread and most food plants have been selected for bland tastes (few secondary chemicals), are spice use and morning sickness just evolutionary anachronisms with no adaptive value today? If Sherman and Flaxman are correct, then these behaviors should be widespread, yet no other mammals are known to regularly spice their foods and morning sickness has been described only in dogs and rhesus monkeys. How can we apply the comparative approach to understanding spice use, morning sickness, or other behaviors exhibited primarily or exclusively by humans?

Although whether a comparative analysis of the behaviors of living species, including human beings, can serve as a window into the evolutionary history of those behaviors is left as an open question, the right kinds of comparisons among existing species are clearly instructive. And any approaches that offer ways to determine whether such fascinating behaviors as nest-weaving in ants, bower-building by bowerbirds, docility of domestic animals, and spice use in cooking are useful today, and how they may have come into being, are approaches well worth knowing about!

Bert Hölldobler
Edward O. Wilson

The Evolution of Communal Nest-Weaving in Ants

Steps that may have led to a complicated form of cooperation in weaver ants can be inferred from less advanced behavior in other species

One of the most remarkable social phenomena among animals is the use of larval silk by weaver ants of the genus *Oecophylla* to construct nests. The ants are relatively large, with bodies ranging up to 8 mm in length, and exclusively arboreal. The workers create natural enclosures for their nests by first pulling leaves together (Fig. 1) and then binding them into place with thousands of strands of larval silk woven into sheets. In order for this unusual procedure to succeed, the larvae must cooperate by surrendering their silk on cue, instead of saving it for the construction of their own cocoons. The workers bring nearly mature larvae to the building sites and employ them as living shuttles, moving them back and forth as they expel threads of silk from their labial glands.

The construction of communal silk nests has clearly contributed to the success of the *Oecophylla* weaver ants. It permits colonies to attain populations of a half million or more, in spite of the large size of the

The authors take pleasure in dedicating this article to Caryl P. Haskins, fellow myrmecologist and distinguished scientist and administrator, on the occasion of his seventy-fifth birthday and his retirement from the chairmanship of the Board of Editors of American Scientist.

Bert Hölldobler is Alexander Agassiz Professor of Zoology at Harvard University. After completing his Dr. rer. nat. at the University of Würzburg in 1965 and his Dr. habil. at the University of Frankfurt in 1969, he served at the latter institution as privatdocent and professor until 1973, when he joined the Harvard faculty. Edward O. Wilson is Frank B. Baird Jr. Professor of Science and Curator in Entomology of the Museum of Comparative Zoology, Harvard University. He received his Ph.D. from Harvard in 1955, held a Junior Fellowship in the Society of Fellows during 1953–56, and has served on the faculty continuously since that time. Address: Museum of Comparative Zoology, Harvard University, Cambridge, MA 02138.

workers, because the ants are freed from the spatial limitations imposed on species that must live in beetles' burrows, leaf axils (the area between the stems of leaves and the parent branch), and other preformed vegetative cavities. This advance, along with the complex recruitment system that permits each colony to dominate up to several trees at the same time, has helped the weaver ants to become among the most abundant and successful social insects of the Old World tropics (Hölldobler and Wilson 1977a, b, 1978; Hölldobler 1979). A single species, *O. longinoda*, occurs across most of the forested portions of tropical Africa, while a second, closely related species, *O. smaragdina*, ranges from India to Queensland, Australia, and the Solomon Islands. The genus is ancient even by venerable insect standards: two species are known from Baltic amber of Oligocene age, about 30 million years old (Wheeler 1914). *O. leakeyi*, described from a fossil colony of Miocene age (approximately 15 million years old) found in Kenya, possessed a physical caste system very similar to that of the two living forms (Wilson and Taylor 1964).

Our recent studies, building on those of other authors, have revealed an unexpectedly precise and stereotyped relation between the adult workers and the larvae. The larvae contribute all their silk to meet the colony's needs instead of their own. They produce large quantities of the material from enlarged silk glands early in the final instar rather than at its end, thus differing from cocoonspinning ant species, and they never attempt to construct cocoons of their own (Wilson and Hölldobler 1980). The workers have taken over almost all the spinning movements from the larvae, turning them into passive dispensers of silk.

It would seem that close attention to the exceptional properties of *Oecophylla* nest-weaving could shed new light on how cooperation and altruism operate in ant colonies, and especially on how larvae can function as an auxiliary caste. In addition, a second, equally interesting question is presented by the *Oecophylla* case: How could such extreme behavior have evolved in the first place? As is the case with the insect wing, the vertebrate eye, and other biological prodigies, it is hard to conceive how something so complicated and efficient in performance might be built from preexisting structures and processes. Fortunately, other phyletic lines of ants have evolved communal nest-weaving independently and to variably lesser degrees than *Oecophylla*, raising the prospect of reconstructing the intermediate steps leading to the extreme behavior of weaver ants. These lines are all within the Formicinae, the subfamily to which *Oecophylla* belongs. They include all the members of the small Neotropical genus *Dendromyrmex*, the two Neotropical species *Camponotus* (*Myrmobrachys*) *senex* and *C.* (*M.*) *formiciformis*, which are aberrant members of a large cosmopolitan genus, and various members of the large and diverse Old World tropical genus *Polyrhachis*.

Two additional but doubtful cases have been reported outside the Formicinae. According to Baroni Urbani (1978), silk is used in the earthen nests of some Cuban species of *Leptothorax*, a genus of the subfamily Myrmicinae. However, the author was uncertain whether the material is obtained from larvae or from an extraneous source such as spider webs. Since no other myrmicine is known to produce silk under any circumstances, the latter alternative seems the more probable.

Similarly, the use of silk to build nests was postulated for the Javan ant *Technomyrmex bicolor textor*, a member of the subfamily Dolichoderinae, in an early paper by Jacobson and Forel (1909). But again, the evidence is from casual field observations only, and the conclusion is rendered unlikely by the fact that no other dolichoderines are known to produce silk.

During the past ten years we have studied the behavior of both living species of *Oecophylla* in much greater detail than earlier entomologists, and have extended our investigations to two of the other, poorly known nest-weaving genera, *Dendromyrmex* and *Polyrhachis*. This article brings together the new information that resulted from this research and some parallel findings of other authors, in a preliminary characterization of the stages through which the separate evolving lines appear to have passed.

In piecing together our data, we have utilized a now-standard concept in organismic and evolutionary biology, the phylogenetic grade. The four genera of formicine ants we have considered are sufficiently distinct from each other on anatomical evidence as to make it almost certain that the communal nest-weaving displayed was in each case independently evolved. Thus it is proper to speak of the varying degrees of cooperative behavior and larval involvement not as the actual steps that led to the behavior of *Oecophylla* but as grades, or successively more advanced combinations of traits, through which autonomous evolving lines are likely to pass. Other combinations are possible, even though not now found in living species, and they might be the ones that were actually traversed by extreme forms such as *Oecophylla*. However, by examining the behavior of as many species and phyletic lines as possible, biologists are sometimes able to expose consistent trends and patterns that lend convincing weight to particular evolutionary reconstructions. This technique is especially promising in the case of insects, with several million living species to sample. Within this vast array there are more than 10,000 species of ants, most of which have never been studied, making patterns of ant behavior exceptionally susceptible to the kind of analysis we have undertaken and are continuing to pursue on communal nest-building.

The highest grade of cooperation

The studies conducted on *Oecophylla* prior to our own were reviewed by Wilson (1971) and Hemmingsen (1973). In essence, nest-weaving with larval silk was discovered in *O. smaragdina* independently by H. N. Ridley in India and W. Saville-Kent in Australia, and was subsequently described at greater length in a famous paper by Doflein (1905). Increasingly detailed accounts of the behavior of *O. longinoda*, essentially similar to that of *O. smaragdina*, were provided by Ledoux (1950), Chauvin (1952), Sudd (1963), and Hölldobler and Wilson (1977a).

The sequence of behaviors by which the nests are constructed can be summarized as follows. Individual workers explore promising sites within the colony's territory, pulling at the edges and tips of leaves. When

Figure 1. To make a nest out of leaves and larval silk, worker ants of the species *Oecophylla smaragdina*, the Australian green tree ant, first choose a pliable leaf. They then form a row and pull in unison, as shown in the photograph, until they force two leaves to touch or one leaf to curl up on itself. (All photographs are by Bert Hölldobler unless otherwise indicated.)

Figure 2. If a single ant cannot bridge the gap between two leaves to be used in a nest, *O. smaragdina* workers arrange themselves in chains and pull together to close the gap. The photograph at the left indicates how many individuals can be involved in this stage of the work; at the right, several parallel chains of ants are shown in more detail.

a worker succeeds in turning a portion of a leaf back on itself, or in drawing one leaf edge toward another, other workers in the vicinity join the effort. They line up in a row and pull together (Fig. 1), or, in cases where a gap longer than an ant's body remains to be closed, they form a living chain by seizing one another's petiole (or "waist") and pulling as a single unit. Often rows of chains are aligned so as to exert a powerful combined force (Fig. 2). The formation of such chains of ants to move objects requires intricate maneuvering and a high degree of coordination. So far as is known, it is unique to *Oecophylla* among the social insects.

When the leaves have been maneuvered into a tentlike configura-tion, workers carry larvae out from the interior of the existing nests and use them as sources of silk to bind the leaves together (Figs. 3, 4). Our previous studies (Wilson and Hölldobler 1980) showed that the *O. longinoda* larvae recruited for this purpose are all in the final of at least three instars, and have heads in excess of 0.5 mm wide. However, their bodies (exclusive of the rigid head capsule) are smaller than those of the larvae at the very end of the final instar, which are almost ready to turn into prepupae and commence adult development. Thus the larvae used in nest-weaving are well along in development and possess large silk glands, but they have not yet reached full size and hence are more easily carried and manipulated by the workers.

In *O. longinoda*, all the workers we observed with spinning larvae have been majors, the larger adults that possess heads between 1.3 and 1.8 mm in width. Hemmingsen (1973) reported that majors of *O. smaragdina* perform the weaving toward the exterior, while minor workers—those with heads 1–1.2 mm wide—weave on the inner surfaces of the leaf cavities. We have observed only major workers performing the task in *O. longinoda*, but admittedly our studies of interior activity have been limited. Hemmingsen also recorded that exterior weaving is rare during the daytime but increases sharply at night, at least in the case of *O. smaragdina* working outdoors in Thailand. We have seen frequent exterior weaving by *O. lon-*

ginoda during the day in a well-lit laboratory, as well as by *O. smaragdina* outdoors in Queensland.

In recent studies reported here for the first time, we followed the spinning process of *O. longinoda* through a frame-by-frame analysis of 16-mm motion pictures taken at 25 frames per second. The most distinctive feature of the larval behavior, other than the release of the silk itself, is the rigidity with which the larva holds its body. There is no sign of the elaborate bending and stretching of the body or of the upward thrusting and side-to-side movements of the head that characterize cocoon-spinning in other formicine ant larvae, particularly in *Formica* (Wallis 1960; Schmidt and Gürsch 1971). Rather, the larva keeps its body stiff, forming a straight line when viewed from above but a slightly curved, **S**-shaped line when seen from the side, with its head pointing obliquely downward as shown in Figure 4. Occasionally the larva extends its head for a very short distance when it is brought near the leaf surface, giving the impression that it is orienting itself more precisely at the instant before it releases the silk. The worker holds the larva in its mandibles between one-fourth and one-third of the way down the larva's body from the head, so that the head projects well out in front of the worker's mandibles.

The antennae of the adult workers are of an unusual conformation that facilitates tactile orientation along the edges of leaves and other vegetational surfaces. The last four segments are shorter relative to the eight segments closest to the body than in other ants we have examined, including even communal silk-spinning formicines such as *Camponotus senex* and *Polyrhachis acuta*. They are also unusually flexible and can be actively moved in various directions in a fashion seen in many solitary wasps.

As the worker approaches the edge of a leaf with a larva in its mandibles, the tips of the antennae are brought down to converge on the surface in front of the ant. For 0.2 ± 0.1 sec ($\bar{x} \pm$ SD, n = 26, involving a total of 4 workers), the antennae play along the surface, much in the manner of a blindfolded person feeling the edge of a table with his hands. Then the larva's head is touched to the surface and held in contact with

Figure 3. The nest of the African *Oecophylla longinoda*, the most sophisticated in design of weaver ants' nests, is formed basically of living leaves and stems bound together with larval silk. Some of the walls and galleries are constructed entirely of the silk.

the leaf for 1 sec (0.9 ± 0.2 sec, n = 26). During this time, the tips of the worker's antennae are vibrated around the larva's head, stroking the leaf surface and touching the larva's head about 10 times (9.2 ± 3.6, n = 26). At some point the larva releases a minute quantity of silk, which attaches to the leaf surface.

About 0.2 sec before the larva is lifted up again, the worker spreads and raises its antennae. Then it carries the larva directly to the edge of the other leaf, causing the silk to be drawn out as a thread. While moving between leaves, the worker holds its antennae well away from the head of the larva. When it reaches the other leaf, it repeats the entire procedure exactly, except that the larva's head is held to the surface for only 0.5 sec (0.4 ± 0.01 sec, n = 26); during this phase the worker's antennae touch the larva about 5 times (5.2 ± 2.4, n = 26). In other words, the workers alternate between a longer time spent at one leaf surface and a shorter time at the opposing surface.

To summarize, the weaving behavior of the *Oecophylla* worker is even more complicated, precise, and distinctive than realized by earlier investigators. The movements are rigidly stereotyped in form and sequence. The antennal tips are used

for exact tactile orientation, a "topotaxis" somewhat similar to that employed by honeybee workers to assess the thickness of the waxen walls of the cells in the comb (Lindauer and Martin 1969). The worker ant also appears to use its flexible antennal tips to communicate with the larva, presumably to induce it to release the silk at the right moment. Although we have no direct experimental proof of this effect, we can report an incidental observation consistent with it. One worker we filmed held the larva upside down, so that the front of the larva's head and its silk-gland openings could not touch the surface or be stroked by the antennal tips. The worker went through the entire sequence correctly, but the larva did not release any silk.

For its part, the larva has evolved distinctive traits and behaviors that serve communal weaving. It releases some signal, probably chemical, that identifies it as being in the correct phase of the final instar. When a worker picks it up, it assumes an unusual **S**-shaped posture. And when it is held against the surface of a leaf and touched by a worker's antennae, it releases silk, in a context and under circumstances quite out of the ordinary for most immature insects.

Figure 4. Once leaves have been pulled together to form a nest, the workers hold them in place with larval silk. A simple form of weaving is practiced by the workers of an Australian *Polyrhachis* species similar to *doddi*; at the top, one worker holds a larva above the surface, allowing it to perform most of the weaving movements. The most sophisticated type of weaving has been developed by *O. smaragdina*; in the bottom photograph, *O. smaragdina* workers perform almost all the movements while the larvae serve principally as passive shuttles.

Intermediate steps

The existence of communal nest-weaving in *Polyrhachis* was discovered in the Asiatic species *Polyrhachis* (*Myrmhopla*) *dives* by Jacobson (Jacobson and Wasmann 1905). However, few details of the behavior of these ants have been available until a recent study by Hölldobler, reported here for the first time.

A species of *Polyrhachis* (*Cyrtomyrma*), tentatively classified near *doddi*, was observed in the vicinity of Port Douglas, Queensland, where its colonies are relatively abundant. The ants construct nests among the leaves and twigs of a wide variety of bushes and trees (Fig. 5). Most of the units are built between two opposing leaves, but often only one leaf serves as a base or else the unit is entirely constructed of silk and is well apart from the nearest leaves.

Polyrhachis ants have never been observed to make chains of their own bodies or to line up in rows in the manner routine for *Oecophylla*. Occasionally a single *Polyrhachis* worker pulls and slightly bends the tip or edge of a leaf, but ordinarily the leaves are left in their natural position and walls of silk and debris are built between them.

The weaving of *Polyrhachis* also differs markedly from that of *Oecophylla*. The spinning larvae are considerably larger and appear to be at or near the end of the terminal instar (Fig. 4). The workers hold them gently from above, somewhere along the forward half of their body, and allow the larvae to perform all of the spinning movements. In laying silk on the nest wall, the larvae use a version of the cocoon-spinning movements previously observed in the larvae of *Formica* and other formicine ants. Like these more "typical" species, which do not engage in communal nest-building, *Polyrhachis* larvae begin by protruding and retracting the head relative to the body segments while bending the forward part of the body downward. Approximately this much movement is also seen in *Oecophylla* larvae prior to their being touched to the surface of a leaf.

The *Polyrhachis* larvae are much more active, however, executing most of the spinning cycle in a sequence very similar to that displayed by cocoon-spinning formicines. Each larva begins with a period of bending and stretching, then returns to its original position through a series of arcs directed alternately to the left and right; in sum, its head traces a rough figure eight. Because the larvae are held by the workers, the movements of their bodies are restricted. They cannot complete the "looping-the-loop" and axial rotary movements described by Wallis (1960), by which larvae of other formicine ants move around inside the cocoon to complete its construction. In fact, the *Polyrhachis* larvae do not build cocoons. They pupate in the naked state, having contributed all their expelled silk to the communal nest. In this regard they fall closer to the advanced *Oecophylla* grade than to the primitive *Dendromyrmex* one, discussed below.

Polyrhachis ants are also intermediate between *Oecophylla* and *Dendromyrmex* in another important respect. The *Polyrhachis* workers do not move the larvae constantly like living shuttles as in *Oecophylla*, nor do they hold the larvae in one position for long periods of time or leave them to spin on their own as in *Dendromyrmex*. Rather, each spinning larva is held by a worker in one spot or moved slowly forward or to the side for a variable period of time (range 1–26 sec, mean 8 sec, SD 7.1 sec, n = 29). After each such brief episode the larva is lifted up and carried to another spot inside the nest, where it is permitted to repeat the stereotyped spinning movements. While the larva is engaged in spinning, the worker touches the substrate, the silk, and the front half of the larva's body with its antennae. However, these antennal movements are less stereotyped than in *Oecophylla*.

The product of this coordinated activity is an irregular, wide-meshed network of silk extending throughout the nest. The construction usually begins with the attachment of the silk to the edge of a leaf or stem. As the spinning proceeds, some workers bring up small particles of soil and bark, wood chips, or dried leaf material that the ants have gathered on the ground below. They attach the detritus to the silk, often pushing particles into place with the front of their heads, and then make the larvae spin additional silk around the particles to secure them more tightly to the wall of the nest. In this way a sturdy outside shell is built, consist-

ing in the end of several layers of silk reinforced by solid particles sealed into the fabric. The ants also weave an inside layer of pure silk, which covers the inner face of the outer wall and the surfaces of the supporting leaves and twigs. Reminiscent of wallpaper, this sheath is thin, very finely meshed, and tightly applied so as to follow the contours of the supporting surface closely. When viewed from inside, the nest of the *Polyrhachis* ant resembles a large communal cocoon (Fig. 5).

A very brief description of the weaving behavior of *Polyrhachis*

(*Myrmhopla*) *simplex* by Ofer (1970) suggests that this Israeli species constructs nests in a manner similar to that observed in the Queensland species. The genus *Polyrhachis* is very diverse and widespread, ranging from Africa to tropical Asia and the Solomon Islands. Many of the species spin communal nests, apparently of differing degrees of complexity, and further study of their behavior should prove very rewarding.

A second intermediate grade is represented by *Camponotus* (*Myrmobrachys*) *senex*, which occurs in moist forested areas of South and Central

Figure 5. The nest of the Australian *Polyrhachis* species (*top*) is at an intermediate level of complexity, consisting of sheets of silk woven between leaves and twigs and reinforced by soil and dead vegetable particles. The interior of this type of nest (*bottom*) has a layer of silk tightly molded to the supporting leaf surfaces.

America. It is one of only two representatives of the very large and cosmopolitan genus *Camponotus* known to incorporate larval silk in nest construction (although admittedly very little information is available about most species of this genus), and in this respect must be regarded as an evolutionarily advanced form. The most complete account of the biology of *C. senex* to date is that of Schremmer (1972, 1979a, b).

Unlike the other weaver ants, *C. senex* constructs its nest almost entirely of larval silk. The interior of the nest is a complex three-dimensional maze of many small chambers and connecting passageways. Leaves are often covered by the silken sheets, but they then die and shrivel, and thereafter serve as no more than internal supports. Like the Australian *Polyrhachis*, *C. senex* workers add small fragments of dead wood and dried leaves to the sheets of silk along the outer surface. The detritus is especially thick on the roof, where it serves to protect the nest from direct sunlight and rain.

As Schremmer stressed, chains of worker ants and other cooperative maneuvers among workers of the kind that characterize *Oecophylla* do not occur in *C. senex*. The larvae employed in spinning are relatively large and most likely are near the end of the final instar. Although they contribute substantial amounts of silk collectively, they still spin individual cocoons—in contrast to both *Oecophylla* and the Australian *Polyrhachis*. Workers carrying spinning larvae

can be most readily seen on the lower surfaces of the nest, where walls are thin and nest-building unusually active. During Schremmer's observations they were limited to the interior surface of the wall and consequently could be viewed only through the nascent sheets of silk. Although numerous workers were deployed on the outer surface of the same area at the same time, and were more or less evenly distributed and walked slowly about, they did not carry larvae and had no visible effect on the workers inside. Their function remains a mystery. They could in fact be serving simply as guards.

Although Schremmer himself chose not to analyze the weaving behavior of *C. senex* in any depth, we have been able to make out some important details from a frame-by-frame analysis of his excellent film (Schremmer 1972). In essence, *C. senex* appears to be very similar to the Australian *Polyrhachis* in this aspect of their behavior. Workers carry the larvae about slowly, pausing to hold them at strategic spots for extended periods. They do not contribute much to the contact between the heads of the larvae and the surface of the nest. Instead, again as in *Polyrhachis*, the larvae perform strong stretching and bending movements, with some lateral turning as well. When held over a promising bit of substrate, larvae appear to bring the head down repeatedly while expelling silk. We saw one larva perform six "figure eight" movements in succession, each time touching its

head to the same spot in what appeared to be typical weaving movements. The duration of the contact between its head and the substrate was measured in five of these cycles; the range was 0.4–1.5 sec and averaged 0.8 sec. During the spinning movements the workers play their antennae widely over the front part of the body of the larva and the adjacent substrate.

The nest-weaving of *C. senex*, then, is the same as that of the Australian *Polyrhachis*. The only relevant difference between the two is that *C. senex* larvae construct individual cocoons and *Polyrhachis* larvae do not.

The simplest type of weaving

A recent study of the tree ants *Dendromyrmex chartifex* and *D. fabricii* has revealed a form of communal silk-weaving that is the most elementary conceivable (Wilson 1981). The seven species of *Dendromyrmex* are concentrated in Brazil, but at least two species (*chartifex* and *fabricii*) range into Central America. The small colonies of these ants build oblong carton nests on the leaves of a variety of tree species in the rain forest (Weber 1944).

The structure of the nests is reinforced with continuous sheets of larval silk (Fig. 6). When the nest's walls are deliberately torn to test their strength, it can be seen that the silk helps hold the carton together securely. Unlike *Oecophylla* larvae, those of *Dendromyrmex* contribute silk only at the end of the final instar, when they are fully grown and ready to pupate. Moreover, only part of the silk is used to make the nest. Although a few larvae become naked pupae, most enclose their own bodies with cocoons of variable thickness. Workers holding spinning larvae remain still while the larvae perform the weaving movements; in *Oecophylla*, the larvae are still and the workers move. Often the larvae add silk to the nest when lying on the surface unattended by workers. Overall, their nest-building movements differ from those of cocoon-spinning only by a relatively small change in orientation. And, not surprisingly, this facultative communal spinning results in a smaller contribution to the structure of the nest than is the case in *Oecophylla* and other advanced weaver ants.

Figure 6. *Dendromyrmex chartifex*, of Central and South America, makes the simplest type of woven nest, a carton-like structure of chewed vegetable fibers reinforced with larval silk. (Photograph from Wilson 1981.)

Anatomical changes

The behavior of communally spinning ant larvae is clearly cooperative and altruistic in nature. If general notions about the process of evolution are correct, we should expect to find some anatomical changes correlated with the behavioral modifications that produce this cooperation. Also, the degree of change in the two

kinds of traits should be correlated to some extent. And finally, the alterations should be most marked in the labial glands, which produce the silk, and in the external spinning apparatus of the larva.

These predictions have generally been confirmed. *Oecophylla*, which has the most advanced cooperative behavior, also has the most modified external spinning appara-

tus. The labial glands of the spinning larvae of *Oecophylla* and *Polyrhachis* are in fact much larger in proportion to the size of the larva's body than is the case in other formicine ant species whose larvae spin only individual cocoons (Karawajew 1929; Wilson and Hölldobler 1980). On the other hand, *C. senex* larvae do not have larger labial glands than those of other *Camponotus* larvae. Schrem-

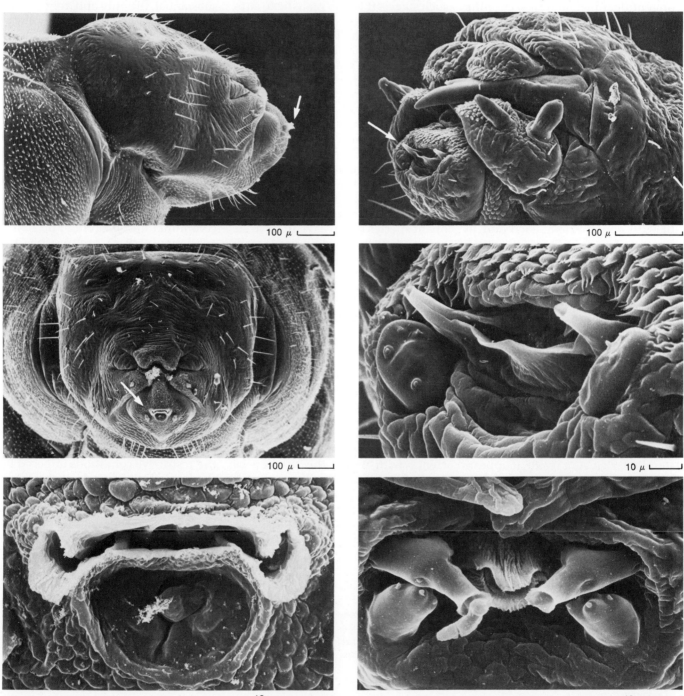

Figure 7. Scanning electron micrographs reveal adaptations in the spinning apparatus of ant larvae in *Oecophylla*. At the left, the head of an *O. longinoda* larva is shown from the side (*top*) and front (*middle*); the arrows indicate the slit-shaped opening of the silk glands, which is modified substantially from the more primitive forms at the right. The reduced lateral nozzles in *O. longinoda* and the larger central nozzle are clearly visible at the bottom left. In *Nothomyrmecia macrops*, a living Australian ant thought to be similar to the earliest formicines, there is no central nozzle and the lateral nozzles are much more prominent; the arrow at the top right points to the area enlarged at the middle right. The silk-gland opening of the Australian weaver ant, a species of *Polyrhachis* (*bottom right*), is similar in structure to that in *Nothomyrmecia*. (Micrographs by Ed Seling.)

mer (1979a) tried to fit this surprising result into the expected pattern by suggesting that the *C. senex* larvae produce silk for longer periods of time than other species that weave nests communally, and therefore do not need larger glands. This hypothesis has not yet been tested.

Until recently, little was known about the basic structure of the spinning apparatus of formicines. Using conventional histological sectioning of larvae in the ant genera *Formica* and *Lasius*, Schmidt and Gürsch (1970) concluded that the silk glands open to the outside by three tube-like projections, or nozzles. They were indeed able to pull three separate silken strands away from the heads of larvae with forceps. However, our studies, which combine histology and the use of the scanning electron microscope, have led us to draw a somewhat different picture.

In general, the labial gland opens to the outside through a small slit with one nozzle at each end, as shown in Figure 7. This is the structure found in the Australian *Nothomyrmecia macrops*, which is considered to be the living species closest to the ancestors of the Formicinae, as well as in a diversity of formicines themselves. Among the formicines examined, including those engaged in communal nest-weaving, only *Oecophylla* has a distinctly different external spinning apparatus. The labial-gland slit of these extremely advanced weaver ants is enlarged into a single nozzle, incorporating and largely obliterating the lateral nozzles. As a result, it appears that each *Oecophylla* larva is capable of expelling a broad thread of silk—the kind of thread needed to create the powerful webs binding an aboreal nest together.

The uncertain climb toward cooperation

In order to summarize existing information on the evolution of communal spinning, the grades in Table 1 are defined according to the presence or absence of particular traits associated with communal nest-weaving. We believe that it is both realistic and useful to recognize three such stages. It is also realistic to suppose that the most advanced weaver ants, those of the genus *Oecophylla*, are derived from lines that passed through lower grades similar to, if not identical with, those exemplified by *Dendromyrmex*, *Polyrhachis*, and *Camponotus senex*.

On the other hand, we find it surprising that communal nest-weaving has arisen only four or so times during the one hundred million years of ant evolution. Even if new cases of this behavior are discovered in the future, the percentage of ant species that weave their nests communally will remain very small. It is equally puzzling that the most advanced grade was attained only once. The separate traits of *Oecophylla* nest-weaving provide seemingly clear advantages that should predispose arboricole ants to evolve them. The remarkable cooperative maneuvers of the workers allow the colony to arrange the substrate in the best positions for the addition of the silk bonds and sheets. By taking over control of the spinning movements from the larvae, the workers enormously increase the speed and efficiency with which the silk can be applied to critical sites. For their part the larvae have benefited the colony by moving the time when they produce silk forward in the final instar, thus surrendering once and for all

the ability to construct personal cocoons but allowing workers to carry and maneuver them more effectively because of their smaller size.

The case of *Dendromyrmex* is especially helpful in envisioning the first steps of the evolution in behavior that culminated in the communal nest-weaving of *Oecophylla*. Although the contribution of the larvae to the structure of the nest is quite substantial, the only apparent change in their behavior is a relatively slight addition to their normal spinning cycle, so that the larva releases some silk onto the floor of the nest while weaving its individual cocoon. It is easy to imagine such a change occurring with the alteration of a single gene affecting the weaving program. Thus, starting the evolution of a population toward communal weaving does not require a giant or otherwise improbable step.

There is another line of evidence indicating the general advantage of communal nest-weaving and hence a relative ease of progression. We discovered that both male and female larvae contribute silk to the nest in the case of *Oecophylla* (Wilson and Hölldobler 1980) and *Dendromyrmex* (Wilson 1981); male contribution has not yet been investigated in *Polyrhachis* and *Camponotus*. Because cooperation and altruism on the part of male ants is rare, it is always worthy of close examination. Bartz (1982) has recently shown that in social Hymenoptera, natural selection will favor the evolution of either male workers or female workers, but not both, and the restrictive conditions imposed by the haplodiploid mode of sex determination—used by all Hymenoptera—favor all-female worker castes. In fact, the sterile workers of hymenopterous societies are always female

Table 1. Grades of communal nest-weaving

	Larvae contribute silk to nest	Workers always hold spinning larvae	Larvae no longer make individual cocoons	Workers repeatedly move larvae	Workers cooperate in adjusting substrate	Workers perform most spinning movements	Silk is produced before end of final instar
Grade 1							
Dendromyrmex spp.	+	−	−	−	−	−	−
Grade 2							
Polyrhachis ?doddi	+	+	+	+	−	−	−
Camponotus senex	+	+	−	+	−	−	−
Grade 3							
Oecophylla spp.	+	+	+	+	+	+	+

(Oster and Wilson 1978). In boreal carpenter ants of the genus *Camponotus*, where the males do contribute some labor to the colony, it is in the form of food-sharing, an apparent adaptation to the lengthy developmental cycle of *Camponotus*. The males are kept in the colonies from late summer or fall to the following spring, and it benefits both the colony and the individual males to exchange liquid food (Hölldobler 1966).

The contribution of silk by male weaver-ant larvae is a comparable case. When the queens of *Oecophylla* and *Dendromyrmex* die, some of the workers lay eggs, which produce males exclusively (Hölldobler and Wilson 1983). Such queenless colonies can last for many months, until the last of the workers have died. During this period it is clearly advantageous for male larvae to add silk to the nest, for their own survival as well as that of the colony as a whole.

In summary, then, weaver ants exemplify very well an important problem of evolutionary theory: why so many intermediate species possess what appear to be "imperfect" or at least mechanically less efficient adaptations. Two hypotheses can be posed to explain the phenomenon that are fully consistent with the manifest operation of natural selection in such cases. The first is that some species remain in the lower grades because countervailing pressures of selection come to balance the pressures that favor the further evolution of the trait. In particular, the tendency for larvae to collaborate in the construction of nests could be halted or even reversed in evolution if surrendering the ability to make cocoons reduces the larvae's chance of survival. In other words, the lower grade might represent the optimum compromise between different pressures.

The second, quite different hypothesis is that the communal weavers are continuing to evolve—and will eventually attain or even surpass the level of *Oecophylla*—but species become extinct at a sufficiently high rate that most such evolutionary trends are curtailed before they are consummated. Even a moderate frequency of extinction can result in a constant number of species dispersed across the various evolutionary grades.

At present we see no means of choosing between these two hypotheses or of originating still other, less conventional evolutionary explanations. The greatest importance of phenomena such as communal nest-weaving may lie in the prospects they offer for a deeper understanding of arrested evolution, the reasons why not all social creatures have attained what from our peculiar human viewpoint we have chosen to regard as the pinnacles of altruistic cooperation.

References

Baroni Urbani, C. 1978. Materiali per una revisione dei *Leptothorax* neotropicali appartenenti al sottogenere *Macromischa* Roger, n. comb. (Hymenoptera: Formicidae). *Entomol. Basil.* 3:395–618.

Bartz, S. H. 1982. On the evolution of male workers in the Hymenoptera. *Behav. Ecol. Sociobiol.* 11:223–28.

Chauvin, R. 1952. Sur la reconstruction du nid chez les fourmis Oecophylles (*Oecophylla longinoda* L.). *Behaviour* 4:190–201.

Doflein, F. 1905. Beobachtungen an den Weberameisen (*Oecophylla smaragdina*). *Biol. Centralbl.* 25:497–507.

Hemmingsen, A. M. 1973. Nocturnal weaving on nest surface and division of labour in weaver ants (*Oecophylla smaragdina* Fabricius, 1775). *Vidensk. Meddr. Dansk Naturh. Foren.* 136:49–56.

Hölldobler, B. 1966. Futterverteilung durch Männchen im Ameisenstaat. *Z. Vergl. Physiol.* 52:430–55.

———. 1979. Territories of the African weaver ant (*Oecophylla longinoda* [Latreille]): A field study. *Z. Tierpsychol.* 51:201–13.

Hölldobler, B., and E. O. Wilson. 1977a. Weaver ants. *Sci. Am.* 237:146–54.

———. 1977b. Weaver ants: Social establishment and maintenance of territory. *Science* 195:900–02.

———. 1978. The multiple recruitment systems of the African weaver ant *Oecophylla longinoda* (Latreille) (Hymenoptera: Formicidae). *Behav. Ecol. Sociobiol.* 3:19–60.

———. 1983. Queen control in colonies of weaver ants (Hymenoptera: Formicidae). *Ann. Entomol. Soc. Am.* 76:235–38.

Jacobson, E., and A. Forel. 1909. Ameisen aus Java und Krakatau beobachtet und gesammelt von Herrn Edward Jacobson, bestimmt und beschrieben von Dr. A. Forel. *Notes Leyden Mus.* 31:221–53.

Jacobson, E., and E. Wasmann. 1905. Beobachtungen ueber *Polyrhachis dives* auf Java, die ihre Larven zum Spinnen der Nester benutzt. *Notes Leyden Mus.* 25:133–40.

Karawajew, W. 1929. Die Spinndrüsen der Weberameisen (Hym. Formicidae). *Zool. Anz.* (Wasmann Festband) 1929:247–56.

Ledoux, A. 1950. Recherche sur la biologie de la fourmi fileuse (*Oecophylla longinoda* Latr.). *Ann. Sci. Nat.* (Paris), ser. 11, 12:313–416.

Lindauer, M., and H. Martin. 1969. Special sensory performances in the orientation of the honey bee. In *Theoretical Physics and Biology*, ed. M. Marois, pp. 332–38. North-Holland.

Ofer, J. 1970. *Polyrhachis simplex*: The weaver ant of Israel. *Ins. Soc.* 17:49–81.

Oster, G. F., and E. O. Wilson. 1978. *Caste and Ecology in the Social Insects*. Princeton Univ. Press.

Schmidt, G. H., and E. Gürsch. 1970. Zur Struktur des Spinnorgans einiger Ameisenlarven (Hymenoptera, Formicidae). *Z. Morph. Tiere* 67:172–82.

———. 1971. Analyse der Spinnbewegungen der Larve von *Formica pratensis* Retz. (Form. Hym. Ins.). *Z. Tierpsychol.* 28:19–32.

Schremmer, F. (prod.). 1972. *Die südamerikanische Weberameise* Camponotus senex (*Freilandaufnahmen*). 16 mm film. Distributed by Inst. Wissenschaft, Göttingen. Film no. W1161.

———. 1979a. Die nahezu unbekannte neotropische Weberameise *Camponotus* (*Myrmobrachys*) *senex* (Hymenoptera: Formicidae). *Ent. Gen.* 5:363–78.

———. 1979b. Das Nest der neotropischen Weberameise *Camponotus* (*Myrmobrachys*) *senex* Smith (Hymenoptera, Formicidae). *Zool. Anz.* 203:273–82.

Sudd, J. H. 1963. How insects work in groups. *Discovery* (London), June, 15–19.

Wallis, D. I. 1960. Spinning movements in the larvae of the ant, *Formica fusca*. *Ins. Soc.* 7: 187–99.

Weber, N. A. 1944. The tree-ants (*Dendromyrmex*) of South and Central America. *Ecology* 25:117–20.

Wheeler, W. M. 1914. The ants of the Baltic amber. *Schrift. Phys.-Ökon. Ges. Königsberg* 55:1–142.

Wilson, E. O. 1971. *The Insect Societies*. Harvard Univ. Press.

———. 1981. Communal silk-spinning by larvae of *Dendromyrmex* tree-ants (Hymenoptera: Formicidae). *Ins. Soc.* 28:182–90.

Wilson, E. O., and B. Hölldobler. 1980. Sex differences in cooperative silk-spinning by weaver ant larvae. *PNAS* 77:2343–47.

Wilson, E. O., and R. W. Taylor. 1964. A fossil ant colony: New evidence of social antiquity. *Psyche* (Cambridge) 71:93–103.

Stingless-Bee Communication

Searching for a proto-dance language reveals possible stages in the evolution of methods by which experienced foragers lead others to food

James C. Nieh

At the close of World War I in the spring of 1919, an Austrian scientist, Karl von Frisch, sat in a former Jesuit cloister feeding honeybees. He had just designed a glass-walled observation hive, and, as he wrote in *The Dance Language and Orientation of Bees*, he was delighted to observe one of his returning foragers performing "a round dance in which the … bees sitting nearby showed lively interest. They tripped along after the dancer, and then left the hive to hasten to the feeding station." Scientists as early as Aristotle had alluded to this recruitment behavior, in which one bee finds a source of food and then advertises its location to nestmates. Nevertheless, von Frisch was the first systematically to study it and crack its code.

The discovery of the honeybee's so-called waggle dance served as the core of von Frisch's achievements. In this dance, a returning forager makes a short, straight run, during which it performs a waggling motion, and then circles off to the right or left to make another straight run. This dance involves multiple sensory cues that communicate the distance and direction of a food source. It also provided the first

James C. Nieh is a Harvard Junior Fellow in the Department of Organismic and Evolutionary Biology at Harvard University. He earned his B.A. in 1991 at Harvard and his Ph.D. in 1997 at Cornell University in the Section of Neurobiology and Behavior. The stingless-bee research reported in this paper was conducted as part of his doctoral dissertation. He held a postdoctoral fellowship from 1997–1998 at the University of Würzburg in the Lehrstuhl für Verhaltensphysiologie und Soziobiologie. His research focuses on honeybee and stingless-bee foraging communication systems. Address: Museum of Comparative Zoology, Harvard University, Cambridge, MA 02138. Internet: nieh@oeb.harvard.edu

animal example of sophisticated representational communication: the ability to encode and transmit spatial information. Although Adrian Wenner of the University of California at Santa Barbara later challenged many of these claims, subsequent research showed that honeybee recruits are able to perceive and use information provided in the waggle dance to find a food source at the correct distance and direction.

Finding this dance in just one species of bee intrigued von Frisch, and he wondered how such a complex system had evolved. Therefore, von Frisch urged his student Martin Lindauer, later of the Universität Würzburg, to study other honeybee species and the stingless bees. In 1985 Lindauer quoted von Frisch as saying, "we cannot believe that the bee dance of the European bees has come from heaven as it is and since the Indian honeybees and the stingless bees there live in a more primitive social organization, we should expect some phylogenetically primitive stages of the bee dance." Lindauer discovered that the dwarf honeybee, *Apis florea*, performs waggle dances on the horizontal surface of its comb, directly orienting toward the sun. Nevertheless, this was the most primitive variant, and it could not explain how the waggle dance evolved.

Lindauer's work on the stingless bees proved more fruitful and is the starting point of my story. The stingless bees are a large monophyletic group of more than 450 species spread through 30 to 50 genera. Although their greatest diversity is in the Neotropics, they are also found in Africa, Asia and Australia. Consequently, they probably evolved prior to the breakup of Gondwanaland. Stingless bees attracted von Frisch's attention because they were

thought to be the most closely related sister group to honeybees. Although this relationship is now uncertain, stingless bees and honeybees share much in common because they are the only highly social bees.

Lindauer and Warwick Kerr of the Universidade de São Paulo discovered a wide variety of communication systems among stingless bees, ranging from random search strategies, which do not communicate food's location, to systems employing scent trails and undirected excitatory "dance" motions inside the nest. In 1965, Harald Esch of the University of Notre Dame discovered that recruiting foragers in stingless bees of the genus *Melipona* produce longer sound pulses inside the nest for food sources that are more distant from it. A similar correlation between sound pulses and distance was known in honeybees, so von Frisch's hope of finding a proto-dance language seemed closer to fulfillment.

Unfortunately, the stingless-bee discoveries failed to ignite further research and laid dormant for decades, despite raising intriguing new questions. Results from recent work, however, should renew interest in stingless bees. As I shall show, one species of stingless bee uses a wide range of communication techniques, from simply watching where other bees go to encoding information in sounds. A careful examination of this behavior suggests how food-recruitment communication might have evolved.

Direction, Distance and Height

In 1991, my doctoral advisor at Cornell University, Tom Seeley, suggested that I take up this story and carefully analyze the communication system of a stingless bee. I chose the *Melipona* because

Ed Ross

of Esch's previous work, and I went to Panama to collaborate with David Roubik of the Smithsonian Tropical Research Institute. Confident that I would be able to record recruitment sounds, I came during the Panamanian dry season armed with a videocamera and microphones to record behavior in Roubik's observation colonies. For my experiments, I selected *Melipona panamica* because it is one of the most common species of *Melipona* and is relatively unaggressive.

Although I spent a long, hot month working in Panama, little came from my efforts. I could barely train bees to my artificial feeder, observed scarcely any recruitment behavior even for natural food sources and discovered that intriguing sounds or movements were generally concealed inside a waxy funnel built by the bees from the nest entrance to the food-storage pots. Any attempts to remove this funnel disrupted foraging. Foragers were also extremely sensitive to light and air currents. Consequently, I could not easily follow recruiting foragers with a moving microphone as I had done before with honeybees. The colony needed to be sealed and placed under special lighting. A month of frustration convinced me that I could not take a shortcut ap-

proach. I needed to choose a time when I could train bees to artificial feeders and to resolve several technical issues in order to observe recruitment behavior inside a nest.

On my next trip, I came at the end of the rainy season, when the fewest natural food sources are available. In addition, I had designed a new observation nest that consisted of a chamber for the central brood comb, a smaller chamber for honey- and pollen-storage pots and an unloading platform—a flattened, triangular chamber that mimicked the natural nest's waxy in-

Figure 1. Stingless bees *(above)* are a large category of insects encompassing more than 450 species spread through 30 to 50 genera. Like honey bees, stingless bees are highly social, and many species are able to communicate the locations of food sources. Foragers of *Melipona panamica*, the subject of the author's study *(left)*, lead others to sources of food through various forms of communication, including sounds, smells and visual cues. By testing food-source locations that vary in height, distance or direction, and then studying the resulting behavior by bees, the author reveals that these bees use a variety of signals to communicate where other bees in the nest can find food. In addition, his work suggests how food-recruitment communication might have evolved. (Except where noted, all photographs by the author.)

ternal funnel. I covered all the chambers with glass and illuminated the nest with a fiber-optic lamp to localize the lighting and reduce the effect of lamp heat. To my surprise, the new observation nest worked well. Bees unloaded food and danced normally on the unloading platform. Microphones inserted along the sides of the unloading platform allowed me to record recruitment sounds.

My first goal was to determine if *Melipona panamica* foragers could actually communicate food location. To test for direction, David Roubik and I

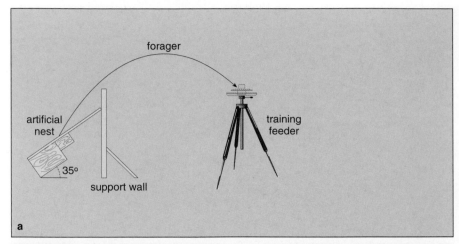

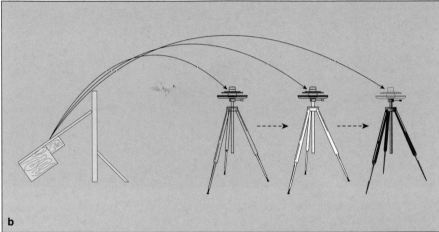

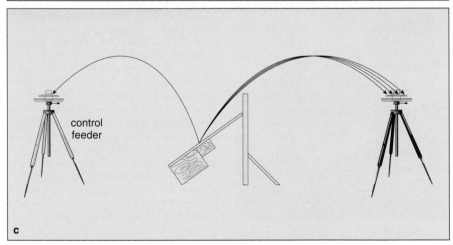

Figure 2. Foragers communicate the direction to a food source to other bees. The author revealed this by training bees from an artificial nest to visit a nearby training feeder *(red)*, which contained sugar water *(a)*. Foragers visited the feeder and then returned to the nest to recruit other bees. Next, the author slowly increased the distance between the training feeder and the nest *(b)*. Once the bees had been trained to the desired distance, the investigators placed a control feeder *(gray)* at the same distance but in the opposite direction *(c)*. Nearly all inexperienced foragers visited the training feeder, indicating that experienced foragers had communicated its direction.

placed two sucrose-solution feeders equal distances from the nest but in opposite directions. We trained bees to a feeder by placing it close to the nest, waiting for bees to feed and then moving it in small steps toward a goal location. All bees landing on this feeder were marked with unique tags or paint dabs. Once this training feeder reached its final site, we set up a control feeder with the same scented-sucrose solution but located in the opposite direction.

Assistants immediately captured all bees coming to this control feeder, so that none could return to the nest and communicate its location.

If experienced foragers cannot communicate direction, then newcomers, which can be identified as unmarked bees, should search randomly for a feeder and arrive in equal numbers at both. On the other hand, if experienced foragers can communicate direction, then significantly more newcomers should arrive at the training feeder. In order to control for potential site bias, we also switched or changed the location of the training feeder in half of the trials. Through these experiments, we discovered that these bees can communicate a food source's direction. Related experiments showed that *M. panamica* can also communicate a food source's distance and height. The communication of height was a welcome surprise because von Frisch and Lindauer reported that the European honeybee cannot communicate food-source height.

How Do They Do It?

At this stage, simple mechanisms could account for all of our data. For example, a recruiter might leave a scent trail between the hive and a feeder, so that newcomers could follow it. Newcomers might also simply follow an experienced forager to the food. A recruiter might even mark a feeder with a pheromone. We needed to test these possibilities.

Roubik and I tested the scent-trail idea by training bees across a water gap, where a scent trail would "disappear" by diffusing immediately into the water. By placing a feeder in a canoe, we trained bees to the opposite shore. As is typical in field research, we encountered unexpected problems. For one thing, other animals ate our carefully trained foragers! After finally building a toad- and basilisk lizard–proof enclosure, we managed to train 20 foragers over the water gap. After that, we observed that significantly more newcomers still arrived at the training feeder than at the control feeder. Recruits apparently did not need a scent trail to find the right feeder.

Recruits also did not appear to follow experienced foragers directly to a food source, because recruits and foragers rarely arrived at a feeder together.

To see whether experienced foragers marked a feeder with a pheromone, I

Figure 3. Artificial nest *(left)* allowed the author not only to observe stingless-bee behavior and record the sounds the bees produce but also to capture and tag *(above)* experienced foragers and control which bees left the nest.

trained bees to a feeder 100 meters from the nest and allowed them to feed for several hours. After recruitment had begun, I placed a suction tube over the nest entrance and began capturing all exiting bees. Meanwhile, my assistant captured all remaining bees at the feeder, sealed them in plastic bags and placed an identical but unvisited control feeder 1 meter to the right or the left of the training feeder. Next, I released potential newcomers one by one from the nest. Significantly more newcomers chose the training feeder, even when it was moved to a new location that experienced foragers had never visited. So the experienced foragers had apparently deposited some odor cue on the food source to guide newcomers. Subsequent experiments showed that the scent beacon has a maximum effective radius of 6 to 12 meters from a food source. The source of the scent beacon, however, remains a mystery.

Throughout my previous experiments, I had heard intriguing sounds through the glass cover of the observation hive and saw recruiting foragers running and spinning inside of the nest. Now, I wondered whether these bees could communicate direction, distance or height inside the nest.

To test each of these parameters, I designed experiments that forced newcomers to search for food based only on information received inside the nest. In each experiment, David Roubik and I allowed bees to forage freely for one hour, and then released bees that had never visited a feeder but that had been exposed to returning for-

agers. In the direction experiment, equal numbers of newcomers arrived at both the control and the training feeder. So recruits apparently do not receive directional information inside the nest. Nevertheless, significantly more newcomers did find the training feeder in the height and distance experiments. Consequently, these experiments suggested that foragers communicate direction outside the nest and height and distance inside it.

Trilling and Twisting

After three years of groundwork, I was prepared to revisit the intriguing behaviors of recruiting foragers inside the nest. An experienced forager returning from a rich food source enters the nest and often begins producing loud pulsed sounds while unloading her food to other bees. As she unloads, other bees cluster around her and often hold their antennae over her vibrating wings. Unloading typically lasts for 20 seconds. Just as she stops unloading, one sometimes hears a short series of pulses, like a slow trill, followed by a dance. The dance phase typically lasts 10 seconds and consists of a recruiter making rapid clockwise and counterclockwise turning motions—about one complete turn per second—while continuing to produce pulsed sounds.

There is no correlation between the degree of turning, rate of turning or order of turning and the direction to the food source. Other bees attend to the dancer during the turns and continue to cluster around her, but they do not follow her every turn in the manner of

honeybee followers. Nevertheless, the followers shift their body positions to face the dancer's general location. Occasionally, they beg a food sample from her. Between turns, a particularly excited forager will sometimes run around the nest, contacting other bees and giving them brief food samples. The followers usually do not keep up with her during these excursions.

Analyzing videotapes of recruiters trained to known feeder locations revealed no correlations between dance movements and the direction to a food source. In addition, recruiter sounds did not correlate with a food source's direction. This was not surprising, because previous experiments suggested that these bees communicate direction outside the nest. However, David Roubik and I did find unloading sounds that corresponded to the height of a food source and dance sounds that corresponded to the distance to a food source.

Foragers recruiting for a feeder 40 meters up in the canopy, for example, produced shorter unloading sound pulses than foragers recruiting for a feeder at the base of the same canopy. This difference was particularly clear for the longest unloading pulse per performance. However, David Roubik and I repeated these measurements with the same colony and feeder locations in 1995 and 1996 and were surprised to observe a change in the unloading pulses. Although the higher feeder consistently elicited shorter unloading pulses, the absolute magnitude of these pulses was not fixed. The durations of 1995 unloading pulses were

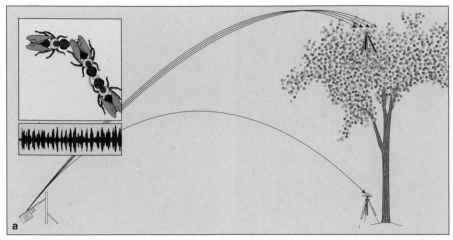

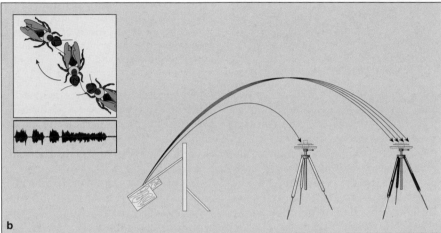

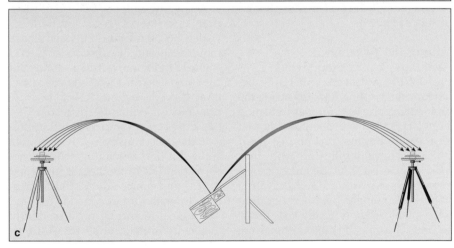

Figure 4. Information about the location of a food source is communicated inside and outside a nest. This can be shown by training bees to a feeder, letting experienced foragers interact with inexperienced bees (ones that have not visited the feeder) inside the nest, and then letting only the inexperienced bees leave the nest and attempt to find a new feeder placed at the same final training location. (The original feeder was removed to eliminate forager-deposited scent marks.) Under these conditions, more of the inexperienced bees find the training feeder *(red)* high in the forest canopy than find the control feeder *(gray)* at the base of the canopy *(a)*. Thus height information is communicated inside the nest, potentially by sounds during the dance phase *(inset panel)*, which involves clockwise and then counterclockwise rotations. If a control feeder is placed between the nest and the training feeder, most of the new bees go to the latter *(b)*. Thus foragers convey distance information inside the nest *(inset panel)*, potentially through sounds made during food unloading. When the author put the control feeder in the opposite direction from the training feeder, however, about equal numbers of new bees arrived at both feeders, indicating that directional information is conveyed outside the nest.

generally longer than the durations of 1996 unloading pulses. Thus a 1996 forager transported back in time to 1995 would mistakenly assume that both feeders were located high in the canopy if she relied solely on unloading pulse durations. Conversely, a 1995 forager transported to 1996 would mistakenly assume that both feeders were located on the forest floor.

Thus the mechanism of height communication presents an intriguing problem. Part of the variation in unloading-pulse durations might come from a poor understanding of what constitutes a good recruitment performance. Recruiters varied in the "strength" of their performances and the number of followers that they attracted. During the seemingly weakest performances, recruiters produced very brief, sporadic sound pulses and did not perform dance turns. Because no one knows precisely what constitutes a strong versus a weak recruitment performance, we collected data on all recruitment performances and thereby averaged the weak and the strong ones. Future studies will focus on teasing out those parameters of the recruitment performance that are most relevant to recruits.

The mechanism of distance communication is clearer. As bees begin making the swift clockwise and counterclockwise turns of the dance phase, they continue to produce pulsed sounds. The duration of these sounds correlates closely with the distance to the food source, with longer pulses indicating greater distances. The longest dance pulse per performance provides the best indication of distance. In fact, in our experiments, distance explained 96 percent of the variance in the average duration of the longest dance pulse. As with the unloading pulses, dance-pulse duration varied considerably for a given food location. However, the variance in dance-pulse duration increased linearly with increasing distance to a food source. In other words, the mechanism was accurate to a fixed percentage of the total distance. If the error was 10 percent, for example, then it would be ±10 meters for a food source 100 meters away and ±100 meters for a feeder 1 kilometer away. Recent studies indicate that this type of error also applies to honeybee distance estimation, but other mechanisms appear to compensate for it so that the absolute error does not increase as the distance to the food source increases.

Because the recruiter apparently communicates both height and distance with sound durations, the recruits must have some way of knowing when she is communicating height and when she is communicating distance. One key might be the relatively motionless state of the recruiter during the unloading phase versus her swift spinning motions during the dance phase. The sounds generated by a bee's vibrating wings are highly directional, so a stationary listener should detect a relatively stable sound field from an unloading bee and a rapidly changing one from a spinning bee. Perhaps this is why follower bees gradually shift their positions to remain close to the dancer but do not closely pursue her as she spins.

The communication of direction remains a mystery. Perhaps recruits gain the directional vector from following the foragers outside the nest for a brief distance. Esch described an intriguing behavior in Brazilian *Melipona* species where recruiters in front of the nest appear to fly in a zigzag pattern pointing in the direction of a food source.

The Evolution of Communication

The initial hope for discovering "missing links" in the form of species with simpler forms of the honeybee-dance language has not been directly fulfilled. The stingless bees might be the closest sister group to the honeybees, but the stingless-bee communication systems probably evolved independently into the wonderful forms and diversity that we see today. Nonetheless, the recruitment communication system of *Melipona panamica* provides fascinating insights into the evolution of sophisticated representational communication systems in highly social bees.

For one thing, both honeybees and *Melipona panamica* foragers use a series of multiply redundant communication channels, what Bert Hölldobler of the Universität Würzburg called "multimodal signals." For example, both honeybees and *Melipona panamica* mark their food sources with scents, and these marks assist orientation near the food source. Long-distance orientation, on the other hand, is guided by cues given at the nest: direction and distance with the honeybee waggle dance and height and distance with the sounds of *Melipona panamica*'s recruit-

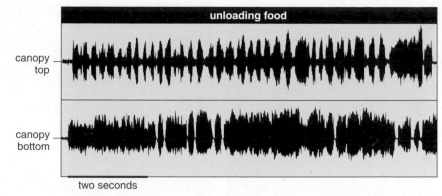

Figure 5. Sounds made during the unloading of food communicate the height of a food source. For example, bees trained to a feeder at the top of the canopy made short unloading pulses, whereas bees trained to a feeder at the bottom of the forest canopy made much longer unloading pulses.

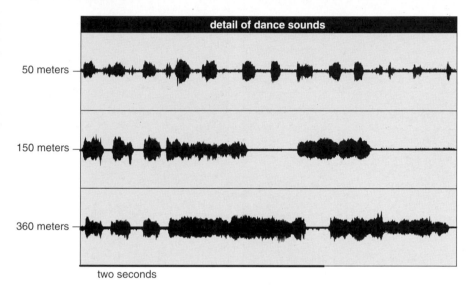

Figure 6. Sounds made during dancing appear to communicate a food source's distance from the nest. The length of sound pulses goes up with increasing distance between the nest and the feeder, as shown here for distances of 50, 150 and 360 meters.

ment performance. Moreover, honeybee and *Melipona panamica* foragers return with the scent of a food source adhering to the their bodies, provide direct food samples and give acoustic cues during their recruitment performances. In other words, both provide olfactory, gustatory, acoustic and perhaps even tactile cues. In some honeybee species, vision might also help followers orient toward dancers. Vision might play a stronger role in *Melipona panamica*, which communicates direction outside the nest, possibly by recruits following the recruiter for a short distance. These findings show that these bees exploit many sensory modalities for the vital task of communicating food-source location. Moreover, multi-modality probably characterizes the recruitment systems of all highly social bees.

Still, one wonders how and why this system has evolved. The manner in which *Melipona panamica* divides up the communication task is revealing. Directly following a forager to the food source is presumably the simplest and most primitive state of a directional communication system. The disadvantages of direct following remain unclear, but they might include problems with tracking a lead forager for long distances, especially in the dim light and dense growth of the forest understory. One backup system found in many stingless bee species is a scent trail, a series of pheromone marks deposited along the substrate. Lindauer and Kerr reported that recruits losing track of an experienced forager would drop close to ground and begin searching for this trail. Yet a scent trail can also guide foragers from other colonies

Figure 7. Food-source communication probably evolved in response to pressures from the local habitat and competitors. In the simplest scenario, an experienced bee *(1)* might lead an inexperienced one *(2)* to a food source. Nevertheless, an obstacle (such as dense foliage) between a nest and food might cause the inexperienced bee to lose track of the leader *(left)*. To solve that problem, bees could lay down a scent trail from a nest to food *(middle)*; some species appear to do this. A scent trail, however, could lead other colonies from the same or other species of bees to the same food source. To solve both the obstacle and competiton problems, bees could communicate a food-source location through information provided inside and near the nest *(right)*, just as *Melipona panamica* does.

and even different species to a food source. Stingless bees face intense inter- and intra-specific competition for limited food sources, and some species apparently specialize on taking over food sources discovered by others. This potential espionage might have driven the evolution of concealed communication at the nest. Once the transfer of spatial information is restricted to the nest, the transfer must become representational. As a result, espionage might have selected for complex representational communication in stingless bees. The potential role of espionage remains speculative for now but is one focus of my current research.

One hypothesis of waggle-dance evolution suggests that living on an open comb was a primitive state in honeybees, and that the waggle dance

ritualized outbound and inbound flights because it was oriented horizontally with the waggle run facing the direction of the food source. The buzzing sounds produced by foragers during the waggle run were thought to have evolved from the sounds produced when foragers warm up their flight muscles prior to take-off. Although the status of open nesting as a primitive state is now uncertain, all honeybees, including cavity nesters, retain the "primitive" ability to orient to a sun cue if the comb is tilted until it becomes a horizontal surface. The simplicity of the hypothesis that the waggle run represents ritualized flight is attractive, despite its speculative nature.

In *Melipona panamica*, the communication of direction outside the nest might, therefore, reflect evolutionary

history, but this remains pure conjecture until comparative data on other *Melipona* species are gathered. Some recent evidence suggests that a few Brazilian *Melipona* species possess a more primitive variant of *Melipona panamica*'s communication system, because a few of these Brazilian species communicate direction well but communicate distance poorly.

We remain at an early stage of understanding *Melipona panamica*'s communication system, and several important questions remain. The most interesting questions concern the basic configuration of the recruitment system. Apart from evolutionary history, what is the advantage of communicating direction outside the nest and height and distance inside the nest? In other words, what are the benefits of a complex symbolic communication system? With *Melipona panamica* alone we cannot resolve the question of which traits are actually primitive. However, this genus contains more than 50 species, and the stingless bees in general contain between 450 to 800 species, which exhibit a broad range of communication complexity. This phylogenetic and behavioral diversity provides the necessary raw material to trace out the evolution of advanced representational recruitment systems. The comparative approach should also enable us to understand how species with simple communication systems can coexist with more sophisticated ones.

As von Frisch observed, "we cannot believe that the bee dance of the European bees has come from heaven," and indeed this representational communication system has now been joined by the communication system of *Melipona panamica*. Nevertheless, two examples out of hundreds of highly social bee species beg the larger question: Why is advanced representational communication so rare? Perhaps we have not yet studied sufficient species, but I believe that the answer to this question might be one of the most important outcomes of the search for a proto-dance language. We could do well by thinking deeply about the precise costs and advantages of such representational communication systems.

Acknowledgements
The author thanks Tom Seeley for his patience, valuable insights and suggestions through the years, David Roubik for introducing him to the mysteries of the stingless-bee world, and his doctoral committee—Kern Reeve, Ron Hoy and Cole Gilbert—for its support and advice. His research would not have been possible without the help of many research assistants: Barrett Klein, Querube Fuenmayor, Elena Schott, Kimberly Denslow, David Stein, Steve Armstrong, Lori Unruh and Diana Nash. He is greatly indebted to the Smithsonian Tropical Research Institute and its staff, especially Oris Acevedo, Daniel Milan, Gloria Maggiori and Georgina DeAlba.

Bibliography

Esch, H., I. Esch and W. E. Kerr. 1965. Sound: an element common to communication of stingless bees and to dances of the honey bee. *Science* 149:320–321.

Lindauer, M. 1985. The dance language of the honeybees: The history of a discovery. *Experimental Behavioral Ecology* 31:129–140.

Lindauer, M., and W. E. Kerr. 1960. Communication between workers of stingless bees. *Bee World* 41:29–41, 65–71.

Nieh, J. C. 1998. The food recruitment dance of the stingless bee, *Melipona panamica. Behavioral Ecology and Sociobiology* 43:133–145.

Nieh, J. C. 1998. The role of pheromones in the communication of food location by the stingless bee, *Melipona panamica. Behavioral Ecology and Sociobiology* 43:47–58.

Nieh, J. C., and D. W. Roubik. 1995. A stingless bee (*Melipona panamica*) indicates food location without using a scent trail. *Behavioral Ecology and Sociobiology* 37:63–70.

Nieh, J. C., and D. W. Roubik. 1998. Potential mechanisms for the communication of height and distance by a stingless bee, *Melipona panamica. Behavioral Ecology and Sociobiology* 43:387–399.

von Frisch, K. 1967. *The Dance Language and Orientation of Bees.* Leigh E. Chadwick, trans. Cambridge: Belknap Press.

Why Do Bowerbirds Build Bowers?

Females prefer to visit courtship areas that provide easy avenues of escape, thereby protecting them from forced copulations

Gerald Borgia

Male bowerbirds of Australia and New Guinea clear and decorate courts and build bowers at display sites where they mate. Bowerbird species, however, differ in several characteristics, including the type and color of court decorations and the form of a bower, if one is even built. Moreover, some male bowerbirds possess bright crest and body plumages, and others do not. Charles Darwin's observations of satin bowerbirds—in the Blue Mountains of Australia during his round-the-world journey on the HMS *Beagle*—contributed to the then-controversial central element of his theory of sexual selection called female choice. The highly sculptured structure of a bower and a male's use of brightly colored decorations suggested to Darwin that female bowerbirds might shop for the most attractive bower, thereby directing the evolution of these display traits.

Nevertheless, several other mechanisms could have driven the evolution of bowers. The so-called good-genes model, for instance, suggests that male-display traits, including bowers, might indicate a male's vigor and, ultimately, his quality as a sire. That is, more vigorous males might have better bowers. A bower could even directly benefit a female, perhaps protecting her from threats, including predators that might attack her during mating or

Gerald Borgia is professor of zoology at the University of Maryland. He holds a long interest in mate choice, which he studied in insects for his doctoral research at the University of Michigan. He began studying bowerbirds 15 years ago, while he was a postdoctoral fellow at the University of Melbourne. This year he and his students will return to his original bowerbird study site for further investigations of how male display affects mate choice. Address: Department of Zoology, University of Maryland, College Park, MD 20742.

males that might try to force her to copulate. Bower building could even arise from an arbitrary or pre-existing female preference, such as an attraction to nest-like structures.

I have used Darwin's method of comparisons of related species to reconstruct the evolution of bower building. My work on several species of bowerbirds confirms the existence of female preferences for males with well-built and highly decorated bowers. The origins of bower building, however, can be best explained as a trait that attracts females because of the protection it provides them from forced copulation by bower owners.

Evaluating Bower-Building Hypotheses
Picking one model of bower-building evolution over another proves difficult because of several problems. One cannot always reconstruct what happened long ago, especially for display behavior that leaves no fossil record. Moreover, bower building may have evolved over a period of time, and different stages of its evolution may have served different functions. Although experiments can show the plausibility of a particular evolutionary process, understanding the origins of traits can best be accomplished by careful comparisons between species whose relationships are known.

Such an analysis depends on accurate and detailed descriptions of bowers and how they are used in courtship in modern species. Obtaining detailed quantitative information on courtship and mating through direct observation proves nearly impossible, because bowers are separated widely, the mating period may last several months and a large proportion of males do not or rarely copulate. Remote-controlled cameras aimed at bowers where males perform their dis-

plays and mate, however, have allowed intensive monitoring of more than 30 bowers for a single species through an entire mating season. That information has provided a direct measurement of male attractiveness and detailed information on how males and females use a bower during courtship. For most of the species that my colleagues and I have studied, we were the first to see these bowerbirds perform successful courtships, which ended in copulation.

By comparing mitochondrial DNA sequences, my colleagues, Robert Kusmierski and Ross Crozier, and I have developed a highly reliable bowerbird phylogeny, which shows evolutionary relationships among species. Our phylogeny indicates, in contrast to some earlier speculation, that all 18 species of bowerbirds evolved from a single ancestral species. Three species branched off from others long ago, and they employ the predominant avian pattern of monogamy with both parents caring for their offspring. All other bowerbirds are polygynous (males mate with more than one female), and they create elaborately decorated display courts. All but two of the polygynous species build bowers. The second major divergence developed between species that build avenue bowers—two vertical stick walls, separated by a central avenue—and those that build maypole bowers—sticks woven around a sapling to create a decorated pillar. It appears that bower building evolved once and then diverged into two types of bowers. The two species that do not build bowers, toothbill and Archbold's bowerbirds, apparently lost bower-building behavior, but they do clear and decorate display courts. Comparisons of bowers and relationships among living species suggest that a

Figure 1. Male bowerbirds build bowers, where they court and mate with females. Some species, such as this satin bowerbird, build avenue bowers, made of two freestanding stick walls. Others build maypole bowers, in which sticks are placed around a central sapling (*Figure 3*). In addition, a male may decorate his bower with a variety of objects, including the pieces of blue plastic shown here. This complex behavior of building and decorating prompts a fundamental question: How did it evolve?

decorated sapling—similar to a simple maypole bower—may represent the ancestral bower type.

Several criteria can be used to evaluate hypotheses for bower evolution based on mating behavior and the evolutionary relationships among species. To the extent that these criteria are met, we can identify the likely initial causes of bower building. First, the proposed function of incipient bowers should be consistent with the design of the supposed ancestral bower. That is, the bower type that appears most consistent with the ancestral bower type should be capable of functioning in accordance with the hypothesized cause of bower origins. Second, the proposed function of the earliest bowers should be consistent with the design of modern bower types. The persistence of bower building among the polygynous species suggests that ancestral functions may remain important. If a consistent function exists for modern bowers, it would be a likely candidate for the ancestral function. Third, species that do not build bowers should possess alternative solutions to the prob-

lem solved by a bower. These species should possess compensatory behaviors, which work in the absence of a bower to protect females from forced copulations by the courting male.

Avenue-Bower Builders

The group of avenue-bower builders consists of three genera and eight species, including the satin bowerbird. Satins inhabit rain forests along the

Figure 2. Inside an avenue bower, a female satin bowerbird observes a male's courtship display. A male flits back and forth across the avenue opening, flicks his wings, mimics the calls of other birds and performs other displays. When a male runs to the rear entrance of the avenue to mate with a female, she either waits to copulate or departs through the front opening.

Figure 3. Macgregor's bowerbird *(left)* **constructs the simplest maypole bower** *(right)*. **A male selects a thin sapling, stacks sticks around it and covers a surrounding display court with a compressed-moss mat. In addition, he decorates the court.**

eastern fringe of Australia. A male aligns his bower along a north-south line, with a display court at the north end. He decorates his display court with blue, yellow and white objects including feathers, flowers, leaves, snail shells and, where available, plastic and paper, over a background of yellow straw. The male trims leaves from above the court, and the northern orientation causes the sun to illuminate the decorated site, perhaps making it more attractive. Males of several species destroy each other's bowers and steal decorations.

A visiting female usually lands in cover south of the bower and then moves rapidly into the avenue between the two stick walls. On the display court, a male makes vocalizations, including guttural chortles and squeaks that progress into a typical call sequence: initial mechanical buzzing followed by mimicking a kookaburra, a Lewin's honeyeater and less frequently a crow. During the buzzing, a male moves swiftly across the northern bower entrance and rapidly flicks one or both wings. When he begins mimicking other birds, he stops at one side of the bower entrance, puffs up his body feathers, holds his wings at his side,

faces the female with a small decoration—usually a yellow leaf—in his mouth and performs a series of knee bends. After that, he usually moves away from the bower, makes several harsh calls and then returns to the bower for more displaying.

In courtships that lead to copulation, a female in the bower avenue crouches deeply as courtship progresses, and a slight lifting of her tail signals her willingness for mating. A male circles around to the opposite end of the bower and mounts her for a three-second copulation. After mating, a female shakes and flaps her wings in or near the bower for a few minutes before leaving. Although a female may visit several bowers, she usually mates with only one male. The average courtship lasts about four minutes. If a male moves to the southern end of a bower before the female is ready, she escapes through the northern exit.

Females exert strong preferences in mating, and only a small proportion of males achieve most of the matings. Males with high quality bowers—with symmetrical walls formed from thin, densely packed sticks—and many decorations on their courts mate most often. Although only nine percent of

satin courtships lead to copulation, the most attractive males mate in 25 percent of their courtships. The higher rate of courtship success by specific males, the significant effect of small decorations and the fine details of bowers on mating success, and the changes in a female's behavior that indicate her readiness to mate after she arrives at the male bower indicate that a female makes her mating decisions after she arrives at the court.

Maypole-Bower Builders

The other major group of bower builders make maypole bowers—a central "pole" surrounded by a circular display court. Some maypole builders cover part of the display court with a hut-like structure. The simplest structure, however, comes from Macgregor's bowerbird, which lives at high elevations in the mountains of central and eastern New Guinea. It decorates a sapling with sticks and moss. This bower may be most similar to the ancestral one for all bowerbird species.

A male Macgregor's bowerbird selects a thin sapling, usually from three to six centimeters in diameter, and surrounds it with horizontal piles of sticks, which increase the pole's diameter to about 25 centimeters. He covers the lower part of the maypole and the court floor with a fine compressed-moss mat that rises up to form a circular rim about 40 centimeters from the pole. He decorates the court with small objects including seeds, and he hangs regurgitated fruit pulp near the ends of the maypole sticks. On the court's rim and nearby logs, he adds woody black fungi.

We have observed young male Macgregor's bowerbirds clearing courts around the naked trunks of small trees. A selected tree's diameter usually approaches that of a fully developed maypole, which is much larger than the saplings selected by an adult male. This shows a functional correspondence between the trees used by young males and the size of maypoles built by adults.

A female arriving for courtship on an adult male's bower usually lands on the maypole and then hops down to the court. The male moves to the opposite side of the maypole with his chest close to it. He calls, and as the female moves around the maypole, he makes a counter move to keep the maypole between them. Calling, moving and counter moving go on for one to two minutes. Then the female stops moving, and the male expands his bright orange head plume and shakes his head from side to side, giving the female a view of rapid orange flashes on alternating sides of the maypole. While shaking his head, the male moves toward the female to copulate. In some cases, the male may charge the female without prolonged head shaking, causing her to escape around the opposite side of the maypole.

Bowerbirds without Bowers

Two species of bowerbirds, toothbill and Archbold's, clear and decorate courts, but they have lost their bower-building behavior. The unique mating tactics of each of these species suggest functional alternatives to bower building.

Archbold's bowerbird has lost bower-building behavior. Instead, it clears a display court that is about four meters long and 2.5 meters wide, and it covers the court with a thick mat of ferns. A male decorates his court with beetle wings, dark fruit, King of Saxon (a bird of paradise) head plumes and snail shells, and he places smaller decorations in piles near a court's edge and on limbs that overhang a court. A male also drapes orchid vines on numerous

Figure 4. Macgregor's bowerbirds dodge around the maypole during courtship. A male calls, a female moves around the pole and the male makes a countermove to keep the pole between them. If a female wants to mate, she stops moving, allowing a male to approach her.

overhanging limbs, making a set of curtains that crisscross and nearly touch the display court.

Archbold's bowerbird courtship begins with a male chasing a court-visiting female. He flies and hops low, close to the court surface, beneath the vine curtains. After repeated chases, a female stops moving, apparently signaling the male that he may approach her. The male then presses his body close to the fern mat and moves toward the female. With his head near the ground, the male faces the female and makes a chattering call, during which he moves his head rapidly with slight side-to-side movements and occasionally jerks up his head and tail. If the female remains stationary after that frontal display, the male moves behind her, staying near the ground, and then rises rapidly to perform a brief copulation. The low position of the male held throughout courtship, in part assured by the low-hanging vine curtains, reduces his opportunity for forced copulations by jumping on the female.

Male toothbills clear courts that are about two meters in diameter. A court encompasses several small trees, and their bases are cleaned meticulously. Unlike other bowerbirds, a toothbill decorates its court with large objects: fresh leaves turned upside down so that their light undersides are showing. Although not visible from adjacent

courts, the courts of different males are aggregated, in a so-called *lek*, and are often less than 30 meters apart allowing them to interact through loud calls. Dominant males interrupt the calls of males on adjacent courts. In addition, toothbills spend little time on the ground and far less time on their courts than do other species. Males at the center of an aggregation—the birds that preliminary studies show to be dominant in vocal interactions—have the highest mating success.

Figure 5. Toothbill bowerbirds clear courts but do not build bowers. A male decorates his court with large leaves, which he turns upside down. The rectangular pieces of paper on this court were added in an experiment to test the colors of artificial objects males would used. (Photograph courtesy of the author.)

Figure 6. Archbold's bowerbirds also build bowerless courts. A male approaches a court-visiting female with low-level flights or hops. He also makes calls and jerks his head from side to side. The female either remains stationary for mating or escapes.

During courting, a female arrives on the court and stands still, as if waiting for a male. After little or no display, the male aggressively mounts the female. The longest observed courtship lasted just 3.8 seconds. Toothbill copulation, however, lasts longer and appears violent compared with the brief and cooperative mating of other bowerbirds. During mating, a male toothbill makes low buzzing calls and beats his wings. After mating, a female leaves immediately. The use of exceptionally loud calls and large decorations and evidence of a female preference for central males on leks suggests that toothbill females may assess males before arriving on the court. If a female chooses a mate before arriving on a court,

B. Chudleigh, Vireo

Figure 7. Great bowerbirds build avenue bowers and decorate them with green objects and shells. A male great bowerbird organizes the decorations strategically, such that they contrast with his lilac crest during his display. The combination of decorations and behavior may enhance a female's interest.

there is little need for a bower's protection from forced copulation.

Why Build Bowers?

Although avenue and maypole bowers differ in form, observations of courtship behavior at bowers show that both provide a barrier that protects a visiting female from forced copulations by a courting male. Both avenue- and maypole-building males perform prolonged and active courtship displays. A male watches a female until she signals her readiness for mating, then he moves behind her to copulate. A female not prepared to mate can escape while an approaching male moves around a barrier created by a bower wall or maypole. A bower also allows a female to observe court decorations from close range with a reduced threat of forced copulation. The freedom from forced copulation offered by the bower may explain the high degree of elaboration of the decorated ground that has evolved in this group, including the use of small decorations on a ground court.

The two species that build courts without bowers offer alternative solutions to the problem of restricted mate choice because of forced copulation. Toothbill females select desirable mates before arriving on a court, so they do not need the protection of a bower. The low position of an Archbold's bowerbird male while courting allows a female to escape an unwanted copulation.

Males of many species gain reproductive success through forced copulation, so why would male bowerbirds build a structure that limits their opportunities for this behavior? The protection of a bower probably attracts females and increases their visitation, which more than compensates a male for losing forced copulations. Given that female bowerbirds choose the courts they visit, they should prefer the ones that provide protection from forced copulations. A female that freely chooses her mate should also be less likely to mate with another male. That combination of behaviors provides bower-building males with increased visitation by females and a high chance of being a female's only mate.

In some bowerbird species, males attack visiting females during courtship, and a bower might protect a female from such a threat. A maypole bower could serve that purpose. At avenue bowers, on the other hand, a male faces a female during courtship, so the bower offers no protection. Moreover, a female confined inside avenue-bower walls makes a susceptible target; she could only escape by moving backward, because the walls prevent her from moving sideways or turning around. If bowers served originally as protection against aggressive attacks from courting males, the evolution of the avenue bowers would require the loss of that function and replacement with others. Although the prostrate position taken by a courting male Archbold's may provide protection from attacks, no such behavior has been observed in toothbills. Overall, it seems unlikely that bowers initially evolved to protect females from attacks by males.

The good-genes hypothesis gains support from some observations, including the tendency of females to choose vigorous males and the intense, athletic displays of males in species with widely separated bowers. These characteristics, however, may derive from the origins of male courtship "dances" and vocal behavior rather than from bower building. In some modern species, a female might assess a male's genetic quality from his ability to maintain his bower in the face of destruction by rivals, but such a process seems unlikely early in the evolution of bower building, when only a few males had bowers. Assessing male quality by his bower probably arose as a secondary function after bower building evolved.

The so-called runaway model suggests that female preferences and male traits evolve together, driven by a mating advantage gained by males that possess an extreme version of a trait, such as bower building. No evidence, though, suggests that males with large bowers mate more. In addition, recent versions of the runaway model expect high costs for a trait that confers a strong mating advantage. In an intense study of satin bowerbirds, I found no evidence of high cost, despite strong effects of bower quality on male mating success.

Several other hypotheses also lack support. The predation hypothesis seems unlikely, in part, because no example of predation on females or males appeared during more than 100,000 hours of monitoring bowerbird display courts in 10 species. That result proves especially relevant given that males of most species are not protected from predators during courtship. In addition, neither major bower type protects a female from behind, where a predator or a marauding male seeking forced copulation might approach.

Evolution of Bower Building

Determining the evolution of many traits requires an explanation of how incipient stages could be used. Our work suggests that the first bowers consisted of a sapling on a display court. If that is correct, why might a female prefer a male that has a sapling on his display court over ones with other attractive attributes that might benefit her or her offspring? Maybe females simply sought protected courts. A court with a natural barrier, such as a sapling, could separate a female from a courting male and allow her to closely observe the male's display and decorations without committing to mating.

By placing sticks around a sapling a male would be less constrained by sapling size and location. He could utilize a much wider range of saplings and ones in particularly suitable locations, by enhancing the diameter of a maypole to an appropriate size with a stick covering. In addition, the soft edge created by a stick maypole allows males and females to observe each other and anticipate each other's moves, which would be more difficult around a tree of equal diameter. In this scenario, stick-built bowers would have been an improvement on a previous practice of using natural barriers on courts. It might have begun with the rearrangement of fallen sticks that were already present to enhance a court's protective qualities and led to a simple maypole bower.

Once the tendency toward stick-built bowers evolved, two trends could emerge. First, bower form could diversify to serve other functions. Second, free-standing stick barriers would allow males even more freedom in selecting bower sites and in concentrating decorations in advantageous locations. The transition to avenue bowers required losing the use of a sapling as a bower support and the addition of a different barrier. The two-walled barrier oriented a female toward parts of a court where a male could concentrate his decorations on a well-lit stage and arrange the decorations to his best advantage. Many avenue-building males separate decoration types in zones around a bower in an apparently functional manner. Male avenue-building great bowerbirds, for example, place green objects beneath the spot where they display their lilac crest. The decorations are a complementary color to a male's crest and probably increase the contrast of his display. The hut-like cover on some maypole-bower courts also orients a female to a male's display.

No one knows whether bower building or decorating came first, but it appears that the development of complex bowers may have strongly influenced the use of decorations and the evolution of male plumage. The late E. Thomas Gilliard of the American Museum of Natural History argued that the degree of male head-crest elaboration correlates inversely with bower size and the number of decorations in maypole bowers. He suggested that plumage characteristics were transferred to the bower and its decorations, but offered no explanation for the transfer.

Observations of how bower shape constrains a male's display may reveal a relationship between plumage, bowers and display areas. Around the simple maypole of Macgregor's bowerbirds or the large bowerless court of Archbold's bowerbirds, decorations are spread widely around the bower. The males of both species possess well-developed crests, which they use actively during displays. That behavior contrasts with most avenue builders, which have either a reduced or no crest and build more complex bowers that orient a female toward a more limited area where decorations can be concentrated and kept in her view. In most polygynous avian species, including bowerbirds with simple or no bowers, the position of males and females varies during displays, and for females to see bright colors males must carry bright plumage. In bowerbirds with more complex bowers, which focus a female's attention on concentrations of decorations, costly bright plumage may be replaced by strategically located arrays of decorations.

The combination of analyzing bowerbird behavior and constructing a phylogeny produces an unexpectedly coherent picture of bower function, despite the diversity in structural form. All types of bowerbird behavior indicate that females seek protection from unwanted mating. No other hypothesis proves consistent with current bower function, the function of a presumed ancestral bower and novel behavior in derived bowerless species. The significance of protection from a courting male suggests an important role for models that predict direct benefits that females might gain from elaborate male traits.

Acknowledgments

This research was supported by the National Science Foundation and the University of Maryland. The New South Wales and Queensland National Parks, The Australian Bird and Bat Banding Scheme, and the PNG Wildlife and Conservation Department provided permits. R. Crozier, J. Dimuda, G. Harrington, I., J., N. and M. Hayes, J. Lauridsen, M. J. Littlejohn, J. Kikkawa, M. Raga, J. Hook, and M. and J. Turnbull provided important support. C. Depkin, D. Bond, K. Collis, R. Condit, A. Day, J. Helms, I. Kaatz, C. Loffredo, J. Morales and D. Sejkora participated as team leaders and/or co-investigators. More than 100 volunteers provided excellent field assistance.

Bibliography

Borgia, G. 1995. Threat reduction as a cause of differences in bower architecture, bower decoration and male display in two closely related bowerbirds *Chlamydera nuchalis* and *C. maculata. Emu* 95:1–12.

Borgia, G. 1985. Bowers as markers of male quality. Test of a hypothesis. *Animal Behavior* 35:266–271.

Borgia, G., and U. Mueller. 1992. Bower destruction, decoration stealing, and female choice in the spotted bowerbird (*Chlamydera maculate*). *Emu* 92:11–18.

Kusmierski, R., G. Borgia, R. Crozier and B. Chan. 1993. Molecular information on bowerbird phylogeny and the evolution of exaggerated male characters. *Journal of Evolutionary Biology* 6:737–752.

Early Canid Domestication: The Farm-Fox Experiment

Foxes bred for tamability in a 40-year experiment exhibit remarkable transformations that suggest an interplay between behavioral genetics and development

Lyudmila N. Trut

Lyudmila N. Trut is head of the research group at the Institute of Cytology and Genetics of the Siberian Department of the Russian Academy of Sciences, in Novosibirsk. She received her doctoral degree in 1980. Her current research interests are the patterns of evolutionary transformations at the early steps of animal domestication. Her research group is developing the problem of domestication as an evolutionary event with the use of experimental models, including the silver fox, the American mink, the river otter and the wild gray rat. Address: Institute of Cytology and Genetics of the Russian Academy of Sciences, 630090 Novosibirsk 90, Russia. Internet: iplysn@bionet.nsc.ru.

When scientists ponder how animals came to be domesticated, they almost inevitably wind up thinking about dogs. The dog was probably the first domestic animal, and it is the one in which domestication has progressed the furthest—far enough to turn *Canis lupus* into *Canis familiaris*. Evolutionary theorists have long speculated about exactly how dogs' association with human beings may have been linked to their divergence from their wild wolf forebears, a topic that anthropologist Darcy Morey has discussed in some detail in the pages of this magazine (July–August 1994).

As Morey pointed out, debates about the origins of animal domestication tend to focus on "the issue of intentionality"—the extent to which domestication was the result of deliberate human choice. Was domestication actually "self-domestication," the colonization of new ecological niches by animals such as wolves? Or did it result from intentional decisions by human beings? How you answer those questions will determine how you understand the morphological and physiological changes that domestication has brought about—whether as the results of the pressure of natural selection in a new niche, or as deliberately cultivated advantageous traits.

In many ways, though, the question of intentionality is beside the point. Domestication was not a single event but rather a long, complex process. Natural selection and artificial selection may both have operated at different times or even at the same time. For example, even if prehistoric people deliberately set out to domesticate wolves, natural selection would still have been at work. The selective regime may have changed drastically when wolves started living with people, but selective pressure continued regardless of anything *Homo sapiens* chose to do.

Another problem with the debate over intentionality is that it can overshadow other important questions. For example, in becoming domesticated, animals have undergone a host of changes in morphology, physiology and behavior. What do those changes have in common? Do they stem from a single cause, and if so, what is it? In the case of the dog, Morey identifies one common factor as *pedomorphosis*, the retention of juvenile traits by adults. Those traits include both morphological ones, such as skulls that are unusually broad for their length, and behavioral ones, such as whining, barking and submissiveness—all characteristics that wolves outgrow but that dogs do not. Morey considers pedomorphosis in dogs a byproduct of natural selection for earlier sexual maturity and smaller body size, features that, according to evolutionary theory, ought to increase the fitness of animals engaged in colonizing a new ecological niche.

The common patterns are not confined to a single species. In a wide range of mammals—herbivores and predators, large and small—domestication seems to have brought with it strikingly similar changes in appearance and behavior: changes in size, changes in coat color, even changes in the animals' reproductive cycles. Our research group at the Institute of Cytology and Genetics in Novosibirsk, Siberia, has spent decades investigating such patterns and other questions of the early evolution of domestic animals. Our work grew out of the interests and ideas of the late director of our institute, the geneticist Dmitry K. Belyaev.

Like Morey, Belyaev believed that the patterns of changes observed in domesticated animals resulted from genetic changes that occurred in the course of selection. Belyaev, however, believed that the key factor selected for was not size or reproduction, but behavior—specifically amenability to domestication, or *tamability*. More than any other quality, Belyaev believed, tamability must have determined how well an animal would adapt to life among human beings. Because behavior is rooted in biology, selecting for tameness and against aggression means selecting for physiological changes in the systems that govern the body's hormones and neurochemicals. Those changes, in turn, could have had far-reaching effects on the development of the animals themselves, effects

Figure 1. In the late 1950s, the Russian geneticist Dmitry K. Belyaev began a decades-long effort to breed a population of tame foxes. Belyaev, then director of the Institute of Cytology and Genetics of the U.S.S.R. (now Russian) Academy of Sciences in Novosibirsk, Siberia, hoped to show that physical and morphological changes in domestic animals such as dogs could have resulted from selection for a single behavioral trait, friendliness toward people. Fourteen years after his death, the experiment continues, and the results appear to support Belyaev's hypothesis. (All photographs courtesy of the author.)

that might well explain why different animals would respond in similar ways when subjected to the same kinds of selective pressures.

To test his hypothesis, Belyaev decided to turn back the clock to the point at which animals received the first challenge of domestication. By replaying the process, he would be able to see how changes in behavior, physiology and morphology first came about. Of course, reproducing the ways and means of those ancient transformations, even in the roughest outlines, would be a formidable task. To keep things as clear and simple as possible, Belyaev designed a selective-breeding program to reproduce a single major factor, strong selection pressure for tamability. He chose as his experimental model a species taxonomically close to the dog but never before domesticated: *Vulpes vulpes*, the silver fox. Belyaev's fox-breeding experiment occupied the last 26 years of his life. Today, 14

years after his death, it is still in progress. Through genetic selection alone, our research group has created a population of tame foxes fundamentally different in temperament and behavior from their wild forebears. In the process we have observed some striking changes in physiology, morphology and behavior, which mirror the changes known in other domestic animals and bear out many of Belyaev's ideas.

Belyaev's Hypothesis

Belyaev began his experiment in 1959, a time when Soviet genetics was starting to recover from the anti-Darwinian ideology of Trofim Lysenko. Belyaev's own career had suffered. In 1948 his commitment to orthodox genetics had cost him his job as head of the Department of Fur Animal Breeding at the Central Research Laboratory of Fur Breeding in Moscow. During the 1950s he continued to conduct genetic re-

appearance of dwarf and giant varieties	all
piebald coat color	all
wavy or curly hair	sheep, poodles, donkeys, horses, pigs goats, mice, guinea pigs
rolled tails	dogs, pigs
shortened tails, fewer vertebrae	dogs, cats, sheep
floppy ears	dogs, cats, pigs, horses, sheep, goats, cattle
changes in reproductive cycle	all except sheep

Figure 2. Early in the process of domestication, Darwin noted long ago, animals often undergo similar morphological and physiological changes. Because behavior is rooted in biology, Belyaev believed that selection for behavior implied selection for physiological characteristics that would have broader effects on the animals' development. These effects might explain patterns in the responses of various animals to domestication.

search under the guise of studying animal physiology. He moved to Novosibirsk, where he helped found the Siberian Department of the Soviet (now Russian) Academy of Sciences and became the director of the Department's Institute of Cytology and Genetics, a post he held from 1959 until his death in 1985. Under his leadership the institute became a center of basic and applied research in both classical genetics and modern molecular genetics. His own work included ground-breaking investigations of evolutionary change in animals under extreme conditions (including domestication) and of the evolutionary roles of factors such as stress, selection for behavioral traits and the environmental photoperiod, or duration of natural daylight. Animal domestication was his lifelong project, and fur bearers were his favorite subjects.

Early in the process of domestication, Belyaev noted, most domestic animals had undergone the same basic morphological and physiological changes. Their bodies changed in size and proportions, leading to the appearance of dwarf and giant breeds. The normal pattern of coat color that had evolved as camouflage in the wild altered as well. Many domesticated animals are piebald, completely lacking pigmentation in specific body areas. Hair turned wavy or curly, as it has done in Astrakhan sheep, poodles, domestic donkeys, horses, pigs, goats and even laboratory mice and guinea pigs. Some animals' hair also became longer (Angora type) or shorter (rex type).

Tails changed, too. Many breeds of dogs and pigs carry their tails curled up in a circle or semicircle. Some dogs, cats and sheep have short tails resulting from a decrease in the number of tail vertebrae. Ears became floppy. As Darwin noted in chapter 1 of *On the Origin of Species*, "not a single domestic animal can be named which has not in some country drooping ears"—a feature not found in any wild animal

except the elephant. Another major evolutionary consequence of domestication is loss of the seasonal rhythm of reproduction. Most wild animals in middle latitudes are genetically programmed to mate once a year, during mating seasons cued by changes in daylight. Domestic animals at the same latitudes, however, now can mate and bear young more than once a year and in any season.

Belyaev believed that similarity in the patterns of these traits was the result of selection for amenability to domestication. Behavioral responses, he reasoned, are regulated by a fine balance between neurotransmitters and hormones at the level of the whole organism. The genes that control that balance occupy a high level in the hierarchical system of the genome. Even slight alterations in those regulatory genes can give rise to a wide network of changes in the developmental processes they govern. Thus, selecting animals for behavior may lead to other, far-reaching changes in the animals' development. Because mammals from widely different taxonomic groups share similar regulatory mechanisms for hormones and neurochemistry, it is reasonable to believe that selecting them for similar behavior—tameness—should alter those mechanisms, and the developmental pathways they govern, in similar ways.

For Belyaev's hypothesis to make evolutionary sense, two more things must be true. Variations in tamability must be determined at least partly by an animal's genes, and domestication must place that animal under strong selective pressure. We have looked into both questions. In the early 1960s our team studied the patterns and nature of tamability in populations of farm foxes. We cross-bred foxes of different behavior, cross-fostered newborns and even transplanted embryos between donor and host mothers known to react differently to human beings. Our studies showed that about 35 percent of the variations in the foxes' defense response to the experimenter are genetically determined. To get some idea of how powerful the selective pressures on those genes might have been, our group has domesticated other animals, including river otters *(Lutra lutra)* and gray rats *(Rattus norvegicus)* caught in the wild. Out of 50 otters caught during recent years, only eight of them (16 percent) showing weak defensive behavior made a genetic contribution to the next generation. Among the gray rats, only 14 percent of the wild-caught yielded offspring living to adulthood. If our numbers are typical, it is clear that domestication must place wild animals under extreme stress and severe selective pressure.

The Experiment

In setting up our breeding experiment, Belyaev bypassed that initial trauma. He began with 30 male foxes and 100 vixens, most of them from a commercial fur farm in Estonia. The founding foxes were already tamer than their wild relatives. Foxes had been farmed since the beginning of this century, so the earliest steps of domestication—capture, caging and isolation from other wild foxes—had already left their marks on our foxes' genes and behavior.

From the outset, Belyaev selected foxes for tameness and tameness alone, a criterion we have scrupulously followed. Selection is strict; in recent years, typically not more than 4 or 5 percent of male offspring and about 20 percent of female offspring have been allowed to breed. To ensure that their tameness results from genetic selection, we do not train the foxes. Most of them spend their lives in cages and are allowed only brief "time dosed" contacts with human beings. Pups are caged with their mothers until they are 1½ to 2 months old. Then they are caged with their litter mates but without their mothers. At three months, each pup is moved to its own cage.

To evaluate the foxes for tameness, we give them a series of tests. When a pup is one month old, an experimenter offers it food from his hand while trying to stroke and handle the pup. The pups are tested twice, once in a cage and once while moving freely with other pups in an enclosure, where they can choose to make contact either with the human experimenter or with another pup. The test is repeated monthly until the pups are six or seven months old.

At seven or eight months, when the foxes reach sexual maturity, they are scored for tameness and assigned to one of three classes. The least domesticated foxes, those that flee from experimenters or bite when stroked or handled, are assigned to Class III. (Even Class III foxes are tamer than the calmest farm-bred foxes. Among other things, they allow themselves to be hand fed.) Foxes in Class II let themselves be petted and handled but show no emotionally friendly response to experimenters. Foxes in Class I are friendly toward experimenters, wag-

Figure 3. Piebald coat color is one of the most striking mutations among domestic animals. The pattern is seen frequently in dogs (border collie, *top right*), pigs, horses and cows. Belyaev's hypothesis predicted that a similar mutation he called *Star*, seen occasionally in farmed foxes, would occur with increasing frequency in foxes selected for tamability. The photograph above shows a fox in the selected population with the *Star* mutation.

ging their tails and whining. In the sixth generation bred for tameness we had to add an even higher-scoring category. Members of Class IE, the "domesticated elite," are eager to establish human contact, whimpering to attract attention and sniffing and licking experimenters like dogs. They start displaying this kind of behavior before they are one month old. By the tenth generation, 18 percent of fox pups were elite; by the 20th, the figure had reached 35 percent. Today elite foxes make up 70 to 80 percent of our experimentally selected population.

Now, 40 years and 45,000 foxes after Belyaev began, our experiment has achieved an array of concrete results. The most obvious of them is a unique population of 100 foxes (at latest count), each of them the product of between 30 and 35 generations of selection. They are unusual animals, docile, eager to please and unmistakably domesticated. When tested in groups in an enclosure, pups compete for atten-

Figure 4. In typical silver foxes, such as those in the founding population of Belyaev's breeding experiment, ears are erect, the tail is low slung and the fur is silver-black, save for the tip of the tail. (All drawings of foxes were made from the author's photographs.)

tion, snarling fiercely at one another as they seek the favor of their human handler. Over the years several of our domesticated foxes have escaped from the fur farm for days. All of them eventually returned. Probably they would have been unable to survive in the wild.

Physical Changes

Physically, the foxes differ markedly from their wild relatives. Some of the differences have obvious links to the changes in their social behavior. In dogs, for example, it is well known that the first weeks of life are crucial for forming primary social bonds with human beings. The "window" of bonding opens when a puppy becomes able to sense and explore its surroundings, and it closes when the pup starts to fear unknown stimuli. According to our studies, nondomesticated fox pups start responding to auditory stimuli on day 16 after birth, and their eyes are completely open by day 18 or 19. On average, our domesticated fox pups respond to sounds two days earlier and open their eyes a day earlier than their nondomesticated cousins. Nondomesticated foxes first show the fear response at 6 weeks of age; domesticated ones show it after 9 weeks or even later. (Dogs show it at 8 to 12 weeks, depending on the breed.) As a result, domesticated pups have more time to become incorporated into a human social environment.

Moreover, we have found that the delayed development of the fear response is linked to changes in plasma levels of corticosteroids, hormones concerned with an animal's adaptation to stress. In foxes, the level of corticosteroids rises sharply between the ages of 2 to 4 months and reach adult levels by the age of 8 months. One of our studies found that the more advanced an animal's selection for domesticated behavior was, the later it showed the fear response and the later came the surge in its plasma corticosteroids. Thus, selection for domestication gives rises to changes in the timing of the postnatal development of certain physiological and hormonal mechanisms underlying the formation of social behavior.

Other physical changes mirror those in dogs and other domesticated animals. In our foxes, novel traits began to appear in the eighth to tenth selected generations. The first ones we noted were changes in the foxes' coat color, chiefly a loss of pigment in certain areas of the body, leading in some cases to a star-shaped pattern on the face similar to that seen in some breeds of dog. Next came traits such as floppy ears and rolled tails similar to those in some breeds of dog. After 15 to 20 generations we noted the appearance of foxes with shorter tails and legs and with underbites or overbites. The novel traits are still fairly rare. Most of them show up in no more than a few animals per 100 to a few per 10,000. Some have been seen in commercial populations, though at levels at least a magnitude lower than we recorded in our domesticated foxes.

Alternative Explanations

What might have caused these changes in the fox population? Before discussing Belyaev's explanation, we should consider other possibilities. Might rates and patterns of changes observed in foxes be due, for example, to inbreeding? That could be true if enough foxes in Belyaev's founding population carried a recessive mutant gene from the trait along with a dominant normal gene that masked its effects. Such mixed-gene, or *heterozygous*, foxes would have been hidden carriers, unaffected by the mutation themselves but capable of passing it on to later generations.

As Morey pointed out, inbreeding might well have been rampant during the early steps of dog domestication. But it certainly cannot explain the novel traits we have observed in our foxes, for two reasons. First, we designed the mating system for our experimental fox population to prevent it. Through outbreeding with foxes from commercial fox farms and other standard methods, we have kept the inbreeding coefficients for our fox population between 0.02 and 0.07. That means that whenever a fox pup with a novel trait has been born into the herd, the probability that it acquired the trait through inbreeding (that is, by inheriting both of its mutant genes from the same ancestor) has varied between only 2 and 7 percent. Second, some of the new traits are not recessive: They are controlled by dominant or incompletely dominant genes. Any fox with one of those genes would have shown its effects; there could have been no "hidden carriers" in the original population.

Another, subtler possibility is that the novelties in our domesticated population are classic by-products of strong selection for a quantitative trait. In genetics, quantitative traits are characteristics that can vary over a range of possibilities; unlike Gregor Mendel's peas, which were either smooth or wrinkly with no middle ground, quantitative traits such as an animal's size, the amount of milk it produces or its overall friendliness toward human beings can be high, low or anywhere in between. What makes selecting for quantitative traits so perilous is that they (or at least the part of them that is genetic) tend to be controlled not by single genes but by complex systems of genes, known as *polygenes*. Because polygenes are so intricate, anything that tampers with them runs the risk of upsetting other parts of an organism's genetic machinery. In the case of our foxes, a breeding program that alters a poly-

Figure 5. Foxes in Belyaev's experimental group were selected to breed depending on how they reacted to their human keepers. Vicious foxes *(top left)* were excluded from the experimental population. Foxes showing slight fear and no viciousness toward humans were used in cross-breeding for the next generation *(top right)*. Their offspring *(photograph, bottom)* were calm and showed no negative emotional responses to people.

gene might upset the genetic balance in some animals, causing them to show unusual new traits, most of them harmful to the fox. Note that in this argument, it does not matter whether the trait being selected for is tameness or some other quantitative trait. Any breeding program that affects a polygene might have similar effects.

The problem with that explanation is that it does not explain why we see the particular mutations we do see. If disrupted polygenes are responsible, then the effects of a selection experiment ought to depend strongly on which mutations already existed in the population. If Belyaev had started with 130 foxes from, say, North America, then their descen-

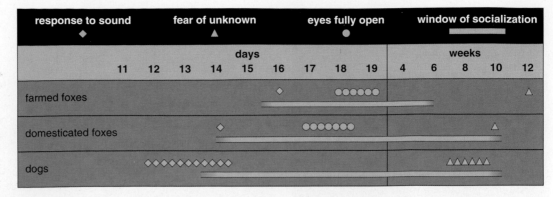

Figure 6. Dogs begin forming social bonds with human beings as soon as they can see and hear; they stop bonding once they start showing fear of the unknown. In foxes bred for tamability, the window of bonding opens earlier and closes later than it does in ordinary farmed foxes.

dants today would have ended up with a completely different set of novelties. Domesticating a population of wolves, or pigs, or cattle ought to produce novel traits more different still. Yet as Belyaev pointed out, when we look at the changes in other domesticated animals, the most striking things about them are not how diverse they are, but how similar. Different animals, domesticated by different people at different times in different parts of the world, appear to have passed through the same morphological and physiological evolutionary pathways. How can that be?

According to Belyaev, the answer is not that domestication selects for a *quantitative* trait but that it selects for a *behavioral* one. He considered genetic transformations of behavior to be the key factor entraining other genetic events. Many of the polygenes determining behavior may be regulatory, engaged in stabilizing an organism's early development, or *ontogenesis*. Ontogenesis is an extremely delicate process. In principle, even slight shifts in the sequence of events could throw it into chaos. Thus the genes that orchestrate those events and keep them on track have a powerful role to play. Which genes are they? Although numerous genes interact to stabilize an organism's development, the lead role belongs to the genes that control the functioning of the neural and endocrine systems. Yet those same genes also govern the systems that control an animal's behavior, including its friendliness or hostility toward human beings. So, in principle, selecting animals for behavioral traits can fundamentally alter the development of an organism.

As our breeding program has progressed, we have indeed observed changes in some of the animals' neurochemical and neurohormonal mechanisms. For example, we have measured a steady drop in the hormone-producing activity of the foxes' adrenal glands. Among several other roles in the body, the adrenal cortex comes into play when an animal has to adapt to stress. It releases hormones such as corticosteroids, which stimulate the body to extract energy from its reserves of fats and proteins.

After 12 generations of selective breeding, the basal levels of corticosteroids in the blood plasma of our domesticated foxes had dropped to slightly more than half the level in a control group. After 28 to 30 generations of selection, the level had halved again. The adrenal cortex in our foxes also responds less sharply when the foxes are subjected to emotional stress. Selection has even affected the neurochemistry of our foxes' brains. Changes have taken place in the serotonin system, thought to be the leading mediator inhibiting animals' aggressive behavior. Compared with a control group, the brains of our domesticated foxes contain higher levels of serotonin; of its major metabolite, 5-oxyindolacetic acid; and of tryptophan hydroxylase, the key enzyme of serotonin synthesis. Serotonin, like other neurotransmitters, is critically involved in shaping an animal's development from its earliest stages.

Selection and Development
Evidently, then, selecting foxes for domestication may have triggered profound changes in the mechanisms that regulate their development. In particular, most of the novel traits and other changes in the foxes seem to result from shifts in the rates of certain ontogenetic processes—in other words, from changes in timing. This fact is clear enough for some of the novelties mentioned above, such as the earlier eye opening and response to noises and the delayed onset of the fear response to unknown stimuli. But it also can explain some of the less obvious ones. Floppy ears, for example, are characteristic of newborn fox pups but may get carried over to adulthood.

Even novel coat colors may be attributable to changes in the timing of embryonic development. One of the earliest novel traits we observed in our domesticated foxes was a loss of pigment in parts of the head and body. Belyaev determined that this piebald pattern is governed by a gene that he named *Star*. Later my colleague Lyudmila Prasolova and I discovered that the *Star* gene affects the migration rate of *melanoblasts*, the embryonic precursors of the pigment cells *(melanocytes)* that give color to an animal's fur. Melanocytes form in the embryonic fox's neural crest and later move to various parts of the embryo's epidermis. Normally this migration starts around days 28 to 31 of the embryo's development. In foxes that carry even a single copy of the *Star* gene, however, melanoblasts pass into the potentially depigmented areas of the epidermis two days later, on average. That delay may lead to the death of the tardy melanoblasts, thus altering the pigmentation in ways that give rise to the distinctive Star pattern.

One developmental trend to which we have devoted particular attention has to do with the growth of the skull. In 1990 and 1991, after noticing abnormal developments in the skulls and jaws of some of our foxes, we decided to study variations in the animals' cranial traits. Of course, changes in the shape of the skull are among the most obvious ways in which dogs differ from wolves. As I mentioned earlier, Morey believes that they are a result of selection (either natural or artificial) for reproductive timing and smaller body size.

In our breeding experiment, we have selected foxes only for behavior, not size; if anything, our foxes may be slightly longer, on average, than the ones Belyaev started with 40 years ago. Nevertheless, we found that their skulls have been changing. In our domesticated foxes of both sexes, cranial height and width tended to be smaller, and snouts tended to be shorter and wider, than those of a control group of farmed foxes.

Another interesting change is that the cranial morphology of domesticated adult males became somewhat "feminized." In farmed foxes, the crania of males tended to be larger in volume than those of females, and various other proportions differed sharply between the sexes. In the domesticated foxes the sexual dimorphism decreased. The differences in volume remained, but in other respects the skulls of males became more like those of females. Analysis of cranial allometry showed that the changes in skull proportions result either from changes in the timing of the first appearance of particular structures or from changes in their growth rates. Because we studied the skulls only of adult foxes, however, we cannot judge whether any of these changes are pedomorphic, as Morey believes they are in dogs.

The most significant changes in developmental timing in our foxes may be the smallest ones: those that have to do with reproduction. In the wild, foxes reach sexual maturity when they are about 8 months old. They are strict seasonal breeders, mating once a year in response to changes in the length of the day (in Siberia the mating season runs from late January to late March) and giving birth to litters ranging from one to thirteen pups, with an average of four or five. Natural selection has hard-wired these traits into foxes with little or no genetic varia-

Figure 7. Changes in the foxes' coat color were the first novel traits noted, appearing in the eighth to tenth selected generations. The expression of the traits varied, following classical rules of genetics. In a fox homozygous for the *Star* gene, large areas of depigmentation similar to those in some dog breeds are seen *(top)*. In addition some foxes displayed the brown mottling seen in some dogs, which appeared as a semirecessive trait.

tion. Fur farmers have tried for decades to breed foxes that would reproduce more often than annually, but all their attempts have failed.

In our experimental fox population, however, some reproductive traits have changed in a correlated manner. The domesticated foxes reach sexual maturity about a month earlier than nondomesticated foxes do, and they give birth to litters that are, on average, one pup larger. The mating season has lengthened. Some females breed out of season, in November–December or April–May, and a few of them have mated twice a year. Only a very small number of our vixens have shown such

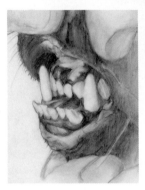

characteristic	animals per 100,000 with trait		increase in frequency
	domesticated population	nondomesticated population	(percent)
depigmentation (*Star*)	12,400	710	+1,646
brown mottling	450	86	+423
gray hairs	500	100	+400
floppy ears	230	170	+35
short tail	140	2	+6,900
tail rolled in circle	9,400	830	+1,033

Figure 8. Foxes in the domesticated population show an unusually high incidence of certain other changes, including *(clockwise from top left)* floppy ears, shortened legs and tails, tails curled upward like dogs', and underbites and overbites. The rates of some common aberrations are compared in the table. In addition to the *Star* depigmentation pattern, the increased incidence of doglike tail characteristics was most marked.

unusual behavior, and in 40 years, no offspring of an extraseasonal mating has survived to adulthood. Nevertheless, the striking fact is that, to our knowledge, out-of-season mating has never been previously observed in foxes experiencing a natural photoperiod.

Lessons Learned

Forty years into our unique lifelong experiment, we believe that Dmitry Belyaev would be pleased with its progress. By intense selective breeding, we have compressed into a few decades an ancient process that originally unfolded over thousands of years. Before our eyes, "the Beast" has turned into "Beauty," as the aggressive behavior of our herd's wild progenitors entirely disappeared. We have watched

new morphological traits emerge, a process previously known only from archaeological evidence. Now we know that these changes can burst into a population early in domestication, triggered by the stresses of captivity, and that many of them result from changes in the timing of developmental processes. In some cases the changes in timing, such as earlier sexual maturity or retarded growth of somatic characters, resemble pedomorphosis.

Some long-standing puzzles remain. We believed at the start that foxes could be made to reproduce twice a year and all year round, like dogs. We would like to understand why this has turned out not to be quite so. We are also curious about how the vocal repertoire of foxes changes under domestication. Some of the calls of our adult foxes resemble those of dogs and, like those of dogs, appear to be holdovers from puppyhood, but only further study will reveal the details.

The biggest unanswered question is just how much further our selective-breeding experiment can go. The domestic fox is not a domestic dog, but we believe that it has the genetic potential to become more and more doglike. We can continue to increase that potential through further breeding, but the foxes will realize it fully only through close contact with human beings. Over the years, other investigators and I have raised several fox pups in domestic conditions, either in the laboratory or at home as pets. They have shown themselves to be good-tempered creatures, as devoted as dogs but as independent as cats, capable of forming deep-rooted pair bonds with human beings—mutual bonds, as those of us who work with them know. If our experiment should continue, and if fox pups could be raised and trained the way dog puppies are now, there is no telling what sort of animal they might one day become.

Whether that will happen remains to be seen. For the first time in 40 years, the future of our domestication experiment is in doubt, jeopardized by the continuing crisis of the Russian economy. In 1996 the population of our breeding herd stood at 700. Last year, with no funds to feed the foxes or to pay the salaries of our staff, we had to cut the number to 100. Earlier we were able to cover most of our expenses by selling the pelts of the foxes culled from the breeding herd. Now that source of revenue has all but dried up, leaving us increasingly dependent on outside funding at a time when shrinking budgets and changes in the grant-awarding system in Russia are making long-term experiments such as ours harder and harder to sustain. Like many other enterprises in our country, we are becoming more entrepreneurial. Recently we have sold some of our foxes to Scandinavian fur breeders, who have been pressured by animal-rights groups to develop animals that do not suffer stress in captivity. We also plan to market pups as house pets, a commercial venture that should lead to some interesting, if informal, experiments in its own right. Many avenues of both applied and basic research remain for us to pursue, provided we save our unique fox population.

Acknowledgments

This article is dedicated to the memory of Dmitry K. Belyaev. The research was supported by grants RBD000 and RBD300 from the International Scientific Funds, grants 93-04-06936 and 96-04-49972 from the Russian Fund of Fundamental Research, and grant 1757 of the Russian University Fund. The author expresses her gratitude to Anna Fadeeva for translation of the manuscript from Russian into English. She is also grateful to Irina Plysnina for help during preparation of the manuscript and to Yekaterina Omelchenko for technical assistance.

Figure 9. Forty years into the experiment, between 70 and 80 percent of the foxes bred for tameness are members of the human-friendly "domesticated elite." When raised as pets, they are devoted, affectionate and capable of forming strong social bonds with people (here, with technical assistant Marina Nurgalieva). The foxes seek out human contact and lick experimenters' hands and faces. The friendly behavior is evident before the fox pups are a month old.

Bibliography

Belyaev, D. K. 1969. Domestication of animals. *Science Journal* (U.K.) 5:47–52.

Belyaev, D. K. 1979. Destabilizing selection as a factor in domestication. *The Journal of Heredity* 70:301–308.

Belyaev, D. K., A. O. Ruvinsky and L. N. Trut. 1981. Inherited activation-inactivation of the star gene in foxes. *The Journal of Heredity* 72:264–274.

Belyaev, D. K., and L. N. Trut. 1982. Accelerating evolution. *Science in the USSR* 5:24–29, 60–64.

Falconer, D. S. 1981. *Introduction to Quantitative Genetics.* New York: Longman.

Logvinenko, N. S., P. M. Krass, L. N. Trut, L. N. Ivanova and D. K. Belyaev. 1979. Genetics and phenogenetics of hormonal characteristics of animals. V. Influence of domestication on ontogenesis of estrogen- and progesterone-secreting functions of ovaries and adrenals in silver fox females. *Genetica* 15:320–326 (in Russian).

Morey, D. F. 1994. The early evolution of the domestic dog. *American Scientist* 82:336–347.

Raff, R. A., and T. C. Kaufman. 1983. *Embryos, Genes and Evolution.* New York: Macmillan Publishing Company.

Scott, J. P., and J. L. Fuller. 1965. *Genetics and the Social Behavior of the Dog.* Chicago: University of Chicago Press.

Trut, L. N. 1988. The variable rates of evolution transformations and their parallelism in terms of destabilizing selection. *Journal of Animal Breeding and Genetics* 105:81–90.

Trut, L. N. 1996. Sex ratio in silver foxes: effects of domestication and the star gene. *Theoretical and Applied Genetics* 92:109–115.

Wayne, R. K. 1993. Molecular evolution of the dog family. *Trends in Genetics* 9:218–224.

Zeuner, F. T. 1954. Domestication of animals. In *A History of Technology*, Volume I, ed. C. Singer, E. J. Holmyard, A. R. Hall and T. I. Williams. Oxford: Oxford University Press.

Links to Internet resources for further exploration of "Early Canid Domestication: The Farm-Fox Experiment" are available on the *American Scientist* Web site:

http://www.amsci.org/ amsci/articles/ articles99/trut.html

Protecting Ourselves from Food

*Spices and morning sickness may shield us
from toxins and microorganisms in the diet*

Paul W. Sherman and Samuel M. Flaxman

Although a well-prepared meal may be an unalloyed pleasure, the act of eating is one of the most dangerous things many of us do every day. Ingesting bits and pieces of the outside world provides a free pass to the bloodstream for whatever lurks within. Microorganisms and toxins are pervasive—benign or not, they are almost always present at some level in the food we eat. Indeed most of us have experienced the unpleasant consequences of "food poisoning." Likewise, we often read newspaper accounts of deaths and illnesses associated with contaminated foods. And all of this occurs despite our modern practices of refrigeration and hygienic food preparation.

One can imagine a time in the past, however, when the absence of refrigeration and the scarcity of food encouraged people to eat meals that might not have met the highest standards of sanitation. The dangers of eating leftovers in those days must have been great indeed. Certainly there must have been many more deaths and illnesses in the past. What did people do?

As evolution-minded biologists, we were intrigued by traditions associated with preparing and eating food. Might they provide some means of protecting us from the dangers of contamination? Consider the humble cookbook. Traditional versions contain

recipes that have been prepared for hundreds, perhaps even thousands, of years. Surely one reason for a recipe's success must be its palatability, but could a recipe also have some adaptive value? That is, might it be adaptive to find certain recipes palatable and others not? Cooking food not only increases its flavor in certain instances, but also has the added value of killing microorganisms that are potentially harmful. Might some of the ingredients in a recipe do the same?

Spices, for example, are used almost universally throughout the world. We wondered whether there might be a relation between their use and their effectiveness in protecting us from foodborne microorganisms. In the first part of this article we discuss research that seeks to test the validity of this idea.

In the second half of this article we approach a seemingly unrelated subject: the adaptive value of "morning sickness"—nausea and vomiting during early pregnancy. About two-thirds of pregnant women experience some degree of morning sickness. From an evolutionary perspective, the prevalence of the phenomenon raises two questions: Why do so many women feel nauseated, as opposed to some other symptom, early in pregnancy? And how does "morning sickness" affect the outcome of pregnancy? We believe that the answers to these questions suggest that morning sickness is another mechanism, this time a physiological one, that protects us from the inherent dangers of eating.

Cooking with Spice

Spices are plant products. They come from various woody shrubs, vines, trees and aromatic lichens, as well as the roots, flowers, seeds and fruits of

herbaceous plants. Each spice has a unique aroma and flavor that derives from "secondary compounds," chemicals that are secondary (not essential) to the plant's basic metabolism. Most spices contain dozens of them. These phytochemicals evolved in plants to protect them from being eaten. Phytochemicals are legacies of multiple coevolutionary races between plants and their enemies—parasites, pathogens and herbivores. These chemical cocktails are the plants' recipes for survival.

People have made use of plant secondary compounds in food preparation and embalming for thousands of years. So valuable were spices that when Alarich, a Gothic leader, laid siege to Rome in 408 A.D. he demanded as ransom various precious metals and 3,000 pounds of pepper! In the Middle Ages, the importance of spices was underscored by the willingness of seafarers like Marco Polo, Ferdinand Magellan and Christopher Columbus to undertake hazardous voyages to establish routes to spice-growing countries. Spice trade was so crucial to national economies that rulers repeatedly mounted costly raids on spice-growing countries, and struggles for their control precipitated several wars. In modern times the spice trade still flourishes. Black pepper, for example, is the world's most widely used spice even though *Piper nigrum* grows naturally only in the Old World tropics.

What accounts for the enduring value of spices? The obvious answer is that they enhance the flavor, color and palatability of food. This *proximate*, or immediate-cause, explanation is true, but it does not address the *ultimate*, or long-term, questions of why people find foods more appealing when they contain pungent plant products, why

Paul W. Sherman is a professor at Cornell University, where he studies the behavioral ecology of various mammals and birds, and teaches animal social behavior and Darwinian medicine. Samuel M. Flaxman was an undergraduate student when he began studying morning sickness. He is now a graduate student at Cornell. Address for both authors: Department of Neurobiology and Behavior, Mudd Hall, Cornell University, Ithaca, New York 14853. Internet: pws6@cornell.edu, smf7@cornell.edu

Figure 1. Spices are ancient commodities that have been cherished by many of the world's cultures throughout history for the flavors they impart to the native cuisine. Enjoying these flavors probably is adaptive because spices may protect us from food-borne illnesses by inhibiting the growth of bacteria. The authors explore this possibility by examining the antimicrobial properties of spices and their use in traditional recipes of cultures worldwide. Here a 19th-century painting captures the active trade in spices in a northern Indian bazaar.

some secondary compounds are tastier than others and why preferences for these chemicals differ among cultures. Answers to proximate and ultimate questions are complementary rather than mutually exclusive. A full understanding of spice use, or any other trait, requires explanations at both these levels of analysis.

The ultimate reason plants possess secondary compounds is to protect themselves from their natural enemies. This may be a clue to the ultimate reason we use these chemicals. Our foods also are attacked by bacteria and fungi, often the same ones that afflict the spice plants. If spices were to kill microorganisms or inhibit their growth or production of toxins, then spice use could protect us from food-borne illnesses and food poisoning.

The Antimicrobial Hypothesis

In a series of recent studies, one of us (Sherman), along with Jennifer Billing and Geoffrey Hash, set out to test this "antimicrobial hypothesis." We located 107 "traditional" cookbooks from 36

countries, representing every continent and 16 of the world's 19 major linguistic groups. Written primarily as archives of the native cuisine, these cookbooks are artifacts of human behavior. Our cookbook database contains information on the use of 42 spices in 4,578 meat-based recipes and 2,129 vegetable-based recipes (containing no meat). Although salt has antimicrobial properties, and it has been used in food preparation and preservation for centuries, it was not included in our analyses because it is a mineral, not a plant product.

The cookbooks confirm what every gastronome knows—across the world there is tremendous variability in the use of different spices. Black pepper and onion are the most frequently used, each appearing in more than 60 percent of the meat-based recipes, with garlic, chili pepper and lemon/lime juices following close behind (*Figure 3*). Most spices, however, are rarely used. In our sample, about 80 percent of spices are used in fewer than 10 percent of the meat-based recipes, with

horseradish, fennel and savory at the bottom of the list.

To test the antimicrobial hypothesis, we developed five critical predictions and examined them by combining information from the microbiology literature with analyses of traditional recipes.

Prediction 1: Spices used in cooking should exhibit antimicrobial activity. Microbiologists and food scientists have challenged various food-borne bacteria and fungi with spice chemicals. Although the data are heterogeneous—owing to differences in laboratory techniques, phytochemical concentrations and definitions of microbial inhibition—there is, nonetheless, overwhelming evidence that most spices have antimicrobial properties (*Figure 3*).

Inhibition of bacteria is especially important because they are more common in outbreaks of food-borne diseases and food poisoning than are fungi. Of the 30 spices for which information was available, all inhibited or killed at least one-quarter of the

Chicken with Latitude

Norway:
Chicken a la Manor

2 spring chickens	3 tablespoons butter
1 teaspoon salt	1 lump sugar
parsley	1 cup chicken stock
½ cup cream	

Recipes from different parts of the world differ widely in their use of spices, as exemplified by these two dishes from Norway and India. If the use of spices is partly due to their antimicrobial properties, then there should be a systematic difference in their use in hot and cold climates, since bacteria grow faster and are more diverse in hot climates. The authors tested this hypothesis by examining the use of 42 spices in 6,707 recipes from 36 countries around the globe.

India:
Badaami Murgh (chicken smothered in aromatic herbs and almonds)

1 chicken cut in pieces	1½ teaspoons lemon juice
2 teaspoons salt	6 tablespoons vegetable oil
3 tablespoons slivered almonds	2 cups chopped onions
1 tablespoon chopped garlic	1 tablespoon chopped
1 stick cinnamon	ginger root
8 green cardamom pods	4 whole cloves
1 teaspoon ground cumin	1 teaspoon ground coriander
½ teaspoon turmeric	2–3 tablespoons almond butter
1 cup chopped ripe tomatoes	1–2 tablespoons chopped
½ teaspoon red pepper	fresh coriander leaves

bacterial species on which they were tested, and half the spices inhibited or killed three-quarters of these bacteria. The four most potent spices—garlic, onion, allspice and oregano—killed every bacterial species tested. Most of the bacteria that were tested are widely distributed and are frequently implicated in food-borne illness.

The antimicrobial hypothesis assumes that the amounts of spices called for in a recipe are sufficient to produce the desirable effects and that cooking does not destroy the active chemicals. Although the efficacy of spices in prepared meals has not been evaluated directly, both assumptions are reasonable. The minimum concentrations of purified phytochemicals necessary to inhibit growth of food-borne bacteria *in vitro* are well within the range of spice concentrations used in cooking. Phytochemicals are generally thermostable, and spices containing those that are not, such as cilantro and parsley, are typically added after cooking, so their antimicrobial effects are not lost.

Prediction 2: Use of spices should be greatest in hot climates, where unrefrigerated foods spoil quickly. Uncooked foods and leftovers, particularly meat products, can build up massive bacterial populations if they are stored at room temperature for more than a few hours, especially in the tropics where temperatures are highest and the diversity of food-borne pathogens also is the greatest. Under the antimicrobial hypothesis, there should be a positive relationship between annual temperatures and spice use, assuming that traditional recipes predate the availability of refrigeration.

In the 36 countries sampled, the average annual temperatures range from 2.8 degrees Celsius in Norway to 27.6 degrees in Thailand. Consistent with this prediction, the use of spices is greater in countries with comparatively high temperatures (*Figure 4*). In particular, the fraction of both meat- and vegetable-based recipes that called for at least one spice, the mean number of spices per recipe and the number of different spices used all were greater in the warmer countries. These trends were especially strong for the "highly inhibitory" spices—those that reduced the growth of 75 percent or more of the bacterial species tested—including chili peppers, garlic, onion, cinnamon, cumin, lemongrass, bay leaf, cloves and oregano. Only the use of dill and parsley was negatively correlated with temperature, and neither is highly inhibitory.

In five of the six hottest countries (India, Indonesia, Malaysia, Nigeria and Thailand) every meat-based recipe called for at least one spice, whereas in the two coldest countries (Finland and Norway) more than a third of the meat-based recipes did not call for any spices at all. In India, 25 different spices were used, and the average meat-based recipe called for about nine of them, whereas the Norwegians used only 10 spices in total and usually less than two spices per recipe. In general, the most powerful spices are the most popular and they become increasingly popular as the climate gets hotter.

Prediction 3: The spices used in each country should be particularly effective against the local bacteria. Spices are only useful if they inhibit or kill the indigenous microbes. Unfortunately, comprehensive lists of native food-borne bacteria do not exist for any country. To work around this problem, we decided to look at the effectiveness of native recipes in killing or inhibiting 30 common food-borne bacteria, including such well-known nasties as *Clostridium botulinum, Escherichia coli, Salmonella pullorum, Staphylococcus aureus* and *Streptococcus faecalis.* Most of these are distributed worldwide, and they often are implicated in outbreaks of food-borne disease. Consistent with this prediction, as annual temperatures increased among countries, the estimated percentage of bacteria that would be inhibited by the spices in an average recipe from each country also rose *(Figure 4)*.

Prediction 4: Within a country, meat recipes should be spicier than vegetable recipes. Unrefrigerated meats are more often associated with food-borne disease outbreaks and food poisoning than are vegetables. This may be because dead plants are naturally better protected from microbial invasions. As we noted earlier, many contain inhibitory phytochemicals, but plants also have specially modified cell walls that contain cellulose and lignin, which are difficult to decompose for most aerobic microorganisms. Furthermore, the *p*H of plant cells (ranging from 4.3 to 6.5) is below the ideal growth conditions (between *p*H 6.6 and 7.5) for most bacteria.

In contrast, the cells of dead animals are relatively unprotected chemically and physically, and their internal *p*H is usually greater than 5.6. The primary defense of animal tissues against bacteria is the immune system, which ceases to function at death. For all these reasons, any relation between spoilage and spice use should be more apparent in meat-based recipes.

Indeed, traditional meat-based recipes from all 36 countries combined called for an average of about 3.9 spices per recipe, significantly more than the 2.4 spices in the average vegetable-based recipe. This is even true within the collection of recipes from each of the 36 countries, indicating that the availability of spice plants in different countries cannot account for the global distinctions we observed *(Figure 5)*.

Prediction 5: Within a country, recipes from lower latitudes and altitudes should be spicier because of the presumably greater microbial diversity and growth rates in these regions. Although we could not find altitude-specific cookbooks, there were large samples of traditional recipes from different latitudes in China and the United States. The results were consistent with this prediction. Recipes from the southern latitudes used a greater variety of spices and used individual spices more often in their recipes. Moreover, the typical southern recipe contained an assortment of spices that was more likely to kill or inhibit bacteria.

Alternative Hypotheses
One can imagine other reasons why spices may have become so common in the human diet. Perhaps the most prominent of these suggests that spices disguise the smell and taste of spoiled foods. Although this idea is consistent with the greater use of spices in hotter climates, it ignores the potential dangers of ingesting bacteria-infested foods. Even a very hungry person would do better to err on the conservative side by passing up rather than covering up the taste of contaminated foods that would be deadly to someone in a weakened condition.

Another popular idea suggests that spices might serve as medicaments. It is true that many spice plants have pharmacological uses in traditional societies. This includes use as topical or ingested antimicrobials, as aids to digestion, as treatment for high blood pressure, as sources of micronutrients and as aphrodisiacs. However, the use of spices in food preparation differs from that in traditional medicine in a number of ways. In cooking, spices are routinely added to specific recipes in relatively small quantities and regardless of the diner's health status. By contrast, in medicinal usage phytochemi-

Figure 2. Colors, shapes and textures of these 20 common spices hint that they will provide the diner with a wide range of flavors and aromas, but offer no obvious clues about their antimicrobial properties. Displayed here *(in a clockwise spiral from upper left)*: onion, pepper, garlic, chili pepper (whole varieties and powder), lemon/lime, parsley, ginger, bay, coriander (seed and leaves), cinnamon, cloves, nutmeg (whole and powder), thyme (fresh and dried), paprika, sweet pepper, cumin, celery, turmeric, allspice and mustard seed.

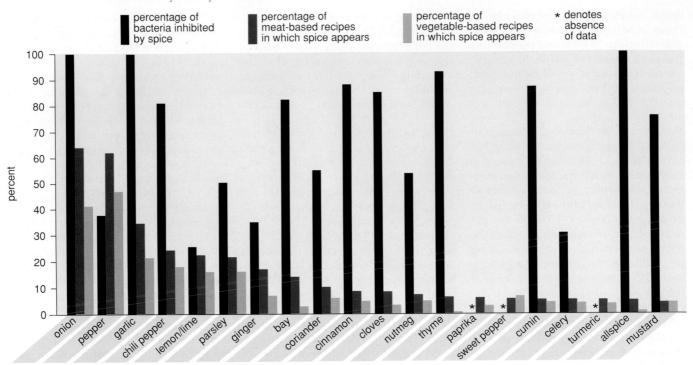

Figure 3. Most commonly used spices are potent inhibitors of bacterial growth. Most spices also are used more often in meat-based recipes than in vegetable-based recipes, which is consistent with their presumptive role in inhibiting food-borne bacteria.

cals are taken occasionally, usually in response to particular maladies, in much larger quantities and not necessarily in association with food. Use of a spice as a medicament is more like taking a pill than preparing a meal.

It has also been suggested that spicy foods might be preferred in hot climates because they increase perspiration and thus help cool the body. Indeed, chilis and horseradish do cause some people to sweat. However, the use of many other spices also increases with temperature, and these do not increase perspiration. In general, physiological mechanisms of temperature regulation operate to keep us cool without the necessity of finding, eating and dealing with the side effects of phytochemicals.

Finally, it could be that people use whatever aromatic plants grow locally just because they taste good. Under this proximate-level hypothesis, spice chemicals should be highly palatable, and spice-use patterns should correspond to availability. Neither prediction is fully supported. There is no relation between the number of countries in which each spice plant grows and either the number of countries in which it is used or their annual temperatures. Second, pungent spices like garlic, ginger, anise and chilis are initially distasteful to most people. For

most unpalatable substances, an initial negative response is sufficient to maintain avoidance for a lifetime. Yet most people come to like spicy foods, often as a result of urging by family members and friends—another sign that spice use is beneficial.

Morning Sickness

There also are costs to using spices. When eaten in sufficient amounts, many phytochemicals act as allergens, mutagens, carcinogens, teratogens and abortifacients. This may suggest why pre-adolescent children typically dislike spicy foods: Children are particularly susceptible to mutagens because some of their tissues are undergoing rapid cell division. Their alternative is to avoid foods that might contain pathogens or phytochemicals—perhaps this is why children have acquired a reputation as "picky eaters."

Rapid cell division also takes place within the body of a pregnant woman. Moreover, pregnant women are especially susceptible to food-borne illnesses and infectious diseases because their cell-mediated immune response is depressed—lest the woman's body reject the foreign tissue that is her baby-to-be. The risks for the mother create even greater dangers for the embryo. Miscarriages and birth defects can result if a pregnant woman contracts an illness,

especially during the first trimester. *Toxoplasma gondii,* for example, is a common food-borne parasite that can be acquired by handling or eating raw or undercooked meat. It is usually only dangerous to those whose immune systems are compromised, such as pregnant women. The parasite has been linked to congenital neurological birth defects, spontaneous abortions and neonatal diseases.

What's a mother to do? She might consider cooking with powerful spices, but this would expose the embryo to phytotoxins. We believe that natural selection has provided another type of answer: morning sickness.

Morning sickness is the common term for nausea and vomiting during pregnancy. It is actually a complete misnomer: Symptoms occur throughout waking hours, not just in the morning, and "sickness" implies pathology whereas healthy women experience the symptoms and bear healthy babies. For these reasons, the medical community uses the acronym NVP, short for "nausea and vomiting of pregnancy." Symptoms first appear about five weeks after the last menstrual period, peak during weeks 8 to 12, and gradually decline thereafter *(Figure 7)*. Some women experience symptoms severe enough to dramatically disrupt their daily lives, and about one out of a hundred experi-

that the highest reported rates of NVP in the world also occur in Japan.

The possibility that NVP may be adaptive in the 21st century is supported by the relation between the occurrence of NVP and reduced chances of miscarriage that has been documented multiple times in the past two decades *(Figure 10)*. This does not mean that pregnant women should eliminate animal products and vegetables from their diets. These foods contain essential vitamins and nutrients. Rather, the message is that NVP is not a "disease" in the normal sense of that word, but rather serves a useful function. There are good evolutionary reasons why so many women respond adversely to the smells or tastes of particular foods. Uncomplicated nausea and vomiting will not hurt the embryo, and may even help to protect it. We hope pregnant women will derive some comfort from this analysis, knowing that NVP indicates the evolved "wisdom" of their bodies rather than their frailty.

Acknowledgments

The authors thank Jennifer Billing and Geoffrey A. Hash for their enthusiastic participation in the spice study, and Jennifer Billing and Mark E. Hauber for comments on preliminary drafts. Funding was provided by the National Science Foundation, the Howard Hughes Foundation, the Olin Foundation, and the College of Agriculture and Life Sciences at Cornell University.

Bibliography

Billing, J., and P. W. Sherman. 1998. Antimicrobial functions of spices: why some like it hot. *Quarterly Review of Biology* 73:3–49.

Deutsch, J. A. 1994. Pregnancy sickness as an adaptation to concealed ovulation. *Rivista di Biologia* 87:277–295.

Flaxman, S. M., and P. W. Sherman. 2000. Morning sickness: A mechanism for protecting mother and embryo. *Quarterly Review of Biology* 75:113–148.

Haig, D. 1993. Genetic conflicts in human pregnancy. *Quarterly Review of Biology* 68:495–532.

Hook, E. B. 1976. Changes in tobacco smoking and ingestion of alcohol and caffeinated beverages during early pregnancy: Are these consequences, in part, of feto-protective mechanisms diminishing maternal exposure to embryotoxins? In *Birth Defects: Risks and Consequences*, ed. S. Kelly, E. B. Hook, D. T. Janerich and I. H. Porter. New York: Academic Press, pp.173–183.

Huxley, R. R. 2000. Nausea and vomiting in early pregnancy: its role in placental development. *Obstetrics and Gynecology* 95:779–782.

Inoue, S., A. Nakama, Y. Arai, Y. Kokubo, T. Maruyama, A. Saito, T. Yoshida, M. Terao, S. Yamamoto and S. Kumagai. 2000. Prevalence and contamination levels of *Listeria monocytogenes* in retail foods in Japan. *International Journal of Food Microbiology* 59:73–77.

Lee, W.-C., T. Sakai, M.-J. Lee, M. Hamakawa, S.-M. Lee and I.-M. Lee. 1996. An epidemiological study of food poisoning in Korea and Japan. *International Journal of Food Microbiology* 29:141–148.

Profet, M. 1992. Pregnancy sickness as adaptation: A deterrent to maternal ingestion of teratogens. In *The Adapted Mind: Evolutionary Psychology and the Generation of Culture*, eds. J. H. Barkow, L. Cosmides and J. Tooby. New York: Oxford University Press, pp. 327–365.

Rodin, J., and N. Radke-Sharpe. 1991. Changes in appetitive variables as a function of pregnancy. In *Chemical Senses. Volume 4: Appetite and Nutrition*, eds. M. I. Friedman, M. G. Tordoff and M. R. Kare. New York: Marcel Dekker, pp. 325–340.

Seto, A., T. Einarson and G. Koren. 1997. Pregnancy outcome following first trimester exposure to antihistamines: Meta-analysis. *American Journal of Perinatology* 14:119–124.

Sherman, P. W. 1988. The levels of analysis. *Animal Behaviour* 36:616–619.

Sherman, P. W., and J. Billing. 1999. Darwinian gastronomy: Why we use spices. *BioScience* 49:453–463.

Sherman, P. W., and G. A. Hash. In press. Why vegetable recipes are not very spicy. *Evolution and Human Behavior*.

Stein, Z., and M. Susser. 1991. Miscarriage, caffeine, and the epiphenomena of pregnancy: The causal model. *Epidemiology* 2:163–167.

Weigel, R. M., and M. M. Weigel. 1989. Nausea and vomiting of early pregnancy and pregnancy outcome. A meta-analytical review. *British Journal of Obstetrics and Gynaecology* 96:1312–1318.

Links to Internet resources for "Protecting Ourselves from Food" are available on the *American Scientist* Web site:

http://www.americanscientist.org/ articles/01articles/sherman.html

PART IV

The Adaptive Value of Reproductive Behavior

The seven articles included here illustrate how behavioral biologists analyze questions about the adaptive value of reproductive behavior, rather than the historical sequence of events underlying the evolution of the behavior as was discussed in Part III. All the authors in Part IV concern themselves primarily with how traits advance the reproductive success of individual members of living species. Here we shall see how they use an adaptationist approach in their work, an approach based on the central tenet that of all the major forces of evolution—mutation, migration, drift, and natural selection—only natural selection leads to adaptation. Therefore, if a behavioral trait is an adaptation, we expect individuals to use it in ways that enhance the transmission of their genes—an unconscious goal that is generally achieved by maximizing lifetime reproduction. Tests of this prediction in particular cases can involve experimental manipulations, long-term observations of individual animals in their natural habitats, and particular kinds of comparisons among living species.

In the opening selection, Randy Thornhill and Darryl Gwynne employ all these techniques in attempting to understand why males and females nearly always differ dramatically in the ways they go about reproducing. Generally, males are extremely eager to court and mate, but expend relatively little time and energy on rearing young, whereas females devote much more effort to choosing their mates and caring for their offspring. These "traditional" sex roles are far from universal, however, as is seen in numerous insects. Female Mormon crickets, for example, often compete vigorously for access to mates, while the males are sexually coy, sometimes rejecting willing partners. Thornhill and Gwynne believe that, in general, whichever sex invests less in offspring will exhibit "male-like" behavior. Thus, males behave in the "traditional" fashion when they are not as parental as females, but when the tables are turned and males do more for their progeny than their mates, they behave like "traditional" females.

Try to develop an alternative hypothesis on sex role differences based on the relative rates at which males and females can produce offspring, rather than their relative investment in offspring. What is the logic behind this alternative, and how is parental investment related to the rate at which males and females can produce offspring?

Thornhill and Gwynne note that sexual reproduction is not always a cooperative affair. When male and female reproductive interests differ, an evolutionary "battle of the sexes" ensues. One outcome, discussed by

William Eberhard in the next article, is the varied and bizarre morphology of animal genitalia. Eberhard proposes that males engage in "copulatory courtship behavior" to encourage choosy females to accept and use their sperm to fertilize their eggs. According to this hypothesis, through evolutionary time female choosiness has resulted in the elaboration of male intromittent organs and behaviors that deliver stimulating sensory messages to females.

Eberhard's article presents numerous other adaptationist hypotheses to explain the shapes and sizes of intromittent organs, their spines, whorls, and horns, and the speed and frequency of copulations. A useful exercise would be to list these alternative hypotheses and then develop ways to discriminate among the competing ideas.

But let us imagine that Eberhard is correct about the role of sexual selection in molding morphological attributes associated with copulatory courtship. Many questions arise. For example, under what circumstances would females be expected to possess penis-like structures? Why do males in many species of rodents and ungulates have bones in their penises, whereas most primates lack penis bones? Why do male birds generally lack penises altogether, whereas most male mammals have large and elaborate organs? And why are there exceptions among the birds, with intromittent organs being found in ostriches, rheas, tinamous, ducks, geese, and swans?

The two previous chapters discussed the now widely accepted view that animal reproductive behavior has been shaped by both within-sex interactions in competition for mates (usually among males) and by interactions between the sexes. Females (typically) exert selection in favor of some males over others on the basis of the males' resources, anticipated parental abilities, or physical attributes (including genitalia, as Eberhard suggests). Tim Birkhead's article extends that perspective on reproduction by examining what is happening among the sperm as they maneuver inside a female's reproductive tract, when the female in question has mated with more than one male. Under these circumstances, the sperm of one male are in competition with the sperm of another male in a fertilization sweepstakes. However, it is not just male against male, but rather male against female or vice versa because females usually have some control over who wins the sperm competition contests inside their bodies.

Birkhead's article explores both the proximate mechanisms by which some males gain an advantage in the competition among sperm to fertilize a given female's eggs and the evolutionary reasons why females might mate with several males, creating the conditions that lead to sperm competition in the first place. You should be able to identify and list the alternative proximate hypotheses tested by Birkhead as well as the different functional explanations that he discusses. Can you come up with any additional hypotheses that Birkhead does not mention?

That sperm competition is widespread among birds and other animals tells us something about the frequency of multiple mating by females (polyandry). At one time monogamy was thought to be the standard mating system for birds in particular, but now we know better. Although most birds pair off, one male to one female, creating what is called social monogamy, members of both sexes may mate with other individuals with the result that broods of mixed paternity are common. The realization that social associations do not necessarily indicate exclusive mating relationships has revolutionized the way behavioral biologists think about mating systems generally and monogamy particularly. The adaptive value for females and males that choose to mate with individuals other than their social partner is a topic of intense interest and debate. After reading Birkhead's article you should be able to advance some ideas about why social and sexual monogamy often do not coincide due to the attempts of both females and males to maximize their genetic success.

But there are cases in which a socially monogamous pair also is genetically monogamous (with females producing broods sired only by their social partner). Lowell Getz and C. Sue Carter describe a possible case of "true" monogamy in a diminutive mammal, the prairie vole. Although these mouse-sized rodents sometimes live in in mixed-sex groups, they not infrequently forms pairs, and when this happens the female's social partner refuses to respond sexually to other receptive virgin females. Why is this remarkable, and what explanation can you provide for the males' behavior? Other interesting aspects of reproduction in prairie voles include the (1) refusal of females to mate with familiar males, (2) frequent failure of incestuous matings to result in pregnancy, and (3) reproductive suppression of virgin females that smell a pheromone produced by reproductively active females in their group. What is surprising about each of these phenomena? Can you develop testable (i.e., falsifiable) hypotheses at different levels of analysis (see the article by Holekamp and Sherman in Part II) that might explain why females possess these attributes?

Getz and Carter believe that the variable social system of prairie voles is an adaptive response to seasonal variation in the amount of available food. An alternative hypothesis is that the voles' social system varies in response to predation risk, which might be greater in the winter, when a brown vole against a snow-covered background would be an easy target for a fox or hawk. How would you test these alternative explanations?

In fishes, the diversity of mating systems reaches a level of complexity unknown among mammals and birds. In fact, as Robert Warner shows, even "maleness"

and "femaleness" in reef fishes may be transitory. Individuals in many species start out their reproductive lives as females, but later become brilliantly colored males. In other species, individuals can change from male to female, while in a few deep-water species, individuals produce both sperm and eggs throughout life. Warner's article develops a general theory to explain how sexual selection and mating system diversity underlie the evolution of these different kinds of hermaphroditism among fishes.

Warner's article offers an opportunity to identify and contrast hypotheses for sex reversal at multiple levels of analysis. Given the fitness benefits that Warner believes derive from the ability to switch sex under appropriate circumstances, why do you suppose sex reversal is a relatively rare phenomenon, even in fishes, and is unknown in birds, mammals, reptiles, and insects? Can you develop a nonadaptationist explanation for the irregular distribution of sex reversal? What are the reproductive costs of sex reversal and hermaphroditism?

Next, David Buss applies an adaptationist approach to human sexual behavior—or rather, to the psychological foundations of human sexual behavior. In Western society, men appear to be much more interested in casual sex—or short-term matings, in Buss's terminology—than are unmarried women, who appear eager to acquire a committed partner or long-term mate. Furthermore, men generally find youthfulness in a potential partner attractive, whereas women generally consider wealth and high social status to be sexually appealing in a man. One general theory to account for these differences is popular among social scientists, namely, that the sexual preferences of men and women are the learned products of tradition, which depends purely on the particular culture that these men and women belong to. How does Buss, a social psychologist, use comparisons among modern cultures to test this theory?

Buss presents an alternative explanation for male and female sexual preferences ("sexual strategies theory") that is based on arguments about the minimum parental investments of men and women. Is this theory different from the one presented by Thornhill and Gwynne? If so, how does it differ?

In essence, Buss proposes that male and female sexual psychology differ adaptively, helping men and women reproduce maximally in a social and biological environment in which the obstacles to producing surviving descendants differ for the two sexes. This hypothesis assumes that evolved psychological reactions are still adaptive when they are produced today because the current human environment is similar enough to the ancestral human environment. This assumption is shared by Sherman and Flaxman (Part III) in their discussion of why spices taste good. Is the assumption reasonable for the behaviors being studied? Is it reasonable for all behaviors of contemporary humans?

Buss tests his argument by generating and evaluating nine hypotheses derived from sexual strategies theory. But if a sexual strategies hypothesis is a causal explanation for why men or women follow a particular mating strategy, then are some of Buss's "hypotheses" really predictions taken from sexual strategies theory? What is the difference between a hypothesis and a prediction? Can you reorganize Buss's article, starting with a description of key differences in male and female attitudes about casual sex, followed by hypotheses, predictions, and tests of the predictions?

Buss used written and oral questionnaires to gather his data. When humans reply to questions, however, they do not always tell the truth. Therefore, how can students of behavior use answers to questionnaires as evidence of people's real emotions and behavioral actions? How does Buss deal with this problem? Would direct observations of humans in mate-choice situations be preferable? Would this approach be practical?

The last article in Part IV, by Douglas Mock, Hugh Drummond, and Christopher Stinson, shifts the focus from reproductive behavior to parental care. A common product of reproductive behavior is progeny, and in some species parents are highly solicitous toward their young. But in a surprising number of predatory bird species, parents tolerate the elimination of one chick by another, rarely interfering in this siblicidal aggression, even when one chick is near death. Adaptationists were justifiably puzzled by parental acceptance of siblicide when it was first discovered because the trait seemed impossible to explain via natural selection. In fact, some persons have argued that the apparently "stupid" behavior of the parent birds occurs because of a physical constraint on brain size. Since birds must economize on weight if they are to get off the ground, they can afford to carry only a limited amount of brain tissue—even if

this means that they are intellectually handicapped. If so, siblicide should be rampant in hummingbirds (it is not).

Avian siblicide will enable you to contrast the "bird-brained" argument with an adaptationist alternative. Mock and his colleagues hypothesize that fitness benefits can flow to the parents of siblicidal offspring. These benefits may arise when food shortages make it impossible for parents to provision a full clutch all the way to fledging. The authors test their idea using a clever combination of comparative and experimental approaches. Consider how examples of convergent evolution (the independent production of similar traits in unrelated species) and divergent evolution (the production of different traits within a lineage of closely related species) were used in this article to generate comparative tests of the adaptive siblicide hypothesis. Finally, consider why, if food is the important limiting factor, cannibalism does not occur—siblicidal chicks never eat the siblings whose deaths they have caused.

Taken together, the articles in Part IV show how behavioral biologists have attempted to make evolutionary sense of some challenging puzzles about reproductive behavior. Why do the sexes differ in eagerness to mate and criteria for choosing partners? Why do some species exhibit reversal of sex roles, and even of individuals' sex? Why do parents tolerate, and sometimes even encourage, the murder of certain offspring? The success that behavioral ecologists have had in answering these questions by studying Mormon crickets, reef fishes, prairie voles, and even humans testifies to the range and power of an evolutionary approach to animal behavior.

Randy Thornhill
Darryl T. Gwynne

The Evolution of Sexual Differences in Insects

The ultimate cause of sexual differences in behavior may be the relative contribution of the sexes to offspring

Evolutionary biologists strive to understand the diversity of life by the study of the evolutionary processes that produced it. Among the more fundamental of these processes is sexual selection, which has probably been a major form of selection in the evolutionary background of all organisms with two sexes. An important question for evolutionary biologists, as one theorist (Williams 1975) has put it, is "Why are males masculine and females feminine and, occasionally, vice versa?" This question focuses on evolutionary causation by natural or sexual selection rather than on the proximate causes of sexual differences, such as genetic influences, hormones, or development.

Sexual selection is distinguished from natural selection in terms of how the differential in the reproduction of individuals is brought about (Darwin 1874). Sexual selection is differential reproduction of individuals in the context of competition for mates. Natural selection is differential reproduction of individuals due to differences in survival. Since reproduction is necessary for selection

Randy Thornhill is Professor of Biology at the University of New Mexico. He received his B.S. and M.S. from Auburn University and his Ph.D. (1974) from the University of Michigan. His research interests include sexual selection, the evolution of sexual differences, insect mating systems, and human social behavior. Darryl Gwynne attended the University of Toronto and Colorado State University (Ph.D. 1979). He was a postdoctoral fellow at the University of New Mexico and is currently a Research Fellow at the University of Western Australia. His interests include sexual selection and communication in insects. The research reported here was supported by grants from the NSF, a Queen Elizabeth II Fellowship (Australia), and the Australian Research Grants Scheme. Address for Dr. Thornhill: Department of Biology, University of New Mexico, Albuquerque, NM 87131.

to act, natural selection also includes differential reproduction of individuals in the context of reproductive acts, such as obtaining a mate of the right species, proper fertilization, and so on. Although both forms of selection involve competition between individuals for genetic representation, competition for mates is a key factor for distinguishing sexual from natural selection. The competition among members of one sex (usually males) for the opposite sex may take the form of attempting to coax choosy individuals to mate, leading to intersexual selection, or may involve striving to obtain access to already receptive individuals who are willing to mate with any individual of the opposite sex, leading to intrasexual selection.

In his treatise on sexual selection, Darwin (1874) compiled an encyclopedic volume of comparative support for the crucial role of the process in the evolution of morphological and behavioral traits important in sexual competition. Current studies of sexual selection involve several approaches. Some researchers seek to describe the consequences and nature of sexual selection by the study of behavioral and morphological traits important in sexual competition as well as by observing the types of mating associations (monogamy and polygyny, for example) in animals and plants (e.g., Bradbury and Vehrencamp 1977; Emlen and Oring 1977; Thornhill and Alcock 1983). Another approach is to measure the intensity of sexual selection, focusing on variation in the reproductive success of individuals in nature (e.g., Payne 1979; Wade and Arnold 1980; Thornhill 1981, 1986; Gwynne 1984a). Other studies are attempting to elucidate how sexual selection works, and there are several compet-

ing hypotheses (reviewed in Thornhill and Alcock 1983). Subtle forms of female mate choice and male-male competition for females have been discovered and are under investigation (Parker 1970; Thornhill 1983; Smith 1984). The area of study we will address in this paper concerns factors that control the operation of sexual selection—factors that govern the extent to which one sex competes for the other.

The evidence of sexual selection in nature raises a number of questions about these factors. When reproducing, why are males usually more competitive and less discriminating of mates than females and thus subject to greater sexual selection? And why in a few exceptional species is the intensity of sexual selection on the sexes apparently reversed, with females competing for males and males discriminating among mates? Moreover, why does the extent of sexual selection vary in the same species?

We will address these questions using evidence from insects, which, as one of the most diverse groups of animals, exhibit a variety of different reproductive biologies and thus provide a wealth of comparative information with which to examine the theory of sexual selection. We will first discuss theory concerning the control of sexual selection and then examine the theory in light of what is known about insects.

Control of sexual selection

Bateman (1948) argued that males typically are more sexually competitive than females primarily because of the sexual asymmetry in gamete size. He noted that female fertility is limited by the production of large,

Figure 1. The female in this mating pair of hangingflies (*Hylobittacus apicalis*) is feeding on a blow fly captured and provided to her by the male. This nuptial feeding by the male enhances the female's fecundity, reduces the risks that must be taken by the female to obtain food, and therefore promotes the male's reproductive success. (From Thornhill 1980.)

costly gametes; for the same amount of reproductive effort, a male produces vastly more gametes. Therefore, the reproductive success of males is limited by their success at inseminating females and not, as in females, by their ability to produce gametes. Bateman's work with fruit flies, *Drosophila melanogaster,* also demonstrated empirically that the intensity of sexual selection is greater on males than on females, and the difference is due to greater variation in mating success among males.

Williams (1966) and Trivers (1972) built a more comprehensive theory than Bateman in noting that what controls the intensity of sexual selection and explains the evolution of sexual differences in reproductive strategy is not just prezygotic investment by the sexes in gametes but all goods and services that contribute to the next generation—that determine the number and survival of offspring. The large disparity in gamete size itself predicts neither the reduced sexual differences seen in monoga-

mous species nor the competitive females and choosy males seen in species with reversals in sex roles.

Material contributions by each sex to the next generation determine the reproductive rate of the population; therefore, sexual competition is ultimately for these contributions because the greater the amount obtained, the higher the reproductive success of a competitor (Thornhill 1986). Simply put, when one sex (usually the females) contributes more to the production and survival of offspring, sexually active members of this sex are in short supply and thus become a limiting resource for the reproduction of the opposite sex; and the extent of sexual competition in the sex contributing least should correspond with the degree to which the opposite sex exceeds it in this contribution. Furthermore, the sex with the greater contribution is subject to a greater loss of fitness if it makes an improper mate choice, because its contribution represents a large fraction of its total reproductive

contribution. This asymmetry, coupled with the availability of the sex investing less, is expected to favor mate choice by the sex contributing more.

An important point to note is that not all forms of effort expended by the sexes in reproduction are expected to control the extent of sexual selection. Energetic expenditures by males that are used in obtaining fertilizations but that do not allow greater reproduction by females or do not promote the fitness of offspring are excluded (Trivers 1972; Thornhill and Alcock 1983; Gwynne 1984b).

Male efforts that are expected to control the extent of sexual selection encompass an array of resources that are of material benefit to females, offspring, or both. Such resources include nutrition, protection, and, under certain conditions, genes in spermatozoa. Bateman argued persuasively that female reproductive success is rarely limited by sperm per se. However, a high variance in the genetic quality of males available as mates may result in female competition for the best mates if genetic variation among males affects offspring survival and if males of high genetic quality limit female reproductive success. Although there is little information with which to examine the influence of high-quality sires on the operation of sexual selection, there are a number of studies, which we will now discuss, that examine sexual selection in species in which males provide immediate, material services.

Contributions to offspring

Both males and females can provide for offspring in a variety of ways. Females invest directly in their offspring both through the material investment in eggs and zygotes as well as through maternal care. The better-known examples of male investment concern direct paternal care of offspring; this has been observed in a large number of species in a variety of animal groups (Ridley 1978). However, males can also contribute indirectly to offspring by providing benefits to their mates, both before and after mating. Examples include the nutritional benefits of "courtship feeding," which is observed in certain birds (Nisbet 1973) and insects, such as the hangingfly *Hylobittacus apicalis* shown in Figure 1 (Thornhill 1976). Protection of the mate is another example of such benefits (Gwynne 1984b; Thornhill 1984). The nature of selection leading to the evolution of benefit-providing males is still poorly understood (Alexander and Borgia 1979; Knowlton 1982). But regardless of the evolved function of the phenomenon, with the evolution of benefit-providing males there may be a change in the action of sexual selection on the sexes.

Contributions by one sex that affect the number and survival of offspring potentially limit the reproduction of the opposite sex. Thus the relative investment of the sexes in these sorts of reproductive efforts should determine the extent to which each sex competes for the opposite sex. This hypothesis can be tested by comparing species or populations with differing investment patterns or by directly manipulating resources that limit the reproduction of the population. Examples we will review use these methods.

There have been few attempts to estimate the extent of sexual selection on the sexes in nature. However, variance in the reproductive success of the sexes has been estimated for *Drosophila melanogaster* (Bateman 1948), a damselfly (Finke 1982), red-winged blackbirds (Payne 1979), and red deer (Clutton-Brock et al. 1982). In these species, the parental contribution by the male is smaller than that of the female; as predicted, all species show greater sexual selection on the males.

Observed sexual differences, the consequences of sexual selection, typically serve as evidence for the relative intensity of sexual selection on the sexes in the evolutionary past. In most species, females provide a large amount of parental contribution and males little, and it is primarily the males that show secondary sexual traits of morphology and behavior which function in competition for mates. It is also well known that sexual differences are greatly reduced under monogamy. This is as expected, because both sexes of monogamous species engage in similar levels of parental care. However, for this comparative test to succeed, a sex reversal in the courtship and competitive roles should be observed in species in which males provide a greater portion of the total contribution affecting offspring number and survival.

Parental care provided only by the male is found throughout the vertebrates, particularly in frogs and toads, fishes, and birds, and is likely to represent a limiting resource for female reproduction. In certain seahorses and pipefishes (Syngnathidae), males care for eggs in a specialized brood pouch (Breder and Rosen 1966). In these fishes, male parental care appears to limit female reproductive success, and females are larger and more brightly colored than males, as well as being more competitive in courtship (Williams 1966, 1975; Ridley 1978). In frogs of the genus *Colostethus,* it is the male in one species and the female in another that provide parental care by carrying tadpoles on their backs. In both species, as predicted by theory, it is the sex emancipated from parental duties that defends long-term mating territories and has a higher frequency of competitive encounters (Wells 1980). In species of birds in which males provide most of the parental care, the roles in courtship behavior are reversed; females compete for mates and sometimes are the larger or more brightly colored sex (reviewed by Ridley 1978).

Exclusive paternal care of eggs or larvae is restricted to about 100 species of insects, all of which are within the order Hemiptera, or true bugs (Smith 1980). In the giant water bugs (Hemiptera: Belostomatidae), females adhere eggs to the wing covers of their mates, and the males aerate the eggs near the water surface and protect them from predators. For *Abedus herberti,* Smith (1979) provides evidence that male back space is a limiting resource for females and that male parental care is essential for offspring survival; females actively approach males during courtship, and males reject certain females as mates.

Although direct investment in offspring through parental care is uncommon within the insects, indirect paternal contributions, with males supplying the females with nutrition or other services such as guarding, is widespread in a number of taxa.

The guarding of the female by the male after mating is usually thought of as functioning to prevent other males from inseminating the female (Parker 1970). However, an alternative hypothesis is that guarding evolved in the context of supplying protection for the female and that it thereby enhances male reproductive success (Gwynne 1984b; Thornhill 1984). Male guarding is known to benefit the female in several species: in damselflies (*Calopteryx maculata*), guarding by males after mating allows females to oviposit undisturbed by other males (Waage 1979); in waterstriders (*Gerris remigis*), harassment of guarded females by other males is similarly reduced, allowing females much longer periods during which to forage for food (Wilcox 1984). At present there is no information for these species concerning whether certain males protect females better than others, which would lead to female competition for more protective males. If female competition occurs, selection should favor mate choice by males.

Mate choice by males has been observed in species in which males provide protection or other services to females. In brentid weevils (*Brentus anchorago*) males prefer large females as mates and are known to assist their mates in competition for

oviposition sites by driving away nearby ovipositing females (Johnson and Hubbell 1984). Male lovebugs (*Plecia nearctica*), so named for their two- to three-day-long periods of copulation, also prefer to mate with large females (Hieber and Cohen 1983). Lengthy copulation in this species may be beneficial for females in that copulating pairs actually fly faster than unattached lovebugs (Sharp et al. 1974). Similar benefits may be obtained by paired amphipod crustaceans (*Gammarus pulex*); pairs in which males are larger than females have a superior swimming performance that minimizes the risk of being washed downstream (Adams and Greenwood 1983). Perhaps these sorts of services supplied by male crustaceans explain the presence of male choice of mates seen in certain groups (e.g., Schuster 1981).

Males can supply nutrition in several ways. Our research has dealt with courtship feeding, where food items such as prey or nutritious sperm packages (spermatophores) are eaten by females, and we discuss this behavior in detail below for male katydids and scorpionflies. There are also more subtle forms of contribution; in several insect species spermatophores or other ejaculatory nutrients are passed into the female's genital tract at mating (Thornhill 1976; Gwynne 1983; Thornhill and Alcock 1983). A number of researchers have done some interesting work on a similar phenomenon in crustaceans. Electrophoretic studies of proteins in the ovaries and the male accessory gland of a stomatopod shrimp (*Squilla holoschista*) strongly suggest that a specialized protein from the male's accessory glands is transferred with the ejaculate into the female's gonopore and then is translocated intact into the developing eggs; females of this species usually initiate mating and will mate repeatedly (Deecaraman and Subramoniam 1983). In a detailed study of another

Figure 2. A female katydid (*Requena verticalis*) just after mating shows the large spermatophore that has been attached by the male to the base of her ovipositor (*top*). The female grasps the nutritious spermatophylax (*middle*) and eats it (*bottom*), leaving the sperm ampulla portion of the spermatophore in place. After insemination, the ampulla is also eaten. The nutrients in the spermatophore represent a considerable material contribution by the male in the reproductive success of the female. (Photos by Bert Wells.)

Figure 3. While this female Mormon cricket (*Anabrus simplex*) is atop the male, the male apparently weighs his potential mate and will reject her if she is too light. Males select mates among females that compete for access to them, preferring females that are larger and therefore more fecund. This represents a reversal of the sex roles much more commonly found in nature. (Photo by Darryl Gwynne.)

stomatopod, *Pseudosquilla ciliata*, Hatziolos and Caldwell (1983) report a reversal in sex roles, with females courting males that appear reluctant to mate; in the absence of obvious male parental contribution, these researchers cite work with insects in suggesting that male *Pseudosquilla* may provide valuable nutrients in the ejaculate.

Studies with several butterfly species have used radiolabeling to show that male-produced proteins are incorporated into developing eggs as well as into somatic tissues of females (e.g., Boggs and Gilbert 1979). Lepidopteran spermatophores potentially represent a large contribution by the male (up to 10% of body weight), and proteins ingested by females are likely to represent a limiting resource for egg production in these insects that feed on nectar as adults (Rutowski 1982). Preliminary experiments by Rutowski (pers. com.) with alfalfa butterflies (*Colias* spp.) indicate that females receiving larger spermatophores lay more eggs. Consistent with theory, there is evidence of males choosing females and of competition by females for males. For example, in the checkered white butterfly (*Pieris protodice*), males prefer young, large (and thus more fecund) females to older, small-

er individuals (Rutowski 1982). And in *Colias*, certain females were observed to solicit courtship by pursuing males; these females may have had reduced protein supplies, as they were shown to have small, depleted spermatophores in their genital tracts (Rutowski et al. 1981). Although there is variation between species of butterflies in the size of the male spermatophore, this variation apparently does not result in large differences between species in the male contribution; a review of the reproductive behavior of several butterfly species did not show consistent differences in courtship when species with small spermatophores were compared to those with large spermatophores (Rutowski et al. 1983). However, as shown by Marshall (1982) and confirmed by our studies described below, spermatophore size is not always a useful measure of the importance of the male nutrient contribution.

Reversal of courtship roles in katydids

Katydids (Orthoptera: Tettigoniidae) are similar to butterflies in that males transfer spermatophores to their mates, and, as shown by radiolabeling, spermatophore nutrients are

used in egg production (Bowen et al. 1984). In contrast to the mated female butterfly, the katydid female ingests the spermatophore by eating it (Fig. 2). The spermatophore consists of an ampulla which contains the ejaculate and a sperm-free mass termed the spermatophylax (Gwynne 1983). Immediately after mating, the female first eats the spermatophylax; while this is being consumed, insemination takes place, after which the empty sperm ampulla is also eaten (Gwynne et al. 1984). However, the katydid spermatophylax appears not to function as protection of the ejaculate from female feeding. The spermatophylax of the katydid *Requena verticalis* is more than twice the size necessary to allow the transfer both of the spermatozoa and of substances that induce a four-day nonreceptive period in females (Gwynne, unpubl.).

Spermatophore nutrients are important to the reproductive success of the female katydid. Laboratory experiments have shown that as consumption of spermatophylax increases, both the size and the number of eggs that females subsequently lay also increase (Gwynne 1984c). Furthermore, the increase in the size of eggs appears to be determined only by male-provided nutrient; an increase in protein in the general diet increases egg number but does not affect egg size (Gwynne, unpubl.).

The size of the spermatophore produced by male katydids varies from less than 3% of male body weight in some species to 40% in others (Gwynne 1983). Differences in the size of the male contribution conform with the predictions of sex-difference theory: in two species of katydids that make very large investment in each spermatophore (25% or more of male weight) and that have been examined in detail—the Mormon cricket (*Anabrus simplex*) and an undescribed species (*Metaballus* sp.) from Western Australia—there is a complete reversal in sex roles, with females competing aggressively for access to males that produce calling sounds, and males selecting mates, preferring large, fecund females (Gwynne 1981, 1984a, 1985). Figure 3 illustrates this reversal in the Mormon cricket. There is no evidence of such a reversal of courtship roles in species with smaller spermatophores; in these species males compete for mating territories and females select mates (Gwynne 1983).

It is evident, however, that a complete estimate of the contribution to offspring requires more than a simple measure of the relative contribution by the sexes to offspring such as the weight of the spermatophore relative to the weight of a clutch of eggs. Both species of katydids showing a role reversal in courtship behavior also had populations that showed no evidence of the reversed roles. For the Mormon cricket, the simple measure of relative contribution did not show a higher contribution by males at the sites of role reversal (Gwynne 1984a). However, these sites had very high population densities, with individuals of both sexes competing vigorously for food in the form of dead arthropods and certain plants. These observations suggest the hypothesis that the limited food supplies at these sites resulted in few spermatophores being produced and that spermatophore nutrients were thus a limiting resource for female reproduction. Food did not appear to be scarce at sites of low population density where the reversal in courtship roles was not observed. Support for the hypothesis that food is a limiting resource at high-density sites came from dissections of the reproductive accessory glands that produce the spermatophore in a sample of males from each of the sites. Only the few calling males at sites of high density had glands large enough to produce a spermatophore, whereas most males at the low-density site had enlarged glands. This difference between the males at the high- and low-density sites was not a result of a higher number of matings by males at the high-density site.

Differences between individuals from the two sites indicate that sexual selection on females at high-density sites was intense compared to the low-density site. (Sexual selection is measured by variance in mating success; see Wade and Arnold 1980.) Some females were very successful at obtaining spermatophores. These tended to be large females that were preferred by males as mates. The evolutionary consequences of the apparently greater sexual selection on high-density females was not only aggressive female behavior in competition for calling mates but also a

larger female body size at this site relative to males. This sexual dimorphism was not seen at sites of low density.

Variation in the expression of sexual differences within the same katydid species suggested that behavior might be flexible; that is, females become competitive and males choosy when they encounter certain environments. This hypothesis was examined using the undescribed *Metaballus* species of katydid from Western Australia, which is similar to the Mormon cricket in that only certain populations show female competition for mates and show males that reject smaller, less fecund females. In this species, discriminating males call females by producing a broken "zipping" song from deep in the vegetation. Sites where courtship roles are reversed consist of mainly the zipping male song, whereas at sites of male competition, males produce continuous songs that appear to be louder. An experiment was conducted which involved shifting a number of males and females from a site where role reversal was not observed to one in which it was noted. The behavior of the males that were moved to the role-reversed site changed to resemble that of the local males: their song changed from a continuous to a zipping song, the duration of courtship increased (possibly to assess the quality of their mates), and they even rejected females as mates. Thus, sexual differences in behavior are plastic; courtship roles of the sexes appear to be dependent on the environment encountered.

For the Mormon cricket, it is likely that the relative contributions of the sexes is the factor controlling sexual selection. A comparison of the weights of spermatophores and egg clutches is undoubtedly a poor estimate of relative contribution by the sexes; spermatophores seem to be important to the reproduction of females at both sites (Gwynne 1984a). However, if food supplies limit spermatophore production at sites where reversals in courtship roles are observed, and if females cannot obtain spermatophore nutrients from other food sources, then spermatophores are likely to have a greater influence on female fecundity and thereby are more valuable to female reproduction at these sites. Thus, the total

contribution from the males at these sites is probably larger than that of the females.

Nuptial feeding in scorpionflies

Most of the evidence supporting the hypothesis that the relative contribution of the sexes to offspring is an important factor controlling the extent of sexual selection has been derived from comparisons between or, in the katydid work, within species. In contrast, studies were conducted in which the relative contribution of males was manipulated to determine its effect on the extent of sexual selection (Thornhill 1981, 1986). This research has focused on scorpionflies of the genus *Panorpa* (Panorpidae), in which males use either dead arthropods or nutritious products of salivary glands to feed their mates.

Males must feed on arthropod carrion, for which they compete through aggression, before they can secrete a salivary mass. Males in possession of a nuptial gift release pheromone that attracts conspecific females from some distance. Females can obtain food without male assistance, but doing so is risky because of exposure to predation by web-building spiders. Movement in the habitat required to find dead arthropods exposes females and males to spider predation, and dead arthropods unattended by males are frequently found in active spider webs. The gift-giving behavior of males is an important contribution, because dead arthropods needed by females to produce eggs are limited both in the absolute sense and in terms of the risks in obtaining them.

In a series of experiments, individually marked male and female *Panorpa latipennis* were placed in field enclosures, and variances in mating success of the sexes were determined in order to estimate the relative intensity of sexual selection. Dead crickets taped to vegetation represented the resource that males defended from other males and to which females were attracted. In one experiment, three treatments were established in which equal numbers of males and females were added to each enclosure and the number of dead crickets varied—two, four, or six crickets per enclosure. As predicted, competition among males was greatest in the enclosures with two crickets; the intensity of sexual selection, calculated by variance in male mating success, was greatest in this treatment and was lowest in the treatment with six crickets.

Variance in female mating success was low and was not significant across cricket abundances over the seven days of the experiment. Sexual selection on females probably often arises from female-female competition for the best mates regardless of the number of mates. *Panorpa* females prefer males that provide large, fresh nuptial gifts of dead arthropods over males that provide salivary masses, and males only secrete saliva when they cannot compete successfully for dead arthropods (Thornhill 1981, 1984). This female mating preference is adaptive in that females mating with arthropod-providing males lay more eggs than females mating with saliva-providing males. Thus an accurate measure of sexual selection on female *Panorpa* would include the variation in egg output by females in relation to the resource provided by the mates of females. This information is not available at present.

However, the results on males from this experiment clearly support the hypothesis that sexual selection is determined by the relative contribution of the sexes. As food is a limiting resource for reproduction by female scorpionflies, the total contribution of food by males in enclosures with more crickets was greater than in enclosures with fewer crickets, and the intensity of sexual selection on males declined as males contributed relatively more.

Such studies of the factors controlling the operation of sexual selection are important for two major reasons. The first is simply that sexual selection has been such an important factor in the evolution of life. Sexual selection seems inevitable in species with two sexes, because, as Bateman (1948) first pointed out, the relatively few large female gametes will be the object of sexual competition among the males, whose upper limit to reproductive success is set by the number of ova fertilized rather than by production of the relatively small, energetically cheap sperm. The role of sexual selection in the history of life can best be explored when such controlling factors are fully understood. The second reason is related to the first: the difference in the operation of sexual selection on the sexes may ultimately account for all sexual differences. Only sexual selection acts differently on the sexes per se (Trivers 1972). Natural selection may act on and may even magnify sexual differences in behavior and morphology, but probably only after these differences already exist as a result of the disparate action of sexual selection.

The insight of Williams (1966) and Trivers (1972) is that the relative contribution of materials and services by the sexes in providing for the next generation is the most important factor controlling the operation of sexual selection. In insects, contributions supplied by males to their mates include not only the paternal care of young, a well-studied phenomenon in vertebrates, but also other services such as courtship feeding, subtle forms of nutrient transfer via the reproductive tract, and "beneficial" guarding of mates.

References

Adams, J., and P. J. Greenwood. 1983. Why are males bigger than females in precopula pairs of *Gammarus pulex*? *Behav. Ecol. Sociobiol.* 13:239–41.

Alexander, R. D., and G. Borgia. 1979. On the origin and basis of the male-female phenomenon. In *Sexual Selection and Reproductive Competition in the Insects*, ed. M. S. Blum and N. A. Blum, pp. 417–40. Academic.

Bateman, A. J. 1948. Intrasexual selection in *Drosophila*. *Heredity* 2:349–68.

Boggs, C. L., and L. E. Gilbert. 1979. Male contribution to egg production in butterflies: Evidence for transfer of nutrients at mating. *Science* 206:83–84.

Bowen, B. J., C. G. Codd, and D. T. Gwynne. 1984. The katydid spermatophore (Orthoptera: Tettigoniidae): Male nutrient investment and its fate in the mated female. *Aust. J. Zool.* 32:23–31.

Bradbury, J. W., and S. L. Vehrencamp. 1977. Social organization and foraging in emballonurid bats. III. Mating systems. *Behav. Ecol. Sociobiol.* 2:1–17.

Breder, C. M., and D. E. Rosen. 1966. *Modes of Reproduction in Fishes*. Nat. Hist. Press.

Clutton-Brock, T. H., F. E. Guinness, and S. D. Albon. 1982. *Red Deer: Behavior and Ecology of Two Sexes*. Univ. of Chicago Press.

Darwin, C. 1874. *The Descent of Man and Selection in Relation to Sex*, 2nd ed. New York: A. L. Burt.

Deecaraman, M., and T. Subramoniam. 1983. Mating and its effect on female reproductive physiology with special reference to the fate of male accessory sex gland secretion in the stomatopod, *Squilla holoschista*. *Mar. Biol.* 77:161–70.

Emlen, S. T., and L. W. Oring. 1977. Ecology, sexual selection, and the evolution of mating systems. *Science* 197:215–22.

Finke, O. M. 1982. Lifetime mating success in a natural population of the damselfly *Enallagma hageni* (Walsh) (Odonata: Coenagrionidae). *Behav. Ecol. Sociobiol.* 10:293–302.

Gwynne, D. T. 1981. Sexual difference theory: Mormon crickets show role reversal in mate choice. *Science* 213:779–80.

———. 1983. Male nutritional investment and the evolution of sexual differences in the Tettigonidae and other Orthoptera. In *Orthopteran Mating Systems: Sexual Competition in a Diverse Group of Insects*, ed. D. T. Gwynne and G. K. Morris, pp. 337–66. Westview.

———. 1984a. Sexual selection and sexual differences in Mormon crickets (Orthoptera: Tettigoniidae, *Anabrus simplex*). *Evolution* 38:1011–22.

———. 1984b. Male mating effort, confidence of paternity, and insect sperm competition. In Smith 1984, pp. 117–49.

———. 1984c. Courtship feeding increases female reproductive success in bushcrickets. *Nature* 307:361–63.

———. 1985. Role-reversal in katydids: Habitat influences reproductive behavior (Orthoptera: Tettigoniidae, *Metaballus* sp.). *Behav. Ecol. Sociobiol.* 16:355–61.

Gwynne, D. T., B. J. Bowen, and C. G. Codd. 1984. The function of the katydid spermatophore and its role in fecundity and insemination (Orthoptera: Tettigoniidae). *Aust. J. Zool.* 32:15–22.

Hatziolos, M. E., and R. Caldwell. 1983. Role-reversal in the stomatopod *Pseudosquilla ciliata* (Crustacea). *Anim. Behav.* 31:1077–87.

Hieber, C. S., and J. A. Cohen. 1983. Sexual selection in the lovebug, *Plecia nearctica*: The role of male choice. *Evolution* 37:987–92.

Johnson, L. K., and S. P. Hubbell. 1984. Male choice: Experimental demonstration in a brentid weevil. *Behav. Ecol. Sociobiol.* 15:183–88.

Knowlton, N. 1982. Parental care and sex role reversal. In *Current Problems in Sociobiology*, ed. King's College Sociobiology Group, pp. 203–22. Cambridge Univ. Press.

Marshall, L. D. 1982. Male nutrient investment in the Lepidoptera: What nutrients should males invest? *Am. Nat.* 120:273–79.

Nisbet, I. C. T. 1973. Courtship-feeding, egg size, and breeding success in common terns. *Nature* 241:141–42.

Parker, G. A. 1970. Sperm competition and its evolutionary consequences in the insects. *Biol. Rev. Cambridge Philos. Soc.* 45:525–67.

Payne, R. B. 1979. Sexual selection and intersexual differences in variance of breeding success. *Am. Nat.* 114:447–66.

Ridley, M. 1978. Paternal care. *Anim. Behav.* 26:904–32.

Rutowski, R. L. 1982. Mate choice and lepidopteran mating behavior. *Fla. Ent.* 65:72–82.

Rutowski, R. L., C. E. Long, and R. S. Vetter. 1981. Courtship solicitation by *Colias* females. *Am. Midl. Nat.* 105:334–40.

Rutowski, R. L., M. Newton, and J. Schaefer. 1983. Interspecific variation in the size of the nutrient investment made by male butterflies during copulation. *Evolution* 37:708–13.

Schuster, S. M. 1981. Sexual selection in the Socorro Isopod *Thermosphaeroma thermophilum* (Cole) (Crustacea: Peracarida). *Anim. Behav.* 29:698–707.

Sharp, J. L., N. C. Leppala, D. R. Bennett, W. K. Turner, and E. W. Hamilton. 1974. Flight ability of *Plecia nearctica* in the laboratory. *Ann. Ent. Soc. Am.* 67:735–38.

Smith, R. L. 1979. Paternity assurance and altered roles in the mating behaviour of a giant water bug, *Abedus herberti* (Heteroptera: Belostomatidae). *Anim. Behav.* 27:716–25.

———. 1980. Evolution of exclusive postcopulatory paternal care in the insects. *Fla. Ent.* 63:65–78.

———, ed. 1984. *Sperm Competition and the Evolution of Animal Mating Systems.* Academic.

Thornhill, R. 1976. Sexual selection and paternal investment in insects. *Am. Nat.* 110:153–63.

———. 1980. Sexual selection in the black-tipped hangingfly. *Sci. Am.* 242:162–72.

———. 1981. *Panorpa* (Mecoptera: Panorpidae) scorpionflies: Systems for understanding resource-defense polygyny and alternative male reproductive effort. *Ann. Rev. Ecol. Syst.* 12:355–86.

———. 1983. Cryptic female choice in the scorpionfly *Harpobittacus nigriceps* and its implications. *Am. Nat.* 122:765–88.

———. 1984. Alternative hypotheses for traits believed to have evolved in the context of sperm competition. In Smith 1984, pp. 151–78.

———. 1986. Relative parental contribution of the sexes to offspring and the operation of sexual selection. In *The Evolution of Behavior,* ed. M. Nitecki and J. Kitchell, pp. 10–35. Oxford Univ. Press.

Thornhill, R., and J. Alcock. 1983. *The Evolution of Insect Mating Systems.* Harvard Univ. Press.

Trivers, R. L. 1972. Parental investment and sexual selection. In *Sexual Selection and the Descent of Man, 1871–1971,* ed. B. Campbell, pp. 136–79. Aldine.

Waage, J. K. 1979. Adaptive significance of postcopulatory guarding of mates and non-mates by *Calopteryx maculata* (Odonata). *Behav. Ecol. Sociobiol.* 6:147–54.

Wade, M. J., and S. J. Arnold. 1980. The intensity of sexual selection in relation to male sexual behaviour, female choice, and sperm precedence. *Anim. Behav.* 28:446–61.

Wells, K. D. 1980. Social behavior and communication of a dendrobatid frog (*Colostethus trinitatis*). *Herpetologica* 36:189–99.

Wilcox, R. S. 1984. Male copulatory guarding enhances female foraging in a water strider. *Behav. Ecol. Sociobiol.* 15:171–74.

Williams, G. C. 1966. *Adaptation and Natural Selection.* Princeton Univ. Press.

———. 1975. *Sex and Evolution.* Princeton Univ. Press.

Animal Genitalia and Female Choice

William G. Eberhard

When I was a senior in college I took a course in ichthyology and learned to enjoy thumbing through taxonomic drawings, which displayed the fascinating theme-and-variations patterns that are so common in nature. The various species in a genus were basically similar, but each had a set of seemingly senseless and often surprising and aesthetically pleasing differences. Later that year I became interested in spiders, and I can still remember my disappointment upon finding that similar drawings of whole spiders did not accompany papers on spider taxonomy. Instead, illustrations in spider papers were limited to male and female genitalia, which are generally extremely complex structures lacking the elegant sweep of fish profiles. Even closely related species of spiders can usually be distinguished by the genitalia alone.

This was my first encounter with a major pattern in animal evolution: among closely related species that employ internal fertilization, the genitalia—especially male genitalia—often show the clearest and most reliable morphological differences. For some reason, the genitalia of most spiders have evolved rapidly, becoming distinct even in recently diverged lines. In contrast, animals that employ external fertilization, such as most fish, do not have species-specific genital morphology.

These trends are widespread. Groups in which intromittent genitalia (for placing gametes inside the mate) are often useful for distinguishing species include flatworms, nematodes, oligochaete worms, insects, spiders, millipedes, sharks and rays, some lizards, snakes, mites, opilionids, crustaceans, molluscs, and mammals (including rodents, bats, armadillos, and primates). In contrast, groups that employ external fertilization all lack species-specific genitalia; they include echinoderms, most polychaete worms, hemichordates, brachiopods, sipunculid worms, frogs, birds, a few insects, and most

Rapid evolutionary divergence of male genitalia may be explained by the ability of females to choose the paternity of their offspring

fish. In such cases, both males and females have only a simple opening through which gametes are released. Even within groups that have recently switched from external to internal fertilization, for example guppies, whose males use a modified anal fin to introduce sperm into the female, the intromittant organs are often useful for distinguishing species.

Rapid and divergent genital evolution also occurs in species in which the male, rather than penetrating the female himself, introduces a spermatophore, or package of his sperm, into the female. In many octopuses, squids, scorpions, some pseudoscorpions, some snails and slugs, some arrow worms, and pogonophoran worms, it is the spermatophore, rather than male genital structures, that is morphologically complex and species-specific.

Why do male mating structures possess such a bewildering diversity of forms? Surely the transfer of a small mass of gametes does not require the elaborate genital structures carried by the males of many groups. Two explanations were proposed some time ago: lock-and-key and pleiotropy. Neither is particularly convincing. According to the lock-and-key hypothesis, females have evolved under selection favoring those individuals that avoided wasting eggs by having them fertilized by sperm of other species. Elaborate, species-specific female genitalia (locks) admit only the genitalia of conspecific males (keys), enabling females to avoid mistakes in fertilization.

Originally proposed nearly 150 years ago for insects (see Nichols 1986), the lock-and-key idea fell into disrepute when it was established that locks are too easily picked. Studies of groups in which females have complex genitalia showed that the female genitalia could not exclude the genitalia of males of closely related species (for a summary of evidence see Shapiro and Porter 1989). The lock-and-key hypothesis is inapplicable in many other groups in which the female genitalia are soft and mechanically incapable of excluding incorrect keys while the male genitalia or spermatophores are nevertheless species-specific in form (flatworms, nematodes, arrow worms, annelid worms, sharks and rays, guppies and their relatives, snakes, lizards, snails and slugs, octopuses and squids, and many insects).

A species-isolation function, whether mechanical or otherwise, is improbable for several reasons. In some

William Eberhard is a member of the staff of the Smithsonian Tropical Research Institute and a professor at the University of Costa Rica. He received both undergraduate and graduate degrees from Harvard University. His research interests include the behavior and ecology of web-spinning spiders, functional morphology of beetle horns and earwig forceps, evolutionary interactions between subcellular organelles and plasmids and the cells that contain them, and the evolution of animal genitalia. Address: Escuela de Biología, Universidad de Costa Rica, Ciudad Universitaria, Costa Rica.

groups with species-specific genitalia, males and females exchange species-specific signals during courtship, and probably seldom if ever reach the point of making genital contact with members of other species. For example, some female moths attract males with species-specific blends of pheromones, and the males, after finding the females, court them with additional species-specific pheromones before beginning to copulate. Yet the male moths have species-specific genitalia (Baker and Cardé 1979). Species-specific genitalia have even evolved in situations in which mistaken, cross-specific matings are essentially impossible, for example in island-dwelling species isolated from all close relatives, or parasitic species that mate on hosts which never harbor more than one species of the parasite. In some of these groups, such as the pinworms of primates, male genitalia provide the best morphological characteristics known to distinguish closely related species (Inglis 1961).

The alternative hypothesis, pleiotropy, is no more satisfying. It holds that genital characteristics are chance effects of genes that code primarily for other characteristics, such as adaptations to the environment. But this idea fails to explain why incidental effects should consistently occur on genitalia and not elsewhere. Nor does it explain why incidental effects fail to occur in species employing external rather than internal fertilization. It also cannot account for the genital morphology of a number of groups, such as spiders and guppies, in which organs (e.g., a pedipalp or anal fin) other than the primary male genitalia acquire the function of introducing sperm into the female; these other organs consistently become subject to the putative "incidental" effects while the primary genitalia do not.

So, until recently, a pattern widespread in animal evolution was left without a plausible explanation. Recently, however, a resurgence of interest in Darwin's ideas on sexual selection and advances in evolutionary theory to encompass male-female conflicts have stimulated new hypotheses.

Two mechanisms proposed by Darwin are potentially involved: male-male competition and female choice. Recent hypotheses are that male genitalia function to remove or otherwise supersede sperm introduced in previous matings of the female (Waage 1979; Smith 1984), and that male genitalia are often used as "internal courtship devices"—inducing the female to use a male's sperm—and thus are under sexual selection by female

Figure 1. The male of this pair of *Altica* beetles performs complex courtship behavior after copulation has begun. Within the first two minutes of copulation, he inflates inside the female a sac which emerges from the tip of the brown, cylindrical basal portion of his genitalia *(visible in the top photo)*, and passes his sperm into the female. During the rest of the approximately 20-minute copulation, he periodically thrusts forward as far as he can *(middle photo)* in a more or less stereotyped pattern. The thrusts do not move his genitalia deeper into the female's genital tract; rather they stretch the walls of the entire basal portion of the tract as it is displaced forward within her body. Between bouts of thrusting the male often rubs his rear tarsi gently but persistently near the tip of the female's abdomen *(bottom photo)*. After he has withdrawn his genitalia, the male sometimes gives her additional, more vigorous rubs with his hind legs. (Photos by D. Perlman.)

Figure 2. The penis of *Notomys mitchelli*, an Australian rodent, is extremely elaborate. The movements of such an organ certainly do not go unnoticed by the female while it is within her. (From Breed 1986; courtesy of W. G. Breed.)

choice (Eberhard 1985). While sperm removal and displacement have been documented in several cases, these are unlikely to be general explanations for the trend of genitalia to evolve and diverge rapidly, because males of many groups with species-specific genitalia do not penetrate deep enough into the female to reach sites where sperm from previous copulations are stored. In the remainder of this article I present some of the evidence supporting the female-choice hypothesis, and show how it calls into question some basic and intuitively "obvious" notions about animal behavior and morphology.

Copulatory courtship and genital stimulation

The obvious function of a male's courtship behavior is to induce the female to mate. Yet males of many species of insects appear to court the female even after they have achieved genital coupling. Their behavior includes typical courtship movements such as waving antennae or colored legs, stroking, tapping, rubbing, or biting the female's body, buzzing the wings in stereotyped patterns, rocking the body back and forth, and singing (Eberhard, unpubl.). A survey of studies of copulation behavior in insects showed that, in just over one-third of 302 species, the male performs behavior apparently designed to stimulate the female (Eberhard, unpubl.). In some species, such as the beetles shown in Figure 1, the male combines movements of body parts such as antennae and hind feet with more or less stereotyped movements of his genitalia. Both genital and non-genital behavior patterns differ in closely related species.

Many male mammals move their genitalia in and out of the female in more or less stereotyped movements prior to insemination (Dewsbury 1972). Some also perform post-ejaculatory intromissions which differ from

the earlier ones and which increase the likelihood that the mating will result in the female becoming pregnant (Dewsbury and Sawrey 1984). Male goldeneye ducks perform four different displays after copulation (Dane and van der Kloot 1962).

Other observations also indicate that male genitalia themselves perform copulatory courtship. A study by Lorkovic (1952) showed that the genitalic "claspers" of some male butterflies are rubbed gently back and forth on the sides of the female's abdomen during copulation. Some snails thrust genitalic darts into the female during courtship or copulation (Fretter and Graham 1964), while some moths have elongate, sharp-pointed scales on their penes which are designed to fall off inside the female (Busck 1931), probably delivering stimuli to the female after the male has left. In a variety of groups, ranging from *Drosophila* flies to mice and sheep, copulatory behavior persists even after the male has exhausted his supply of sperm (Dewsbury and Sawrey 1984). In marmosets, male stimulatory effects during copulation have been documented by showing that some female responses disappear when the female's reproductive tract is anesthetized (Dixson 1986). Male cats and some male rodents have backwardly directed spines on their penes which make stimulation of the female inevitable (Fig. 2); ovulation in female cats is known to be induced by mechanical stimulation of the vagina (Greulich 1934; DeWildt et al. 1978). These stimulation devices probably represent mechanical equivalents of the visual displays of erect, brightly colored male penes in some lizards and primates (Bohme 1983; Eckstein and Zuckerman 1956; Hershkovitz 1979).

In some groups, mechanical stimulation and sperm transfer are particularly clear because they are performed separately. Male spiders and millipedes generally transfer sperm to modified structures on their pedipalps or legs, then use these secondary genitalia to introduce sperm into the female. In some species copulation always occurs first with the secondary genitalia empty; the male then withdraws, loads the secondary genitalia with sperm, and copulates again (Austad 1984). A number of species, including beetles, wasps, and rodents, perform a series of preliminary or extra intromissions which apparently do not result in sperm transfer (Cowan 1986; Schincariol and Freitag 1986; Dewsbury, in press). Mallards, one of the few bird groups having intromittent organs, frequently copulate during pair formation, six months before egg-laying, when the male gonads are repressed and sperm are not produced (McKinney et al. 1984).

Copulatory courtship behavior is understandable given the perspective that copulation is only one of a series of events which must occur if a male is to sire offspring. The female must remain still, or at least not actively attempt to terminate copulation prematurely. In rats and fleas, for instance, genitalic stimulation increases a female's tendency to stand still (Rodriguez-Sierra et al. 1975; Humphries 1967). The male's sperm must be transported to the storage site or fertilization site; this process seems to depend to a large extent on female peristalsis or other transport movements rather than on the motility of the sperm (see Overstreet and Katz 1977 on mammals; Davey 1965 on insects). Females of many species have, associated with their sperm-storage organs, glands which must be activated to help keep sperm alive and healthy. In some mammals and arthropods, sperm must be "activated" once inside the female in order to become capable of fertilization (Hamner et al. 1970; Leopold and Degrugillier 1973; Brown 1985). In some species, such as roaches and cats, ovulation and maturation of the eggs are induced by mechanical stimuli associated with copulation (Roth and Stay 1961; Greulich 1934). Brood care in earwigs is thought to be induced by mating (Vancassel cited in Lamb 1976).

Another critical female response often associated with copulation is the lack of further sexual receptivity. Sperm from subsequent matings can offer dangerous competition because it is extremely rare for internal fertilization of eggs to occur immediately following copulation. In some bees and wasps, postcopulatory courtship appears to reduce the frequency of remating by females (van den Assem and Visser 1976; Alcock and Buchmann 1985). Stimuli from both copulation (without spermatophore transfer) and the spermatophore itself reduce further sexual receptivity in the female butterfly

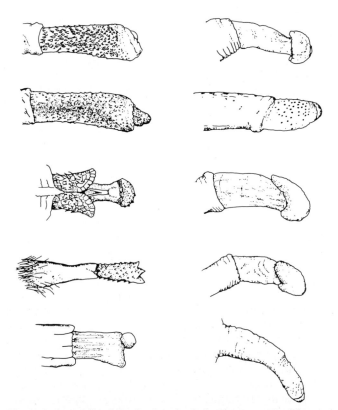

Figure 3. As predicted by the female-choice theory, more elaborate penis morphology occurs in species in which the female is more likely to mate with more than one male, thus being able to choose the father of her offspring. The male genitalia above are from different primate species. Those on the right belong to species in which a single male usually monopolizes the matings of a female during a period of estrous; those on the left belong to species in which receptive females can be mated by more males. (On the left proceeding down: *Galago crassicaudatus, G. garnettii, Arctocebus calabarensis, Euoticus elegantulus, Nycticebus coucang.* On the right proceeding down: *Colobus guereza, Callithrix jacchus, Mandrillus sphinx, Erythrocebus patas, Saguinus oedipus.*)(After Dixson 1987; drawn to different scales.)

Pieris rapae (Obara et al. 1975; Sugawara 1979). Finally, in mice, copulation can inhibit transport of sperm from previous matings (Dewsbury 1985). In sum, a male's reproductive success can be greatly affected by his ability to induce females to perform any of several critical post-coupling activities. Copulatory courtship behavior by males probably serves this end.

Number of mates and coyness

The female-choice hypothesis predicts that genital morphology of males should be under stronger sexual selection in those species in which females mate with more males. Dixson (1987) recently tested this prediction using a sample of 130 primate species. His data show, even when corrected for possible effects of common ancestors, that males of species in which sexually active females are not monopolized by single males have relatively longer genitalia, more highly developed hard spines on the penis, more complex shapes at the tip of the penis, and a more developed baculum (penis bone)(Fig. 3). In addition, males of species in which receptive females are not monopolized display more elaborate copulatory behavior, with prolonged and multiple intromissions. A similar trend occurs in *Heliconius* butterflies; males of species in which females remate more often tend to have more distinctive genitalia (Eberhard 1985).

If reproductive processes in females are triggered by male stimuli after coupling, then sexual selection theory predicts that it will be advantageous for a female to avoid having each copulation result in the fertilization of all her available eggs. This is because females able to favor males proficient at stimulation will have sons that are superior reproducers because the sons will be, on average, proficient stimulators (Fisher 1958). This type of selection can, in theory, give rise to a runaway process in which males develop increasingly elaborate apparatus and females become increasingly discriminating. Along with the probable advantages to females of controlling the timing of fertilization, female choice may help explain the tortuous and complex morphologies of many female reproductive tracts (Fig. 4, 5) and the rarity of designs in which males simply place their sperm at the site of fertilization. Female genitalia may be designed not only to facilitate fertilization, but also to prevent it under certain circumstances.

Perhaps the most dramatic and well-documented case supporting this idea is that of bedbugs and their allies (Carayon (1966). Some male bedbugs have evolved a hypodermic penis which can be inserted at a variety of sites on the female's body; the sperm are injected into her blood, and they migrate to the ovaries where they accumulate in huge masses even after only a single copulation. In some groups the females have responded by evolving a new genital system, complete with an opening on the top of the abdomen, a storage organ, and ducts to the oviduct. This new system would seem unnecessary for sperm transport, and both its developmental origin and mode of action suggest that it serves instead to selectively prevent fertilization. The cells of the new female system are derived from types used to combat infections, and only a very small proportion of the sperm that enter ever reach the ovaries.

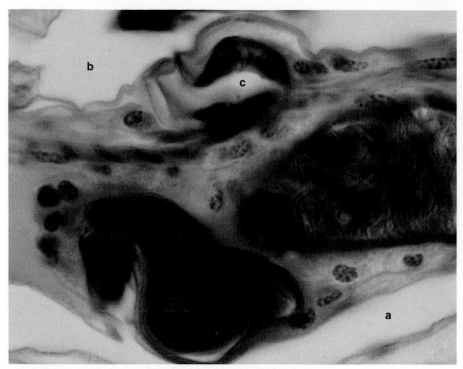

Figure 4. This transverse section of a portion of the female reproductive organs of the velvet water bug, *Hebrus ruficeps*, makes it clear that the female has control over the fate of sperm inside her. The male inserts his phallus along the shaft of her ovipositor into her genital chamber *(a)*, where he then everts a sac at the tip of the phallus into a farther chamber, the gynatrial sac *(b)*. He depends on the female to draw his sperm—each of which is longer than her entire body—out of his genitalia with a special set of dilator muscles. The female must then pump the sperm into the proximal end of her storage organ, the spermatheca, using the spermathecal pump *(c)* and spermathecal muscles. (From Heming-van Battum and Heming 1986; courtesy of B. Heming.)

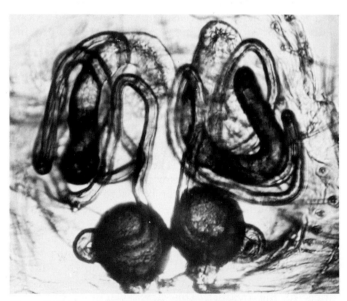

Figure 5. Tortuous ducts lead from the external copulatory opening *(circular forms at bottom)* to the internal sperm storage cavity in a female *Goyenia sylvatica* spider. The male has a pair of long, slender structures which presumably traverse at least part of this maze of ductwork. Females may test the mechanical properties of male genitalia with such hard-to-reach structures. (Photo by R. Foster.)

Such complicated coyness is not limited to insects. In pigs and several rodents, the rapid buildup of sperm in the uterus after mating is followed by the arrival of leucocytes which engulf and digest the sperm (Overstreet and Katz 1977). Other female mammals have barriers to sperm transport: as much as 80% of the sperm in the ejaculate of a rabbit is expelled from the vagina within minutes of coitus; an additional 15% probably does not migrate upward from the vagina; most of the remaining 5% is lost or retained along the way (Overstreet and Katz 1977). The cervix of a female human moves away from rather than toward the vagina during sexual arousal (Masters and Johnson 1966). Among males, some individuals are apparently better than others at overcoming these female barriers (Overstreet and Katz 1977).

Reconsidering old ideas

These observations suggest that the concept of courtship needs to be expanded to include behavior designed to induce the female not only to accept a male's copulation, but also to accept, care for, and use his sperm rather than that of other males. By using the sperm of some males and not others with whom she has copulated, the female can exercise what Thornhill (1983) has termed "cryptic female choice." If this is the case, then evolutionary biologists cannot measure the selective effects of male behaviors and morphologies by simply counting copulations. A male's reproductive success is not necessarily a simple function of the number of females with whom he copulates (Fig. 6).

Recognizing the possible role of sexual selection forces one to rethink the functions of intromission and copulation and the evolutionary origin of external male genitalia. Consider the fact that intromittent organs are more common in males than females. If mating is a cooperative effort by the male and female to fertilize eggs, and if intromittent genitalia have evolved because of the advantage of protecting gametes and zygotes from environmental vicissitudes (both generally accepted notions), then why is it that the male—the sex with relatively small, energetically cheap gametes—nearly always has the intromittent organ even though internal fertilization has evolved many times independently?

Use of a classic comparative technique—looking for an exception to the rule and checking that case for other unusual traits—suggests that sexual selection is involved. A well-documented exception occurs in sea horses and their relatives. The female sea horse has a functional penis, called an ovipositor, which she inserts into the male to transfer her eggs into his pouch, where they are fertilized. Sea horses are unusual in another respect: the males make large investments in the offspring, brooding the eggs and in some cases nourishing the offspring in the pouch before they leave. Females of some species are limited reproductively by a lack of access to males who will brood their eggs (Berglund et al. 1989). The coincidence of a reversal both in sexual morphology and in which of the two sexes represents a "bottleneck" or limiting resource in the reproduction of the other, is expected if sexual selection is involved. In more typical cases, male reproduction is limited by the number of mates, whereas female reproduction is limited not by males but by other factors such as food availability. Sexual selection theory predicts that the sex whose reproduction is more limited by access to the other sex will be selected to distribute its gametes to as many members of the opposite sex as possible. The sea horses thus suggest that intromittent organs in general are competitive devices for placing gametes in favorable sites and inducing even not completely willing partners to accept and use the gametes.

This implies that the notion that internal fertilization evolved to protect gametes and zygotes (Hinton 1964) may have to be reevaluated. Protection in some cases may have been an incidental consequence of competitive maneuvers for placing gametes in positions where they were more likely to achieve fertilization (Parker 1970). Sexual selection in this context may have had other important incidental consequences. Perhaps some lineages were able to move from aquatic to terrestrial habitats partly because ancestral males had evolved intromittent organs.

Another classical notion that must be modified is the distinction, made by Darwin himself, between primary sexual characteristics, which are supposed to function strictly for gamete transfer, and secondary sexual characteristics, such as horns or bright colors, which function in the context of sexual selection (as in fights between males over females or the attraction of females). Correcting Charles Darwin on a point of natural selection theory is an event so unusual that it inspires trepidation, even when the point in question is only a footnote in one of his books.

Good genes or propaganda?

Studies of genital function inspired by new evolutionary theory may in turn help advance the theory. Evolutionary theorists are currently divided over whether or not female choice necessarily favors males that are superior in other contexts, as in the ability to resist parasites or capture more prey (Bradbury and Andersson 1987). The "good genes" hypothesis interprets elaborate morphologies such as the tails of peacocks, which are classically

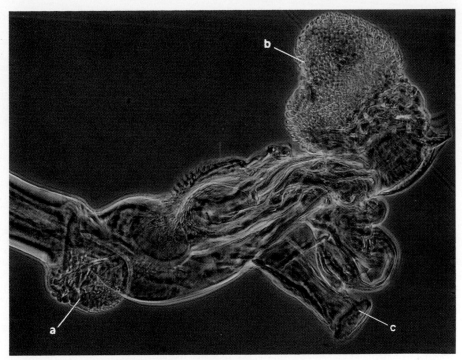

Figure 6. The complex tip of the genitalia of a male medfly is at the end of a long, snakelike structure which is about 40% of the length of the male's body. He inserts his genitalia by folding them and threading them up the shaft of the female's ovipositor. Then he gradually unfolds them inside her. They must go still deeper, however, and the male apparently uses an inching movement of an expandable finger-like structure *(a)* that pulls them farther inward. When the tip finally reaches the upper end of the vagina, it must hit a deeper target; the male uses a second expandable sac *(b)* to drive a tubular structure *(c)* into a cone in the wall of the vagina, and from there into a cylindrical organ attached to the cone's tip. The female has muscles in the ducts leading to the sperm storage sites (spermathecae), and apparently these must contract in order for the sperm to be pumped into the spermathecae. **(Photo by M. Vargas and H. Camacho).**

associated with sexual selection by female choice, as indicators of overall male quality. Females may judge the health of the male by his ability to produce such extravagant features (Zahavi 1987; Hamilton and Zuk 1982). On the other hand, as Ronald Fisher argued over 50 years ago, male signaling ability per se should be advantageous once females begin to use male signals as criteria for mating, or, as we have seen here, as triggers for other postcoupling processes leading to fertilization. A male with a greater ability to capture the female's attention and induce her to accept and use his sperm for fertilization could be favored whether or not that talent for "advertisement" is correlated with other aspects of male fitness. The ability to produce effective signals would, in and of itself, be a component of the male's fitness. Clarification of the causes of genital evolution may help resolve this debate, because typical genital characteristics probably have little or no relationship with overall male fitness in other contexts. If the female-choice hypothesis for genital evolution is correct, then the choosing of males on the basis of advertisement rather than good genes must be widespread.

The study of genitalia probably represents the last major frontier of an old and very successful branch of biology: functional morphology. It can be difficult to observe genitalia in action, and special techniques such as freezing pairs at different stages of copulation are often needed. However, the payoffs in increased understanding that result when genitalic events are included in studies of the selective consequences of different male and female behaviors are already evident (Simmons 1986; 1987). In addition to cryptic female choice, direct competition between males can occur inside females, as with the removal or repositioning of sperm from previous copulations in odonates (Waage 1986; Siva-Jothy 1988). Other possibilities are chemical trickery, in which sperm from previous males are induced to leave the storage site and become diluted by the sperm of a second male (Gilbert et al. 1984), and shuffling or selective discarding of sperm by the female as a result of genital stimulation (Otronen 1988). Studies of such phenomena, combined with more information on unknown behavior of the sperm inside the female, will probably constitute a major area of interest in evolutionary studies of behavior and morphology. There are many exciting surprises in store.

References

Alcock, J., and S. L. Buchmann. 1985. The significance of post-insemination display by male *Centris pallida* (Hymenoptera: Anthophoridae). *Zeitschrift für Tierpsychologie* 68:231–43.

Assem, J. van den, and J. Visser. 1976. Aspects of sexual receptivity in the female *Nasonia vitripennis* (Hymenoptera: Pteromalidae). *Biol. Behav.* 1:37–56.

Austad, S. N. 1984. Evolution of sperm priority patterns in spiders. In *Sperm Competition and the Evolution of Animal Mating Systems*, ed. R. L. Smith, pp. 223–49. Academic Press.

Baker, T. C., and R. T. Cardé. 1979. Courtship behavior of the oriental fruit moth *(Grapholitha molesta)*: Experimental analysis and consideration of the role of sexual selection in the evolution of courtship pheromones in the Lepidoptera. *Ann. Entom. Soc. Am.* 72:173–88.

Berglund, A., G. Rosenquist, and I. Svensson. 1989. Reproductive success of females limited by males in two pipefish species. *Am. Natur.* 133:506–16.

Bohme, T. 1983. The Tucano Indians of Colombia and the lizard *Plica plica*: Ethnological, herpetological and ethological implications. *Biotropica* 15:148–50.

Bradbury, J., and M. Andersson. 1987. *Sexual Selection: Testing the Alternatives.* Wiley.

Breed, W. G. 1986. Comparative morphology and evolution of the male reproductive tract in the Australian hydromysine rodents (Muridae). *J. Zool.* A209:607–29.

Brown, S. 1985. Mating behavior of the golden orb-weaving spider, *Nephila clavipes*: II. Sperm capacitation, sperm competition, and fecundity. *J. Comp. Psychol.* 99:167–75.

Busck, A. 1931. On the female genitalia of the microlepidoptera and their importance in the classification and determination of these moths. *Bull. Brooklyn Entom. Soc.* 26:119–216.

Carayon, J. 1966. Traumatic insemination and the paragenital system. In *Monograph of Cimicidae*, ed. R. Usinger, pp. 81–166. Entom. Soc. Am.

Cowan, D. P. 1986. Sexual behavior of eumenid wasps (Hymenoptera: Eumenidae). *Proc. Entom. Soc. Wash.* 88:531–41.

Dane, B., and W. van der Kloot. 1962. Analysis of the display of the goldeneye duck *(Bucephala clangula* L.) *Behavior* 22:282–328.

Davey, K. G. 1965. *Reproduction in Insects.* Oliver and Boyd.

DeWildt, D. E., S. C. Guthrie, and S. W. J. Seager. 1978. Ovarian and behavioral cyclicity of the laboratory-maintained cat. *Hormones and Behav.* 10:251–57.

Dewsbury, D. 1972. Patterns of copulation behavior in mammals. *Quart. Rev. Biol.* 47:1–33.

———. 1985. Interactions between males and their sperm during multi-male copulatory episodes of deer mice *(Peromyscus maniculatus). Animal Behav.* 33:1266–74.

———. In press. Copulatory behavior as courtship communication. *Ethology.*

Dewsbury, D., and D. K. Sawrey. 1984. Male capacity as related to sperm production, pregnancy initiation, and sperm competition in deer mice *(Peromyscus maniculatus). Behav. Ecol. Sociobiol.* 16:37–47.

Dixson, A. F. 1986. Genital sensory feedback and sexual behaviour in male and female marmosets *(Callithrix jacchus). Physiol. Behav.* 37:447–50.

———. 1987. Observations on the evolution of genitalia and copulatory behavior in primates. *J. Zool.* 213:423–43.

Eberhard, W. G. 1985. *Sexual Selection and Animal Genitalia.* Harvard Univ. Press.

Eckstein, P., and S. Zuckerman. 1956. Morphology of the reproductive tract. In *Marshall's Physiology of Reproduction*, ed. A. S. Parkes, vol. 1, pp. 43–155. Longmans, Green and Co.

Fisher, R. 1958. *The Genetical Theory of Natural Selection.* Dover.

Fretter, V., and A. Graham. 1964. Reproduction. In *Physiology of Mollusca*, ed. K. M. Wilbur and C. M. Yonge, pp. 127–64. Academic Press.

Gilbert, D. G., R. C. Richmond, and K. B. Sheehan. 1984. Studies of esterase 6 in *Drosophila melanogaster*. V. Progeny production and sperm use in females inseminated by males having active or null alleles. *Evolution* 38:24–37.

Greulich, W. W. 1934. Artificially induced ovulation in the cat *(Felis domestica). Anat. Rec.* 58:217–23.

Hamilton, W. D., and M. Zuk. 1982. Heritable true fitness and bright birds: A role for parasites? *Science* 218:384–87.

Hamner, C. E., L. L. Jennings, and N. J. Skojka. 1970. Cat *(Felis catus* L.) spermatozoa require capacitation. *J. Reprod. Fert.* 23:477–80.

Heming-van Battum, K. E., and B. S. Heming. 1986. Structure, function and evolution of the reproductive system in females of *Hebrus pusillus* and *H. ruficeps* (Hemiptera, Gerromorpha, Hebridae). *J. Morph.* 190:121–67.

Hershkovitz, P. 1979. *Living New World Monkeys (Platyrrhini) with an Introduction to Primates*, vol. 1. Univ. Chicago Press.

Hinton, H. E. 1964. Sperm transfer in insects and the evolution of haemocoelic insemination. In *Insect Reproduction*, ed. K. C. Highnam, pp. 95–107. Royal Entom. Soc. London.

Humphries, D. A. 1967. The mating behaviour of the hen flea *Ceratophyllus gallinae* (Shrank) (Siphonaptera: Insecta). *Animal Behav.* 15:82–90.

Inglis, W. G. 1961. The oxyurid parasites (Nematoda) of primates. *Proc. Zool. Soc. London* 136:103–22.

Lamb, R. J. 1976. Parental behavior in the Dermaptera with special reference to *Forficula auricularia* (Dermaptera: Forficulidae). *Canadian Entom.* 108:609–19.

Leopold, R. A., and M. E. Degrugillier. 1973. Sperm penetration of housefly eggs: Evidence for involvement of a female accessory secretion. *Science* 181:555–57.

Lorkovic, A. 1952. L'accouplement artificiel chez les Lépidoptères et son application dans les recherches sur la fonction de l'appareil génital des insectes. *Physiol. Comp. Oecol.* 3:313–19.

Masters, W. H., and V. E. Johnson. 1966. *Human Sexual Response.* Little Brown.

McKinney, F., K. M. Cheng, and D. J. Bruggers. 1984. Sperm competition in apparently monogamous birds. In *Sperm Competition and the Evolution of Animal Mating Systems*, ed. R. L. Smith, pp. 523–45. Academic Press.

Nichols, S. W. 1986. Early history of the use of genitalia in systematic studies of Coleoptera. *Quaest. Entom.* 22:115–41.

Obara, Y., H. Tateda, and M. Kuwabara. 1975. Mating behavior of the cabbage white butterfly, *Pieris rapae crucivora* Boisduval. V. Copulatory stimuli inducing changes of female response patterns. *Zool. Mag.* (Tokyo) 84:71–76.

Otronen, M. 1988. Studies on reproductive behavior in some carrion insects. D. Phil. Thesis, Univ. of Oxford.

Overstreet, J. W., and D. F. Katz. 1977. Sperm transport and selection in the female genital tract. In *Development in Mammals*, ed. M. H. Johnson, vol. 2, pp. 31–65. North Holland Pub. Co.

Parker, G. A. 1970. Sperm competition and its evolutionary consequences in the insects. *Biol. Rev.* 45:525–67.

Rodriguez-Sierra, J. F., W. R. Crowley, and B. R. Komisaruk. 1975. Vaginal stimulation in rats induces prolonged lordosis responsiveness and sexual receptivity. *J. Comp. Physiol. Psychol.* 89:79–85.

Roth, L. M., and B. Stay. 1961. Oocyte development in *Diploptera punctata* (Eschscholtz) (Blattaria). *J. Ins. Physiol.* 7:186–202.

Shapiro, A. M., and A. H. Porter. 1989. The lock and key hypothesis: Evolutionary and biosystematic interpretation of insect genitalia. *Ann. Rev. Entom.* 34:231–45.

Schincariol, L. A., and R. Freitag. 1986. Copulatory locus, structure and function of the flagellum of *Cicindela tranquebarica* Hergst (Doleoptera: Cicindelidae). *Int. J. Invert. Reprod. Devel.* 9:333–38.

Simmons, L. 1986. Female choice in the field cricket *Gryllus bimaculatus* (DeGreer). *Animal Behav.* 35:1463–70.

———. 1987. Sperm competition as a mechanism of female choice in the field cricket *Gryllus bimaculatus*. *Behav. Ecol. Sociobiol.* 21:197–202.

Siva-Jothy, M. 1988. Sperm repositioning in *Crocothemis erythraea*, a libellulid dragonfly with a brief copulation. *J. Ins. Behav.* 1:235–45.

Smith, R. L., ed. 1984. *Sperm Competition and the Evolution of Animal Mating Systems.* Academic Press.

Sugawara, T. 1979. Stretch reception in the bursa copulatrix of the butterfly *Pieris rapae crucivora*, and its role in behaviour. *J. Comp. Physiol.* 130:191–99.

Thornhill, R. 1983. Cryptic female choice and its implications in the scorpionfly *Harpobittacus nigriceps*. *Am. Nat.* 122:765–88.

Waage, J. K. 1979. Dual function of the damselfly penis: Sperm removal and transfer. *Science* 203:916–18.

———. 1986. Evidence for widespread sperm displacement ability amongst Zygoptera and the means for predicting its presence. *Biol. J. Linn. Soc.* 28:285–300.

Zahavi, A. 1987. The theory of signal selection and some of its implications. In *International Symposium Biological Evolution*, ed. V. P. Delfino, pp. 305–27. Adriatica Editrice.

Mechanisms of Sperm Competition in Birds

When a female bird mates with an extra-pair male, timing and the number of sperm inseminated by each male determine which fathers the most offspring

Tim R. Birkhead

Figure 1. Zebra finches, like many other bird species, practice extra-pair matings more often than common wisdom would have you believe. In fact, in many bird species a greater number of offspring are produced through extra-pair matings than through copulations with the female's true mate. The different, genetically determined coloration of zebra finches can be exploited experimentally to determine paternity. A homozygous wild-type gray male produces only wild-type

The springtime image of birds pairing up has often been used as a metaphor for human mating behavior. Birds have long been thought of as forming life-long monogamous partnerships, an ideal to which many people aspire. But appearances can be deceptive. Far from being models of fidelity, it has been clear for some time that many birds are sexually promiscuous.

The first inklings that this might be the case arose with the earliest stirrings of behavioral ecology in the early 1970s. The traditional view of monogamy was that each partner benefited from mutual cooperation, but behavioral ecology's focus on individual selection raised the notion that even within a monogamous relationship both males and females should behave selfishly. In a landmark paper published in 1972 Robert Trivers, then at Harvard, predicted that males should

attempt to inseminate the females of other males and hence parasitize their paternal care. By doing so, they could increase their reproductive success—and, because sperm were believed to be cheap, at minimal expense. Females on the other hand were predicted to be coy, since they need to copulate only once to ensure that their eggs are fertilized. Moreover, it was thought that infidelity was potentially rather costly for a female since her partner would retaliate by reducing his investment in her offspring. At that time there had been relatively few observations of extra-pair behavior of birds, and those that had been made were explained away as nonadaptive (and in highly anthropomorphic terms)—for example, the males were "sexually dissatisfied at home."

My own involvement in this field had its beginnings in 1970 during a

memorable undergraduate lecture in which we were presented the heady cocktail of Trivers's ideas and the results of University of Liverpool ecologist Geoff Parker's studies of multiple mating in the yellow dung fly *Scatophaga stercoraria*. Parker showed that the sperm competition that results from males copulating with already mated females creates powerful and conflicting selection pressures on males. Male dung flies have a suite of adaptations to help them secure multiple matings and a suite of counter-adaptations to help them avoid being cuckolded. These results hung so elegantly on their theoretical framework, and the whole subject area had such a feel of revolution about it, that I knew at that moment that this was what I wanted to work on. With a long-term interest in ornithology, I specifically wondered if these ideas were applicable to birds.

Tim Birkhead is professor of behavioral ecology at the University of Sheffield in England. He was awarded his D. Phil. from the Edward Grey Institute, University of Oxford, in 1976. He has spent many summers studying seabirds in the Arctic and sperm competition in birds in warmer parts of the world, including zebra finches in Australia. He has written and edited a number of books, including The Cambridge Encyclopedia of Ornithology, *which he co-edited with M. de L. Brooke and which was awarded the McColvin medal for the best reference book. Address: Department of Animal and Plant Sciences, The University, Sheffield, S10 2TN, UK.*

offspring when mating with a fawn-colored female *(left).* **Two fawn birds, on the other hand, produce only fawn offspring** *(right).* **If a fawn female is inseminated by a male of each type, the paternity of each male's offspring is apparent at hatching. Experiments where females are mated sequentially with two differently colored males have demonstrated that the second male fathers more offspring than does the first. (All paintings and photographs courtesy of the author.)**

Monogamy Myths

The first field studies of sperm competition in birds were essentially observational. They sought to determine whether extra-pair copulations take place in monogamous birds and whether paired males attempt to protect their partner from the sexual advances of other males. Both prove true: Males routinely attempt to copulate with the fertile partners of other males, while at the same time remaining very close to their own female when she is fertile to avoid being cuckolded. It seems hard to believe now, but it was not even known in the mid-1970s whether multiple mating by female birds in the wild could result in multiple paternity. While these observational studies were in full swing, the new technique of DNA fingerprinting was developed in the mid-1980s, for the first time enabling researchers to assess parentage unambiguously. Luckily for those of us who study birds, Terry Burke of the University of Leicester and David Parkin of the University of Nottingham were quick to realize the potential of fingerprinting for studies of the mating systems of birds. To date, paternity analyses have been conducted on approximately 100 species of bird, revealing levels of extra-pair paternity that even Trivers and Parker would not have predicted. In some species, such as the superb fairy wren *Malurus cyaenus*, more than half of all offspring come from extra-pair matings. True monogamy, with zero extra-pair paternity, is relatively rare but is evident in some seabirds, such as the fulmar petrel *Fulmarus glacialis*. These studies show that extra-pair copulations do increase the reproductive success of certain males—sometimes substantially. For example, in the purple martin *Progne subis*, a bird familiar to many North Americans, successful cuckolders can double the number of offspring they father.

But two major questions remained. First, why do females engage in extra-pair copulations? And, second, when they do, what determines which male fathers the offspring?

The first of these questions is one that many behavioral ecologists are currently trying to answer. In the true tradition of behavioral ecology it is a functional question and concerns the adaptive significance of females copulating with more than one male during a single reproductive cycle. Although there are plenty of ideas, so far there are no clear answers. A frequent explanation is that females mate with additional males to ensure fertilization, but there is no evidence for this. The answer currently attracting the most attention, some evidence and not a little controversy is that a female gains better genes for her offspring by performing one or more extra-pair copulations with a male that is of better quality than her partner. We still have a long way to go, however, to demonstrate this unequivocally and in enough species for it to constitute a general explanation for female infidelity.

The second question, concerned with the fertilization success of copulations, however, is one where some clear answers are now emerging. What is more, this is a mechanistic question, concerned not with the evolutionary significance of phenomena but with underlying physiological processes. After a long and productive period in which they focused almost exclusively on functional aspects, behavioral ecologists have started to appreciate that by broadening their horizons to include mechanisms, they can look forward to a highly productive marriage between these complementary levels of explanation. The mechanism of sperm

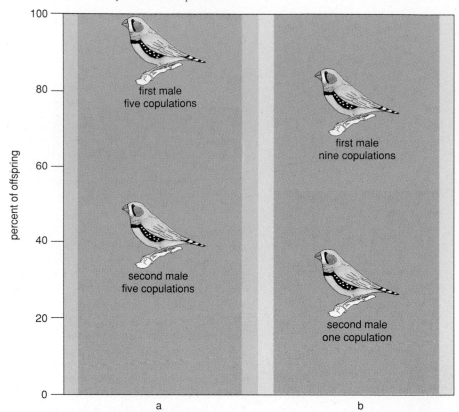

Figure 2. Experiments demonstrate the advantage the second male has in fathering offspring. Fawn female copulates with a male of each coloration for the same number of times (a), in this experiment, five. Nevertheless, the second male fathers almost 75 percent of the offspring. Second-male precedence persists (b), even when the odds are skewed to favor the first male. In an experiment identical in design with a, the first male copulates with the female nine times, whereas the second male copulates with her only once. In spite of the 9-to-1 ratio, the second male fertilizes more than 50 percent of the eggs.

competition in birds provides a nice example of this fruitful approach.

Last-Male Precedence

In many organisms, the second of two males to copulate with a female typically fertilizes most of her eggs. This phenomenon, known as second- (or last-) male sperm precedence, has been reported in insects, crustaceans and birds. A study on chickens conducted in the 1970s reported a particularly strong last-male effect: When females were artificially inseminated just four hours apart with equal numbers of sperm from two genotypes, the second insemination fathered about 77 percent of the offspring—regardless of which genotype's sperm was inseminated first. The proposed mechanism was that the sperm from the two successive inseminations remained stratified within the female's sperm store; the last sperm in were the first out and hence fertilized most eggs.

Investigating the mechanism in detail meant following sperm along the obstacle course from insemination to fertilization. After insemination, only 1 percent of the several millions of sperm ejaculated traverse the intensely hostile vagina and enter the female's sperm-storage tubules. Many of the remaining sperm are ejected with fecal material within minutes of insemination. The sperm that get into the storage tubules remain quiescent and can fertilize eggs up to 30 days later. Over this time, however, sperm leak out of the tubules at a constant rate and are carried up the oviduct to the infundibulum, where fertilization takes place. Fertilization in birds is sequential: Each egg of a clutch is fertilized separately and, in the case of most small birds, at 24-hour intervals. The ovum is shed from the ovary, usually in the early morning, into the infundibulum. There is then a 15- to 30-minute window during which fertilization must take place. After that time the protein-rich albumen is laid down around the ovum, preventing further access by sperm. The ovum spends the next 23 hours in the oviduct having water, membranes and shell added before the fully formed egg is laid early the next morning. Within an hour of laying, the next ovum is shed from the

ovary, and the cycle repeats itself. Between the laying of one egg and the fertilization of the next ovum, the infundibulum is replenished with sperm from the sperm-storage tubules.

Last In, First Out

When I started to investigate the mechanisms of sperm competition in birds, I needed a species that would breed readily in captivity—allowing me to control natural matings—but also one that I could observe in the wild to determine whether sperm competition takes place naturally. The zebra finch *Taeniopygia guttata* was an ideal species. An inhabitant of Australia's arid regions and an opportunistic breeder, the zebra finch breeds throughout the year in captivity, given the right conditions.

The observations I made of individually marked zebra finches at Richard Zann's study colony in northern Victoria, Australia, confirmed that extra-pair copulation was frequent. Subsequent DNA fingerprinting, performed in collaboration with Terry Burke and other colleagues, revealed that extra-pair copulations resulted in extra-pair paternity. Despite the technological elegance of DNA fingerprinting, for laboratory work in Sheffield I needed a more rapid technique to assign paternity and decided on the low-tech but tried-and-trusted method of genetic plumage markers. Zebra-finch breeders have developed a number of different color forms whose mode of inheritance is well known. The fawn mutation is sex linked and recessive to the wild (gray) type, which means that when a fawn female copulates with a fawn male they produce only fawn offspring. But when a fawn female is fertilized by a homozygous gray (wild-type) male, they produce only gray-plumaged offspring. The beauty of this method is that when a fawn female mates with a male of each genotype, the paternity of the offspring is obvious as soon as they hatch.

I performed two experiments, both designed to mimic situations that take place naturally in the wild. The first was a mate-switching experiment in which two males replaced each other. Despite each securing a similar number of copulations, the second male fertilized most (75 percent) eggs. This suggested a last-male effect, so the second experiment was designed to test simultaneously for this and to determine the efficacy of a single extra-pair copulation. The first male secured about nine copulations on average, but the second male was extraordinarily successful and fertilized over half the eggs with just a single insemination. These results confirmed a last-male advantage, and demonstrated the potency of a single extra-pair copulation. Since my results were similar to those reported for chickens, I started to wonder about the "last-in, first-out" system as an explanation for last-male precedence. Because the sperm competition experiments I was conducting were extremely time-consuming, however, I realized that I could easily spend years conducting empirical tests to figure out the mechanism of last-male sperm precedence.

Kate Lessells, then at the University of Sheffield and now at the Netherlands Institute of Ecology, and I took a shortcut and built mathematical models of sperm competition to identify the most plausible mechanisms for last-male precedence. Because more was known at that time about the reproductive biology of the chicken than of the zebra finch, we used information from the chicken studies to test our models. Almost our first finding was that the last-in, first-out system could not account for the 77 percent last-male effect. This stratification model, as we called it, predicted that with increasing time since insemination the first male's sperm should fertilize more and more offspring, but the empirical data showed that the ratio of offspring remained constant. In addition, we now know that it takes considerably longer than four hours for sperm to occupy the sperm storage tubules, providing further evidence for the implausibility of this mechanism. Our next model, referred to as the passive sperm-loss model, predicted a last-male effect simply because by the time the second insemination took place some sperm from the initial insemination had already died or been lost from the female tract. In other words, the outcome of sperm competition depended entirely on the relative numbers of sperm from each male present at the point of fertilization. We also rejected this model because the rate at which chickens lose sperm from their reproductive tract is much too slow to explain the observed level of last-male sperm precedence. The only model that came anywhere close to providing a plausible explanation for the 77 percent precedence was a displacement model in which sperm from the second insemination displaced sperm from the earlier insemination. The greater the degree of displacement, the greater the last-male advantage.

The next stage was an empirical test of the displacement model. A simple test would be to inseminate equal numbers of sperm four hours apart and show that after the second insemination fewer of the first male's sperm remained in the female's sperm-storage tubules. But to do this, one needs to be able to distinguish between the sperm of different males inside the sperm-storage tubules. At that time there was only a single vital fluorescent dye available that would label sperm without obviously affecting their viability. When we inseminated females sequentially with labeled and unlabeled sperm, however, the dyed sperm stained the unlabeled sperm, preventing us from distinguishing between the two. It was reassuring but no less frustrating to learn that two other research groups experienced the same problem in trying to resolve this question.

Science is supposed to progress in a logical Popperian fashion, but sometimes it does not work out like that. On coming up against a brick wall, I changed tack and decided to go back to first principles by repeating the sperm-competition experiment using chickens.

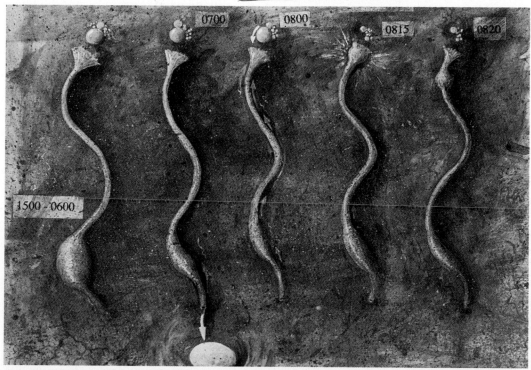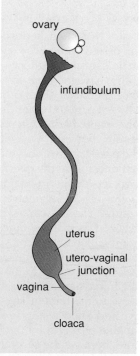

Figure 3. Avian reproductive physiology contains part of the key to the mystery of second-male advantage. From midafternoon onward, the albumen-covered ovum (or yolk), which has already been fertilized, resides in the uterus where water, membranes and hard egg shell are added (*a*). By early the next morning, the egg is complete and is laid at about 7:00 a.m. Once the egg is laid, sperm, which have been stored in tubules at the utero-vaginal junction, travel unimpeded up the oviduct to the infundibulum, where they wait (*b*). Within an hour of laying the first egg, the next ovum is shed from the ovary (*c*). There is a 15- to 30-minute window during which fertilization must take place (*d*). The fertilized ovum (*e*) travels down the oviduct, where it will remain for the next 23 hours, as water, membranes and shell are added around it. It is laid the next morning, and so the cycle repeats itself.

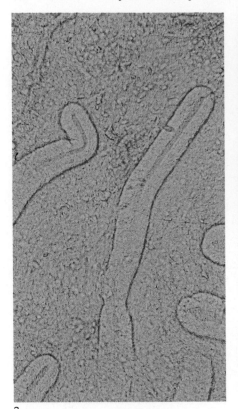

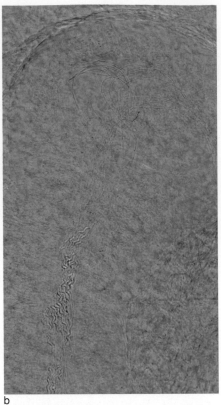

a   b

Figure 4. Sperm-storage tubules seen close-up in a light microscope appear sausage-shaped. Tubules are located at the utero-vaginal junction and retain sperm for several days following copulation, during which time they constantly leak out and fertilize eggs. These are tubules from the Japanese quail *Coturnix japonica (a)* and a single sperm-filled tubule from the Chaffinch *Fringilla coelebs (b)*.

When we did this and inseminated females twice four hours apart, we were surprised to detect virtually no last-male sperm precedence. Thinking we might have made an error, we repeated the experiment, several times in fact, but on each occasion there was no evidence of a marked last-male precedence. Our results, however, were consistent with the passive sperm-loss model. This then left unresolved the question of why the two sets of sperm-competition experiments on chickens gave such different results. The answer, as I subsequently realized, is that the outcome of sperm competition, in the chicken at least, depends on the time at which the inseminations are made. In the original study the initial insemination was made very close to the time of egg laying, a time we now know results in a relatively low uptake of sperm by the female. Therefore, because the second insemination was performed four hours later, it had a considerable advantage and hence the 77 precent precedence. In our chicken experiments, both inseminations were made at least seven hours after egg laying and hence avoided this particular problem. I have subsequently analyzed

published data from other sperm-competition experiments with chickens and turkeys, and these are all consistent with the passive sperm-loss model.

Encouraged by the success of this model, we then examined the results of our zebra finch experiments to see if they too could be explained by the passive loss of sperm from the female tract. To do this, however, we needed two pieces of information: the numbers of sperm inseminated by males and the rate at which sperm are lost or die in the female tract. We devised an empirical method, using a surrogate female, to determine the number of sperm inseminated. Using this technique, we found that the time since the last ejaculation had an important effect on the numbers of sperm transferred. Males that had not copulated for a week or more, referred to as "rested" males, transferred about 8 million sperm. But since they have limited sperm stores and a relatively low rate of sperm production, subsequent ejaculates on the same or the next day were much smaller—about 1 million sperm. In the single extra-pair copulation experiment, we used males that had not copulated for

at least one week, so these males would have produced a single ejaculate containing a large number of sperm. The paired male was also rested when the experiment started, so he would also have produced one large initial ejaculate followed by a series of smaller ones. To estimate the ratio of sperm from the two males at the time of fertilization, however, we also had to take into account that sperm can remain in the sperm-storage tubules for up to 13 days, during which time they constantly leak out and can fertilize new eggs.

To ascertain the rate of loss of sperm from the tract, we developed a technique used initially by poultry biologists to predict the fertility of eggs. This consisted of counting the sperm trapped in the layers surrounding the yolk—the perivitelline layers—of laid eggs. At ovulation the ovum (the yolk) is released from the ovary into the infundibulum where a number of sperm are present. At this stage the ovum is surrounded by the inner perivitelline layer, and within minutes this is penetrated by one or more sperm at the germinal disc. One of these sperm then fuses with the egg's nucleus, thereby combining paternal and maternal DNA. Shortly afterwards the infundibulum starts to secrete the outer perivitelline layer, and as it is laid down around the ovum, it traps the other sperm present in the infundibulum. These sperm can be readily seen and counted in a laid egg—the perivitelline layers are removed from around the yolk, stained with a DNA-specific fluorescent dye and examined microscopically. We counted the number of sperm on successive eggs after a single copulation and found that numbers declined exponentially, indicating that sperm were lost at a constant rate.

Armed with quantitative measures of the number of sperm inseminated and the rate at which they are lost from the female tract, we were able to determine whether they could account for the last-male sperm precedence we recorded in our experiments with zebra finches. Incorporating these values into the passive sperm-loss model, it was reassuring to find that the fit between what was predicted and what we observed was remarkably close. This means that the relative numbers of sperm from different males in the infundibulum at the time of fertilization determines the outcome of sperm competition. Any factor that affects the uptake of sperm by the female tract will therefore influence a

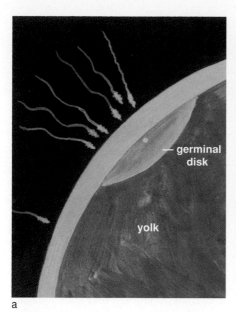
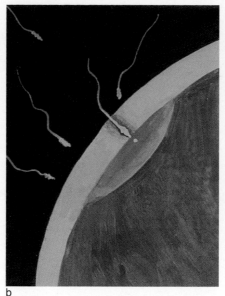
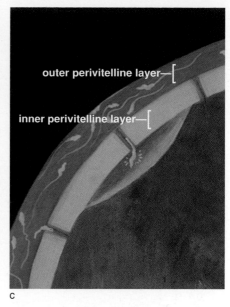

a   b   c

Figure 5. Fertilization triggers the deposition of additional material around the ovum to form the egg. The unfertilized egg (*a*) exists as an ovum surrounded by the inner perivitelline layer. Sperm present in the infundibulum surround the unfertilized egg. Several of these typically penetrate the germinal disc (*b*), and one fuses with the ovum's nucleus, which houses the female's genes. In this way, male and female genes are merged in the germinal disc. Within minutes of fertilization, the infundibulum secretes an outer perivitelline layer around the egg (*c*), which traps remaining sperm present in the infundibulum.

particular male's chances of fertilizing eggs. The chicken experiments showed that one such factor is the timing of insemination relative to egg laying—inseminations close to egg laying were less successful. Another factor that could affect the proportion of sperm reaching the site of fertilization is the quality of the sperm themselves.

A Question of Quality

Breeders of domestic mammals and birds, and those who work in human *in vitro* fertilization (IVF) clinics, have known for a long time that the motility of a semen sample is a reasonable indicator of its likelihood of successfully fertilizing ova. Motility, which usually refers to the proportion of sperm that are actually motile, is, however, a relatively crude measure of quality. More recently the ability to measure the actual velocity of individual sperm, using computer-assisted sperm analyses, has provided a much better predictor of fertilizing potential. In our experiments with surrogate females we found that the ejaculated sperm from rested males moved at twice the speed (33.4 micrometers per second) as those from the next ejaculate made just one hour later (17.4 micrometers per second). In addition, a higher proportion of the sperm in a rested male's ejaculate are morphologically normal than those in subsequent ejaculates. By examining sperm in the male's sperm

store, the seminal glomera, we found that sperm appear to mature and gain the potential for high-speed movement as they travel down the male tract. A similar maturation process takes place in the epididymis of male mammals, but with one important difference: Unlike the zebra finch, male mammals

never ejaculate immature sperm. The reason for this difference remains to be resolved. If faster-swimming sperm are more likely to traverse the female's hostile vagina and enter the sperm-store tubules, it may provide an additional reason why the single copulation from a rested male zebra finch was so

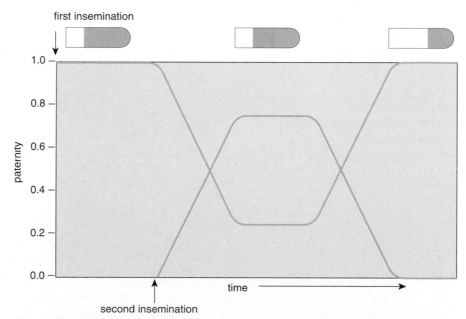

Figure 6. Models have been constructed to explain why the sperm of the second male are more likely to produce offspring than those of the first. All three models—stratification, passive sperm loss and displacement—focus on the disposition of the sperm in the sperm-storage tubules. The stratification model, shown above, relies on a first-in, first-out principle. After a single insemination, the first male fertilizes all of the eggs (*blue*). The second male's sperm (*brown*) is assumed to remain stratified in the sperm-storage tubules in such a way that it is closer to the exit. In this way, the second male's sperm is the first out of the tubule, and fertilizes most of the female's eggs, until it is used up. Then any remaining viable sperm from the first copulation can fertilize eggs.

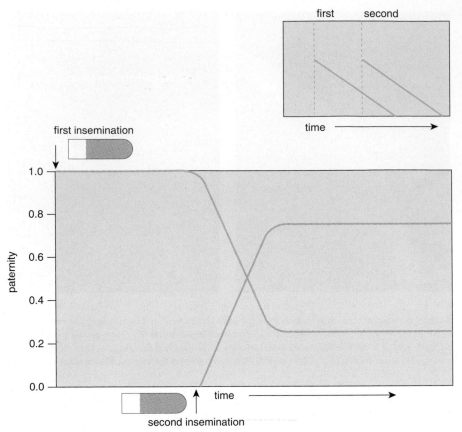

Figure 7. Passive sperm-loss model assumes that sperm from both males are mixed within the sperm-storage tubules. This model assumes that by the time the second male's insemination is made (brown), sperm from the first male (blue) have died or have been lost from the female tract. The second male, therefore, is more likely to fertilize the eggs after this point. This model is illustrated another way in the upper graph. Sperm from each insemination are lost at a constant rate, indicated by the slopes of the lines. Because the ratio of sperm from each male remains constant over time, this model predicts that the relative number of offspring from each male also remains constant over time.

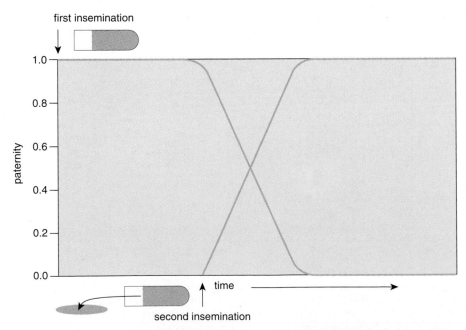

Figure 8. Displacement model suggests that the limited volume of the sperm-storage tubules accounts for the second-male precedence. This model predicts that incoming sperm from the second male (brown) displace sperm previously stored sperm from copulation with the first male (blue).

successful in fertilizing eggs. We are currently investigating this.

Across the animal kingdom the original view was that sperm competition has been driven primarily by selection on males. It is clear that by safeguarding their own paternity and fertilizing other females, males increase their reproductive output. In contrast, the only benefit socially monogamous females were thought to obtain from copulating with more than one male was an increase in the quality of her offspring. The view that selection operates more intensively on males than on females in terms of sperm competition has been questioned in recent years, and there is now considerable focus on female strategies. Although males of a few species (for example, mallard ducks *Anas platyrhynchos*) are able to force copulations on females, the evidence that females generally control extra-pair behavior in birds is incontrovertible. For example, field studies of the purple martin and blue tit *Parus caeruleus*, both socially monogamous species, reveal that females actively seek and initiate extra-pair copulations with males that are of better quality than their partner. Similarly, studies of zebra finches in captivity also indicate that females prefer males with certain characteristics (for example, with a high song rate and a particularly red beak) as extra-pair partners. How they benefit from this behavioral choice of copulation partner is still unknown. We tested the idea that zebra finches with these particular phenotypic traits might also produce fast moving, high-quality sperm but found no evidence for such an effect.

In addition to controlling events behaviorally, females that have mated with several males may also have the physiological ability to favor the sperm of one male over another's. Since fertilization takes place inside the female's body, it seems reasonable that selection will act particularly strongly to give them the ultimate control. Whether females have the physiological ability to determine paternity in this way, referred to as cryptic choice, remains to be seen, but there are some indications that they might.

One of the most remarkable examples concerns an organism about as *un*bird like as you can imagine. In the comb-jelly (ctenophore) beroë, several sperm typically penetrate the ovum (a phenomenon referred to as polyspermy, which also occurs in birds), and the egg's nucleus moves around the cytoplasm visiting

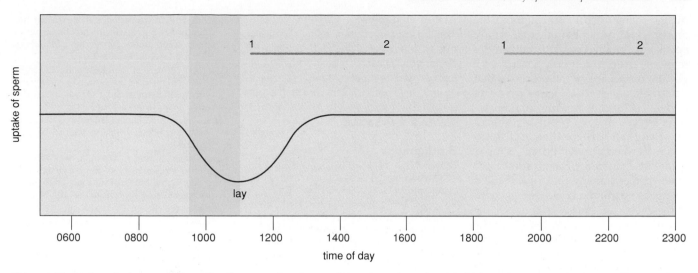

Figure 9. Time of egg-laying can profoundly affect second-male precedence. An early study with chickens demonstrated an unusually strong last-male sperm precedence when inseminations were only four hours apart. It was known at the time that inseminations made up to 90 minutes before egg laying (*yellow zone*), had a low chance of fertilization, but it was assumed that those made soon after laying had as good a chance as those made later. However, the uptake of sperm by the female tract (*red line*) is reduced for approximately 90 minutes on either side of egg laying —hence the reduced success of initial inseminations and the strong last-male effect (*blue line*). In the author's experiments, researchers inseminated females as far away from egg laying as possible (*brown line*). Sperm uptake by females was high for both inseminations. (For the blue and brown lines, "1" indicates the point at which the first insemination is made; "2" indicates the point at which the second is made.)

each of the sperm nuclei before fusing with one. In birds the potential for female choice is considerable: Over 99 percent of the sperm a male inseminates is rejected by the female. By modifying the degree of rejection only slightly one way or the other, a female could change the odds in the fertilization stakes considerably, but whether she has the ability to do this remains to be seen.

All of this raises the perplexing question of why a female should bother to pair up with a male only to go off and mate with another later? Several possible answers have presented themselves.

An attractive possibility is that the extra-pair male helps rear the offspring, but no evidence exists to suggest this is true. A second possibility is that the female gets a valuable item—food, for example—in exchange for services rendered. But this theory also lacks evidence. It has been suggested that females mate with other males when their partner's sperm count is low, but there is no evidence for this.

So it seems that many scientists are left with a single possible explanation. Females are looking for higher quality males to father their offspring. Why doesn't the female just choose the highest-quality male to be her mate in the first place? That is not always possible, as the example of migratory birds, such as swallows, shows. Males generally arrive at the breeding grounds a few days before females do. The first females to return have the pick of all the males and generally select the highest quality. Later-returning females have fewer good choices left to them, and the last females to return have the poorest choices. Rather than foregoing her chance to breed at all that season, the late-returning female takes a poor-quality male as her mate. But she doesn't have to completely forgo the chance to have some higher-quality offspring, and so she engages in extra-pair mating.

It is hard to tell what constitutes high quality for birds. In swallows, it is the length and symmetry of the tail. Studies have shown that females paired to males with short and asymmetric tails seek extra-pair matings from males whose tails are longer and more symmetric.

Studies have also shown that the extra-pair matings are responsible for pro-

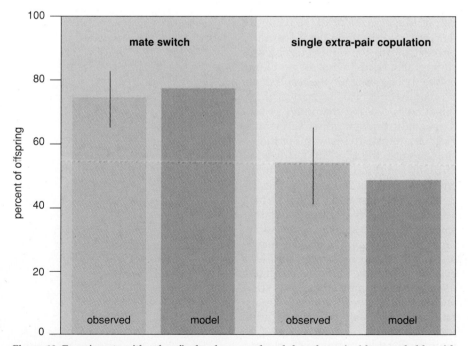

Figure 10. Experiments with zebra finches have produced data that coincide remarkably with those predicted by the passive sperm-loss model. The second male in the experiment described in Figure 2a is predicted to father about 75 percent of the offspring, which is very similar to the observed value (*left two bars*). In the experiment depicted in Figure 2b, the second male is predicted to father approximately 50 percent of the offspring after just one extra-pair mating, and observed values are similar (*right two bars*). The error bars on the gray columns are the 95-percent confidence intervals.

ducing many offspring throughout the animal kingdom. Females of many species routinely copulate with more than one male, but in only a small proportion of species are the processes that determine which males father offspring understood. Just as in birds, in many insects the last male to copulate fertilizes most eggs. In dragonflies and damselflies the mechanism is brutally simple; before inseminating his own sperm, the male scrapes out any previously stored sperm using minute hooks on his penis. In the ghost crab *Inachus phalangium*, the last male also has precedence, but he uses a different trick. Before transferring sperm to a female, he introduces highly modified seminal fluid that sets hard and seals off previously inseminated sperm at the back of the female's sperm store. In the majority of invertebrates, however, the mechanisms by which last-male precedence is achieved are more complicated. In the yellow dungfly it is thought that the incoming sperm displace and flush out some of the sperm from previous matings.

Molecular studies of paternity have demonstrated that sperm competition also takes place in mammals—including apparently monogamous species like ourselves. Studies of the inheritance of human disease have revealed a number of mismatches—many of which can be explained only by extrapair fertilization. None of the paternity data for people have been formally published, but the figures apparently suggest that between one in ten to one in twenty offspring are the result of extra-pair matings. The mechanics of fertilization success in mammals is fundamentally different from that in birds. First, if a female ovulates more than a single ovum, the eggs are fertilized at virtually the same time (not on different days, as in birds). Second, there is no consistent last-male effect. In fact, all combinations of effects occur— sometimes the first male wins, sometimes it is the second male. The reason for this is that in mammals, timing is everything. Unlike birds, the fertilizing lifespan of mammalian sperm is short and they also have to undergo capacitation, the physiological preparation for fertilization. Because ova also remain fertilizable for only a short time, to have a good chance of fertilization, a male must time his copulations so that his sperm have undergone capacitation just at the time the female ovulates and the ovum is ready to be fertilized.

Acknowledgments

This article was written while the author was in receipt of a Leverhulme Research Fellowship. The author also gratefully acknowledges the support of the Biotechnology and Biological Sciences Research Council and The Royal Society for the research described in this article.

Bibliography

Birkhead, T. R. 1996. Sperm competition: evolution and mechanisms. *Current Topics in Developmental Biology* 33:103–158.

Birkhead, T. R., T. Burke, R. Zann, F. M. Hunter and A. P. Krupa. 1990. Extra-pair paternity and intraspecific brood parasitism in wild zebra finches *Taeniopygia guttata*, revealed by DNA fingerprinting. *Behavioral Ecology and Sociobiology* 27:315–324.

Birkhead, T. R., K. Clarkson and R. Zann. 1988. Extra-pair courtship, copulation and mate guarding in wild zebra finches *Taeniopygia guttata*. *Animal Behavior* 35:1853–1855.

Birkhead, T. R., and F. Fletcher. 1995. Male phenotype and ejaculate quality in the zebra finch *Taeniopygia guttata*. *Proceedings of the Royal Society of London (Biology)* 262:329–334.

Birkhead, T. R., F. Fletcher, E. J. Pellatt and A. Staples. 1995. Ejaculate quality and the success of extra-pair copulations in the zebra finch. *Nature* 377:422–423.

Birkhead, T. R., F. M. Hunter and J. E. Pellatt. 1989. Sperm competition in the zebra finch, *Taeniopygia guttata*. *Animal Behavior* 38:935–950.

Birkhead, T. R., and A. P. Møller. 1992. *Sperm Competition in Birds: Evolutionary Causes and Consequences*. London : Academic Press.

Birkhead, T. R., J. E. Pellatt and F. M. Hunter. 1988. Extra-pair copulation and sperm competition in the zebra finch. *Nature* 334:60–62.

Birkhead, T. R., G. J. Wishart and J. D. Biggins. 1995. Sperm precedence in the domestic fowl. *Proceedings of the Royal Society of London (Biology)* 261:285–292.

Colegrave, N., T. R. Birkhead and C. M. Lessells.

1995. Sperm precedence in zebra finches does not require special mechanisms of sperm competition. *Proceedings of the Royal Society of London (Biology)* 259:223–228.

Compton, M. M., H. P. Van Kery and P. B. Siegel. 1978. The filling and emptying of the uterovaginal sperm-host glands in the domestic hen. *Poultry Science* 57:1696–1700.

Davies, N. B. 1992. *Dunnock Behaviour and Social Evolution*. Oxford, England: Oxford University Press.

Kempenaers, B., and A. Dhondt. 1993. Why do females engage in extra-pair copulations? A review of hypotheses and their predictions. *Belgian Journal of Zoology* 123:93–103.

Kempenaers, B., G. R. Verheyen, M. V. D. Broeck, T. Burke, C. V. Broeckhoven and A. A. Dhondt. 1992. Extra-pair paternity results from female preference for high-quality males in the blue tit. *Nature* 357:494–496.

Lessells, C. M., and T. R. Birkhead. 1990. Mechanisms of sperm competition in birds: mathematical models. *Behavioral Ecology and Sociobiology* 27:325–337.

Lifjeld, J. T., P. O. Dunn and D. F. Westneat. 1994. Sexual selection by sperm competition in birds: male male competition or female choice? *Journal of Avian Biology* 25:244–250.

Lifjeld, J. T., and R. J. Robertson. 1992. Female control of extra-pair fertilization in tree swallows. *Behavioral Ecology and Sociobiology* 31:89–96.

Møller, A. P. 1994. Sexual Selection and the Barn Swallow. Oxford, England: Oxford University Press.

Møller, A. P., and T. R. Birkhead. 1994. The evolution of plumage brightness in birds is related to extra-pair paternity. *Evolution* 48:1089–1100.

Sheldon, B. C. 1994. Male phenotype, fertility, and the pursuit of extra-pair copulations by female birds. *Proceedings of the Royal Society of London (Biology)* 257:25–30.

Westneat, D. F., P. W. Sherman and M. L. Morton. 1990. The ecology and evolution of extra-pair copulations in birds. *Current Ornithology* 7:331–369.

Prairie-Vole Partnerships

This rodent forms social groups that appear to have evolved as an adaptation for living in a low-food habitat

Lowell L. Getz and C. Sue Carter

If you part the vegetation and closely examine the surface of almost any grassy field in central North America, you may discover a network of small, tunnel-like runways. Bare soil serves as the floor of such a runway, because the grass gets trampled by countless tiny feet. The dead litter and grass surrounding a runway are chewed away, leaving smooth and rounded walls that allow unhindered movement by a runway's occupant. If you follow a runway far enough, you will find three or four burrows, each about three centimeters in diameter, that lead underground. The burrows regroup at an underground nest constructed of tightly interwoven and finely shredded grass that fills a chamber, which is from 10 to 15 centimeters in diameter and lies a dozen or so centimeters below the surface. You might also discover a few surface nests of similar size and construction.

Prairie voles—small rodents that weigh between 35 and 45 grams as adults—build these elaborate runway systems and nests. The prairie voles represent one of the most abundant small mammals in the North American grasslands, from the midwestern United States to the Canadian prairie provinces. An adult prairie vole may live alone or in a group, composed either of relatives or genetic strangers. Re-

Lowell L. Getz is professor of ecology, ethology and evolution at the University of Illinois at Urbana-Champaign. He earned a Ph.D. from the University of Michigan in 1960. He has been studying the social organization and demography of prairie voles for 30 years. C. Sue Carter is professor of zoology at the University of Maryland at College Park. She earned her Ph.D. from the University of Arkansas in 1969. She continues to study the physiological substrates of monogamy in prairie voles. Address for Getz: University of Illinois, 515 Morrill Hall, 505 South Goodwin Avenue, Urbana, IL 61801–3799.

gardless of the social groupings, prairie-vole populations experience high-amplitude fluctuations in numbers. A population's density varies from lows of only one or two animals per hectare (about 2.5 acres)—often disappearing completely—to highs that surpass 600 voles per hectare. Some populations reach high densities every three to four years, but others vary erratically. These fluctuations in population density develop from a combination of a high reproductive rate and heavy mortality from predators.

In many cases, male and female prairie voles practice monogamy—living and mating with a single partner. Many recent articles have examined the hormonal and neurobiological bases for such behavior. By monitoring more than 850 free-living social groups, we found that the social and mating systems of prairie voles depend on other factors as well, including the quantity of available food and the population density. In this article, we shall describe the results of 25 years of laboratory and field studies on the social behavior and population dynamics of prairie voles in east-central Illinois.

Social Groups

In addition to building complex living quarters, prairie voles forge equally complicated living arrangements. Three different types of breeding units, or social groups, can be found in prairie-vole populations. A unit may be a male-female pair, a single female or a communal group.

A male-female pair consists of an adult male, an adult female and, perhaps, newborn young. Such adults display traits that are associated with behavioral monogamy. For example, a male and female share a nest and an associated home range, or territory. More-

over, the male of a pair expends considerable time and effort in constructing and maintaining a nest and in providing care to his young, including grooming and retrieval of pups that wander from the nest. Once paired, a male and a female tend to stay together. In fact, three-quarters of these pairings persist until one member dies.

If a male from such a pair dies, the female rarely acquires a new mate. Instead, she remains alone. Most single-female breeding units consist of a survivor from a male-female pair that was disrupted, usually by the death of a partner. A single-female breeding unit may include juveniles but not an adult male. If the female of a pair dies, the male typically leaves the nest to wander as an unsettled male.

A communal group of prairie voles usually forms from an extended family of a male-female pair or a single female. In such cases, offspring remain in their natal nest, even as adults, and such animals are called *philopatric*. In fact, most prairie voles can be described as philopatric, because 68 percent of the males and 73 percent of the females born to male-female pairs or single females remain at their natal nests through adulthood. As these offspring mature, they help care for younger offspring. When two or more philopatric offspring reach adulthood, unrelated males and females may join an extended family. Nevertheless, communal groups emerge largely from philopatric behavior, because 70 percent of the additions to an original breeding unit come from offspring that stay at home.

Not all adult voles, however, join a social group. At any given time in a population of voles, approximately 45 percent of the males and 24 percent of the females are not permanent residents of a nest. Most of the roaming males

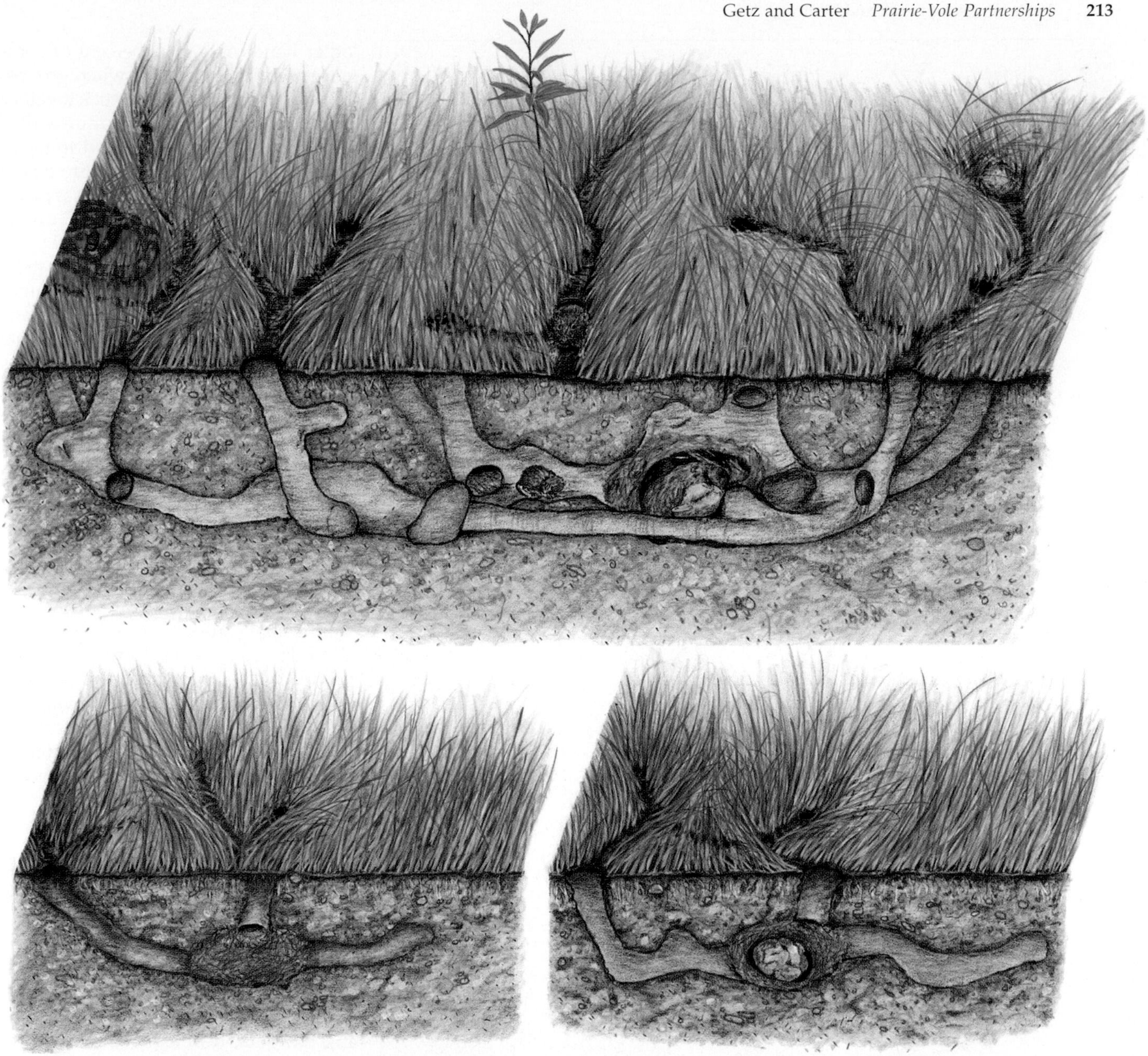

Figure 1. Prairie voles build systems of runways and nests in many North American grasslands. Inside these systems, these small rodents search for food, find mates and raise litters in several forms of families: communal groups *(top)*, single females *(left)* and male-female pairs *(right)*. Although a single population of prairie voles usually supports all three family forms, the predominant one depends primarily on the mortality of adult males of pairs and nestlings.

tend to stay that way, wandering through the population but making frequent visits to social groups. Socially unattached females, on the other hand, soon settle into a nest, either singly or as part of a male-female pair or a communal group. The percentage of wandering males in a population remains the same even when population density varies; so a high-density population contains more wandering males than does a low-density one.

Seasonal Sociability

The distribution of social groups in a population of voles varies seasonally. In an average population between March and mid-October in central Illinois, for instance, about 36 percent of the groups are male-female pairs, 37 percent are single females and 27 percent are small communal groups, made up of three individuals on average. In mid-October, communal living increases and remains high throughout the winter, when 69 percent of the social groups are communal, 24 percent are male-female pairs and seven percent are single females. From late October through February, an average communal group consists of eight voles, and groups made up of more than a dozen appear frequently.

These two general seasons will be called spring–early autumn and late autumn–winter.

The stability of communal groups also varies with season. In spring–early autumn, communal groups stay together an average of 20 days; late autumn–winter groups stay together an average of 87 days. So stability increases in late autumn–winter communal groups. The stability of a communal group depends largely on individual vole behavior. In spring–early autumn, 21 percent of the members of communal groups move from one group to another. During late autumn–winter, on the

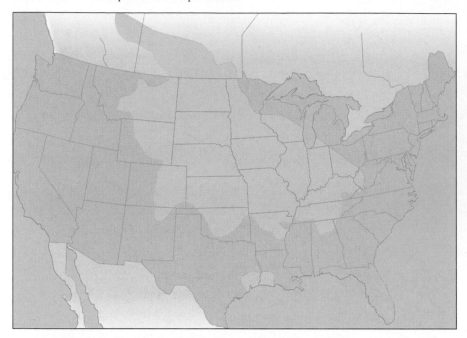

Figure 2. Prairie voles inhabit North American grasslands from the midwestern United States to the Canadian prairie provinces. (Adapted from Hall 1981.)

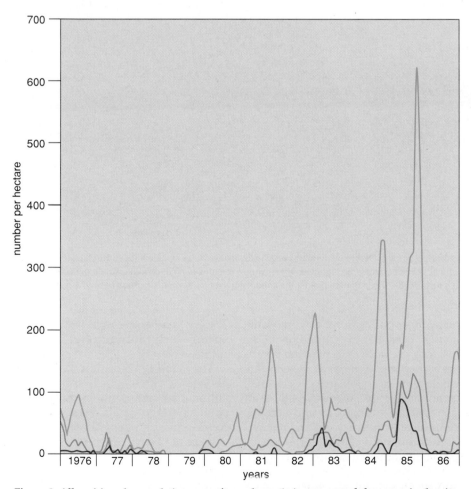

Figure 3. All prairie-vole populations experience dramatic increases and decreases in density. This graph depicts more than 10 years worth of population densities of prairie voles in three different habitats: high food or alfalfa (*green*), medium food or bluegrass (*blue*) and low food or tall-grass (*red*). Prairie-vole densities rise and fall—sometimes every few years and at other times less predictably—in each habitat, but the most significant changes exist when food is plentiful.

other hand, only four percent of communal-group members switch groups. In fact, communal groups stick together so tightly during late autumn-winter that they have been observed to move more than 20 meters as a unit to a new nest location—something that has never been observed in a communal group during spring-early autumn.

As winter progresses, however, mortality decreases the size of communal groups. By spring, survivors from various communal groups form male-female pairs. Nearly two-thirds of those pairs will be composed of individuals from different late autumn-winter communal groups. Moreover, fewer than two percent of the pairs consist of related voles. Although prairie voles tend to live throughout most of a year in communal groups comprised primarily of family members, each spring brings a thorough mixing of a breeding population.

Communal Cause

Why do prairie voles form communal groups? Other mammals nest communally for various reasons, including enhanced exploitation of a resource, such as food, protection against predators or to conserve energy during cold periods. None of these explanations, however, applies to prairie voles.

Prairie voles do not gain food, safety or warmth from group living. No apparent result of group living should increase an individual vole's efficiency at feeding on green vegetation. In terms of predation, even a large number of voles could not fight off most of their enemies, especially the large ones, including badgers, domestic cats, coyotes, foxes, hawks, mink, owls, raccoons, skunks, snakes and weasels. Group warming also fails to explain communal living in prairie voles. First, communal groups, albeit small ones, make up one-fourth of the social groups during summer. Second, prairie voles begin forming larger communal groups in mid-October, even though low temperatures do not become significant for another month. Third, nearly half of the winter social groups consist of fewer than three voles—essentially the same number as in summer. Fourth, when natural causes during winter reduce a large communal group to three or fewer voles, nearly half of the survivors never leave the nest. Moreover, voles that do leave nests tend to join smaller groups.

Although many species of rodents form communal groups only during

Figure 4. Prairie voles often mate monogamously. Once a pair forms, a male and a female tend to stay together until one member dies. (Except where noted, all photographs courtesy of Lisa L. Davis and Lowell L. Getz.)

nonreproductive periods, the same cannot be said of prairie voles. In fact, two or more concurrent litters exist in many prairie vole–communal groups. Moreover, all spring–early autumn–communal groups include at least one reproductive adult. In addition, 73 percent of these groups include reproductive, unrelated males, and 86 percent include unrelated, adult females. Unrelated, reproductive adults can also be found regularly in winter communal groups.

Perhaps the cause of communal-group forming can be found by asking: Why are communal groups most prevalent during late autumn–winter? Although the groups arise from young voles remaining in their natal nest, philopatric behavior does not vary seasonally. Seasonal variation in nestling mortality provides a more obvious reason for the prevalence of communal groups during late autumn-winter. During spring-early autumn, an average of only 0.3 offspring per litter survive 30 days, which leaves too few offspring to form extended families, thereby making communal groups rare. By mid-October, however, an average of 1.3 offspring per litter survive 30 days—an increase of more than four times—which leads to most breeding units becoming communal groups.

From spring through early autumn, the most probable cause of increased nestling mortality in central Illinois is believed to be snakes, which are active during that time and do prey on prairie voles, especially nestlings. When snakes were excluded from part of our study site, communal groups of voles formed in August and September, and those groups were similar in size and composition to those normally observed only in late autumn and winter. When snakes are active, it appears that most social groups remain as male-female pairs or single females, and the few communal groups that do form are small. After snakes hibernate in mid-October, nestling survival increases and large communal groups predominate. So communal groups form in the absence of nest predation.

Mating Triggers

The social organization of a population of prairie voles describes the grouping of individuals in space and time, but it does not reveal which individuals reproduce. Understanding the selective advantage of communal nesting in prairie voles depends on knowing the mating system within social groups. The secrecy of voles, however, prevents observations of their mating under natural conditions, so much of the information about prairie-vole mating comes from laboratory observations. A combination of laboratory studies and field work provides a general understanding of mating in prairie voles.

A chemical signal, or pheromone, in a male prairie vole's urine stimulates the first period of estrus, or heat, in a virgin female. This pheromone does not travel through the air, so it's transmission depends on a female sniffing a male's genital area. A female usually acquires this pheromone from an unfamiliar male, because she does not sniff the genitals of familiar males. Within 48 hours of sniffing this substance, and if a female stays in the presence of a male (not necessarily the one she sniffed), she achieves estrus and mates. Her pregnancy lasts about 21 days. Within a day of giving birth, the female enters postpartum estrus and will usually mate immediately if a reproductive male (usually her mate) is nearby.

If a female practices philopatry—living with her extended family—repro-

Figure 5. Male of a male-female pair contributes considerably to rearing offspring. These efforts include nest building and maintenance, grooming young and, as shown here, retrieving wandering pups.

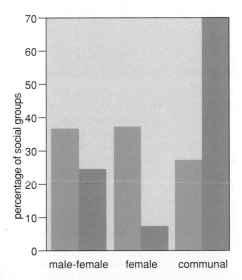

Figure 6. Prairie-vole populations include three social groups: male-female pairs (*left*), single females (*middle*) and communal groups (*right*). From spring to early autumn (*green*), a population consists of nearly equal percentages of male-female pairs and single females and a somewhat lower percentage of communal groups. From autumn through winter (*blue*), communal groups increase dramatically, largely through philopatry—offspring remaining as adults in the nest where they were born.

Figure 7. Virgin female sniffs a male's genital area for exposure to a urinary pheromone. This pheromone triggers reproductive activity in a female. A female that lives with a male becomes reproductively active after a single exposure to this pheromone. If a female lives in a communal group with her extended family, she requires multiple exposures, usually from an unrelated male.

ductive activation becomes more complicated. Reproductively active females in a group produce another urine-based pheromone that suppresses the reproductive activation of young females. So a single bout of male-genital sniffing will not activate a virgin female. Instead, she requires multiple exposures, about a dozen, to the male-urine pheromone over three or four days to achieve estrus.

For the most part, philopatric females do not mate with family members or even familiar males. Laboratory studies indicate that a reproductive, philopatric female will not mate with her brothers or her father if unfamiliar males are present. Without the presence of an unfamiliar male, on the other hand, these females occasionally mate with familiar males, including family members. The presence of a female's mother, however, inhibits a daughter from mating with her father. In any case, matings between philopatric females and their brothers or father rarely lead to pregnancy. As already mentioned, such a female usually becomes reproductive only through frequent visits from unfamiliar males, with which she could mate in the wild.

In addition, philopatric males do not become reproductive in family groups unless they interact with unfamiliar animals, males or females. Moreover, males reared with family members display less than average reproductive behavior when they encounter reproductive females. These behaviors should also reduce matings between related animals in a social group.

Mating Pairs

Given the social structure of a prairie-vole population, opportunities for mating with various partners might exist. Frequent group living, for instance, could provide several potential mating partners for any individual. Moreover, an unfamiliar partner could come from the wandering voles, which include nearly half of the males and one-quarter of the females in a population. The likelihood of a vole mating with more than one partner, however, depends on its type of social group.

In male-female pairs, several factors support monogamy. In such a pair, the male rejects overtures from unfamiliar, virgin females by not allowing them to sniff his genital area. In addition, a male excludes unfamiliar males from the vicinity of his nest, as well as from the periphery of his home range. This mate-guarding behavior works best, as expected, during low-population periods. During high-population periods, unfamiliar males may enter the home range of a male-female pair, which produces an opportunity for the female to mate with another male.

Laboratory observations indicate that a paired, estrous female might mate with unfamiliar males. Nevertheless, DNA fingerprinting of embryos from females of free-living, male-female pairs revealed that the females had mated exclusively with their partners, even at a high population density. It appears that males of such pairs also remain loyal to their mates, because they are rarely found visiting nearby nests.

Single females, on the other hand, receive numerous visits from males, suggesting that litters may be fathered by more than one male. Field evidence reveals that single females could potentially mate with wandering males or neighboring paired males. Nonetheless, wandering males make up more than 90 percent of the male visitors at single-female nests, regardless of the population density. So single females probably mate with wandering males most of the time.

In communal groups, resident males defend the outskirts of their home range against unfamiliar males. At low population densities, unfamiliar males can be excluded, and only 17 percent of the philopatric females become reproductive in a communal group, probably because they are not exposed to unfamiliar males. At high population densities, on the other hand, resident males may fail to exclude all of the wandering males, and 77 percent of the philopatric females become reproductive. Such intruding males may also mate with these females. In fact, we have evidence for multiple paternity in some litters from females that lived in communal groups in high-density populations. The resident males were not mating with multiple partners, so the wandering males must have been mating with a groups' females.

The largely restricted mating interactions in a population of prairie voles bring up a fundamental question: How do prairie voles maintain genetic variability? Populations that undergo periods of low density—so-called population bottlenecks—may experience reduced genetic variability simply because of a reduced number of individuals. Likewise, populations that form genetically distinct subunits, called *demes*, risk even higher losses of variability during a population bottleneck. If a population crashes and an entire deme dies out, unique genes may be lost.

At first, we predicted that prairie voles would be susceptible to a loss of genetic variability. Their social organization includes communal groups formed largely from family members, and every two to five years this species experiences extreme population bottlenecks of 5 to 15 generations. Although this combination could favor reduced genetic variability, prairie voles appar-

1 2 3

Figure 8. Male guards his nest against intruding males. When an unfamiliar male approaches (seen entering the bottom of *1*), the resident male becomes aggressive and attacks the other male (*2* and *3*). In a low-density population, a male can usually guard his mate from other males and keep his daughters from becoming reproductively active. When population density increases, however, increasing numbers of wandering males often make mate guarding fail.

ently avoid that problem. First, hormonal and behavioral mechanisms reduce the probability of mating between related voles. Second, a breeding population of voles gets mixed completely at the beginning of each annual breeding period. Finally, voles that do disperse from their natal nest travel far enough away that they are not likely to mate with a family member. So most voles mate with unrelated partners, which facilitates gene flow and genetic variability in a population, even during bottlenecks.

Communal Evolution
The preceding information suggests an evolutionary answer to an earlier question: Why do prairie voles form communal groups? This animal evolved in tallgrass prairies, which are very low in food. Although these habitats include forbs—broad-leaved herbs with succulent stems and leaves—that are the es-

sential food for prairie voles, these plants provide food only during spring, and even then they are scattered. By early summer, the growth of tall surrounding grasses shades a forb's base, causing the lower leaves to wither and die. At the same time, a forb's stem turns dry. During the winter, little green vegetation of any sort exists in a tallgrass prairie. This low level of food leads to low population densities and widely dispersed females.

We believe that the social organization and mating system of prairie voles might have evolved as an adaptation to low-food habitats. If a low-food habitat induces a low population density among voles and widely dispersed females, then a male might do better by forming a pair with a female. That pair-forming behavior would assure a male of a suitable mate and prevent him from continually expending energy on searching for other females. Once a male selects a mate, he

might increase his fitness by guarding her from other males and by providing care for his offspring. Then all of the young produced by the female would be his and more would survive, because of receiving additional parental care.

In a low-food habitat, any voles leaving their natal nest would face little chance of finding a mate and locating a home range that would provide enough food for successful reproduction. So offspring would do better by staying home. In addition, reproductive activation of philopatric offspring would be selected against, because the habitat would not offer enough food for a mother and her daughters to raise litters on the same home range. Few of the offspring that did leave their natal nest would survive, because of limited food. That would lead to very few wandering males. So a male of a male-female pair could adequately guard his mate and prevent reproductive activation of his daughters. The offspring that remained at their natal nest could increase their own reproductive success, or fitness, by caring for younger siblings, which share common genes with their extended family. That helping, instead of reproducing, would further limit the size of a prairie-vole population.

Although the social organization of prairie voles apparently evolved in response to limited food, several field studies reveal that the present social organization does not vary with food availability. Perhaps the basic behaviors associated with communal nesting in a low-food habitat—behavioral monogamy and philopatry—remain stable even in high-food habitats. Nevertheless, some behaviors, such as mate guarding and reproductive suppression of philopatric females, may not maintain the original mating system in high-food habitats. In fact, we predict that ample amounts of food lead to increased survival for voles that leave their natal nest. That would induce more breeding groups, more wandering males and, in turn, more potential visitors to social groups than in low-food situations.

At some point, a male of a male-female pair could not fend off all of the wandering males, and most of the philopatric females would become reproductive. With a high level of food, however, those females could raise successful litters in their natal nest. Nevertheless, such a disruption of the normal mating system would lead to higher population densities, especially in late autumn, when more nestlings survive. Overall, this change in prairie-vole mating would lead to large-amplitude swings in population in high-food habitats. Therefore, we conclude that there is a clear relationship between the social organization and mating system of prairie voles and the variations in population demography that are observed in different habitats. These conclusions should be tested by comparing the social organization and mating system in high- and low-food habitats across a range of population densities.

Bibliography

Carter, C. S., and L. L. Getz. 1993. Monogamy and the prairie vole. *Scientific American* 268:100–106.

Getz, L. L., D. Gudermuth and S. Benson. 1992. Pattern of nest occupancy of the prairie vole, *Microtus ochrogaster*, in different habitats. *American Midland Naturalist* 128:197–202.

Getz, L. L., B. McGuire, J. Hofmann, T. Pizzuto and B. Frase. 1993. Social organization of the prairie voles (*Microtus ochrogaster*). *Journal of Mammalogy* 74:44–58.

Getz, L. L., N. Solomon and T. Pizzuto. 1990. The effects of predation of snakes on social organization of the prairie vole, *Microtus ochrogaster*. *American Midland Naturalist* 123:365–371.

Hall, E. R. 1981. *The Mammals of North America*. New York: John Wiley & Sons.

Hofmann, J. E., and L. L. Getz. 1988. Multiple exposures to adult males and reproductive activation of virgin female *Microtus ochrogaster*. *Behavioural Processes* 17:57–61.

Hofmann, J. E., L. L. Getz and L. Gavish. 1984. Home range overlap and nest cohabitation of male and female prairie voles. *American Midland Naturalist* 112:314–319.

McGuire, B., and L. L. Getz. 1991. Response of young female prairie voles (*Microtus ochrogaster*) to nonresident males: implications for population regulation. *Canadian Journal of Zoology* 69:1348–1355.

McGuire, B., L. L. Getz, J. Hofmann, T. Pizzuto and B. Frase. 1993. Natal dispersal and philopatry in prairie voles (*Microtus ochrogaster*) in relation to population density, season, and natal social environment. *Behavioural Ecology and Sociobiology* 32:293–302.

McGuire, B., T. Pizzuto and L. L. Getz. 1990. Potential for social interaction in a natural population of prairie voles, *Microtus ochrogaster*. *Canadian Journal of Zoology* 68:391–398.

Mating Behavior and Hermaphroditism in Coral Reef Fishes

The diverse forms of sexuality found among tropical marine fishes can be viewed as adaptations to their equally diverse mating systems

Robert R. Warner

The colorful diversity of coral-reef fishes has long been a source of fascination for both scientists and amateurs. We now know that this diversity of form and color is matched by an immense variety of social behaviors and sexual life histories, including several kinds of functional hermaphroditism. Recent observations and experimental work suggest that the sexual patterns found in fishes may best be viewed as evolutionary responses to the species' mating systems, and much of the evidence I review here bears out this idea. Since most theory in behavioral ecology has been derived from studies of terrestrial vertebrates and insects, which have strictly separate sexes, the relationships between sexual expression and mating behavior in fishes offer new insights into the role of sexuality in social evolution.

Like other vertebrates, most fish species have separate sexes, a condition known as gonochorism. However, fishes are by no means restricted to this pattern: in many species individuals are capable of changing sex, a phenomenon sometimes called sequential hermaphroditism, and in others fishes can be both sexes at the same time, displaying simultaneous hermaphroditism.

This sexual flexibility is quite widespread. At least fourteen fish families contain species that exhibit sex change from female to male, termed protogyny, as a normal part of their life histories (see Policansky 1982 for a recent review). Eleven of these families are common in coral-reef areas; and in the wrasses (Labridae), parrotfishes (Scaridae), and larger groupers (Serranidae) protogyny occurs in the great majority of the species studied (Fig. 1). Changes from female to male are also known to occur in damselfishes (Pomacentridae; Fricke and Holzberg 1974), angelfishes (Pomacanthidae), gobies (Gobiidae), porgies (Sparidae), emporers (Lethrinidae; Young and Martin 1982), soapfishes (Grammistidae), and dottybacks (Pseudochromidae; Springer et al. 1977). However, we have no indication of how common sex change might be in these families, since few species have been carefully investigated. New reports of protogynous species are constantly cropping up, and the phenomenon may be much more frequent than previously imagined.

Change of sex from male to female—protandry—appears to be less common. It is known in eight families of fishes, three of which—porgies (Sparidae), damselfishes (Pomacentridae), and moray eels (Muraenidae; Shen et al. 1979)—are found on coral reefs. The damselfish and porgy families also include species that are protogynous. Such variability within a family offers an important opportunity to test sex-change theory, and deserves further study.

Robert R. Warner is Associate Professor of Marine Biology at the University of California, Santa Barbara. He received his Ph.D. from Scripps Institution of Oceanography in 1973. Before coming to Santa Barbara in 1975, he was a postdoctoral fellow at the Smithsonian Tropical Research Institute at Balboa, Panama. His interests include the interactions of life history and behavioral characteristics, the evolution of sexual dimorphism, the adaptive significance of delayed maturity in males, and the sexual patterns of tropical marine fishes. Address: Department of Biological Sciences, University of California, Santa Barbara, CA 93106.

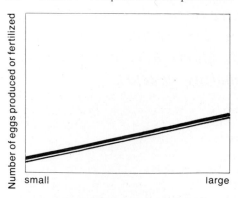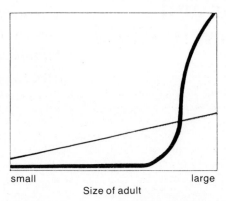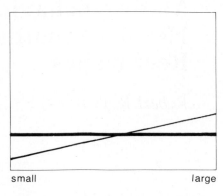

Figure 2. The expected fertility of a female fish (*thin line*), measured as the number of eggs produced, usually depends directly on size and thus shows a steady increase with growth. By contrast, a male's expected fertility (*thick line*), measured as the number of eggs fertilized, is affected by the mating system. When males and females form monogamous pairs matched by size or when males compete with each other to fertilize eggs and thus to produce the most sperm, both sexes show a similar increase of fertility with size (*left*), and no selection for sex change exists.

When large males monopolize mating to the detriment of small males, however, male fertility rises dramatically at a certain point in growth (*center*), and an individual that remains a female when small and changes to a male at a large size will be selectively favored. When mating consists of random pairing (*right*), an individual would do best to function as a male while small (with a chance of fertilizing a larger female) and as a female when large (capitalizing on a high capacity for egg production). (After Warner 1975.)

Species in which individuals are simultaneous hermaphrodites, producing eggs and sperm at the same time, are common among deep-sea fishes (Mead et al. 1964) but quite rare elsewhere. However, several species of small sea basses (Serranidae) found in shallow water in tropical areas are known to be simultaneous hermaphrodites (Smith 1975). The existence of this sexual pattern in an abundant species such as the sea basses poses special evolutionary problems that are discussed below.

Because the warm, clear waters of coral reefs create conditions that are nearly ideal for observing animal behavior, the social and mating systems of tropical marine fishes have been particularly well studied. It is my intention here first to outline a central hypothesis which relates sexual patterns to social behavior, and then to test this hypothesis at several levels, using information available in the literature.

Sex change

Why should natural selection favor sexual patterns different from pure gonochorism? In other words, under what circumstances might we expect hermaphroditism to be adaptive? This question was first dealt with in a comprehensive fashion by Ghiselin (1969, 1974), who suggested the general conditions under which hermaphroditism would be expected to evolve. More recent work has related these generalities to the specific social systems of hermaphroditic species (Warner 1975; Fischer 1980; Charnov 1982). It is best to view sequential her-

maphroditism, or sex change, separately from simultaneous hermaphroditism, because they are distinct phenomena under the influence of very different selective regimes.

Ghiselin proposed the "size-advantage model" to account for many cases of sex change. The concept is simple: if the expected number of offspring (measured, say, as the number of eggs produced or fertilized) differs between the sexes with size, then an individual that changes sex at the right size or age will have more offspring than one that remains exclusively male or female.

What might cause such sexual differences in the distribution of expected fertility? Two factors are important here: the relative number of male and female gametes produced and the characteristic mating behavior of the species. In many cases, a female's fertility is limited by the number of eggs she can hold or manufacture, which in turn is controlled by her size, her store of energy, or both. Thus it makes little difference whether she mates with one male or with many. Male fertility, on the other hand, is often limited not by the number of sperm an individual can produce but rather by the number of females with whom he mates and their fertility. Because of this, the fertility of males is potentially much more variable than that of females, and can reach very high levels in certain circumstances (Williams 1966; Trivers 1972).

While size of gamete production sets the stage for potential differences in the reproductive success of males and females of various sizes, it is often the mating system that determines the actual values (Fig. 2). For example, in monogamous species where both members of a pair are normally about the same size, the fertility of males and females is approximately the same over their entire size range, and changing sex conveys no advantage.

By contrast, many coral-reef fishes have mating systems in which larger males monopolize the spawning of females. In this situation smaller males may not spawn at all, while females of equivalent size have little trouble finding a mate. The spawning rate of small males is thus lower than that of small females, but large males expe-

Figure 1. The parrotfish family Scaridae is one of several common families of coral-reef fishes in which the majority of species include individuals that change from female to male, a phenomenon known as protogyny. In the striped parrotfish, *Scarus iserti*, small females are boldly striped (*top*); as females grow larger, they change sex and adopt the brilliant blue and yellow coloration found in older males (*bottom*). Groups of young females defend territories along the edges of reefs; males and sex-changed females have larger territories which usually encompass several of these harems.

rience relatively high mating success. Since the distribution of fertility differs between the sexes with size, we would expect sex change to be adaptive: an individual that functioned as a female when small and as a male after attaining a large size would have more offspring over its lifetime than one that remained either male or female, and thus protogyny should be favored by natural selection (Warner 1975).

Other mating systems lead to selection for protandry. Males usually produce millions of sperm, and small individuals are physically capable of fertilizing females of almost any size. Thus in mating systems where no monopolization occurs and where mating consists of random pairing, it should be advantageous to be a male when small (since it is probable that any mating will be with larger individuals) and a female when large (thereby taking advantage of a high capacity for egg production).

Because the fertilization of eggs occurs outside the body in many fish species, spawning is not limited to simple pairs: numerous individuals can mate simultaneously in large groups. Although mating occurs more or less at random in such spawning groups, protandry is not necessarily adaptive if many males release sperm simultaneously. In this case, competition among sperm from several males to fertilize eggs creates a situation in which male fertility is limited by the number of sperm produced. Such production should increase with size in a fashion similar to egg production by females, and thus no fertility differential between the sexes exists.

The size-advantage model has been refined over the years to allow for sexual differences both in mortality and in the rate at which fertility changes with size (Warner et al. 1975; Leigh et al. 1976; Jones 1980; Char-

nov 1982; Goodman 1982). Individuals should change sex when the other sex has a higher reproductive value—that is, higher future expected reproduction taking into account the probability of death. This means that individuals may (and do) change sex and suffer an initial drop in reproductive success, but by making the change they increase the probability of attaining a high level of success in the future. These are complications we need not consider here, since they do not affect the general idea that the mating system can determine the adaptive value of various forms of sex change.

Testing an evolutionary idea such as this is difficult, since experimental manipulations are often impossible or exceedingly time-consuming. Typically, one must rely instead on a search for correlations between the hypothesized cause and effect. As long as sufficient variation exists in the traits in question, the search may take place among unrelated species, within a related group, or even within a single species. The wide diversity of sexual patterns and behaviors among coral-reef fishes allows investigation on all these levels. In addition, the fact that sex changes are often direct responses to external cues makes possible experimental study as well.

Mating in sex-changers

In general, the mating systems of sex-changing species are those in which reproductive success varies with sex and size. Larger males tend to monopolize mating, either by defending spawning sites that females visit or by controlling a harem of females. In most of the species, eggs are simply released into the water, and males are free to devote a large amount of time to courtship, spawning, and defense of mating sites.

Figure 3. The sexual pattern and mating system of the bluehead wrasse, *Thalassoma bifasciatium*, is typical of many wrasses. As in most protogynous species, large males play a dominant role. These older individuals—both primary males and sex-changed females—defend spawning grounds, pair-mating with as many as 150 visiting females daily (*left*). Young primary males engage in group spawning (*right*) and in "sneaking," the practice of interfering with the mating of older males by rushing in as sperm is released. Young females and young primary males such as those shown group-spawning are characterized by greenish-black lateral markings, whereas both older primary males and sex-changed females display a distinctive white band bordered with black, like the larger individual at the left. (Photos by S. G. Hoffman.)

Figure 4. The study of three closely related species of the wrasse genus *Bodianus* seems to support the idea that sex change is less common in species in which large males have less opportunity to monopolize mating. Both *Bodianus rufus* (*top*), a species in which large males defend harems of females, and *B. diplotaenia* (*center*), one in which large males defend temporary spawning sites, are protogynous, as might be expected in mating systems where large males dominate. By contrast, *B. eclancheri* (*bottom*), which spawns in groups with no pattern of domination by large males, is functionally gonochoric, with the sexes existing in equal ratios. (Photos by S. G. Hoffman.)

A good example of such a mating system is found in the wrasse *Labrides dimidiatus*, which feeds by cleaning the skin, mouth, and gills of other fishes at specific "cleaning stations" on the reef. The cleaner-wrasses at a station live in a group consisting of a single male and a harem of five or six females. The male actively defends these females and mates with each one every day. This appears to be a system in which there is no advantage to being a small male, and indeed the species is totally protogynous (Robertson 1972).

In the last decade, marine biologists have begun to study the mating behavior of a wide variety of protogynous species in their coral-reef habitats. It is striking that virtually all these species exhibit some form of monopolization of mating by large males, even though they are found in a diverse array of families such as wrasses, parrotfishes, damselfishes, angelfishes, basses, and gobies (see, for example, Moyer and Nakazono 1978a; Robertson and Warner 1978; Cole 1982; and Thresher and Moyer 1983). Haremic mating systems appear to be most common among these protogynous species. Coral-reef fishes are often quite sedentary, and it is not surprising that large males have come to dominate and defend a local group of females in many cases. It is just this kind of situation that evolutionarily favors sex change from female to male.

There are also some apparent exceptions to the trend toward the dominance of the large male in protogynous species. In a number of species of gobies (Lassig 1977), a small bass (Jones 1980), and a wrasse (Larsson 1976), the social system appears to be monogamous, and thus protogyny would not be adaptive. Lassig (1977) suggests that sex change in the gobies he studied is an adaptation to allow reconstitution of a mated pair in case of death. This explanation probably would not apply to the more active wrasses and basses, however.

For other families in which protogyny occurs, such as the groupers and the emperors, we simply lack sufficient knowledge of the mating behavior to state whether the predictions of the size-advantage model hold.

Our knowledge of the mating habits of protandrous fishes is also incomplete. Many of the species known to be protandrous live in large schools not closely associated with the substrate; however, the details of their mating behavior have not been reported. The size-advantage model suggests that mating might consist of haphazard pairing, but this prediction remains to be tested. The anemonefishes, the one group of protandrous species whose mating system is well known, fit the size-advantage model in a precise but unexpected way, as is discussed below.

While most studies of fishes known to change sex lend support to the size-advantage model, approaches from the opposite direction are less satisfactory. For example, sex change is not found in every species in which mating is monopolized by large males. Perhaps this is asking too much of evolution, since an adaptive situation does not guarantee the appearance of a trait. It may simply be that the capacity for sexual flexibility has not yet evolved in some species, or that unknown factors reduce the advantage of sex change. Unfortunately, like many evolutionary arguments, this one is virtually untestable.

Comparisons within families

We can avoid some of the uncertainty inherent in broad comparisons among families by examining the sexual patterns and mating behaviors of a group of species within a family in which we know sex change is widespread. Using this approach, the absence of sex change where it is theoretically adaptive is less easily dismissed as evolutionary lag.

The wrasses (Labridae) and the parrotfishes (Scaridae) are large and well-known families of coral-reef fishes that include many species made up of both primary males—that is, fish that remain males for their entire lives—and protogynous individuals. Both primary males and females can become dominant, territorial males if they grow large enough. The proportion of smaller males is a measure of the degree of sex change present in the species: in cases where small males are absent, sex change is at a maximum, and when they form half the population, the species is essentially gonochoristic.

The diversity of sexual types within the wrasses and parrotfishes is reflected in a diversity of mating behavior (Fig. 3). Small males either interfere with the mating activities of larger males by darting in to join the spawning couple at the moment sperm is released, a practice called "sneaking," or they take over a whole spawning site en masse and group-spawn with the females that appear there. In group spawning a single female releases her eggs in the midst of an aggregation of males, all of whom participate in fertilization. Spawning groups can contain from two to over a hundred males.

For the size-advantage model to hold, variation in the degree of sex change should correspond to differences in the mating systems. Specifically, sex change should be less common or absent in species where large males have less opportunity to monopolize mating. Hoffman's recent work (1980 diss., 1983) on three closely related wrasses of the genus *Bodianus* provides a clear demonstration of this relationship between sexual expression and mating system (Fig. 4). *Bodianus rufus* of the Caribbean is haremic, whereas *B. diplotaenia* of the eastern tropical Pacific defends a spawning site visited by females; thus large males monopolize mating in both species. Correspondingly, small males are absent in both species, and all males are the result of sex change in functional females. On the other hand, the multicolored *B. eclancheri* of the Galapagos Islands is a group-spawner with no apparent pattern of dominance related to size or sex. Hoffman could find no evidence of sex change during adult life in this species. Individuals appear instead to be functional gonochores: change from female to male occurs before maturation, and males are equally common in all size classes of the population. The production of small males through prematurational sex change has also been noted in some species of parrotfishes (Robertson and Warner 1978).

What factors lead to changes in the monopolization of mating by large males? We have found that increased population density around spawning sites plays a role in lowering the ability of a male to defend his harem or territory adequately against smaller males (Warner and Hoffman 1980a). In extreme cases, some spawning sites can be undefendable and may be entirely abandoned to group-spawners (Warner and Hoffman 1980b).

A recent study of the wrasses of the Caribbean (genera *Thalassoma*, *Halichoeres*, *Bodianus*, and *Clepticus*) revealed that among species living in similar habitats, those with low population densities tended to have few or no small males (Warner and Robertson 1978). Regardless of whether the mating system was characterized by harems or spawning-site defense, larger males successfully monopolized mating in these species. In species living at greater densities, the proportion of primary males rose as high as 35%. Among these densely distributed species, group spawning as well as territorial mating was seen, with larger males subject to varying amounts of interference from small primary males.

The most thoroughly studied species in this group is the bluehead wrasse, *Thalassoma bifasciatum*. In this species, large males normally control the spawning sites on smaller reefs where the density of the mating population is low, and small males are nearly absent from these local populations (Fig. 5). On large reefs, where spawning sites are much more crowded, group-spawning aggregations occupy the major sites and small males are relatively common (Warner and Hoffman 1980b). Since individuals arrive on reefs as drifting planktonic larvae, the precise mechanisms leading to the distribution of small males are not known, but this example serves as a useful illustration of how density can affect the sex-changing strategy.

Not surprisingly, the effect on monopolization of mating is most pronounced at extreme densities. *T. lucasanum* of the eastern tropical Pacific is the most densely

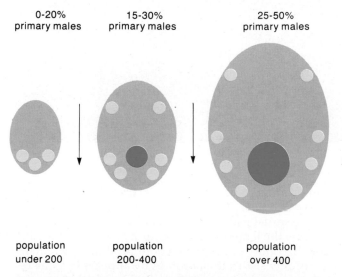

0-20% primary males	15-30% primary males	25-50% primary males

population under 200	population 200-400	population over 400

Figure 5. The mating system of the bluehead wrasse, which spends its adult life on a single reef, varies dramatically with the size of the reef. In this diagram of typical mating configurations, the predominant direction of the current is indicated by arrows; sites at the downcurrent end of the reef are preferred by spawning females. On small reefs (2 to 600 m²), mating occurs exclusively with large males in their territories (*light blue areas*), and primary males are relatively rare. On intermediate-sized reefs (600 to 1,000 m²), territorial males continue to occupy the prime spawning area, but a spawning group of smaller males is active just upcurrent (*dark blue area*). On the largest reefs (above 1,000 m²), a large spawning group of small males occupies the major downcurrent spawning site, and territorial males are relegated to less productive upcurrent sites. Primary males reach their highest concentrations on such reefs, constituting up to 50% of the population.

Figure 6. The most adaptive direction for sex change can depend on the size of the social group. When the group consists of a single pair, both individuals profit if the larger member of the pair is a female, since she could produce more eggs than a smaller individual. Protandry is most adaptive in this case, and is found in the strictly monogamous anemonefishes of the genus *Amphiprion* (*top*). In larger social groups, the combined egg production of smaller members can easily exceed the egg production of the largest individual, and thus his output is maximized by functioning as a male. Protogyny would be expected here, and is found in the group-living damselfishes of the genus *Dascyllus* (*bottom*), which are closely related to anemonefishes. (Photo at top by H. Fricke; photo at bottom by F. Bam.)

distributed wrasse thus far studied; its population is essentially gonochoristic, with about 50% primary males (Warner 1982). Large territorial males are rather rare and only moderately successful in this species, and nearly all mating takes place in groups.

Certain characteristics of the habitat that allow access to spawning sites by small males apt to engage in "sneaking" should also affect mate monopolization. These characteristics are difficult to measure in a quantitative fashion, but some trends are evident. For example, small parrotfishes that live in beds of sea grass near coral reefs have a higher proportion of small males than species that exist in similar densities on the reefs themselves (Robertson and Warner 1978). In one grass-dwelling species, sex change appears to be entirely absent (Robertson et al. 1982). Sea grasses offer abundant hiding places for small fishes, and dominant males in these habitats suffer interference from smaller males in a high proportion of their matings.

Perhaps the most telling variation within a family occurs in the damselfishes (Pomacentridae), where sex change was only recently discovered (Fricke and Fricke 1977). Small damselfishes called clownfishes or anemonefishes (genus *Amphiprion*) live in or near large stinging anemones in reef areas and thus have extremely limited home ranges. They appear to be unaffected by the stinging cells of the anemone, and may enjoy a certain amount of protection from the close association (Allen 1972). An anemonefish society consists of two mature individuals and a variable number of juveniles. The species are protandrous; the largest individual is a female, the smaller adult a male (Fricke and Fricke 1977; Moyer and Nakazono 1978b). The per capita production of fertilized eggs is higher when the larger individual of a mating pair is the female, and protandry is thus advantageous to both adults (Warner 1978).

Note that the advantage of protandry in this case depends on the fact that the social group is rigidly limited to two adults. If more adults were present, the most adaptive sexual pattern could instead be protogyny. This is because the largest individual, as a male, might be able to fertilize more eggs than it could produce as a female. In accordance with this, protogyny appears in some related damselfishes (genus *Dascyllus*) in which the social groups of adults are larger (Fig. 6; Fricke and Holzberg 1974; Swarz 1980 and pers. com.; Coates 1982).

Social control of sex change

Another way of testing the size-advantage model is through an investigation of the dynamics of sex change within a species. So far, I have stressed the importance of the mating system in determining the advantage of a given sex and size. Within a mating system, it is often relative rather than absolute size that determines reproductive expectations. For example, when dominance depends on size, the probable mating success of a particular male is determined by the sizes of the other males in the local population. It would be most adaptive for individuals to be able to change from female to male when their expectations of successful reproduction as a male increase considerably. Thus the removal of a large, dominant male from a population should result in a change of sex in the next largest individual, but no change should be expected in the rest of the local population.

Such social control of sex change has been noted in several species of protogynous coral-reef fishes. Because haremic species exist in small, localized groups, they have proved to be exceptionally good candidates for studies of this kind. In the cleaner-wrasse *L. dimidiatus*, Robertson (1972) found that if the male is removed from the harem, the largest female rapidly changes sex and takes over the role of harem-master. Within a few hours she adopts male behaviors, including spawning with the females. Within ten days this new male is producing active sperm. By contrast, the other females in the harem remain unchanged.

Social control of sex change has also been found in other haremic species (Moyer and Nakazono 1978a; Hoffman 1980; Coates 1982), as well as in species that live in bigger groups with several large males present (Fishelson 1970; Warner et al. 1975; Shapiro 1979; Warner 1982; Ross et al. 1983). In all cases, it is always the largest remaining individuals that undergo sex change when the opportunity presents itself. Even when experimental groups consist entirely of small individuals, sex change can still be induced in the largest individuals present, in spite of the fact that they may be far smaller than the size at which sex change normally occurs (Hoffman 1980; Warner 1982; Ross et al. 1983).

The exact behavioral cues used to trigger sex change appear to differ among species. Ross and his co-workers have shown experimentally that the sex-change response in the Hawaiian wrasse *T. duperrey* depends solely on relative size and is independent of the sex and coloration of the other individuals in a group, whereas Shapiro and Lubbock (1980) have suggested that the local sex ratio is the critical factor in the bass *Anthias squamipinnis*. While it is still unclear how sex change is regulated in fishes that live in large groups, the mechanisms appear to operate with some precision. Shapiro (1980) found that the simultaneous removal of up to nine male *Anthias* from a group led to a change of sex in an equivalent number of females.

Social control of sex change occurs in protandrous fishes as well, and in a pattern consistent with the size-advantage model. A resident male anemonefish will change sex if the female is removed (Fricke and Fricke 1977; Moyer and Nakazono 1978b). One of the juveniles—who apparently are otherwise repressed from maturing—then becomes a functional male and the adult couple is reconstituted.

Simultaneous hermaphroditism

In one sense, the adaptive significance of simultaneous hermaphroditism is obvious: by putting most of their energy into egg production and producing just enough sperm to ensure fertilization, a hermaphroditic mating couple can achieve a much higher output of young than a male-female pair (Fig. 7; Leigh 1977; Fischer 1981). The problem, however, rests with the maintenance of simultaneous hermaphroditism in the face of an alternative male strategy. Consider an individual that fertilizes the eggs of a hermaphrodite, but does not reciprocate by producing eggs of its own. Instead, this individual uses the energy thus saved to find and fertilize other hermaphrodites. This strategy would spread rapidly in a purely hermaphroditic population, effectively forcing it to become gonochoristic. It would therefore appear that where simultaneous hermaphroditism is present, there should exist some means of preventing this kind of "cheating" (Leigh 1977; Fischer 1981).

Among the small coral-reef basses (Serranidae) that are known to be simultaneous hermaphrodites, two types of possible anticheating behavior have been observed. The hamlets, small basses common on Caribbean coral reefs (genus *Hypoplectrus*), appear to ensure that investments in eggs are kept nearly even between the members of a spawning pair by what Fischer (1980) has called "egg trading." In this behavior, a pair alternates sex roles over the course of mating (Fig. 8). Each time an individual functions as a female, it extrudes some, but not all, of its eggs. As a male, it fertilizes the eggs of its partner, who also parcels out eggs in several batches. Thus both individuals are forced to demonstrate their commitment to egg production, and neither has the chance for an unreciprocated fertilization of a large batch of eggs.

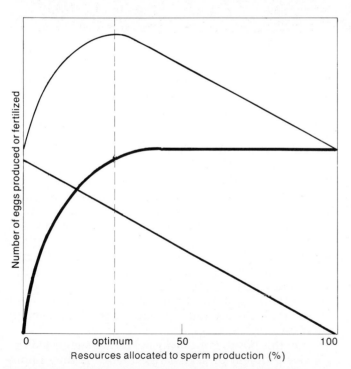

Figure 7. The number of eggs produced by an individual is normally directly related to the amount of energy devoted to their manufacture (*thin line*). In the case of sperm, however, a relatively low output can often produce maximum success (*thick line*) — that is, a small amount of sperm can fertilize all of a partner's eggs, and further investments are superfluous. Natural selection favors the individual with the highest overall reproductive rate combining male and female functions (*black line*), and in this case the optimum result is obtained by putting most of the energy into egg production. A simultaneous hermaphrodite following this strategy has a much higher reproductive success than an individual that is exclusively male or female. (After Fischer 1981.)

Y-axis: Number of eggs produced or fertilized

X-axis: Resources allocated to sperm production (%)
0   optimum   50   100

Figure 8. Simultaneous hermaphrodites among fishes include the hamlet (genus *Hypoplectrus*), a small bass that may alternate sexual roles as many as four times in the course of a single mating, by turns offering eggs to be fertilized and fertilizing its partner's eggs. Here the fish acting as the male curves its body around the relatively motionless female, cupping the upward-floating eggs as he fertilizes them. The strategy of parceling out eggs in a number of batches means that the egg contributions of the two partners are roughly equal, and reduces the rewards of "cheating" by fertilizing a large batch of eggs and then refusing to offer eggs for fertilization in return. (Photo by S. G. Hoffman.)

Another method of preventing desertion is to reduce the opportunities of your partner to find another mate. Some simultaneously hermaphroditic species of the genus *Serranus* delay their mating until late dusk, just before nightfall. These species do not engage in egg-trading, but presumably the onset of darkness means that time is quite limited before shelter must be taken for the night, and thus further mating is impossible (Pressley 1981).

Although these anticheating behaviors are fascinating in their own right, they give us little insight into the origin of the sexual pattern itself. Simultaneous hermaphroditism has been viewed as an adaptation to extremely low population density: if finding mates is difficult, it helps a great deal to be able to mate with whomever you meet (Tomlinson 1966; Ghiselin 1969). Thus many deep-sea fishes, sparsely distributed in their habitat, are simultaneous hermaphrodites (Mead et al. 1964). Perhaps the small serranids evolved from ancestors who lived at low densities, and developed their anticheating behaviors at a later stage when densities were higher. Alternatively, and perhaps more likely, the existence of late-dusk spawning behavior could have allowed the development of hermaphroditism. Unfortunately, we again run up against the problem of untestability and limited predictive power: several other coral-reef fishes mate late in the day, but they are not simultaneous hermaphrodites.

Broader patterns

The contrast between sex change and simultaneous hermaphroditism is intriguing: sex change, particularly protogyny, appears to be a specific adaptation to certain mating systems that happen to be common on coral reefs. These mating systems may be prevalent in coral habitats because clear water, relatively low mobility, and the absence of paternal care allow a greater degree of dominance by large males. Simultaneous hermaphroditism, on the other hand, is theoretically adaptive in a wider variety of circumstances, but is evolutionarily unstable unless male-type cheating can be prevented. While the wide dispersion of deep-sea fishes automati-

cally works against cheating, I can see no reason why coral reefs are particularly good places for such prevention to come about.

Among hermaphroditic groups other than tropical marine fishes, our knowledge of behavior and ecology is generally insufficient to carry out a similar analysis of the relationship of sexual pattern to mating system. Broad surveys of vertebrates (Warner 1978) or organisms in general (Policansky 1982) have shown large-scale tendencies toward hermaphroditism in some groups, sporadic appearance of the phenomenon in others, and a total lack of sexual flexibility in still others. While major features such as the greater complexity of terrestrial reproduction may help to explain the lack of hermaphroditism in some large groups (Warner 1978), many others must await more thorough investigation of the life histories and behavior of the organisms in question.

On this point, Policansky (1982) has suggested that a major problem of sex-change theory is that among closely related species with similar life histories, some change sex and some do not. In light of this review, I am not yet ready to take such a dim view of the size-advantage model. Sexual expression in fishes is extraordinarily adaptable, and closely related species can have quite different sexual patterns that appear to be predictable from their different mating systems. For the fishes, at least, divergent sexual expressions may be no more surprising than differences in coloration. Detailed considerations of the mating systems and life histories of other sexually labile groups are needed to test the hypothesis further.

References

Allen, G. R., 1972. *The Anemonefishes, Their Classification and Biology.* Neptune City, NJ: T. F. H. Publications.

Charnov, E. L. 1982. *The Theory of Sex Allocation.* Princeton Univ. Press.

Coates, D. 1982. Some observations on the sexuality of humbug damselfish, *Dascyllus aruanus* (Pisces, Pomacentridae) in the field. *Z. für Tierpsychol.* 59:7–18.

Cole, K. S., 1982. Male reproductive behavior and spawning success in a temperate zone goby, *Coryphopterus nicholsi. Can. J. Zool.* 10: 2309–16.

Fischer, E. A. 1980. The relationship between mating system and simultaneous hermaphroditism in the coral reef fish *Hypoplectrus nigricans* (Seranidae). *Anim. Beh.* 28:620–33.

———. 1981. Sexual allocation in a simultaneously hermaphroditic coral reef fish. *Am. Nat.* 117:64–82.

Fishelson, L. 1970. Protogynous sex reversal in the fish *Anthias squamipinnis* (Teleostei, Anthiidae) regulated by presence or absence of male fish. *Nature* 227:90–91.

Fricke, H. W., and S. Fricke. 1977. Monogamy and sex change by aggressive dominance in coral reef fish. *Nature* 266:830–32.

Fricke, H. W., and S. Holzberg. 1974. Social units and hermaphroditism in a pomacentrid fish. *Naturwissensch.* 61:367–68.

Ghiselin, M. T. 1969. The evolution of hermaphroditism among animals. *Quart. Rev. Bio.* 44:189–208.

———. 1974. *The Economy of Nature and the Evolution of Sex.* Univ. of California Press.

Goodman, D. 1982. Optimal life histories, optimal notation, and the value of reproductive value. *Am. Nat.* 119:803–23.

Hoffman, S. G. 1980. Sex-related social, mating, and foraging behavior in some sequentially hermaphroditic reef fishes. Ph.D. diss., Univ. of California, Santa Barbara.

———. 1983. Sex-related foraging behavior in sequentially hermaphroditic hogfishes (*Bodianus* spp.). *Ecology* 64:798–808.

Jones, G. P. 1980. Contribution to the biology of the redbanded perch *Ellerkeldia huntii* (Hector), with a discussion on hermaphroditism. *J. Fish Biol.* 17:197–207.

Larsson, H. O. 1976. Field observations of some labrid fishes (Pisces: Labridae). In *Underwater 75*, vol. 1, ed. John Adolfson, pp. 211–20. Stockholm: SMR.

Lassig, B. R. 1977. Socioecological strategies adapted by obligate coral-dwelling fishes. *Proc. 3rd Int. Symp. Coral Reefs* 1: 565–70.

Leigh, E. G., Jr. 1977. How does selection reconcile individual advantage with the good of the group? PNAS 74:4542–46.

Leigh, E. G., Jr., E. L. Charnov, and R. R. Warner. 1976. Sex ratio, sex change, and natural selection. PNAS 73:3656–60.

Mead, G. W., E. Bertelson, and D. M. Cohen, 1964. Reproduction among deep-sea fishes. *Deep Sea Res.* 11:569–96.

Moyer, J. T., and A. Nakazono. 1978a. Population structure, reproductive behavior and protogynous hermaphroditism in the angelfish *Centropyge interruptus* at Miyake-jima, Japan. *Japan. J. Ichthyol.* 25:25–39.

———. 1978b. Protandrous hermaphroditism in six species of the anemonefish genus *Amphiprion* in Japan. *Japan. J. Ichthyol.* 25: 101–6.

Pressley, P. H. 1981. Pair formation and joint territoriality in a simultaneous hermaphrodite: The coral reef fish *Serranus tigrinus. Z. für Tierpsychol.* 56:33–46.

Policansky, D. 1982. Sex change in plants and animals. *Ann. Rev. Ecol. Syst.* 13:471–95.

Robertson, D. R. 1972. Social control of sex reversal in a coral reef fish. *Science* 1977:1007–9.

Robertson, D. R., R. Reinboth, and R. W. Bruce. 1982. Gonochorism, protogynous sex-change, and spawning in three sparasomatinine parrotfishes from the western Indian Ocean. *Bull. Mar. Sci.* 32: 868–79.

Robertson, D. R., and R. R. Warner. 1978. Sexual patterns in the labroid fishes of the western Caribbean, II: The parrotfishes (Scaridae). *Smithsonian Contributions to Zoology* 255:1–26.

Ross, R. M., G. S. Losey, and M. Diamond. 1983. Sex change in a coral-reef fish: Dependence of stimulation and inhibition on relative size. *Science* 221:574–75.

Shapiro, D. Y. 1979. Social behavior, group structure, and the control of sex reversal in hermaphroditic fish. *Adv. Study Beh.* 10:43–102.

———. 1980. Serial female sex changes after simultaneous removal of males from social groups of a coral reef fish. *Science* 209:1136–37.

Shapiro, D. Y., and R. Lubbock. 1980. Group sex ratio and sex reversal. *J. Theor. Biol.* 82:411–26.

Shen, S.-C., R-P. Lin, and F. C-C. Liu. 1979. Redescription of a protandrous hermaphroditic moray eel (*Rhinomuraena quaesita* Garman). *Bull. Instit. Zool. Acad. Sinica* 18(2):79–87.

Smith, C. L. 1975. The evolution of hermaphroditism in fishes. In *Intersexuality in the Animal Kingdom*, ed. R. Reinboth, pp. 295–310. Springer-Verlag.

Springer, V. G., C. L. Smith, and T. H. Fraser. 1977. *Anisochromis straussi*, new species of protogynous hermaphroditic fish, and synonomy of Anisochromidae, Pseudoplesiopidae, and Pseudochromidae. *Smithsonian Contributions to Zoology* 252:1–15.

Swarz, A. L. 1980. Almost all *Dascyllus reticulatus* are girls! *Bull. Mar. Sci.* 30:328.

Thresher, R. E., and J. T. Moyer. 1983. Male success, courtship complexity, and patterns of sexual selection in three congeneric species of sexually monochromatic and dichromatic damselfishes (Pisces: Pomacentridae). *Anim. Beh.* 31:113–27.

Tomlinson, N. 1966. The advantages of hermaphroditism and parthenogenisis. *J. Theoret. Biol.* 11:54–58.

Trivers, R. L. 1972. Parental investment and sexual selection. In *Sexual Selection and the Descent of Man, 1871–1971*, ed. B. Campbell, pp. 136–79. Chicago: Aldine Publishing Co.

Warner, R. R. 1975. The adaptive significance of sequential hermaphroditism in animals. *Am. Nat.* 109:61–82.

———. 1978. The evolution of hermaphroditism and unisexuality in aquatic and terrestrial vertebrates. In *Contrasts in Behavior*, ed. E. S. Reese and F. J. Lighter, pp. 77–101. Wiley.

———. 1982. Mating systems, sex change, and sexual demography in the rainbow wrasse, *Thalassoma lucasanum. Copeia* 1982:653–61.

Warner, R. R., and S. G. Hoffman, 1980a. Population density and the economics of territorial defense in a coral reef fish. *Ecology* 61: 772–80.

———. 1980b. Local population size as a determinant of a mating system and sexual composition in two tropical reef fishes (*Thalassoma* spp.). *Evolution* 34:508–18.

Warner, R. R., and D. R. Robertson. 1978. Sexual patterns in the labroid fishes of the western Caribbean, I: The wrasses (Labridae). *Smithsonian Contributions to Zoology* 254:1–27.

Warner, R. R., D. R. Robertson, and E. G. Leigh, Jr. 1975. Sex change and sexual selection. *Science* 190:633–38.

Williams, G. C. 1966. *Adaptation and Natural Selection.* Princeton Univ. Press.

Young, P. C., and R. B. Martin. 1982. Evidence for protogynous hermaphroditism in some lethrinid fishes. *J. Fish Biol.* 21:475–84.

Avian Siblicide

Killing a brother or a sister may be a common adaptive strategy among nestling birds, benefiting both the surviving offspring and the parents

Douglas W. Mock, Hugh Drummond and Christopher H. Stinson

Occasionally, the pen of natural selection writes a murder mystery onto the pages of evolution. But unlike a typical Agatha Christie novel, this story reveals the identity of the murderer in the first scene. The mystery lies not in "whodunit," but in why.

The case at hand involves the murder of nestling birds by their older siblings. Observers in the field have frequently noted brutal assaults by elder nestmates on their siblings, and the subsequent deaths of the younger birds. The method of execution varies among different species, ranging from a simple push out of the nest to a daily barrage of pecks to the head of the younger, smaller chick. Such killings present a challenge to the student of evolutionary biology: Does siblicide promote the fitness of the individuals that practice it, or is such behavior pathological? In other words, are there certain environmental conditions under which killing a close relative is an adaptive behavior? Moreover, are there other behaviors or biological features common to siblicidal birds that distinguish them from nonsiblicidal species?

Avian siblicide holds a special interest for several reasons. First, because nestling birds are relatively easy to observe, a rich descriptive literature exists based on field studies of many species. Second, because birds tend to

Douglas W. Mock is associate professor of zoology at the University of Oklahoma. He was educated at Cornell University and the University of Minnesota, where he received his Ph. D. in ecology and behavioral biology in 1976. Address: Department of Zoology, University of Oklahoma, Norman, OK 73019. Hugh Drummond is a researcher in animal behavior at the Universidad Nacional Autònoma de México. He was educated at Bristol University, the University of Leeds and the University of Tennessee, where he received his Ph. D. in psychology in 1980. Address: Centro de Ecologia, Universidad Nacional Autònoma de México, AP 70-275, 04510 México, D.F. Christopher H. Stinson was educated at Swarthmore College, the College of William and Mary and the University of Washington, where he received his Ph. D. in 1982. Address: 4005 NE 60th Street, Seattle, WA 98115.

be monogamous, siblicide is likely to involve full siblings. (Although recent DNA studies suggest that birds may not be as monogamous as previously thought, most nestmates are still likely to be full siblings.) Third, young birds require a large amount of food during their first few weeks of development, and this results in high levels of competition among nestlings. The competitive squeeze is exacerbated for most species because the parents act as a bottleneck through which all resources arrive. Fourth, some avian parents may not be expending their maximum possible effort toward their current brood's survival (Drent and Daan 1980, Nur 1984, Houston and Davies 1984, Gustafsson and Sutherland 1988, Mock and Lamey in press). Parental restraint may be especially common in long-lived species, in which a given season's reproductive output makes only a modest contribution to the parents' lifetime success (Williams 1966).

Siblicide—or juvenile mortality resulting from the overt aggression of siblings—is not unique to birds. It is also observed, for example, among certain insects and amphibians; in those groups, however, the behavioral pattern is rather different. Most siblicidal insects and amphibians immediately consume their victims as food, whereas in birds (and mammals) siblicide rarely leads to cannibalism. For example, tadpoles of the spadefoot toad acquire massive dentition (the so-called "cannibal morph") with which they consume their broodmates (Bragg 1954), and fig wasps use large, sharp mandibles to kill and devour their brothers (Hamilton 1979). In contrast, among pronghorn antelopes, one of the embryos develops a necrotic tip on its tail with which it skewers the embryo behind it (O'Gara 1969), and piglet littermates use deciduous eye-teeth to battle for the sow's most productive teats (Fraser 1990). Among birds and mammals it seems that the

goal is to secure a greater share of critical parental care.

Although biologists have known of avian siblicide for many years, only recently have quantitative field studies been conducted. The current wave of such work is due largely to the realization that siblicide occurs routinely in some species that breed in dense colonies; such populations provide the large sample sizes needed for formal testing of hypotheses.

Models of Nestling Aggression

Our examination of siblicidal aggression focuses on five species of birds. Two of these, the black eagle (*Aquila verreauxi*) and the osprey (*Pandion haliaetus*), are raptors that belong to the family Accipitridae. A third species, the blue-footed booby (*Sula nebouxii*) is a seabird belonging to the family Sulidae. We also present studies of the great egret (*Casmerodius albus*) and the cattle egret (*Bubulcus ibis*), both of which belong to the family Ardeidae. Each of these species exhibits a distinct behavioral pattern; the range of variation is important to an understanding of siblicide.

The black eagle is one of the first birds in which siblicide was described. This species, also called Verreaux's eagle, lives in the mountainous terrain of southern and northeastern Africa, as well as the western parts of the Middle East. Black eagles generally build their nests on cliff ledges and lay two eggs between April and June. The eaglets hatch about three days apart; and so the older chick is significantly larger than the younger one. The black eagle is of particular interest for the study of

Figure 1 (right). Two cattle egrets peer down at their recently evicted younger sibling. For several days before the eviction, the elder siblings pecked at the head of their smaller nestmate. Here the younger bird holds its bald and bloodied head out of reach. Soon after the photograph was made, the bird was driven to the ground and perished. (Photograph by the authors.)

siblicide because the elder eaglet launches a relentless attack upon its sibling from the moment the younger eaglet hatches. In one well-documented case, the senior eaglet pecked its sibling 1,569 times during the three-day lifespan of the younger nestling (Gargett 1978).

Among ospreys, sibling aggression is neither so severe nor so persistent as it is among black eagles. Ospreys are widely distributed throughout the world, including the coastal and lacustrine regions of North America. The nests are generally built high in trees or on other structures near water. A brood typically consists of three chicks, which usually live in relative harmony. Nevertheless, combative exchanges between siblings do occur in this species;

comparisons between the fighting and the pacifist populations offer insights into the significance of aggression.

The blue-footed booby lives exclusively on oceanic islands along the Pacific coast from Baja California to the northern coast of Peru. Blue-footed boobies are relatively large, ground-nesting birds that typically form dense colonies near a shoreline. Two or three chicks hatch about four days apart, and this results in a considerable size disparity between the siblings. As in many other siblicidal species, the size disparity predicts the direction of the aggression between siblings.

Young nestmates also differ in size in the two egret species we have studied. The larger of these, the great egret, is distributed throughout the middle

latitudes of the world, and also throughout most of the Southern Hemisphere. Great egrets make their nests in trees or reed beds in colonies located near shallow water. The cattle egret also nests in colonies, but not necessarily close to water. Cattle egrets live in the middle latitudes of Asia, Africa and the Americas. As their name suggests, they are almost always found in the company of grazing cattle or other large mammals, riding on their backs and feeding on grasshoppers stirred up by the movement of the animals. Despite their differences in habitat, great egrets and cattle egrets have a number of behaviors in common. Typically, three or four egret nestlings hatch at one- to two-day intervals, and fighting starts almost as

Figure 2. Aggression in black eagle nestlings almost always results in the death of the younger sibling. Here a six-day-old black eagle chick tears at a wound it has opened on the back of its day-old sibling.

Peter Steyn

soon as the second sibling has hatched. Aggressive attacks lead to a "pecking order" that translates into feeding advantages for the elder siblings (Fujioka 1985a, 1985b; Mock 1985; Ploger and Mock 1986). In about a third of the nests, the attacks culminate in siblicide through socially enforced starvation and injury or eviction from the nest.

Obligate and Facultative Siblicide

It is useful to distinguish those species in which one chick almost always kills its sibling from those in which the incidence of siblicide varies with environmental circumstances. Species that practice obligate siblicide typically lay two eggs, and it is usually the older, more powerful chick that kills its nestmate. The black eagle is a good example of an obligate siblicide species. In 200 records from black eagle nests in which both chicks hatched, only one case exists where two chicks fledged (Simmons 1988). Similar patterns of obligate siblicide have been reported for other species that lay two eggs, including certain boobies, pelicans and other eagles (Kepler 1969; Woodward 1972; Stinson 1979; Edwards and Collopy 1983; Cash and Evans 1986; Evans and McMahon 1987; Drummond 1987; Simmons 1988; Anderson 1989, 1990).

A far greater number of birds are facultatively siblicidal. Fighting is frequent among siblings in these species, but it does not always lead to the death of the younger nestling. There are various patterns of facultative siblicide. For example, in species such as the osprey, aggression is entirely absent in some populations, and yet present in others (Stinson 1977; Poole 1979, 1982; Jamieson et al. 1983). In other species aggression occurs at all nests but differs in form and effect. In the case of the blue-footed booby a chick may hit its sibling only a few times per day for several weeks, and then rapidly escalate to a lethal rate of attack (Drummond, Gonzalez and Osorno 1986). Egret broods tend to have frequent sibling fights—there are usually several multiple-blow exchanges per day—but the birds do not always kill each other (Mock 1985, Ploger and Mock 1986).

Traits of Siblicidal Species

Five characteristics are common to virtually all siblicidal birds: resource competition, the provision of food to the nestlings in small units, weaponry, spatial confinement and competitive disparities between siblings. The first

Figure 3. Blue-footed booby nestlings maintain dominance over their younger siblings through a combination of aggression and threats *(upper photograph)*. The assaults do not escalate to the point of eviction unless the food supply is inadequate. An evicted chick has little chance of survival in the face of attacks from neighboring adults *(lower photograph)*. (Photographs courtesy of the authors.)

four traits are considered essential preconditions for the evolution of sibling aggression; the study of their occurrence may shed some light on the origin of siblicidal behavior. The fifth trait—competitive disparities among nestmates resulting from differences in size and age—is also ubiquitous and important, but it is probably not essential for the evolution of siblicide. In fact, competitive disparities may be a consequence rather than a cause of siblicidal behavior; having one bird appreciably stronger than the other re-

duces the cost of fighting, since asymmetrical fights tend to be brief and it is less likely that both siblings will be hurt during combat (Hahn 1981, Fujioka 1985b, Mock and Ploger 1987).

Of the five traits common to siblicidal species, the competition for resources is probably the most fundamental. Among birds, the competition is primarily for food. Experiments have shown that the provision of additional food often diminishes nestling mortality (Mock, Lamey and Ploger 1987a; Magrath 1989). But "brood reduction"—the general term for nestling deaths brought about by the competition for food—does not necessarily entail direct aggression. Nestlings die even in nonsiblicidal species, but the usual cause of death is starvation; weaker chicks continually lose to their more robust siblings in the scramble for food. What distinguishes siblicidal species is that the competition for food is intensified to the point of overt attack. (In nonavian species, the competition may be over reproductive opportunity. For example, male fig wasps and female "proto-queen" honeybees kill all of their same-sex siblings immediately after hatching in order to gain the breeding unit's single mating slot. In certain species of mammalian social carnivores, one female dominates her sisters, rendering them effectively sterile.)

In avian species, if the source of food cannot be defended, aggression does not appear to be advantageous. The food must come in morsels small enough to be monopolized through combat. In all known species of siblicidal birds, food is presented to the young in small units through direct transfer from parent to chick (Mock 1985). For example, very young raptor chicks take small morsels held in the mother's bill, whereas boobies either reach inside the parent's throat or use their own bills to form a tube with the parent's bill, and egrets scissor the parent's bill crosswise so as to intercept the food as it emerges.

The link between the size of the food and sibling aggression lies in the relation between intimidation and monopolization. From the chick's perspective, food descends from the inaccessible heights of its parent's bill, becoming potentially available only at the moment it arrives within reach. A sibling's share depends primarily on its position relative to its competitors; that position can be enhanced through physical aggression or threat (much as the use of elbows can enhance a basketball player's chance of catching a rebound). For food items that can be taken directly from the parent's bill, the sibling's share should rise in relation to the degree of intimidation achieved. Thus, small food items create incremental rewards for aggression.

A diet of large, cumbersome items that cannot be intercepted by the chicks generally does not give rise to sibling aggression. Although killing all of its siblings would enable a chick to monopolize large items, the rewards for mild forms of aggression are sharply reduced. Thus, when food units are large, sublethal fighting may be less effective than simply eating as fast as possible. The great blue heron (*Ardea herodias*) is developmentally flexible with respect to prey size and aggression. These birds express siblicidal aggression only when the food is small enough to be taken directly from the parent. Great blue heron nestlings in Quebec fight vigorously over small units of food that can be intercepted by aggressive actions (Mock et al. 1987). In contrast, nestlings of the same species in Texas typically receive very large morsels and seldom fight (Mock 1985). Moreover, if the normally nonaggressive Texas herons are raised by great egrets, which feed their young small morsels of food, the herons quickly adopt the direct feeding method and exhibit siblicidal aggression (Mock 1984).

Figure 4. Great egret chicks fight frequently, regardless of food levels. Siblicide occurs in about a third of the nests, through socially enforced starvation and injury or as a result of eviction. As in other species of siblicidal birds, the parents do not interfere with the fights and evictions among their offspring. (Photograph courtesy of the authors.)

Figure 5. Five characteristics are common to virtually all siblicidal birds (*from top left to bottom right*): competition for food, provision of food to the nestlings in small units, weaponry, competitive disparities between siblings and spatial confinement. Four of the traits are considered essential preconditions for the evolution of sibling aggression, whereas competitive disparities between siblings may be a consequence rather than a cause of siblicidal behavior.

A shortage of food, and the ability to defend each unit of food, set the stage for siblicide, but the nestling must also possess some means of carrying out a lethal attack. In this regard it is notable that most siblicidal birds are predatory and have hooked or pointed beaks capable of inflicting serious damage on nestmates. Even so, where obvious weaponry is lacking, other means of siblicide may be possible—such as simply rolling eggs out of the nest cup.

Weaponry aside, effective aggression among nestling birds is also correlated with small nests or nesting territories. Chicks assaulted by their senior siblings do not necessarily have the option of escaping the nest. In tree-nesting species, a chick that leaves its nest risks falling from a narrow limb. Dom-inant chicks of the cliff-nesting kitti-wake (*Rissa tridactyla*) simply drive their siblings off the nest ledge (Braun and Hunt 1983). In the dense colonies of the blue-footed booby, young chicks oppressed at home by their siblings may face even greater persecution from adult neighbors if they leave their natal territory. In a tunnel-nesting bee-eater species, the nestlings have a special hook on their beaks with which they defend the opening to the nest (and the source of the food) against their younger siblings (Bryant and Tat-ner 1990). In each of these cases, the lack of suitable space (either for escape or as an alternative route to food) con-tributes directly to the victim's death.

The competitive disparities com-monly observed among nestlings of siblicidal birds may hold an important clue to the evolution of sibling aggres-sion. Parents usually create such dis-parities by starting to incubate one egg at some point prior to laying the final egg in the clutch. Because eggs are pro-duced at intervals of one or more days, the chick hatched from the egg laid first has an important head start. (Par-ents may also initiate competitive asymmetries by laying different-sized eggs within a clutch or by feeding cer-tain young preferentially, but these mechanisms are less common than asynchronous hatching.)

The Oxford ornithologist David Lack proposed that asynchronous hatching is a behavioral adaptation that allows for a secondary adjustment in brood size to match resource levels

Richard Estes (Photo Researchers, Inc.)

Michael Dick (Animals Animals)

Charles J. Lamphiear (DRK Photo)

Figure 6. Food is presented directly to the chick in small units in all known species of siblicidal birds. This direct method of feeding means that a chick may increase its share of food by physically intimidating, and not just by killing, its competing siblings. The young black eagle (*top*) is fed a piece of hyrax meat by the direct-transfer method, even though the bird is well into the fledgling stage. In the blue-footed booby (*lower left*), the parent transfers small pieces of fish from its mouth directly into the mouth of a chick. An osprey chick (*lower right*) receives a piece of meat from its parent while its sibling waits. Osprey chicks take turns feeding, and will fight only if food becomes scarce.

(Lack 1954). Parents must commit themselves to a fixed number of eggs early in the nesting cycle, before the season's bounty or shortcomings can be assessed. Thus, it is often advantageous for parents to produce an additional egg or two, in case later conditions are beneficent, while reserving the small-brood option by making the "bonus" offspring competitively inferior, in case the season's resources are poor. The production of an inferior sibling may be advantageous, since the senior sibling can then eliminate its younger nestmate with greater ease. In fact, experimentally synchronizing the hatchings of cattle egrets results in an increase in fighting, which reduces the reproductive efficiency of the parents (Fujioka 1985b, Mock and Ploger 1987).

Siblicide as an Adaptation

To understand siblicide, we must understand how the killing of a close relative can be favored by natural selection. At first this may seem a simple matter. Eliminating a competitor improves one's own chance of survival, and thereby increases the likelihood that genes promoting such behavior will be represented in the next generation. According to this simple analysis, natural selection should always reward the most selfish act, and siblicide is arguably the epitome of selfishness.

The trouble with this formulation is that it implies that all organisms should be as selfish as possible, which is contrary to observation. (Siblicide is fairly common, but certainly not universal.) A more sophisticated analysis was provided in the 1960s by the British theoretical biologist William D. Hamilton. In Hamilton's view, the fitness of a gene is more than its contribution to the reproduction of the individual. A gene's fitness also depends on the way it influences the reproductive prospects of close genetic relatives.

This expanded definition of evolutionary success, called inclusive fitness, is a property of individual organisms. An organism's inclusive fitness is a measure of its own reproductive success plus the incremental or decremental influences it has on the reproductive success of its kin, multiplied by the degree of relatedness to those kin (Hamilton 1964). Hamilton's theory is generally invoked to explain apparently altruistic behavior, but the theory also specifies the evolutionary limits of selfishness.

An example will help to clarify Hamilton's idea. Suppose a particular gene predisposes its bearer, X, to help a sibling. Since the laws of Mendelian inheritance state that X and its sibling share, on average, half of their genes, X's sibling has a one-half probability of carrying the gene. From the gene's point of view, it is useful for X to promote the reproductive success of a sibling because such an action contributes to the gene's numerical increase. Therefore, helping a sibling should be of selective advantage. It is in this light that we must understand and explain siblicide. Since selection favors genes that promote their own numerical increase, what advantage might there be in destroying a sibling—an organism with a high probability of carrying one's own genes? The solution to the problem lies in the role played by the "marginal" offspring, which may be the victim of siblicide.

In all siblicidal species studied to date there is a striking tendency for the victim to be the youngest member of the brood (Mock and Parker 1986). The youngest sibling is marginal in the sense that its reproductive value can be assessed in terms of what it adds to or subtracts from the success of other family members. Specifically, the marginal individual can embody two kinds of reproductive value. First, if the marginal individual survives in addition to all its siblings, it represents an extra unit of parental success, or extra reproductive value. Such an event is most likely during an especially favorable season, when the needs of the entire brood can be satisfied. Alternatively, the marginal offspring may serve as a replacement for an elder sibling that dies prematurely. In such instances the marginal individual represents a form of insurance against the loss of a senior sibling. The magnitude of this insurance value depends on the probability that the senior sibling will die.

Among species that practice obligate siblicide, the marginal individual offers no extra reproductive value; marginal chicks serve only as insurance against the early loss or infirmity of the senior chick. In these species, if the senior chick is alive but weakened and in-

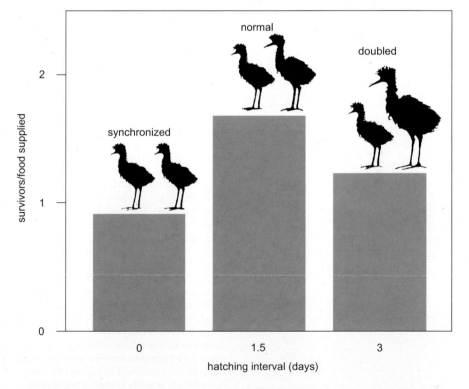

Figure 7. Effect of hatching asynchrony on avian domestic violence was investigated by switching eggs in the nests of cattle egrets. Reproductive efficiency is maximal when chicks hatch at an interval of one and one-half days (as they do under normal conditions). Synchronized hatching (an interval of zero days) increases the amount of fighting between chicks, which results in greater chick mortality. The normal one-and-a-half day interval reduces the amount of fighting since the older chick is able to intimidate the younger chick. Doubling the asynchrony, so that the eggs hatch three days apart, greatly reduces the amount of fighting but exaggerates the competitive asymmetries, so that the youngest nestmates receive little food. The experiments were performed by Douglas Mock and Bonnie Ploger at the University of Oklahoma.

capable of killing the younger chick, the latter may be able to reverse the dominance and kill the senior chick. Such scenarios appear to be played out regularly: In a sample of 22 black eagle nests in which both chicks hatched, the junior chick alone fledged in five of the nests, and the senior chick alone fledged in the remaining 17 cases (Gargett 1977). Similarly, in a sample of 59 nests of the masked booby, the junior chick was the sole fledgling in 13 nests, and the senior chick the sole fledgling in the other 46 nests (Kepler 1969). In

both of these species, the junior chick's chance of being the sole survivor—its insurance reproductive value to the parents—is about 22 percent. Removing the "insurance" eggs results in a reduction in the mean number of fledglings per nest (Cash and Evans 1986). Consequently, the insurance value of the marginal offspring should improve parental fitness if the cost of producing that offspring is reasonable. (In fact, the cost of producing one additional egg seems fairly modest: approximately 2.5 percent of the body weight of the black eagle female.)

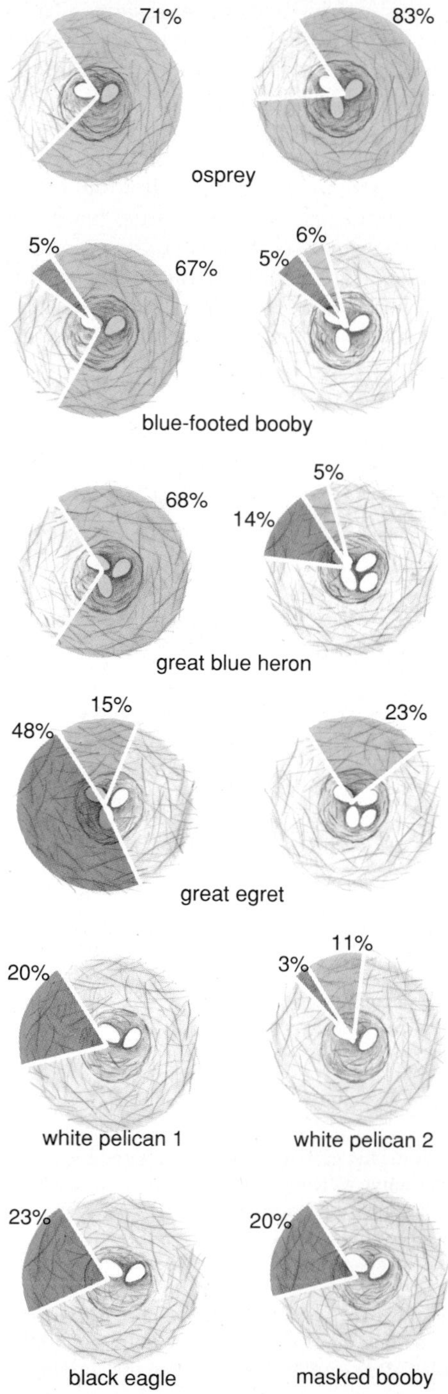

Among species that practice facultative siblicide, the marginal offspring may be a source of insurance but may also provide extra reproductive value. The relative contribution of the marginal offspring to the reproductive success of the parents appears to vary considerably within and between species. For example, among great egrets the proportion of nests in which all nestlings survive—the extra reproductive value—varies from 15 to 23 percent, whereas the proportion of the nests in which at least one senior sibling dies and the youngest sibling

lives—the insurance reproductive value—may vary from 0 to 48 percent (Mock and Parker 1986). The blue-footed booby shows great variation in the extra reproductive value provided by the marginal offspring (5 to 67 percent), whereas the insurance reproductive value is generally quite low (5 to 6 percent). In both of these species the magnitudes of the total reproductive values depend on the size of the brood. In general, the marginal offspring provides a greater total reproductive value to the parents when the brood size is smaller.

The Timing of the Deed

A senior sibling should kill its younger sibling as soon as two conditions are met: (1) the senior sibling's own viability seems secure; and (2) the resources are inadequate for the survival of both siblings. Killing the junior sibling before these conditions are met would waste the potential fitness the junior sibling could offer in the form of extra reproductive value or insurance reproductive value. Delaying much beyond the point at which the conditions are met also has a cost. First, the food eaten by the victim is a loss of resources, and, second, the cost of execution may increase as the victim gains strength and is more likely to defeat the senior sibling.

In obligate siblicide species, the average food supply is presumably inadequate for supporting two chicks at reasonable levels of parental effort, and as a result the second chick is dispatched as soon as possible after it hatches. For example, the mean longevity of the victim in the case of the masked booby is 3.3 days (Kepler 1969), and only 1.75 days for brown boobies (Cohen et al. in preparation).

Among facultative siblicide species, the mean longevity of the victim is usually greater; in the blue-footed booby it is 18 days (Drummond, Gonzalez and Osorno 1986). Although the senior blue-footed booby chick may peck at the head or wrench the skin of its nestmate, the younger sibling is seldom killed by these direct physical assaults. Instead, death typically results from starvation or violent pecking by adult neighbors when the junior chick is routed from the home nest (Drummond and Garcia Chavelas 1989).

The Causes of Siblicide
The evolutionary difference between the obligate and the facultative forms

71%   83%
osprey

5%   67%   5%   6%
blue-footed booby

68%   14%   5%
great blue heron

15%   48%   23%
great egret

20%   11%   3%
white pelican 1   white pelican 2

23%   20%
black eagle   masked booby

Figure 8. Reproductive value of the youngest member of a brood (the usual victim of siblicide) varies across species and brood size. The reproductive value is represented as the proportion of nests (in broods of two, three or four eggs) in which the youngest chick survives. If the youngest chick survives in addition to its elder siblings, it contributes "extra reproductive value" (*blue sections*); when the youngest chick survives as a replacement for an elder sibling that dies early, the junior bird provides "insurance reproductive value" (*red sections*). Among birds that almost always commit siblicide, such as the black eagle and the masked booby, the youngest chick's reproductive value is entirely due to its role as an "insurance policy." In species where siblicide is more occasional, such as the great egret, the youngest chick may provide either form of reproductive value. These estimates of reproductive value are maxima, since they represent survival only part way through the prefledgling period and not recruitment into the breeding population. These data are derived from studies by: Gargett 1977 (black eagle), Stinson 1977 (osprey), Cash and Evans 1986 (white pelican 1), Evans and McMahon 1987 (white pelican 2), Mock and Parker 1986 (great egret and great blue heron), and Drummond (unpublished data on the blue-footed booby). The data on the masked booby are combined from studies by Kepler 1969 and Anderson 1989.

of siblicide may be a function of the risk that a junior chick poses to the welfare of its senior sibling. That risk can be defined both in terms of resource consumption and in terms of the potential for bodily harm. If the resources are adequate only for the survival of a single chick, or if a young chick poses a significant physical threat to an older chick, then the senior sibling might be expected to destroy the younger one. On the other hand, if there is enough food for both chicks, and if the younger sibling can be subjugated so that it does not present a threat, then the survival of the younger sibling is beneficial because it increases the inclusive fitness of the senior sibling. In such circumstances, natural selection should favor a measure of clemency on the part of the senior sibling. Accordingly, we would expect obligate siblicide to evolve in circumstances in which resources are routinely limited and siblings tend to pose a physical threat to one another. In contrast, facultative siblicide should arise in circumstances in which resources are not always limited.

The analysis offered above concerns the inheritance of a long-term predisposition to siblicide. Recent studies suggest that food shortages also act as an immediate stimulus to, or proximal cause of, sibling fighting. A link between the food supply and siblicide was suggested by the finding that brood reductions in the blue-footed booby tend to occur soon after the weight of the senior chick drops about 20 percent below the weight expected at its current age in a good year (Drummond, Gonzalez and Osorno 1986). The relationship between food deprivation and aggression was confirmed by experiments in which the senior chick's neck was taped to prevent it from swallowing food. The experimentally deprived senior chicks pecked their nestmates about three to four times more frequently with the tape in place than without the tape, and they subsequently received a greater share of the food (Drummond and Garcia Chavelas 1989).

In older booby broods, the increase in the amount of aggressive pecking was delayed by about a day after the chick's neck was taped, suggesting that aggression is controlled by a factor that changes progressively over time, such as hunger or growth status. In fact, the increased pecking rate coincided with a 20 percent weight loss by the senior

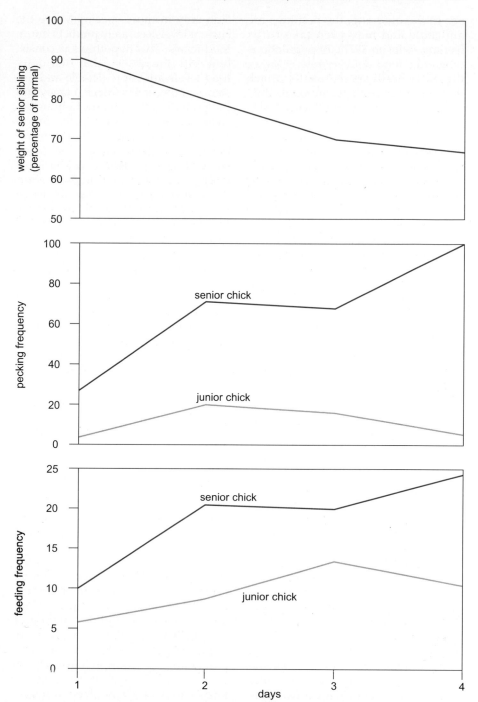

Figure 9. Effect of food deprivation on aggression and food distribution in blue-footed booby nestlings was investigated by taping the senior chick's neck to prevent it from swallowing food. As the weight of the senior chick drops more than 20 percent below normal (*top*), the rate at which it pecks its younger sibling increases more than three-fold (*middle*). The escalating aggression of the elder chick brings it a greater share of food (*bottom*). The experiments were performed by Hugh Drummond and Cecilia Garcia Chavelas at the Universidad Nacional Autònoma de México.

chick. When the tape was removed, the aggressive pecking rate returned toward the baseline level. These results suggest that nestling aggression among certain facultative species is a reversible response that is sensitive to the weight level of the senior chick.

There is also suggestive evidence in other species that practice facultative siblicide that the amount of food available to the nestlings may affect sibling aggression. For example, junior kittiwake siblings are lost from the nest at higher rates following prolonged periods of bad weather, when parental foraging is reduced (Braun and Hunt 1983). Among osprey populations in which there is a high rate of prey delivery to the offspring, the nestlings are amicable and may even take turns feeding (Stinson 1977). In populations where the food delivery rate is lower, the older nestlings frequently attack their younger siblings, although they do not kill them outright (Henny 1988, Poole 1982).

In contrast, the relative abundance of food does not appear to affect the level of aggression in obligate siblicide species. Black eagle nestlings kill their siblings even in the midst of several kilograms of prey, and even while the mother eagle is offering food to the senior sibling. There does not appear to be the same direct relationship between the immediate availability of food and the level of sibling aggression. Since black eagle nestlings require large amounts of food over a period of many weeks, short-term abundance of food may not be an accurate indicator of long-term food levels. As a consequence, aggression and siblicide might be favored in order to obviate any future competition (Anderson 1990, Stinson 1979).

Perhaps the appropriate "sibling aggression policy" is obtained from simple cues available to the chicks from the outset. Assuming that parents deliver food at some optimal rate, then a chick may be able to estimate in advance whether sufficient levels will be available for its own growth. It is interesting to note that the facultatively siblicidal golden eagle (*Aquila chrysaetos*) provides the same amount of food regardless of the number of chicks in the brood (Collopy 1984). If this is typical, then the senior chick may be able to detect whether the food will be enough to support all nestmates. Eagles that practice obligate siblicide generally deliver less food to the nest than facultative species, and consequently no assessment by the chicks is necessary (Bortolotti 1986). The average amount of food provided by the parents may be consistently low enough for natural selection to favor preemptive killing—a system that benefits both the senior chick and its parents. In other words, the insurance policy is canceled.

Even in species that practice facultative siblicide, aggression is sometimes insensitive to food supply. For example, the level of fighting among heron and egret chicks appears to be independent of the amount of food available (Mock, Lamey and Ploger 1987). It may be that the current food level acts as a proximate cue for sibling aggression only in those species where the current level accurately predicts future food levels. This hypothesis is consistent with the observation that daily food levels are unpredictable and unstable among egrets (Mock, Lamey and Ploger 1987). Interestingly, fighting between egrets ceases when the brood size drops from three to two (thus reducing future food demands) and may be reinstated by restoring the third chick (Mock and Lamey in press). Further studies of other species are necessary to determine whether the degree to which food levels fluctuate is related to aggressive behavior.

Future Directions

The study of siblicide as an adaptive strategy is still in its infancy. Much of the work to date has been devoted to identifying the proximate causes of aggressive behavior and documenting its utility for controlling resources. Less is known about the effects of siblicide on the inclusive fitness of the perpetrators. Although many theoretical models of avian siblicide have been proposed (O'Connor 1978; Stinson 1979; Mock and Parker 1986; Parker, Mock and Lamey 1989, Godfray and Harper 1990), the field data are limited.

Several areas of research need to be explored further. We would like to determine the short-term costs of sibling rivalry, perhaps by comparing the energetics of competitive begging and fighting. Likewise we need to know the long-term costs of temporary food shortages; there is particular interest in the relationship between the development of the chick and the amount of food available. Similarly, what is the relation between the amount of effort parents put into supplying food, the resulting chick survival rate and the long-term costs of reproduction among brood-reducing species? Is there any relation between chick gender, hatching order and siblicide—particularly in siblicidal species that have a large degree of sexual dimorphism? Another area of interest is the role of extra-pair copulations, which reduce the relatedness of nestmates and thereby increase the potential benefits of selfishness; it would be useful to know whether chicks have the ability to discriminate half-siblings from full siblings. Finally, why is it that parents appear not to interfere with the execution process in siblicidal species (O'Connor 1978; Drummond, Gonzalez and Osorno 1986; Mock 1987)? Answers to these questions can give us a better understanding of how siblicidal behavior may have evolved.

Bibliography

Anderson, D. J. 1989. Adaptive adjustment of hatching asynchrony in two siblicidal booby species. *Behavioral Ecology and Sociobiology* 25:363-368.

Anderson, D. J. 1990. Evolution of obligate siblicide in boobies. I: A test of the insurance egg hypothesis. *American Naturalist* 135:334-350.

Bortolotti, G. R. 1986. Evolution of growth rates in eagles: sibling competition vs. energy considerations. *Ecology* 67:182-194.

Bragg, A. N. 1954. Further study of predation and cannibalism in spadefoot tadpoles. *Herpetologica* 20:17-24.

Braun, B. M., and G. L. Hunt, Jr. 1983. Brood reduction in black-legged kittiwakes. *The Auk* 100:469-476.

Bryant, D. M., and P. Tatner. 1990. Hatching asynchrony, sibling competition and siblicide in nestling birds: studies of swiftlets and bee-eaters. *Animal Behaviour* 39:657-671.

Cash, K., and R. M. Evans. 1986. Brood reduction in the American white pelican, *Pelecanus erythrorhynchos*. *Behavioral Ecology and Sociobiology* 18:413-418.

Collopy, M. 1984. Parental care and feeding ecology of golden eagle nestlings. *The Auk* 101:753-760.

Drent, R. H., and S. Daan. 1980. The prudent parent: energetic adjustments in avian breeding. *Ardea* 68:225-252.

Drummond, H. 1987. Parent-offspring conflict and brood reduction in the Pelecaniformes. *Colonial Waterbirds* 10:1-15.

Drummond, H., E. Gonzalez and J. Osorno. 1986. Parent-offspring cooperation in the blue-footed booby, *Sula nebouxii*. *Behavioral Ecology and Sociobiology* 19:365-392.

Drummond, H., and C. Garcia Chavelas. 1989. Food shortage influences sibling aggression in the blue-footed booby. *Animal Behaviour* 37:806-819.

Edwards, T. C., Jr., and M. W. Collopy. 1983. Obligate and facultative brood reduction in eagles: An examination of factors that influence fratricide. *The Auk* 100:630-635.

Evans, R. M., and B. McMahon. 1987. Within-brood variation in growth and conditions in relation to brood reduction in the American white pelican. *Wilson Bulletin* 99:190-201.

Fraser, D. 1990. Behavioural perspectives on piglet survival. *Journal of Reproduction and Fertility, Supplement* 40:355-370.

Fujioka, M. 1985a. Sibling competition and siblicide in asynchronously-hatching broods of the cattle egret, *Bubulcus ibis*. *Animal Behaviour* 33:1228-1242.

Fujioka, M. 1985b. Food delivery and sibling competition in experimentally even-aged broods of the cattle egret. *Behavioral Ecology and Sociobiology* 17:67-74.

Gargett, V. 1977. A 13-year population study of the black eagles in the Matopos, Rhodesia, 1964-1976. *Ostrich* 48:17-27.

Gargett, V. 1978. Sibling aggression in the black eagle in the Matopos, Rhodesia. *Ostrich* 49:57-63.

Godfray, H. C. J., and A. B. Harper. 1990. The evolution of brood reduction by siblicide in birds. *Journal of Theoretical Biology* 145:163-175.

Gustafsson, L., and W. J. Sutherland. 1988. The costs of reproduction in the collared flycatcher *Ficedula albicollis*. *Nature* 335:813-815.

Hahn, D. C. 1981. Asynchronous hatching in the laughing gull: Cutting losses and reducing rivalry. *Animal Behaviour* 29:421-427.

Hamilton, W. D. 1964. The genetical evolution of social behaviour. *Journal of Theoretical Biology* 7:1-52.

Henny, C. J. 1988. Reproduction of the osprey. In *Handbook of North American Birds*, ed. R. E. Palmer. Yale University Press.

Houston, A. I., and N. B. Davies. 1985. The evolution of cooperation and life history in the dunnock *Prunella modularis*. In R. Sibly and R. Smith (eds.) *Behavioral Ecology: The Ecological Consequences of Adaptive Behaviour.* Blackwell: Oxford. pp. 471-487.

Jamieson, I. G., N. R. Seymour, R. P. Bancroft and R. Sullivan. 1983. Sibling aggression in nestling ospreys in Nova Scotia. *Canadian Journal of Zoology* 61:466-469.

Kepler, C. B. 1969. Breeding biology of the blue-faced booby on Green Island, Kure Atoll. *Publications of the Nuttal Ornithology Club* 8.

Lack, D. 1954. *The Natural Regulation of Animal Numbers*. Clarendon Press: Oxford.

Magrath, R. 1989. Hatch asynchrony and reproductive success in the blackbird. *Nature* 339:536-538.

Mock, D. W. 1984. Siblicidal aggression and resource monopolization in birds. *Science* 225:731-733.

Mock, D. W. 1985. Siblicidal brood reduction: The prey-size hypothesis. *American Naturalist* 125:327-343.

Mock, D. W. 1987. Siblicide, parent-offspring conflict, and unequal parental investment by egrets and herons. *Behavioral Ecology and Sociobiology* 20:247-256.

Mock, D. W., and T.C. Lamey. In press. The role of brood size in regulating egret sibling aggression. *American Naturalist*.

Mock, D. W., and G.A. Parker. 1986. Advantages and disadvantages of ardeid brood reduction. *Evolution* 40:459-470.

Mock, D. W., and B.J. Ploger. 1987. Parental manipulation of optimal hatch asynchrony in cattle egrets: An experimental study. *Animal Behaviour* 35:150160.

Mock, D. W., T. C. Lamey and B. J. Ploger. 1987. Proximate and ultimate roles of food amount in regulating egret sibling aggression. *Ecology* 68:1760-1772.

Mock, D. W., T. C. Lamey, C.F. Williams and A. Pelletier. 1987. Flexibility in the development of heron sibling aggression: An intraspecific test of the prey-size hypothesis. *Animal Behaviour* 35:1386-1393.

Nur, N. 1984. Feeding frequencies of nestling blue tits (*Parus coeruleus*): Costs, benefits, and a model of optimal feeding frequency. *Oecologia* 65:125-137.

O'Connor, R. J. 1978. Brood reduction in birds: Selection for infanticide, fratricide, and suicide? *Animal Behaviour* 26:79-96.

O'Gara, B. W. 1969. Unique aspects of reproduction in the female pronghorn, *Antilocapra americana. American Journal of Anatomy* 125:217-232.

Parker, G. A., D. W. Mock and T. C. Lamey. 1989. How selfish should stronger sibs be? *American Naturalist* 133:846-868.

Ploger, B. J., and D. W. Mock. 1986. Role of sibling aggression in distribution of food to nestling cattle egrets, *Bubulcus ibis. The Auk* 103:768-776.

Poole, A. 1979. Sibling aggression among nestling ospreys in Florida Bay. *The Auk* 96:415-417.

Poole, A. 1982. Brood reduction in temperate and sub-tropical ospreys. *Oecologia* 53:111-119.

Simmons, R. 1988. Offspring quality and the evolution of Cainism. *Ibis* 130:339-357.

Stinson, C. H. 1977. Growth and behaviour of young ospreys, *Pandion haliaetus. Oikos* 28:299-303.

Stinson, C. H. 1979. On the selective advantage of fratricide in raptors. *Evolution* 33:1219-1225.

Williams, G. C. 1966. Natural selection, the costs of reproduction, and a refinement of Lack's principle. *American Naturalist* 100:687-690.

Woodward, P. W. 1972. The natural history of Kure Atoll, northwestern Hawaiian Islands. *Atoll Research Bulletin* 164.

The Strategies of Human Mating

A theory of human sexual strategies accounts for the observation that people worldwide are attracted to the same qualities in the opposite sex

David M. Buss

David M. Buss is professor of psychology at the University of Michigan. He received his Ph.D. in psychology from the University of California, Berkeley in 1981. He is director of the International Consortium of Personality and Social Psychologists, which conducts cross-cultural research around the world. His most recent book, The Evolution of Desire: Strategies of Human Mating, *was published this year by Basic Books. Address: Department of Psychology, University of Michigan, Ann Arbor, MI 48109-1346.*

What do men and women want in a mate? Is there anything consistent about human behavior when it comes to the search for a mate? Would a Gujarati of India be attracted to the same traits in a mate as a Zulu of South Africa or a college student in the midwestern United States?

As a psychologist working in the field of human personality and mating preferences, I have come across many attempts to answer such questions and provide a coherent explanation of human mating patterns. Some theories have suggested that people search for mates who resemble archetypical images of the opposite-sex parent (à la Freud and Jung), or mates with characteristics that are either complementary or similar to one's own qualities, or mates with whom to make an equitable exchange of valuable resources.

These theories have played important roles in our understanding of human mating patterns, but few of them have provided specific predictions that can be tested. Fewer still consider the origins and functions of an individual's mating preferences. What possible function is there to mating with an individual who is an archetypical image of one's opposite-sex parent? Most theories also tend to assume that the processes that guide the mating preferences of men and women are identical, and no sex-differentiated predictions can be derived. The context of the mating behavior is also frequently ignored; the same mating tendencies are posited regardless of circumstances.

Despite the complexity of human mating behavior, it is possible to address these issues in a single, coherent theory. David Schmitt of the University of Michigan and I have recently proposed a framework for understanding the logic of human mating patterns from the standpoint of evolutionary theory. Our theory makes several predictions about the behavior of men and women in the context of their respective sexual strategies. In particular, we discuss the changes that occur when men and women shift their goals from short-term mating (casual sex) to long-term mating (a committed relationship).

Some of the studies we discuss are based on surveys of male and female college students in the United States. In these instances, the sexual attitudes of the sample population may not be reflective of the behavior of people in other cultures. In other instances, however, the results represent a much broader spectrum of the human population. In collaboration with 50 other scientists, we surveyed the mating preferences of more than 10,000 men and women in 37 countries over a six-year period spanning 1984 through 1989. Although no survey, short of canvassing the entire human population, can be considered exhaustive, our study crosses a tremendous diversity of geographic, cultural, political, ethnic, religious, racial and economic groups. It is the largest survey ever on mate preferences.

What we found is contrary to much current thinking among social scientists, which holds that the process of choosing a mate is highly culture-bound. Instead, our results are consistent with the notion that human beings, like other animals, exhibit species-typical desires when it comes to the selection of a mate. These patterns can be accounted for by our theory of human sexual strategies.

Competition and Choice

Sexual-strategies theory holds that patterns in mating behavior exist because they are evolutionarily advantageous. We are obviously the descendants of people who were able to mate successfully. Our theory assumes that the sexual strategies of our ancestors evolved because they permitted them to survive and produce offspring. Those people who failed to mate successfully because they did not express these strategies are not our ancestors. One simple example is the urge to mate, which is a universal desire among people in all cultures and which is undeniably evolutionary in origin.

Although the types of behavior we consider are more complicated than simply the urge to mate, a brief overview of the relevant background should be adequate to understand the evolutionary logic of human mating strategies.

Figure 1. Species-typical mating preferences are expressed by the American businessman Donald Trump and his new wife Marla Maples, here evoking an image of the ideal family for readers of *Vanity Fair*. The traits of a desirable mate appear to be consistent throughout the world: Men prefer to mate with beautiful young women, whereas women prefer to mate with men who have resources and social status. The author argues that these traits offer evolutionarily adaptive advantages to the opposite-sex mate, which account for their ubiquitous desirability.

As with many issues in evolutionary biology, this background begins with the work of Charles Darwin.

Darwin was the first to show that mate preferences could affect human evolution. In his seminal 1871 treatise, *The Descent of Man and Selection in Relation to Sex*, Darwin puzzled over characteristics that seemed to be perplexing when judged merely on the basis of their relative advantage for the animal's survival. How

type of mating	men's reproductive challenges	women's reproductive challenges
short-term	• partner number • identifying women who are sexually accessible • minimizing cost, risk and commitment • identifying women who are fertile	• immediate resource extraction • evaluating short-term mates as possible long-term mates • attaining men with high-quality genes • cultivating potential backup mates
long-term	• paternity confidence • assessing a woman's reproductive value • commitment • identifying women with good parenting skills • attaining women with high-quality genes	• identifying men who are able and willing to invest • physical protection from aggressive men • identifying men who will commit • identifying men with good parenting skills • attaining men with high-quality genes

Figure 2. Mate-selection problems of men and women differ in short-term mating (casual sex) and long-term mating (a committed relationship) because each gender faces a unique set of reproductive challenges. In short-term mating contexts, a man's reproductive success is constrained by the number of fertile women he can inseminate. A man must solve the specialized problems of identifying women who are sexually accessible, identifying women who are fecund, and minimizing commitment and investment in order to effectively pursue short-term matings. In contrast, a woman's short-term mating strategy involves identifying men who would be good long-term mates, identifying men who have "high-quality" genes (are evolutionarily fit), extracting resources from a short-term mate and cultivating potential backup mates. In long-term mating contexts, men must identify women who have high reproductive value, good parenting skills and "high-quality" genes. Men must also assure that they are the father of their mate's offspring. In contrast, women must identify long-term mates who are willing and able to invest resources, can provide physical protection, have good parenting skills and have "high-quality" genes. Because men and women face different reproductive challenges, each gender has evolved different sexual strategies and is attracted to different qualities in the opposite sex.

could the brilliant plumage of a male peacock evolve when it obviously increases the bird's risk of predation? Darwin's answer was sexual selection, the evolution of characteristics that confer a reproductive advantage to an organism (rather than a survival advantage). Darwin further divided sexual selection into two processes: intrasexual competition and preferential mate choice.

Intrasexual competition is the less controversial of the two processes. It involves competition between members of the same sex to gain preferential access to mating partners. Characteristics that lead to success in these same-sex competitions—such as greater strength, size, agility, confidence or cunning—can evolve simply because of the reproductive advantage gained by the victors. Darwin assumed that this is primarily a competitive interaction between males, but recent studies suggest that human females are also very competitive for access to mates.

Preferential mate choice, on the other hand, involves the desire for mating with partners that possess certain characteristics. A consensual desire affects the evolution of characteristics because it gives those possessing the desired characteristics an advantage in obtaining mates over those who do not possess the desired characteristics. Darwin assumed that

preferential mate choice operates primarily through females who prefer particular males. (Indeed, he even called this component of sexual selection *female choice*.)

Darwin's theory of mate-choice selection was controversial in part because Darwin simply assumed that females desire males with certain characteristics. Darwin failed to document how such desires might have arisen and how they might be maintained in a population.

The solution to the problem was not forthcoming until 1972, when Robert Trivers, then at Harvard University, proposed that the relative parental investment of the sexes influences the two processes of sexual selection. Specifically, the sex that invests more in offspring is selected to be more discriminating in choosing a mate, whereas the sex that invests less in offspring is more competitive with members of the same sex for sexual access to the high-investing sex. Parental-investment theory accounts, in part, for both the origin and the evolutionary retention of different sexual strategies in males and females.

Consider the necessary *minimum* parental investment by a woman. After internal fertilization, the gestation period lasts about nine months and is usually followed by lactation, which in tribal societies typically can last several years. In contrast, a man's minimum parental investment can

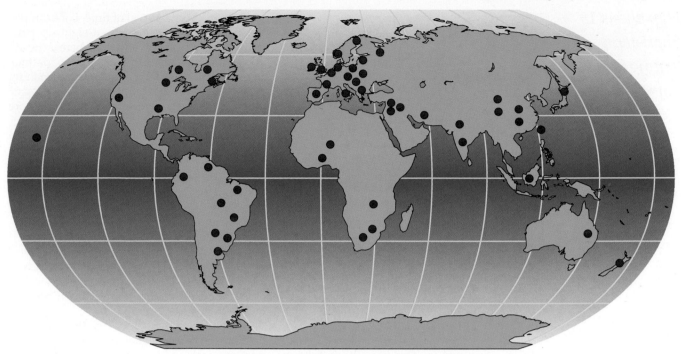

Figure 3. Thirty-seven cultures, distributed as shown above, were examined by the author in his international study of male and female mating preferences. The author and his colleagues surveyed the mating desires of 10,047 people on six continents and five islands. The results provide the largest data base of human mating preferences ever accumulated.

be reduced to the contribution of sperm, an effort requiring as little time as a few minutes. This disparity in parental investment means that the replacement of a child who dies (or is deserted) typically costs more (in time and energy) for women than men. Parental-investment theory predicts that women will be more choosy and selective about their mating partners. Where men can provide resources, women should desire those who are able and willing to commit those resources to her and her children.

Sexual Strategies

Our evolutionary framework is based on three key ingredients. First, human mating is inherently strategic. These strategies exist because they solved specific problems in human evolutionary history. It is important to recognize that the manifestation of these strategies need not be through conscious psychological mechanisms. Indeed, for the most part we are completely unaware of *why* we find certain qualities attractive in a mate. A second component of our theory is that mating strategies are context-dependent. People behave differently depending on whether the situation presents itself as a short-term or long-term mating prospect. Third, men and women have faced different mating problems over the course of human evolution and, as a consequence, have evolved different strategies.

As outlined here, sexual strategies theory consists of nine hypotheses. We can test these hypotheses by making several predictions about the behavior of men and women faced with a particular mating situation. Even though we

make only a few predictions for each hypothesis, it should be clear that many more predictions can be derived to test each hypothesis. We invite the reader to devise his or her own tests of these hypotheses.

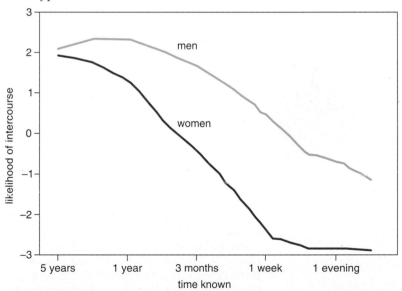

Figure 4. Willingness to have sexual intercourse (measured on a scale from 3, *definitely yes*, to –3, *definitely no*) differs for men and women with respect to the length of time they have been acquainted with their prospective mate. Although men and women are equally likely to engage in sexual intercourse after knowing a mate for five years (both responding with a score of about 2, *probably yes*), women are significantly less inclined to have sex with a prospective mate for all shorter lengths of time. The average man was positive about having intercourse with a woman even after knowing her for only one week, whereas the average women was highly unlikely to have intercourse after such a brief period of time. The data are based on a sample of 148 college students in the midwestern United States. The results support the hypothesis that short-term mating is more important for men than for women.

Hypothesis 1: Short-term mating is more important for men than women.

Figure 5. Stereotypical images of the womanizing male and the marriage-minded female are caricatures of the underlying sexual strategies of men and women. In the television program *Cheers* the character played by the actor Ted Danson exhibited the short-term male sexual strategy of mating with many women. The character played by the actress Shelley Long exhibited the female strategy of seeking a male willing to commit to a long-term relationship. Image courtesy of AP/Wide World Photos.

Hypothesis 1: Short-term mating is more important for men than women. This hypothesis follows from the fact that men can reduce their parental investment to the absolute minimum and still produce offspring. Consequently, short-term mating should be a key component of the sexual strategies of men, and much less so for women. We tested three predictions based on this hypothesis in a sample of 148 college students (75 men and 73 women) in the midwestern United States.

First, we predict that men will express a greater interest in seeking a short-term mate than will women. We asked the students to rate the degree to which they were currently seeking a short-term mate (defined as a one-night stand or a brief affair) and the degree to which they were currently seeking a long-term mate (defined as a marriage partner). They rated their interests on a 7-point scale, where a rating of 1 corresponds to a complete lack of interest and a 7 corresponds to a high level of interest.

We found that although the sexes do not differ in their stated proclivities for seeking a long-term mate (an average rating of about 3.4 for both sexes), men reported a significantly greater interest (an average rating of about 5) in seeking a short-term sexual partner than did women (about 3). The results also showed that at any given time men are more interested in seeking a short-term mate rather than a long-term mate, whereas women are more interested in seeking a long-term mate than a short-term mate.

Second we predict that men will desire a greater number of mates than is desired by women. We asked the same group of college students how many sexual partners they would ideally like to have during a given time interval and during their lifetimes. In this instance men consistently reported that they desired a greater number of sex partners than reported by the

women for every interval of time. For example, the average man desired about eight sex partners during the next two years, whereas the average woman desired to have one sex partner. In the course of a lifetime, the average man reported the desire to have about 18 sex partners, whereas the average woman desired no more than 4 or 5 sex partners.

A third prediction that follows from this hypothesis is that men will be more willing to engage in sexual intercourse a shorter period of time after first meeting a potential sex partner. We asked the sample of 148 college students the following question: "If the conditions were right, would you consider having sexual intercourse with someone you viewed as desirable if you had known that person for *(a time period ranging from one hour to five years)*?" For each of 10 time intervals the students were asked to provide a response ranging from –3 (definitely not) to 3 (definitely yes).

After a period of 5 years, the men and women were equally likely to consent to sexual relations, each giving a score of about 2 (probably yes). For all shorter time intervals, men were consistently more likely to consider sexual intercourse. For example, after knowing a potential sex partner for only one week, the average man was still positive about the possibility of having sex, whereas women said that they were highly unlikely to have sex with someone after knowing him for only one week.

This issue was addressed in a novel way by Russell Clark and Elaine Hatfield of the University of Hawaii. They designed a study in which college students were approached by an attractive member of the opposite sex who posed one of three questions after a brief introduction: "Would you go out on a date with me tonight?" "Would you go back to my apartment with me tonight?" or "Would you have sex with me tonight?"

Of the women who were approached, 50 percent agreed to the date, 6 percent agreed to go to the apartment and none agreed to have sex. Many women found the sexual request from a virtual stranger to be odd or insulting. Of the men approached, 50 percent agreed to the date, 69 percent agreed to go back to the woman's apartment and 75 percent agreed to have sex. In contrast to women, many men found the sexual request flattering. Those few men who declined were apologetic about it, citing a fiancée or an unavoidable obligation that particular evening. Apparently, men are willing to solve the problem of partner number by agreeing to have sex with virtual strangers.

Hypothesis 2: Men seeking a short-term mate will solve the problem of identifying women who are sexually accessible. We can make at least two predictions based on this hypothesis. First, men will value qualities that signal immediate sexual accessibility in a short-term mate highly, and less

Hypothesis 2: Men seeking a short-term mate will solve the problem of identifying women who are sexually accessible.

Alain Evrard (Photo Researchers, Inc.)

Alain Evrard (Photo Researchers, Inc.)

Figure 6. Prostitution is a worldwide phenomenon that is partly a consequence of the short-term mating strategy of males. The relatively rapid exchange of sex and money solves the short-term mating problems of males who can minimize the commitment of resources and quickly identify women who are sexually accessible. Above and on the facing page, interactions on Patpong Street (the prostitution district) in Bangkok play out a scene that is repeated daily in cities around the world.

so in a long-term mate. When we asked men in a college sample of 44 men and 42 women to rate the desirability of promiscuity and sexual experience in a mate, both were significantly more valued in a short-term mate. Although men find promiscuity mildly desirable in a short-term mate, it is clearly undesirable in a long-term mate. It is noteworthy that women find promiscuity extremely undesirable in either context.

We also predict that qualities that signal sexual inaccessibility will be disliked by men seeking short-term mates. We asked men to rate the desirability of mates who have a low sex drive, who are prudish or who lack sexual experience. In each instance men expressed a particular dislike for short-term mates with these qualities. A low sex drive and prudishness are also disliked by men in long-term mates, but less so. In contrast, a lack of sexual experience is slightly valued by men in a long-term mate.

Hypothesis 3: Men seeking a short-term mate will minimize commitment and investment. Here we predict that men will find undesirable any cues

that signal that a short-term mate wants to extract a commitment. We asked the same group of 44 men to rate the variable *wants a commitment* for short-term and long-term mates. Of all the qualities we addressed, this one showed the most striking dependence on context. The attribute of wanting a commitment was strongly desirable in a long-term mate but strongly undesirable in a short-term mate. This distinction was not nearly so strong for women. Although women strongly wanted commitment from a long-term mate, it was only mildly undesirable in a short-term mate.

Hypotheses 4 and 5: Men seeking a short-term mate will solve the problem of identifying fertile women, whereas men seeking a long-term mate will solve the problem of identifying reproductively valuable women. Because these hypotheses are closely linked it is useful to discuss them together. Fertility and reproductive value are related yet distinct concepts. Fertility refers to the probability that a woman is *currently* able to conceive a child. Reproductive value, on the other hand, is defined actuarially in units of expected future

Hypothesis 3: Men seeking a short-term mate will minimize commitment and investment.

Figure 7. Beautiful young women are sexually attractive to men because beauty and youth are closely linked with fertility and reproductive value. In evolutionary history, males who were able to identify and mate with fertile females had the greatest reproductive success. These three young women were photographed at a "fern bar" in Newport Beach, California.

Hypothesis 4: Men seeking a short-term mate will solve the problem of identifying fertile women.

reproduction. In other words, it is the extent to which persons of a given age and sex will contribute, on average, to the ancestry of future generations. For example, a 14-year-old woman has a higher reproductive value than a 24-year-old woman, because her *future* contribution to the gene pool is higher on average. In contrast, the 24-year-old woman is more fertile than the 14-year-old because her *current* probability of reproducing is greater.

Since these qualities cannot be observed directly, men would be expected to be sensitive to cues that might be indicative of a woman's fertility and reproductive value. One might expect that men would prefer younger women as short-term and long-term mates. Again, since age is not something that can be observed directly, men should be sensitive to physical cues that are reliably linked with age. For example, with increasing age, skin tends to wrinkle, hair turns gray and falls out, lips become thinner, ears become larger, facial features become less regular and muscles lose their tone. Men could solve the problem of identifying reproductively valuable women if they attended to physical features linked with age and health, *and* if their standards of attractiveness evolved to correspond to these features.

As an aside, it is worth noting that cultures do differ in their standards of physical beauty, but less so than anthropologists initially assumed. Cultural differences of physical beauty tend to center on whether relative plumpness or thinness is valued. In cultures where food is relatively scarce, plumpness is valued, whereas cultures with greater abundance value thinness. With the exception of plumpness and thinness, however, the physical cues to youth and health are seen as sexually attractive in all known cultures that have been studied. In no culture do people perceive wrinkled skin, open sores and lesions, thin lips, jaundiced eyes, poor muscle tone and irregular facial features to be attractive.

A woman's reproductive success, however, is not similarly dependent on solving the problem of fertility in mates. Because a man's reproductive capacity is less closely linked with age and cannot be assessed as accurately from appearance, youth and physical attractiveness in a mate should be less important to women than it is to men.

Among our sample of American college students we asked men and women to evaluate the relative significance (on a scale from 0, unimportant, to 3, important) of the characteristics *good looking* and *physically attractive* in a short-term and a long-term mate. We found that

Courtesy of The Museum of Modern Art Film Stills Archive

Spencer Grant (Photo Researchers, Inc.)

Jodi Cobb (© National Geographic Society)

Hypothesis 5: Men seeking a long-term mate will solve the problem of identifying reproductively valuable women.

Figure 8. King Hussein' Ibn Talal' (born in 1935) of Jordan and his wife, Queen Noor (formerly Lisa Halaby of Washington, DC, who was born in 1951), provide an example of the general tendency for men to mate with women who are significantly younger than themselves. Sexual-strategies theory holds that men are attracted to younger women as long-term mates because they have a higher reproductive value than older women. The King and Queen are pictured with some of their children.

men's preference for physical attractiveness in short-term mates approached the upper limit of the rating scale (about 2.71). Interestingly, this preference was stronger in men seeking short-term mates than in men seeking long-term mates (about 2.31). The results are a little surprising to us because we did not predict that men would place a greater significance on the physical attractiveness of a short-term mate compared to a long-term mate.

Women also favored physical attractiveness in a short-term mate (2.43) and a long-term mate (2.10). Here again, physical attractiveness was more important in short-term mating than in long-term mating. In both contexts, however, physical attractiveness was significantly less important to women than it is to men.

We also tested these predictions in our international survey of 37 cultures. My colleagues in each country asked men and women to evaluate the relative importance of the characteristics *good looking* and *physically attractive* in a mate. As in our American college population, men throughout the world placed a high value on physical attractiveness in a partner.

In each of the 37 cultures men valued physical attractiveness and good looks in a mate

more than did their female counterparts. These sex differences are not limited to cultures that are saturated with visual media, Westernized cultures or racial, ethnic, religious or political groups. Worldwide, men place a premium on physical appearance.

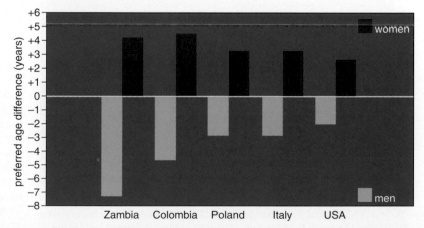

Figure 9. Preferences for an age difference between oneself and one's spouse differ for men and women. Men in each of the 37 cultures examined by the author prefer to mate with younger women, whereas women generally prefer to mate with older men. Here the disparities between the mating preferences of men and women in five countries show some of the cultural variation across the sample.

Hypothesis 6: Men seeking a long-term mate will solve the problem of paternity confidence.

Figure 10. Othello's jealous rage and murder of Desdemona in Shakespeare's tragic play was incited by her presumed sexual infidelity. According to sexual-strategies theory, men have strong emotional responses to a mate's sexual infidelity because the ancestors of modern males who protected themselves against cuckoldry had greater reproductive success. Here the German actor Emil Jannings plays the role of Othello in a movie from the early part of this century.

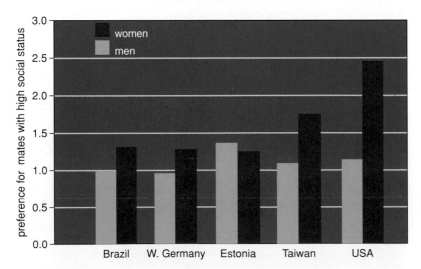

Figure 11. Mate's high social status (ranging from 3, *indispensible*, to 0, *unimportant*) is typically greater for women than for men. Of the 37 cultures examined, only males in Estonia valued the social status of their spouses more than did their female counterparts. (Only a sample of five countries is shown here.) Women generally prefer men with a high social standing because a man's ability to provide resources for her offspring is related to his social status.

A further clue to the significance of reproductive value comes in an international study of divorce. Laura Betzig of the University of Michigan studied the causes of marital dissolution in 89 cultures from around the world. She found that one of the strongest sex-linked causes of divorce was a woman's old age (hence low reproductive value) and the inability to produce children. A woman's old age was significantly more likely to result in divorce than a man's old age.

Hypothesis 6: Men seeking a long-term mate will solve the problem of paternity confidence. Men face an adaptive problem that is not faced by women—the problem of certainty in parenthood. A woman can always be certain that a child is hers, but a man cannot be so sure that his mate's child is his own. Historically, men have sequestered women in various ways through the use of chastity belts, eunuch-guarded harems, surgical procedures and veiling to reduce their sexual attractiveness to other men. Some of these practices continue to this day and have been observed by social scientists in many parts of the world.

Most of these studies have considered three possibilities: (1) the desire for chastity in a mate (cues to *prior* lack of sexual contact with others), (2) the desire for fidelity in mates (cues to no *future* sexual contact with others), and (3) the jealous guarding of mates to prevent sexual contact with other men. We have looked at these issues ourselves in various studies.

In our international study, we examined men's and women's desire for chastity in a potential marriage partner. It proved to be a highly variable trait across cultures. For example, Chinese men and women both feel that it is indispensable in a mate. In the Netherlands and Scandinavia, on the other hand, both sexes see chastity as irrelevant in a mate. Overall, however, in about two-thirds of the international samples, men desire chastity more than women do. Sex differences are especially large among Indonesians, Iranians and Palestinian Arabs. In the remaining one-third of the cultures, no sex differences were found. In no cultures do women desire virginity in a mate more than men. In other words, where there is a difference between the sexes, it is always the case that men place a greater value on chastity.

Although we have yet to examine the desire for mate fidelity in our international sample, in her cross-cultural study Betzig found that the most prevalent cause of divorce was sexual infidelity, a cause that was highly sex-linked. A wife's infidelity was considerably more likely to result in a divorce than a husband's infidelity. Compromising a man's certainty in paternity is apparently seen worldwide as a breach so great that it often causes the irrevocable termination of the long-term marital bond.

We have examined the issue of fidelity among American college students. Indeed,

Photofest

Hypothesis 7: Women seeking a short-term mate will prefer men willing to impart immediate resources.

Figure 12. Female preference for short-term mates who are willing to provide resources was examined in the 1993 film *Indecent Proposal*. Robert Redford played the role of a wealthy older man who offered one million dollars to a younger married woman, played by Demi Moore, in exchange for a short-term mating. Since multiple short-term matings do not directly increase a woman's reproductive success, sexual-strategies theory holds that a woman can increase her reproductive success by acquiring resources.

Schmitt and I found that fidelity is the characteristic most valued by men in a long-term mate. It is also highly valued by women, but it ranks only third or fourth in importance, behind such qualities as honesty. It seems that American men are concerned more about the future fidelity of a mate than with her prior abstinence.

Our studies of jealousy reveal an interesting qualitative distinction between men and women. Randy Larsen, Jennifer Semmelroth, Drew Westen and I conducted a series of interviews in which we asked American college students to imagine two scenarios: (1) their partner having sexual intercourse with someone else, or (2) their partner falling in love and forming a deep emotional attachment to someone else. The majority of the men reported that they would be more upset if their mate had sexual intercourse with another man. In contrast, the majority of the women reported that they would be more upset if their mate formed an emotional attachment to another woman.

We also posed the same two scenarios to another group of 60 men and women, but this time we recorded their physiological responses. We placed electrodes on the corrugator muscle in the brow (which contracts during frowning), on two fingers of the right hand to measure skin conductance (or sweating), and the thumb to measure heart rate.

The results provided a striking confirmation of the verbal responses of our earlier study. Men

became more physiologically distressed at the thought of their mate's sexual infidelity than their mate's emotional infidelity. In response to the thought of sexual infidelity, their skin conductances increased by an average of about 1.5 microSiemens, the frowning muscle showed 7.75 microvolt units of contraction and their hearts increased by about five beats per minute. In response to the thought of emotional infideli-

Hypothesis 8: Women will be more selective than men in choosing a short-term mate.

DWF WRITER/PHOTOGRAPHER—creative, pretty, warm-hearted, independent, energetic, NYC—seeks friendship and romance with good-natured, thoughtful, intelligent, successful, fit SW man 45–60. Please no egoists, philistines, or couch potatoes! Box 3389.

L.A. SCIENTIST, M.D., C.E.O., WORKAHOLIC, 51, Joe Mantegna-type, seeks smart, slim, funny, independent SWF to help him slow down, enjoy travel, jazz. Box 3222.

DETROIT. 52 y/o MWF, happy, leggy, athletic. Fine Arts academician. Husband on 12 month sabbatical. Seeks ruggedly handsome younger man for discreet tryst. Box 3998.

FEMALE FRIEND/LOVER sought by tall, slim, handsome, athletic, married WM Ivy Lawyer/Bus. Man, 60s. Box 4235

BABE WITH LOOKS, BOOKS, AND BRAINS, NOW IN LOS ANGELES needs to meet her match for all things men and women do so well together including, but not limited to, great, safe sex. I'm a widowed lawyer, 50+, lively, slim, and trim. You are handsome, successful, well-educated, amusing, trustworthy. Your chronological age does not concern me but your attitude does. I'm looking for a mature man who is young at heart. Letter, phone, photo. No married men, please. Box 1212

Figure 13. Personal advertisements in newspapers for people seeking short-term mates support the hypothesis that women are generally more selective than men. Women tend to define specific qualities they seek in a man, whereas men tend to define their own qualities—attractiveness, high social status, ambition, and professional standing—that they believe will attract women.

Photofest

Figure 14. Female preferences for long-term mates who are willing to provide resources is parodied in the 1993 film *Addams Family Values*. Joan Cusack *(right)* plays the cultural stereotype of the beautiful "gold-digging" woman, and Christopher Lloyd *(left)* plays the wealthy but unattractive man, Uncle Fester. The character Cousin Itt *(hair in the center)* performs the wedding ceremony.

Hypothesis 9: Women seeking a long-term mate will prefer men who can provide resources for their offspring.

ty, the men's skin conductance showed little change from baseline, their frowning increased by only 1.16 units, and their heart rates did not increase. Women, on the other hand, tended to show the opposite pattern. For example, in response to the thought of emotional infidelity, their frowning increased by 8.12 units, whereas the thought of sexual infidelity elicited a response of only 3.03 units.

Hypothesis 7: Women seeking a short-term mate will prefer men willing to impart immediate resources. Women confront a different set of mating problems than those faced by men. They need not consider the problem of partner number, since mating with 100 men in one year would produce no more offspring than mating with just one. Nor do they have to be concerned about the certainty of genetic parenthood. Women also do not need to identify men with the highest fertility since men in their 50s, 60s and 70s can and do sire children.

In species where males invest parentally in offspring, where resources can be accrued and defended, and where males vary in their ability and willingness to channel these resources, females gain a selective advantage by choosing mates who are willing and able to invest resources. Females so choosing afford their offspring better protection, more food and other material advantages that increase their ability to survive and reproduce. Do human females

exhibit this behavior pattern? If so, we should be able to make a few predictions.

In short-term contexts, women especially value signs that a man will immediately expend resources on them. We asked 50 female subjects to evaluate the desirability of a few characteristics in a short-term and a long-term mate: *spends a lot of money early on, gives gifts early on,* and *has an extravagant lifestyle.* We found that women place greater importance on these qualities in a short-term mate than in a long-term mate, despite the fact that women are generally less exacting in short-term mating contexts.

We would also predict that women will find undesirable any traits that suggest that a man is reluctant to expend resources on her immediately. When we tested this prediction with the same sample population, we found that women especially dislike men who are stingy early on. Although this attribute is undesirable in a long-term mate as well, it is significantly more so in a short-term mate.

Hypothesis 8: Women will be more selective than men in choosing a short-term mate. This hypothesis follows from the fact that women (more than men) use short-term matings to evaluate prospective long-term mates. We can make several predictions based on this hypothesis.

First, women (more than men) will dislike short-term mates who are already in a relationship. We examined the relative undesirability of a prospective mate who was already in a relationship to 42 men and 44 women, using a scale from –3 (extremely undesirable) to 3 (extremely desirable). Although men were only slightly bothered (averaging a score of about –1.04) by this scenario, women were significantly more reluctant to engage in a relationship with such a mate (average score about –1.70).

We would also predict that women (more than men) will dislike short-term mates who are promiscuous. To a woman, promiscuity indicates that a man is seeking short-term relationships and is less likely to commit to a long-term mating. We tested this prediction in the same sample of 42 men and 44 women using the same rating scale as before. Although men found promiscuity to be of neutral value in a short-term mate, women rated the trait as moderately undesirable (an average of about –2.00).

Finally, because one of the hypothesized functions for female short-term mating is protection from aggressive men, women should value attributes such as physical size and strength in short-term mates more than in long-term mates. When we asked men and women to evaluate the notion of a mate being *physically strong,* we found that women preferred physically strong mates in all contexts more than men did, and that women placed a premium on physical strength in a short-term mate. This was true despite the higher standards women generally hold for a long-term mate.

Hypothesis 9: Women seeking a long-term mate will prefer men who can provide resources for her offspring. In a long-term mating context, we would predict that women (more than men) will desire traits such as a potential mate's ambition, earning capacity, professional degrees and wealth.

In one study we asked a group of 58 men and 50 women to rate the desirability (to the average man and woman) of certain characteristics that are indicators of future resource-acquisition potential. These included such qualities as *is likely to succeed in profession*, *is likely to earn a lot of money*, and *has a reliable future career*. We found that in each case women desired the attribute more in a long-term mate than in a short-term mate. Moreover, women valued each of these characteristics in a long-term mate more than men did.

In our international study, we also examined men's and women's preferences for long-term mates who can acquire resources. In this case we looked at such attributes as *good financial prospects*, *social status* and *ambition-industriousness*—attributes that typically lead to the acquisition of resources. We found that sex differences in the attitudes of men and women were strikingly consistent around the world. In 36 of the 37 cultures, women placed significantly greater value on financial prospects than did men. Although the sex differences were less profound for the other two qualities, in the overwhelming majority of cultures, women desire *social status* and *ambition-industriousness* in a long-term mate more than their male counterparts do.

Finally, in her international study of divorce, Betzig found that a man's failure to provide proper economic support for his wife and children was a significant sex-linked cause of divorce.

Conclusion

The results of our work and that of others provide strong evidence that the traditional assumptions about mate preferences—that they are arbitrary and culture-bound—are simply wrong. Darwin's initial insights into sexual selection have turned out to be scientifically profound for people, even though he understood neither their functional-adaptive nature nor the importance of relative parental investment for driving the two components of sexual selection.

Men and women have evolved powerful desires for particular characteristics in a mate. These desires are not arbitrary, but are highly patterned and universal. The patterns correspond closely to the specific adaptive problems that men and women have faced during the course of human evolutionary history. These are the problems of paternity certainty, partner number and reproductive capacity for men, and the problems of willingness and ability to invest resources for women.

It turns out that a woman's physical appear-

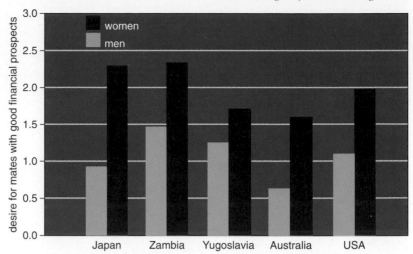

Figure 15. Mate's good financial prospects are consistently more important (measured on a scale from 3, *indispensible*, to 0, *unimportant*) for women than for men throughout the world. In the author's cross-cultural study of 37 countries, women valued the financial prospects of a potential spouse more than their male counterparts did in every culture but one (Spain). A sample of five countries is shown here.

ance is the most powerful predictor of the occupational status of the man she marries. A woman's appearance is more significant than her intelligence, her level of education or even her original socioeconomic status in determining the mate she will marry. Women who possess the qualities men prefer are most able to translate their preferences into actual mating decisions. Similarly, men possessing what women want—the ability to provide resources—are best able to mate according to their preferences.

Some adaptive problems are faced by men and women equally: identifying mates who show a proclivity to cooperate and mates who show evidence of having good parenting skills. Men do not look at women simply as sex objects, nor do women look at men simply as success objects. One of our most robust observations was that both sexes place tremendous importance on mutual love and kindness when seeking a long-term mate.

The similarities among cultures and between sexes implies a degree of psychological unity or species typicality that transcends geographical, racial, political, ethnic and sexual diversity. Future research could fruitfully examine the ecological and historical sources of diversity, while searching for the adaptive functions of the sexual desires that are shared by all members of our species.

Bibliography

Buss, D. 1994. *The Evolution of Desire: Strategies of Human Mating.* New York: Basic Books.

Buss, D. M., et al. 1990. International preferences in selecting mates: A study of 37 cultures. *Journal of Cross-cultural Psychology* 21:5–47.

Buss, D. M., and Schmitt, D. P. 1993. Sexual Strategies Theory: An evolutionary perspective on human mating. *Psychological Review* 100:204–232.

Buss, D. M., Larsen, R., Westen, D., and Semmelroth, J. 1992. Sex differences in jealousy: Evolution, physiology, psychology. *Psychological Science* 3:251–255.

PART V
The Adaptive Value of Social Behavior

Social behavior is particularly interesting to humans, perhaps because we are among the most social of organisms ourselves, but also because the evolution of sociality poses some especially challenging puzzles. How can the competition that we expect should occur between individuals be reduced enough to allow societies to form and social complexity to develop? For example, how can it be adaptive for some animals to share food rather than monopolizing any bonanza they happen to find? Why do members of some species form family groups while others do not? Why in some species do certain family members fail to reproduce, and instead spend most or all of their lives laboring for others? How could such self-sacrificing behavior spread when one would expect natural selection to eliminate fitness-reducing behavioral traits from the populations in which they occur?

Paradoxically, a great deal of "social" behavior consists of rather antisocial behavior, that is, aggressive interactions between individuals who are competing for resources, not living together harmoniously. Recall, for example, the articles on langur infanticide by Hrdy and by Curtin and Dolhinow in Part I. If antisocial behavior has evolved by natural selection, then we can expect that animals will make decisions during their aggressive encounters that generally raise their reproductive success. With this perspective in mind, we have a way to identify interesting Darwinian puzzles, namely actions that appear to reduce rather than raise an individual's chances of passing on its genes to subsequent generations.

In the first article, Michael Mesterton-Gibbons and Eldridge Adams identify three such puzzles in animal contests: the failure of thinner male damselflies to concede quickly to fatter rivals in their aerial disputes over territories, even though the thin males would conserve energy by giving up sooner; the readiness of mantis shrimps to threaten to attack when they are molting their cuticles, even though they would sustain serious damage if physically attacked by a retaliating rival; and the fact that members of a Mexican spider species will abandon a valuable home retreat without resistance if an intruder attempts to enter their hiding place. Each of these observations provides a paradox because it appears that animals behave maladaptively. Mesterton-Gibbons and Adams examine each case using a mathematical tool called "game theory" to resolve the paradox. They argue that this approach provides hypotheses and suggests ways to test them that could not be achieved by any other means.

Although none of the authors in Part IV used game theory formally, in the sense of developing and testing a mathematical model (hypothesis) per se, all discuss examples of behavioral phenomena that can be considered as products of evolutionary games. Return to one or more of these preceding chapters and use your understanding of game theory to identify "contests" that could be amenable to the kind of approach advocated by Mesterton-Gibbons and Adams. For example, don't many of the reproductive interactions between males and females involve conflicts of interest, with the adaptive value of decisions made by members of one sex depending entirely on how members of the opposite sex behave? You may be able to fill out the elements of the "game" you choose following the example set by Mesterton-Gibbons and Adams in their figures 2, 4 and 6.

Social life is a complex mixture of competition and cooperation, as you undoubtedly are well aware from personal experience. Bernd Heinrich and John Marzluff lead off our evolutionary analysis of cooperation with a detailed look at why ravens share food. The authors explain why it is odd that ravens will call others to a spot where they have found a large edible carcass. Why don't the finders just remain silent and keep the location of the food a secret? Heinrich and Marzluff discuss alternative explanations for how some individuals might actually gain by attracting others to carrion. Readers can outline the authors' use of the hypothetico-deductive method (see Woodward and Goodstein's article in Part I), which enabled them to discriminate among these explanations and to conclude that the "selfish" interests of individuals can be served by helping others—under special circumstances.

In addition, ravens sometimes roost communally in the evening, forming another kind of society away from a food source. Heinrich and Marzluff argue that these groups exist because of the advantages that juvenile ravens gain from sharing information about the location of food, and they provide convincing evidence that this is indeed the reason why the young birds are social. They note in passing, however, that the "information center hypothesis" is not the only possible explanation for the formation of nocturnal bird roosts. What alternatives can you develop, and what would you need to know in order to reject or accept these other hypotheses for why individual ravens join a communal roost?

In ravens, cooperators are not closely related. In some species, however, family groups form (see the article by Getz and Carter in Part IV), creating conditions that may be especially favorable for the evolution of certain kinds of self-sacrificing behavior, or altruism. Stephen Emlen, Peter Wrege, and Natalie Demong examine such a family-forming species, the white-fronted bee-eater. This bird, like the Florida scrub-jay (see the article by Schoech in Part II), lives in groups consisting of breeding adults and nonbreeding helpers-at-the-nest. The helpers provide food for chicks that are not their own. Like the jays but unlike the cooperative ravens, helper white-fronted bee-eaters work to increase the number of offspring produced by others while reducing their own offspring production (i.e., direct reproduction) through their altruism.

Altruism poses a special, but not insuperable, problem for evolutionary biologists. Emlen and his coauthors argue that it is possible for nonbreeding helpers to propagate their genes, provided the genetic cost of not reproducing during a breeding season is outweighed by the increased reproductive success of the relatives they assist. The authors explain the concept of inclusive fitness, and show that white-fronted bee-eaters are extremely skillful decision makers in terms of adopting the behavioral option that will maximize their genetic success.

The special power of the inclusive fitness approach is apparent when one realizes that it can be used to understand conflict, as well as cooperation, between offspring and their parents. For example, Emlen and his coauthors also use inclusive fitness theory to explain family disharmony in bee-eater societies. Try your own hand at analyzing the decisions bee-eaters make, incorporating both game theory and inclusive fitness approaches. Do you agree that it is sometimes adaptive for young bee-eaters to behave altruistically (and sometimes not), or do these birds frequently make errors that reduce rather than enhance their fitness?

Rodney Honeycutt's article revisits the issue of altruism and its evolution in a bizarre little subterranean mammal in which helpers-at-the-burrow lead a celibate life—a more extreme altruism than the facultative and generally temporary helping at the nest of white-fronted bee-eaters. Prior to 1981, the only animals known to have permanent nonbreeding helpers—or workers, as they are more commonly known—were insects: the ants, the termites, some bees and some wasps, a few beetles, several aphids, and a handful of thrips. In these insects, workers may be physiologically sterile, unable to reproduce on their own, spending their lives helping their mother produce additional siblings, some of which will reproduce and pass on the genes they share in common with their sterile helpers.

Twenty years ago, the naked mole-rat, a buck-toothed East African rodent, burst on the scene as the first vertebrate known to have a nonbreeding caste of workers. Naked mole-rats are wonderfully odd (and some would say downright ugly) animals that live in colonies of dozens to hundreds of workers, within which only one female and one to three males reproduce. The other members of the colony, both males and females, forage for food and excavate, maintain, and defend the colony's vast subterranean burrow system, but they do not breed. Like ants and termites, and other less dramatically altruistic species such as bee-eaters, naked mole-rat workers aid close relatives, with whom they share a relatively high proportion of their genetic mate-

rial. The breeding female is the largest and most aggressive individual in her colony (after becoming the breeder she grows longer by adding new bone to her vertebral column!). By vigorously shoving colony mates, the breeding female encourages them to work and maintains her pinnacle reproductive position.

But the discovery that helpers, whether permanently sterile or not, assist other family members cannot by itself explain the evolution of helping-at-the-nest or-burrow. After all, in all diploid species, full siblings are on average as closely related to one another as parents are to their offspring, and yet relatively few species of animals have nonbreeding helpers. The evolution of reproductive self-sacrifice must require special ecological and demographic factors, as Honeycutt argues. What are these factors, and might some of the same ecological pressures that favored helping in naked mole-rats and bee-eaters also have contributed to the evolution of non-breeding workers in weaver ants (Part III) and honey bees? In many human families, some offspring help their parents raise other offspring, their younger brothers and sisters. Is such behavior altruistic? In what ways are humans "eusocial," and in what ways are they not?

We conclude with two articles, one by Mark Winston and Keith Slessor and the other by Thomas Seeley, on another magnificent social creature: the now familiar honey bee (see Robinson's article in Part II). The single most striking feature of a honey bee colony from an evolutionary perspective is that, as in naked mole-rats, only one female, the queen, reproduces regularly. Other colony members, which in this case are all females, generally do not directly reproduce, and instead work to maintain the conditions needed for the survival of the queen and her brood. How do the workers know what to do and when to do it? Both articles answer this question by describing the remarkable communication system that enables tens of thousands of individuals to do all the things the colony needs to survive and multiply.

On the one hand, Seeley makes the case that workers exchange a great deal of information without centralized control from the queen. Winston and Slessor, on the other hand, suggest that the queen bee does indeed have much to say about what her workers do. Queen bees are not physically imposing or aggressive, as are naked mole-rat queens, but their mandibular glands produce massive amounts of a potent chemical cocktail. Some workers receive this pheromone orally from the queen and then distribute it to nest-mates, which pass it on to still other nest-mates, all of which are reproductively suppressed as a result. Winston and Slessor used traditional experimental procedures to establish precisely what chemicals—and in what proportions—are passed around by the workers. Their painstaking chemical sleuthing paid off when they finally isolated the "essence of royalty," a mixture of three decenoic acids and two aromatic compounds.

The queen's mandibular pheromone and the other mechanisms of communication within the hive produce a remarkable unity of purpose in the colony. The cohesive nature of the workers' activities in helping their queen makes it possible for Seeley to argue that the colony can be considered a superorganism, essentially a single "individual" composed of thousands of units, all of which labor for the same end: the production of relatives capable of reproducing. If all goes well, a colony will succeed in producing a daughter queen who will inherit about half the colony workforce when the old queen and the other half of the workers swarm away to find a new nest site. According to Seeley, the shared goals of workers and queens make it advantageous for workers to accept the chemical signals they receive from their mother, blocking their own reproduction in order to advance their mother's reproduction and their own genetic interests. The sophisticated nature of information flow in the hive, which among other things, leads foragers to work harder when there are more open honeycomb cells, lends credence to Seeley's view.

Workers are, however, no more closely related genetically to their mother than human offspring are to their parents (although honey bee full-sisters are more closely related than human siblings due to the bees' haplo-diploid genetic system, as explained by Honeycutt). Conflicts occur in honey bee colonies, just as they do in families of other organisms. Indeed, as Winston and Slessor suggest, queens may be attempting to "force" workers to follow orders by giving them chemical propaganda, which the workers consume not to accept these orders, but to digest and destroy the signal,

the better to do what is best for their own genes, rather than advance only their queen mother's fitness. According to this view, the degree of cooperation in a colony is far less complete than it appears on the surface, with the queen trying to control a population that may be attempting to defy her.

How would you test the validity of these two competing views of honey bee sociality? If workers are truly cooperative members of a superorganism, would you expect them to ever reproduce? If workers fight against the reproductive suppressing effects of the queen's pheromone, would you expect them to accept the material from other workers? How should selfish workers behave toward a sister-worker that attempts to lay eggs of her own? Because queens mate with many males, workers can be either full or half-sisters. Can you use this fact to design a test of the alternative superorganism versus selfish individual views of honey bee societies?

Ravens, bee-eaters, naked mole-rats, and honey bees all illustrate the complex mixture of cooperation and conflict that characterize social organisms, which also is why they offer such a range of interesting phenomena for evolutionary biologists to study. The articles in Part V demonstrate how the scientific process has been used to identify major questions about social evolution, and how animal behaviorists have begun the difficult but satisfying task of solving these puzzles.

Animal Contests as Evolutionary Games

Paradoxical behavior can be understood in the context of evolutionary stable strategies. The trick is to discover which game the animal is playing

Michael Mesterton-Gibbons and Eldridge S. Adams

The science of behavioral ecology thrives on paradoxes, baffling inconsistencies between intuition and evidence that engage our attention and stimulate further investigation. Mathematical models can modify our intuition by showing that apparently sensible explanations are actually problematic, or that seemingly outrageous proposals are downright reasonable. This is especially true in the study of animal contests, where the consequences of an interaction between two animals with opposing interests are difficult to guess.

Consider a famous example known as the "handicap principle." The behavioral ecologist Amotz Zahavi of Tel Aviv University argued that animals with conflicting interests should evolve behavioral displays that are costly to the signaler, even if they lower its chances for survival. By showing that it can endure a handicap, the animal reliably indicates its high quality, a message that other animals do well to respect. For example, on sighting a predator, a gazelle may *stott*—that is, jump high in the air on all four legs—several times before fleeing, thereby demonstrating that it is in superb physical condition, and that the predator would only waste time and energy by pursuing it. This hypothesis was initially rejected by many partly because it contradicted the biologist's in-

tuition that evolution should favor signals with low costs to the animal producing them, especially since these costly behaviors will be passed onto the signaler's offspring.

Yet formal models of communication revealed that the handicap principle is logically sound under certain conditions. In particular, the magnitude of the handicap must increase with the intensity of the signal, and the cost must be especially damaging for animals of lower quality. For example, if a weak gazelle stotted as vigorously as a strong one, then it would waste what little strength it had and be unable to flee from a pursuing predator. These models have convinced many biologists that a previously unaccepted idea has broad explanatory power.

Efforts to resolve such questions in animal behavior have relied increasingly on collaborations between biologists and mathematicians, using analytical tools called games. In particular, behavioral ecologists use evolutionary games. A game in this context is a mathematical model of strategic interaction, which arises when the outcome of an individual's actions depends on the actions of others.

A game has three components. First, there are at least two interacting individuals, called players. In an evolutionary game played within a single species, the set of players is an *ecotype*, a population of animals in a given ecological environment. For example, a population of spiders in a grassland habitat and the same species in a riparian habitat form two different ecotypes.

Second, each player has a set of feasible strategies. In an evolutionary game, this set is the same for every player and is constrained by the *information structure* of the interaction. For example, ani-

mals can modify their behavior in different circumstances, such as whether they are owners or intruders, only if they are aware of such roles.

Third, the *pattern of interaction* must be well defined and accompanied by a formula for how each player's reward from the interaction depends on its strategy and on those of the other players. In evolutionary games, the rewards are measured in terms of expected future reproductive success.

For a game to be useful, it must be possible to identify one or more strategies from among those feasible as the "solution" for a given purpose. In our case, the solution is the behavior that can be expected to evolve by natural selection. If a behavior is fixed in a real population, then it must at least be true that every feasible alternative behavior would yield a lower reward, for otherwise the alternative behavior would have spread into the population. The relevant solution concept, which was introduced a quarter of a century ago by John Maynard Smith of the University of Sussex, is that of an *evolutionary stable strategy*, a population strategy that yields a higher reward than any feasible mutant strategy (that is, any newly introduced alternative strategy).

Just as a model population is only a caricature of a real population, so a paradox is only a caricature of real ignorance. So, in terms of realism, a game and a paradox are a perfect match. A game theorist strives to unravel the paradox by establishing conditions for an evolutionary stable strategy to exist in the model population, and by analyzing its properties when it does exist. Now, a paradox arises because evidence fails to support intuition, which (assuming the evidence to be sound) can happen only if the intuition relies on a false

Michael Mesterton-Gibbons is a professor in the Department of Mathematics, Florida State University. Eldridge S. Adams is an assistant professor in the Department of Ecology & Evolutionary Biology at the University of Connecticut, Storrs. Address for Mesterton-Gibbons: Department of Mathematics, Florida State University, Tallahassee, FL 32306–4510. Internet: mmestert@mailer.fsu.edu.

Figure 1. Dueling damselflies will contest a territory until the stronger one prevails, but unlike many mammalian species, these insects cannot observe the relative strength of their opponent. Instead, the damselflies appear to play a war-of-attrition game in which they rely only on a knowledge of their own energy reserves. This strategy was discovered by evolutionary game theorists in an attempt to explain why weaker damselflies did not yield to stronger opponents earlier in a contest. The authors argue that the methods of game theory provide a valuable tool in discovering the strategies that can evolve in contests between animals of the same species. (Photograph courtesy of Jim Marden, Pennsylvania State University.)

assumption, albeit an implicit one. So the way to resolve a paradox is to spot the false assumption.

In other words, if a paradox of animal behavior exists, then we have wrongly guessed which game best models how the real population interacts, and to resolve this paradox we must guess again—if necessary, repeatedly—until eventually we guess correctly. Assuming the validity of our solution concept, that is, assuming that the observed behavior corresponds to some evolutionary stable strategy, there are only three things we could be wrong about: the ecotype, the information structure and strategy set or the pattern of interaction and reward. So there are also only three things to be right about.

Each can be important, as we illustrate with examples from our recent work on animal contests among damselflies, mantis shrimps and spiders.

Information Structure in a Game

A few years ago, Jim Marden of Pennsylvania State University, in collaboration with Jonathan Waage of Brown University, staged a series of territorial contests between male damselflies (*Calopteryx maculata*). The male pairs in these contests had various, unequal reserves of fat, an indicator of strength. A common expectation for such contests (derived from game-theory models) is that each animal compares its own strength to that of its opponent and withdraws when it judges itself to be

the probable loser. This has been confirmed by experimental studies on a variety of animals. The duration of such contests is greatest when the opponents are of nearly equal fighting ability, so that it is more difficult to judge which is likely to win. This is the same reason that many human sporting events are longer and more exciting when teams are evenly matched.

But the duels between the damselflies failed to follow this logic. Although the weaker animal ultimately conceded to its opponent in more than 90 percent of these encounters, Marden found no negative correlation between the duration of a contest and the difference in the relative strengths of the males. Here was a paradox, all the more baffling because the expected relationship had emerged repeatedly in studies of other species. Could there be something amiss in our assumptions about damselfly contests?

One possibly errant assumption is simply the belief that animals can assess one another's strength. Because it is such a good assumption for contests between animals with good visual acuity (such as

bighorn sheep), and because so much of a behavioral ecologist's intuition about contest behavior had developed in that context, it was an easy assumption to overlook. But fat is stored internally in insects, so a damselfly cannot directly observe its opponent's reserves. (Marden also found no correlation between fat reserves and traits that surely are observable, such as body length or wing span.) Thus there emerged the intriguing possibility that damselflies do not assess their opponents' strength at all.

Accordingly, in a collaboration with Marden and Lee Dugatkin of the University of Louisville, one of us (Mesterton-Gibbons) developed a model in which a key assumption is that players know only their own reserves. In the resultant game, known as a "war of attrition," a strategy is the proportion of an animal's initial reserves that it is prepared to expend in a prolonged contest over a disputed site. A second key assumption is that the value of winning is proportional to the winner's remaining reserves. The more an animal has left, the more successful it will be in attracting a mate, finding food or defending its territory (and hence in producing surviving offspring). These assumptions determined the strategy set and the reward formula.

But what about the ecotype? Since a model is merely a caricature of nature, it effectively reduces a real ecological environment to a few parameters, with different values for different ecotypes. In the war-of-attrition game, there are only two such parameters, each a number between zero and one. The first, a *coefficient of variation*, measures the dispersion of energy reserves about their mean. For example, a coefficient of variation of 0.6 means that one standard deviation in fat reserves is 60 percent of the mean. The second parameter, a *cost/benefit ratio*, compares the reproductive cost of a spent unit of fat reserves to the eventual winner's reproductive benefit from a saved unit.

With each component of the game identified, we can look for an evolutionary stable strategy. In general, because an evolutionary stable strategy is a population strategy, the ecological parameters determine whether one exists. In this particular game, an evolutionary stable strategy exists (for reasons described below) if the coefficient of variation exceeds a critical threshold, which increases with the cost/benefit ratio but is never much more than about 0.5. The

---

War-of-Attrition Game

Parameters defining the model ecotype

- Coefficient of variation *(CV)*. Measures variation in fat energy reserves.
- Cost/Benefit Ratio *(CBR)*. Compares the cost of the contest to the benefit of owning the site (both in terms of the expected future reproductive success).

In a real ecotype, the cost/benefit ratio would vary from individual to individual. In the model we assume each individual has the same CBR.

Information structure

An animal knows its own reserves, but not the reserves of its opponent.

Reward structure

A site is worth more to an animal that has greater remaining reserves of energy, which is a safe assumption for a contest over a mating territory.

Pattern of interaction

An opponent's unknown energy reserves are drawn at random.

Strategy set

An animal's strategy is the proportion of its energy reserves that it is willing to expend in contesting a site.

Evolutionary stable strategy

Ecological parameters determine whether an evolutionary stable strategy exists (see graph below). If one exists, then the nature of the strategy depends on those parameters.

$$\text{Evolutionary stable proportion} = \frac{1}{1 + CBR \times f(CV)}$$

$f(CV)$ increases from 0 to 1 as CV increases from 0 to 1, so the proportion exceeds 1/2. If (CV, CBR) is not evolutionarily stable, then any positive proportion of the population can be invaded by a mutant adopting proportion 0 (that is, giving up immediately).

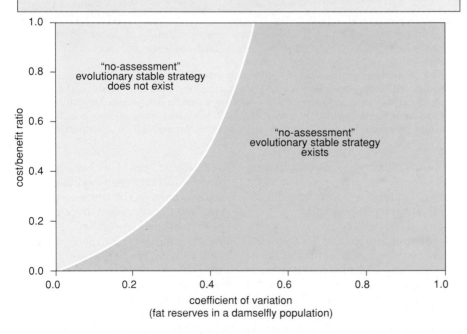

Figure 2. Existence of a "no-assessment" evolutionary stable strategy in the *war-of-attrition* game between male damselflies depends on two ecological parameters: the variation in fat energy reserves in the population and the relative cost of the contest compared to the benefit of owning a mating site. In this instance, an evolutionary stable strategy exists if the coefficient of variation of energy reserves exceeds a critical threshold *(white line)*, which increases with the cost/benefit ratio but is never more than about 0.5. The damselflies do not know their opponents' fat energy reserves but can achieve a balance between the costs and the benefits by responding to the distribution of reserves in the population.

coefficient of variation for Marden's damselflies is 0.51, which would fall below critical if the cost/benefit ratio were between 0.99 and 1. However, in practice the cost/benefit ratio is likely to be considerably less than 0.99. Thus we could assume that fat reserves among damselflies are variable enough that an evolutionary stable strategy involving no assessment of opponents' strength could exist.

At this evolutionary stable strategy, both contestants are prepared to expend at least 50 percent of their initial reserves on the contest, although only the loser actually does so. We can verbally describe the reason for the stability of the no-assessment strategy. Being prepared to expend too small a proportion of initial reserves may mean ceding needlessly to weaker opponents (who would otherwise lose), whereas too high a proportion may mean wasting reserves needlessly against much stronger opponents (who would win in any case). Although animals do not know the reserves of their actual opponents, they achieve a balance between trade-offs by responding to the distribution of reserves among the population.

Marden later staged further damselfly duels in conjunction with Bob Rollins of Pennsylvania State University, and he pooled his data sets. Although the percentage of wins by fatter males fell from 90 percent to 86 percent, it was still remarkably high. Now, judicious approximation is the essence of modeling—effects that are small in a real population are typically absent from a model. For example, we would expect an 86 percent win rate for fatter males in the real world to translate into a 100 percent win rate for fatter males in the model. And this is precisely what happens because both contestants are prepared to deplete their initial reserves by the same proportion. The skinnier one invariably gives up first. In other words, although there is no assessment, the fatter male always wins.

Thus the assessment and no-assessment hypotheses both predict that fatter males always win. But there is also a difference. The assessment hypothesis predicts a negative correlation between strength difference and contest duration. Although, in the pooled data, Marden detected such a correlation in contests exceeding 500 seconds, the variation in strength difference could explain only 14 percent of the variation in contest duration. By contrast, in the

Figure 3. Mantis shrimps weakened by molting will threaten and successfully deter intruders despite their inability to win a fight. The success of the weakened shrimp's bluff raises the question of why all mantis shrimps do not threaten an intruder. The authors argue that the behavior of the strong and weak mantis shrimps can be explained by appealing to game theory. (Photograph courtesy of Roy L. Caldwell, University of California, Berkeley.)

no-assessment model, a contest ends when the loser gives up after using a fixed proportion of its reserves. So the no-assessment hypothesis predicts a positive correlation between final loser reserves and contest duration. And, in contests over 500 seconds, the variation in loser reserves was found to explain 29 percent of the variation in duration.

It is tempting to infer from these results that the no-assessment hypothesis is twice as likely to be correct. But we should desist, because 14 percent and 29 percent are both so much less than 100 percent. The results are really inconclusive. Nevertheless, our attempt to resolve the initial paradox has yielded a valuable new insight on animal contest behavior, namely, that a pattern of victory by stronger animals need not imply that strength is being assessed. Moreover, the analysis indicates the kind of population in which we could expect to find such a no-assessment evolutionary

stable strategy: one with a high coefficient of variation.

Reward Structure in a Game
Several years ago, one of us (Adams), in collaboration with Roy Caldwell at the University of California, Berkeley, observed a series of contests between stomatopods, or mantis shrimps, of the species *Gonodactylus bredini*. These crustaceans inhabit cavities in coral rubble. If one intrudes upon another, the resident often defends its cavity by threatening with a pair of claw-like appendages. These threat displays often deter intruders, so that contests are settled without any physical contact. A surprising observation is that when stomatopods are weakened by molting, so that they are completely unable to fight, they threaten more frequently than animals that are between molts. Furthermore, threats by weaklings often deter much stronger intruders, who

would easily win a fight if there were one. In other words, the weaklings bluff. But if the very weakest members of the population can threaten profitably, then why don't all animals threaten? If the display can be given by animals that cannot back it up, then why do their opponents respect it?

To explore this paradox, we developed a game in which a resident can either threaten or not threaten, and an intruder responds by either attacking or fleeing. We assume that animals vary in weakness, that is, in how much their strength falls short of the maximum strength to be found in the population. We also introduce a reward structure that differs from that assumed by models of "honest" signaling. Specifically, threat displays increase the vulnerability of the signaler to injury inflicted by its opponent. The threat is thus a display of bravado that bears no special cost if the signaler is not attacked, but which adds an additional cost, the *threat cost*, if it is attacked by a stronger opponent. In addition, if there is a fight, both contestants pay a combat cost, consisting of a *fixed cost*, which even the strongest animal pays, and a variable cost that increases with weakness at a constant rate of *marginal cost*.

The threat cost, the fixed cost and the marginal cost together comprise the model ecotype. We also assume that, if there's a fight, the stronger animal wins. Finally, because the molt condition of stomatopods is not externally visible, we assume (as in the war-of-attrition game) that each contestant is unaware of its opponent's strength. So its own strength must determine its behavior.

For this particular reward structure, there is always an evolutionary stable strategy at which the weakest and strongest animals both threaten when resident, whereas those of intermediate strength do not *(Figure 4, left)*. Why such a counterintuitive result? All residents face a similar trade-off: Threats deter some opponents without the necessity of fighting, but if the opponent is stronger and chooses to attack, then the threat increases the vulnerability of the signaler. The weaker the signaling animal, the more likely it is to pay the price of its increased vulnerability. That's why animals of intermediate strength cannot afford to threaten, but stronger animals can. This is essentially an instance of the handicap principle.

Threat Game

Parameters defining model ecotype

- A fixed cost of escalating to physical combat.
- A marginal cost of a unit of weakness.
- A threat cost, which a resident pays if it threatens, is attacked and loses the ensuing fight.

All costs are measured in terms of expected future reproductive success.
Strength is assumed to be uniformly distributed between 0 (weakest) and 1 (strongest).
The combat cost varies between fixed cost for the strongest animal to fixed cost + marginal cost for the weakest. Threat cost can be operationally defined as that part of the total cost of combat which could have been avoided by not theatening.

Information structure and strategy set

An animal knows its own strength, but not the strength of its opponent. A resident's decision to threaten or not precedes an intruder's decision to attack or flee.

Reward structure

A site is worth the same to each animal. Assume that the value of a site is one unit of reproductive success.

Pattern of interaction

An opponent's unknown strength is drawn at random from the uniform distribution for the population.

Evolutionary stable strategy

Provided that the marginal cost is positive and the threat cost is paid only when the resident is attacked and loses, there is a unique evolutionary stable strategy. As usual, the evolutionary stable strategy depends on the ecological parameters (see graphs below, where the strategies are plotted against fixed cost for constant values of marginal cost and threat cost). If the marginal cost is 0 or if the threat cost is paid when the resident is attacked (regardless of whether it wins or loses), then there is no evolutionary stable strategy. Thus, the evolutionary stable strategy is a consequence of a very special reward structure, which the model identifies.

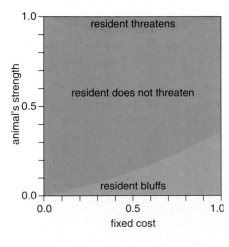
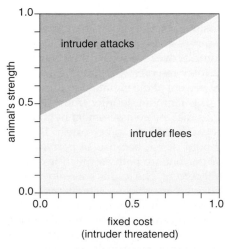
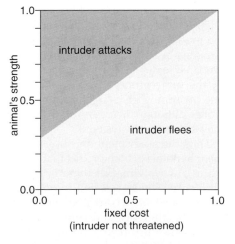

Figure 4. Strategy of a mantis shrimp in the *threat game* depends on its strength and the relative cost of fighting for residents *(left)* and intruders *(middle and right)*. Game theory suggests that residents will threaten (bluff) when they are weak *(green)* because they have little to lose, whereas the strongest residents threaten *(orange)* because they are unlikely to lose. In contrast, residents of intermediate strength *(blue)* do not threaten because the costs of losing exceed the benefits of threatening. The threshold for the intruder's decision to attack or flee depends on whether it is threatened *(middle)* or not *(right)* by the resident. This threshold for the intruder's strength is raised if it is faced with a resident that threatens.

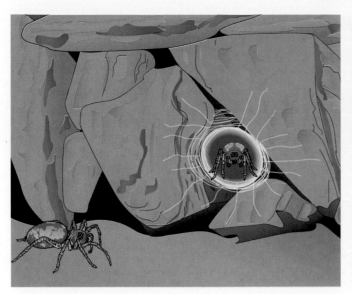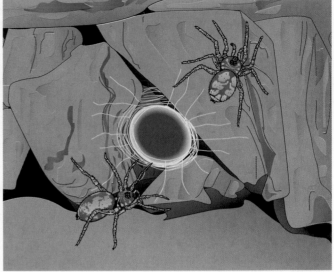

Figure 5. Spiders of the species *Oecobius civitas* will relinquish a hiding place to an intruder despite the advantage that usually accompanies ownership. The evicted spider will similarly displace another spider from its hiding place, potentially initiating a domino effect of evictions until a spider finds an empty hiding place. This paradoxical behavior may be understood in the context of an evolutionary game in which the spider's probability of winning a fight is compared to the probability of surviving a period of searching for another hiding place.

However, the model illustrates a point lacking from previous discussions of handicaps, namely that the benefits of threats are greater for weaker animals, particularly those that have no other means of driving off competitors. These animals would be unlikely to win any of the contests against opponents that are deterred by threats. So the model shows a potential resolution of the paradox: The net costs of bluffing are lower for weak animals than they are for stronger ones. Bluffing therefore forms a part of the evolutionary stable strategy.

The Ecotype in a Game

As the damselflies exemplify, contests between animals of unequal strength are usually won by the stronger one. Suppose, however, that the animals are equally strong. What advantage could either then have? One possibility in competitions for resources is ownership. Indeed evidence abounds, in species as diverse as baboons, butterflies, lions and red deer, that if animals perceive themselves to be in the separate roles of owner and intruder when contesting a permanent resource, then the owner usually retains it with little serious fighting. There is nothing paradoxical about such respect for ownership. But wouldn't it surprise you if animals avoided serious fighting, not because intruders retreated, but rather because owners fled? That is what appears to happen in a species of Mexican spider, *Oecobius civitas*.

According to a study by J. Wesley Burgess, populations of this tiny spider live in darkness on the undersides of rocks. If a spider is disturbed from its hiding place, it is apt to enter the hiding place of another, which, if in residence, does not attack the intruder, but instead departs to find a new retreat. A domino effect can ensue, with retreats successively changing ownership until eventually a spider finds one unoccupied. This behavior is inherently paradoxical because, assuming a home-court advantage, the prize always goes to the contestant likelier to lose an actual fight. How can we make sense of this?

As usual, when a behavior is fixed, we expect to understand it as the evolutionary stable strategy of an evolutionary game, with effects that are small in the real population being absent from a model. Thus, we expect butterfly behavior to correspond to an evolutionary stable strategy at which owners *always* (as opposed to usually) win, and *O. civitas* behavior to an evolutionary stable strategy at which owners *always* lose. And, in both cases, we expect fighting to be absent (as opposed to rare). That is, in Maynard Smith's terminology, we expect butterfly behavior to follow a *Bourgeois* strategy—attack if an owner but be non-aggressive (or display) if an intruder. *O. civitas* behavior reflects an *anti-Bourgeois* strategy—display if an owner but attack if an intruder. Likewise, animals that invariably attack, regardless of whether they are owners or intruders, follow a *Hawk* strategy, and

those that display in either role follow a *Dove* strategy. These four strategies imply an information structure, namely, that animals know their roles.

Using a simple game with these four strategies, Maynard Smith showed that only the Hawk strategy is evolutionarily stable if costs are low enough, whereas the Bourgeois and anti-Bourgeois strategies are evolutionarily stable if costs are high enough. But when two evolutionary stable strategies exist, the population can only be at one of them, and Maynard Smith's game could not say which.

One of us (Mesterton-Gibbons) incorporated Maynard Smith's strategy set into a more elaborate model. Each animal has a finite number of periods in which to find a site, which is essential for reproduction. It does not matter whether the site is acquired early or late, as long as it is owned by the end of the last period. The number of periods is unknown, but all animals have the same probability of surviving predation each period. There are more sites than animals, but the search for a site is random. So a vagrant animal may search in vain, and even if it finds a site it may be occupied. In that case, the animal can either continue to search or contest the site. If it wins, then it becomes an owner, and the previous owner must search again.

There's a catch to this, however. When there's a serious fight, the loser is so seriously injured that it can no longer search, and hence cannot reproduce. Even a winner may suffer injury, and have lower reproductive success

Iterated Hawk-Dove Game

Parameters defining model ecotype

- Probability of surviving a period of searching for a site (a function of predation on the spiders).
- Owner's probability of winning a fight (the resource-holding potential), assumed to exceed 0.5.
- Probability per search period of locating a site.
- Ratio of number of animals to number of sites.
- Probability of injury in winning a fight.
- Ratio of site value for an injured owner to that for an uninjured owner.

Information structure

Animals know whether they are owners or intruders.

Reward structure

A site is worth the same to each animal. Assume that the value of a site is one unit of reproductive success.

Pattern of interaction

Searching animals are equally likely to discover any site.

Strategy set

Hawk: Attack as owner, and attack as intruder.
Bourgeois: Attack as owner, and show a non-aggressive display as intruder.
Anti-Bourgeois: Non-aggressive display as owner, and attack as intruder.
Dove: Non-aggressive display as owner and intruder.

Evolutionary stable strategy

This game differs from the other two in having multiple evolutionary stable strategies. The most interesting dependence of the evolutionary stable strategy is that on survival probability and the resource-holding potential (for fixed values of the other four parameters). This dependence is quite complicated, and so is merely cartooned in the graph below.

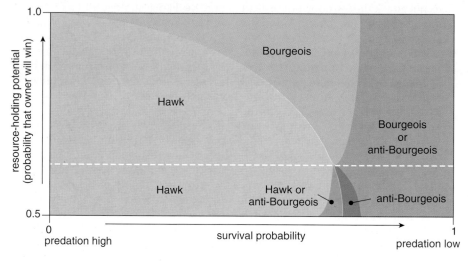

Figure 6. Evolutionary stable strategies in the *iterated hawk-dove game* depend on an animal's chance of surviving predation when searching for a site and the probability that it will win a fight. The general trend in this game is for animal populations to evolve from a fighting to a nonfighting strategy as predation pressure decreases. Above a certain probability of winning a fight as owner (*dashed line*) a population will evolve to use a Bourgeois strategy rather than an anti-Bourgeois strategy. The Mexican spider's anti-Bourgeois strategy—in which it is non-aggressive (or merely displays) as an owner of a site but attacks as an intruder—can be explained in this context if an owner's probability of winning a fight is relatively modest (though above 0.5), at least until the population reaches the blue region.

than an uninjured animal. So there's a trade-off between the uncertainty of search and the risk of injury. The owner's probability of winning an actual fight is called its *resource-holding potential*. It is a number between 0.5 and one, indicating that the owner is more likely to win.

Whether animals fight depends on their strategies. A Bourgeois owner and an anti-Bourgeois intruder, or two Hawks, would always fight; an anti-

Bourgeois owner and a Bourgeois intruder, or two Doves, would merely display; and a Hawk would always scare off a Dove, in either role. This reward and information structure results in a six-parameter ecotype.

Although the evolutionary stable strategy depends on all six ecological parameters, in terms of our paradox it depends critically on the animal's fighting ability and the probability of its survival. From Figure 6 it is evident that only the Hawk strategy is evolutionarily stable if survival is sufficiently low. Despite the risk of injury, it is better to fight than to hope to find a vacant site before the competition ends. At higher probabilities of survival, however, the Bourgeois or the anti-Bourgeois strategy is evolutionarily stable. To see why, suppose that the resource-holding potential remains constant, but that predation decreases slowly with time (so that the probability of survival increases slowly with time) and that the population tracks its changing environment by continually evolving to a new evolutionary stable strategy. Then we think of the population as a point moving horizontally to the right in Figure 6. As this point migrates, the Hawk strategy must eventually cease to be evolutionarily stable.

Why? Consider a lone Bourgeois mutant in an otherwise Hawk population. Because the Bourgeois animal behaves like a Hawk at home, being an owner has no effect on its reward. So a mutant Bourgeois has a higher reward than the population strategy if a Bourgeois intruder does better than an Hawk intruder. Now, a Bourgeois intruder runs from a Hawk owner, hoping to find an empty site eventually, but a Hawk intruder fights, hoping to win a site immediately. If predation is sufficiently low and the resource-holding potential is sufficiently high, however, the first hope is much more likely to be realized than the second. Thus the Bourgeois fares better than the Hawk and so will spread in the population.

Similarly, a mutant anti-Bourgeois in a Hawk population will behave differently only as an owner. It runs from an intruding Hawk, hoping eventually to find a vacant site, whereas a Hawk owner would fight, hoping to win immediately. If predation and the resource-holding potential are both sufficiently low, then the first of these hopes is much more likely to be realized than the second. Thus the anti-Bourgeois fares better than the Hawk and so will spread in the population.

In summary, under decreasing predation, a Hawk population evolves to Bourgeois if the resource-holding potential is relatively large, but the Hawk population evolves to anti-Bourgeois if the resource-holding potential is relatively small (*Figure 6*). If predation falls sufficiently, then the population will remain at either the Bourgeois or the anti-Bourgeois strategy depending on the initial resource-holding potential. In either case, a non-fighting population evolves from a population of fighters.

To the extent that lower predation increases the chance that not fighting and searching will eventually yield a site without injury, these results agree with those of Maynard Smith. But the new model goes further to suggest a resolution of our paradox. If an ownership asymmetry exists, then the resource-holding potential is almost always high enough to keep the population in a Bourgeois strategy under conditions of low predation. However, there are rare environments in which the resource-holding potential is low enough to keep the population in an anti-Bourgeois strategy under conditions of low predation. Does *O. civitas* represent such an exception? Perhaps. But without any empirical evidence that the resource-holding potential is indeed low in *O. civitas*, we can do no more than conjecture.

The Value of Game Theory

It is not unusual for an exercise in game theory to remain partially inconclusive. On the one hand, game-theoretic models are valuable because they suggest ways to test new ideas. The damselfly model suggests a test—whether the reserves of the loser correlate positively with contest duration—for an evolutionary stable strategy in which the contestants cannot assess their opponents' strength. The stomatopod model suggests a test— whether animals that threaten and lose pay higher costs than those that lose without threatening—for the idea of an evolutionary stable strategy with partial bluffing. Finally, the spider model suggests a test—whether an intruder is almost as likely as an owner to win an actual fight—for the idea of an anti-Bourgeois evolutionary stable strategy. On the other hand, suggesting a test is not the same thing as conducting it, and the difficulties of doing so should not be underestimated. For example, the only way to measure resource-holding potential is to precipitate actual fights, which may be far from easy in a species as reclusive as the spider *O. civitas*.

In fact, the difficulties of testing the predictions of evolutionary game theory have led some to question its value. But games are not valuable solely because they suggest ways to test new ideas. They are also valuable because they allow us to explore the logic of a verbal argument rigorously, assuming biologically realistic ecotypes, and to determine when it is true and when it is false. As these cases illustrate, game theory often demonstrates what is difficult to intuit. As a result of exploring damselfly duels, we now understand that victory by stronger animals need not imply that strength is being assessed. As a result of investigating stomatopod strife, we now understand how bluffing can persist at a high frequency. And through analyzing spider spats, we now understand how a non-fighting population of usurpers can evolve, under decreasing predation, from a fighting population that ignores the asymmetry of ownership.

Perhaps non-assessment of strength, high-frequency bluffing and sequential displacement are all remarkably rare in nature. For example, the domino effect has not been observed in colonial spiders other than *O. civitas*. But it is rarely the commonplace that piques our interest. Rather, it is strange behavior that engages our attention and spurs us on to deeper understanding, even of the commonplace. In that regard, evolutionary games have proved their usefulness over and over again. They have become indispensable analytical tools toward progress in behavioral ecology.

Acknowledgment

This work was supported by award #9626609 from the National Science Foundation and by a Fellowship in Science and Engineering from the David and Lucile Packard Foundation.

Bibliography

Adams, E. S., and R. L. Caldwell. 1990. Deceptive communication in asymmetric fights of the stomatopod crustacean, *Gonodactylus bredini*. *Animal Behaviour* 39:706–716.

Adams, E. S., and M. Mesterton-Gibbons. 1995. The cost of threat displays and the stability of deceptive communication. *Journal of Theoretical Biology* 175:405–421.

Burgess, J. W. 1976. Social spiders. *Scientific American*. March, 100–106.

Davies, N. B. 1978. Territorial defence in the speckled wood butterfly (*Pararge aegeria*), the resident always wins. *Animal Behaviour* 26:138–147.

Dugatkin, L. A., and H. K. Reeve. 1998. *Game Theory and Animal Behavior*. New York: Oxford University Press.

Grafen, A. 1990. Biological signals as handicaps. *Journal of Theoretical Biology* 144:517–546.

Hammerstein, P., and S. E. Reichert. 1988. Payoffs and strategies in territorial contests: ESS analyses of two ecotypes of the spider *Agelenopsis aperta*. *Evolutionary Ecology* 2:115–138.

Hodge, M. A., and G. W. Uetz. 1995. A comparison of agonistic behaviour of colonial web-building spiders from desert and tropical habitats. *Animal Behaviour* 50:963–972.

Johnstone, R. A. 1998. Game theory and communication. In *Game Theory and Animal Behavior*, ed. L. Dugatkin and H. Reeve. New York: Oxford University Press, pp. 94–117.

Krebs, J. R., and N. B. Davies. 1993. *An Introduction to Behavioural Ecology*. 3rd edition. Oxford: Blackwell Scientific Publications.

Marden, J. H., and J. K. Waage. 1990. Escalated damselfly territorial contests are energetic wars of attrition. *Animal Behaviour* 39:954–959.

Marden, J. H., and R. A. Rollins. 1994. Assessment of energy reserves by damselflies engaged in aerial contests for mating territories. *Animal Behaviour* 48:1023–1030.

Maynard Smith, J. 1982. *Evolution and the Theory of Games*. Cambridge: Cambridge University Press.

Mesterton-Gibbons, M. 1992. Ecotypic variation in the asymmetric Hawk-Dove game: when is Bourgeois an evolutionarily stable strategy? *Evolutionary Ecology* 6:198–222.

Mesterton-Gibbons, M., J. H. Marden and L. A. Dugatkin. 1996. On wars of attrition without assessment. *Journal of Theoretical Biology* 181:65–83.

Nur, N., and O. Hasson. 1984. Phenotypic plasticity and the handicap principle. *Journal of Theoretical Biology* 110:275–297.

Parker, G. A., and D. I. Rubenstein. 1981. Role assessment, reserve strategy, and acquisition of information in asymmetric animal conflicts. *Animal Behaviour* 29:221–240.

Riechert, S. E. 1986. Spider fights as a test of evolutionary game theory. *American Scientist* 74:604–610.

Riechert, S. E. 1998. Game theory and animal contests. In *Game Theory and Animal Behavior*, ed. L. Dugatkin and H. Reeve. New York: Oxford University Press, pp. 64–93.

Steger, R., and R. L. Caldwell. 1983. Intraspecific deception by bluffing: A defense strategy of newly moulted stomatopods. *Science* 21:558–560.

Zahavi, A. 1977. Reliability in communication systems and the evolution of altruism. In *Evolutionary Ecology*, ed. B. Stonehouse and C. M. Perrins. London: Macmillan, pp. 253–259.

Zahavi, A. 1987. The theory of signal selection and some of its implications. In *International Symposium of Biological Evolution*, ed. V. P. Delfino. Vari: Adriatica Editrice, pp. 305–327.

Zahavi, A., and A. Zahavi. 1997. *The Handicap Principle*. New York: Oxford University Press.

Why Ravens Share

Young ravens eat regularly, even when food is rare, because they direct one another to food bonanzas and fend off adults by feeding in large crowds

Bernd Heinrich and John Marzluff

In a forest in northern New England, a moose dies in a spruce thicket. Coyotes soon find the dead moose and feed on it at night. The next day, a hungry young common raven discovers this bonanza of food. But the raven does not feed: It circles above the carcass, then flies off. A few days afterward, daybreak reveals a raucously calling string of about 40 ravens, flying in for a feast. Within a week, more than 100 ravens have joined in consuming more than 90 percent of the 1,000-pound carcass.

We have observed this scenario, or scenarios much like it, more than 100 times while studying ravens in the Maine forests. Our findings indicate that most of the birds that come to feed on such a treasure—a lifeline during a harsh winter—learned of its location from the raven that made the original discovery. Such communication might be expected within a closely knit group of related individuals. But a feasting flock of ravens hardly fits this definition: It consists of birds that usually defend exclusive domains, and that wander widely before settling down and eventually mating. Among such animals, ecological and evolutionary theory suggests that a large carcass, such as a moose, should be defended, not shared. After all, it provides a source of food that might last an entire winter for a bird lucky enough to find

Bernd Heinrich is professor of biology at the University of Vermont. He earned his Ph.D. in 1970 from the University of California at Los Angeles. He continues to work on Maine ravens. John Marzluff is a research scientist at Sustainable Ecosystems Institute in Boise, Idaho. He earned his Ph.D. in 1987 from Northern Arizona University. He is currently working on western corvids. Address for Heinrich: Department of Biology, Marsh Life Science Building, University of Vermont, Burlington, VT 05405–0086.

and defend it. Sharing a carcass represents altruism—a selfless act—because a bird that shares a heap of food might starve later in winter.

Nevertheless, when a young raven finds a large supply of food, it brings in other ravens from as far away as 30 miles. We wondered how the birds communicate the location of food, and whether sharing proves truly selfless, or advantageous. We suspected, and found, that juvenile ravens possess immediately selfish reasons for this apparently altruistic act. In fact, food sharing turns out to be a successful strategy for maximizing survival in an environment where food is sparsely and unevenly distributed in space and time, and where young birds must cooperate in order to defend and feed on a carcass at the same time.

Origins of Altruism

According to Darwin's original theory of evolution, altruism cannot evolve because it requires an animal to sacrifice its reproductive fitness to generate benefits for other animals. But nature provides many examples in which an animal behaves in a way that benefits other animals. In some cases, an animal might even sacrifice its own life to help others. These examples generate a theoretical problem, because selfless behavior cannot be transmitted genetically.

In general, self-sacrificing behavior, or helping, buys delayed or hidden benefits. In other words, selfishness lies behind seemingly selfless behavior. An organism, for example, may give up resources, such as a big pile of food, if that favor will be repaid in the future, or if food is being shared with a relative who can pass on the sharer's genes to future generations. A termite soldier, for instance, may sacrifice its life in defending its colony against an ant attack. Al-

though such behavior appears altruistic from a soldier's perspective, it proves selfish from a genetic perspective—provided that a soldier's behavior, on average, enhances the survival of its colony, whose members are closely related, and the propagation of its genes by others.

During winter in New England, ravens might share food because they have difficulty finding carcasses on a regular basis. Finding carcasses is difficult because they are rare, can be hidden in thick brush and often become covered with snow. The short daylight hours of winter make searching even more difficult. Ravens must feed on a regular basis to survive. In addition, a raven cannot penetrate the hide of deer or most other large animals that it eats from, so a raven usually feeds on a carcass that has already been torn apart by a mammalian carnivore or another scavenger, which means that some of the food has already been consumed. When a raven does find a carcass, it may provide enough of a meal for a short bout of sharing, but the carcass will not last long, and the raven soon returns to its main task: finding the next carcass.

Ravens might improve their chances of finding spatially and temporally unreliable food by forming alliances composed of many individuals that search independently for carcasses and then get together when a carcass is found. An individual raven that shares an ample-size carcass might be giving up little because, after the ravens eat their fill, the remaining meat might be consumed by waiting carnivores anyway. By sharing finds, ravens reduce the uncertainty of finding another meal. In fact, the larger and rarer carcasses are, the smaller is the cost of sharing and the larger is the benefit of foraging cooperatively. A carcass hidden under snow or brush may

Figure 1. Pairs of adult ravens in the forests of northern New England establish territories and defend their discoveries of food during the harsh winter. But their control may be short-lived: If a juvenile raven comes upon a defended carcass, it communicates the location to many other young ravens, which soon arrive at the carcass and feed as a group. The adult pair's behavior is well explained by evolutionary theory, but the sharing of food by juveniles poses a challenge to the theory. The authors suspected that the seemingly altruistic behavior of juveniles may actually be rooted in self-interest. (Photographs courtesy of Bernd Heinrich.)

escape the attention of one raven, but having more pairs of looking eyes increases the likelihood that all birds will be fed, and on a continuous basis.

The food-sharing scenario, however, suffers from a potential flaw. If an individual raven refuses to share, that bird gains an immediate advantage. If all ravens share their food, on the other hand, the entire population benefits because they all eat regularly. In other words, the system of food sharing among ravens relies on trust—each bird trusting that all other birds will share. Might ravens be the first truly communistic social organization, based on trust and rationality, where individuals give according to their ability and receive according to their needs?

Scout or Squad

Determining whether ravens really share food and why requires knowledge of how they find food and who eats it. Do single individuals or crowds of ravens usually find carcasses? Answering that question took more than 1,000 hours of experimentation: Putting out a carcass (cattle, deer, moose, sheep and so on), obtained from a farmer or game warden, and then watching it from a snow-covered spruce-fir blind until one or more ravens happened to find the food. A single flying raven or two ravens flying together discovered each of the 25 carcasses put out in this experiment.

A crowd of ravens arrived only after one or two ravens discovered the bait. In addition, the crowd usually came at

dawn, and most of birds in the crowd flew in from the same direction. These observations suggest that a crowd of ravens feeding on a carcass does not develop from a flock happening to find it, and a feeding crowd must assemble as a group the night before flying to a carcass. It appears, therefore, that a lone discover or a pair of discoverers brings in the crowds of dozens of ravens from some assembly location. At a carcass, ravens often "yell," or make a special type of vocalization. Playing recorded yells attracts nearby ravens. Clearly, ravens at a carcass do not try to hide the discovery.

Still, we wondered which birds make up a crowd at a carcass. To find out, we captured 463 ravens and marked them with colored and numbered wing tags and radio transmitters. That procedure did not disperse the birds, because the largest number of marked birds could always be found immediately after being released. In addition, we demonstrated that the tags stayed in place for at least two years on birds that lived in a large aviary. Markers on wild birds allowed us to identify ravens feeding at a carcass. On every carcass that we put out, new birds came and "old" birds left on an hourly basis. No two groups of eating ravens consisted of the same individuals, because the composition of a group changed constantly.

Watching these very shy birds from blinds constructed from spruce and balsam-fir boughs in the forest near our baits, we were surprised by the continual turnover of different birds from one day to the next. On some very large baits, for instance, some ravens returned daily for up to six days, but most stayed for only a couple of days, and many stayed for only an hour or less. Up to 500 ravens partic-

Figure 2. Raven yells, attracting its cohorts after discovering a food bonanza. The yell (*see Figure 7*) is different from calls made by infant and fledgling ravens.

Figure 3. Winter forces ravens to survive on a variable supply of food. Carcasses are scarce and can be difficult to locate. The short days of winter reduce the available searching time, and a carcass can be hidden in brush or snow. Furthermore, feeding on a frozen, picked-over carcass does not provide quick satiation. Feeding in crowds may give subadults the ability to retain control of a carcass long enough to satisfy their hunger.

ipated in consuming one "super bonanza," two skinned cows, but usually no more than 50 birds fed at once.

Working with Patricia Parker of Ohio State University and Thomas A. Waite then of Simon Fraser University, we found by DNA fingerprinting that the ravens feeding in a group are not family members that stayed together on a home turf. In fact, ravens that fed together were no more closely related than ravens that fed in separate groups. We concluded that "vagrants"—ravens that have no home domain or that wander over a very large range—make up a crowd of feeding ravens. In fact, one bird that we marked was sighted 200 miles away. Combining those results suggests that sharing food provides little or no opportunity for a raven to promote the propagation of its own genes by sacrificing food to feed a relative. In addition, vagrancy reduces the odds that a raven will receive a future favor by sharing food with an unrelated bird, because the chances of meeting that same bird again are slim.

Individual Eaters

By marking ravens, we discovered two distinct groups: residents and wanderers. The residents, mostly adults, defend any food bonanza located in their domain. The wanderers, mostly juveniles, get little or no access to defended food unless they come in sufficient numbers to neutralize the adults' defenses. In other words, juveniles must recruit larger numbers of birds to get access to a carcass.

Several lines of evidence support the idea that wandering subadults create ad hoc gangs, which allows them to feed at a carcass, even when it is defended by territorial adults. A pair of resident adults may dominate a carcass for weeks and then be swamped by a crowd of young ravens in just one or two days. Second, radio-tagging studies showed that, although residents roost alone, vagrant ravens sleep communally, in a group consisting of from 20 to 60 birds (about the same number that appears at a carcass at one time), and at a site near a carcass. Third, if a juvenile arrives at a carcass, the resident adults make dominance displays and attack the juvenile, which leads to submissive postures and vocalizations by the juvenile. Finally, a juvenile's calls attract other vagrant juveniles.

We tested our gang hypothesis in an

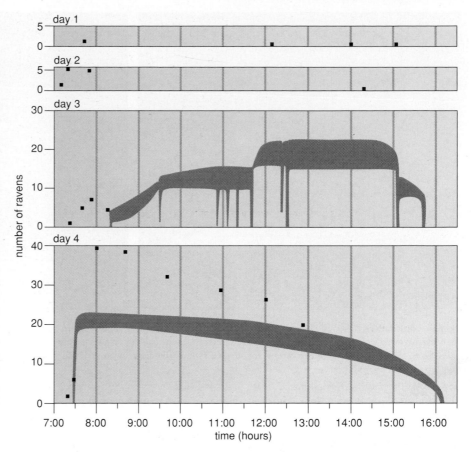

Figure 4. Ravens arrive in a group to feed on a carcass. A sheep carcass, which the authors put out as bait, was discovered by a raven on day 1. (Individual appearances are denoted by *black* squares.) That bird apparently returned several times but did not eat. By day 2, ravens were feeding, but in small numbers. By day 3, a group of ravens (*gray* areas, where thickness indicates variability in the number of ravens in a group) shared the sheep. On day 4, about 20 ravens gathered in the morning and consumed the remainder of the carcass.

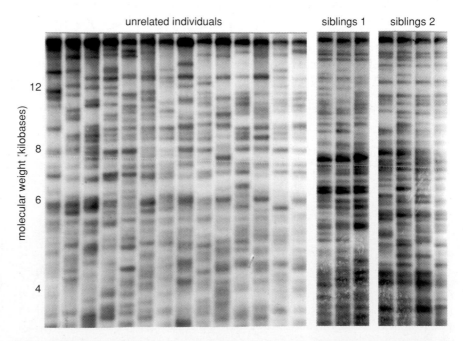

Figure 5. DNA fingerprinting shows that unrelated ravens make up feeding groups. Two groups of siblings possess similar DNA (*right*), but ravens in one feeding group show wide variations in their DNA profiles (*left*), meaning that they are not related. So ravens do not share food in hopes that a feeding relative will pass on shared genes.

Figure 6. Adult ravens *(left)* may control a carcass found in their territory for a week or so, until a juvenile raven flies over and sees the carcass. When the juvenile's signals attract other juveniles, a struggle for control ensues *(right)*. A small group of juveniles cannot overpower a pair of adults. So a juvenile that discovers a carcass must recruit more juveniles, or none of the young birds will eat.

outdoor aviary, which enclosed 7,100 cubic meters. That huge aviary, located at our study site in Maine, consisted of a main area that housed about 20 subadults and side-arms leading to two ancillary aviaries, one of which housed a pair of resident adults that were captured from the wild. From a central observation hut, we could watch the birds through one-way glass, and we controlled access between aviaries by opening and closing gates with guy wires. Our studies in the aviary showed that, as in the wild, adults viciously attacked hungry subadults that attempted to reach food, and it required a gang of at least nine subadults to feed relatively unmolested.

Eating from a frozen, picked-over carcass, however, precludes quick satiation, because a raven cannot rush in and grab a billfull of meat. Instead, a raven can only chip off tiny pieces of frozen meat. Moreover, bones and skin obscure most of the meat. A raven might feed for several hours before becoming satiated. The limited number of choice feeding sites on a carcass might suggest that each feeding bird eats less as the overall number of birds increases. Nevertheless, the average amount of feeding increased with larger group sizes, at least for medium-sized carcasses and larger ones, including deer and cows. In the presence of adults, juveniles benefit even more from a crowd, because the adults attack the dominant subadults, and that distraction allows subordinate

subadults to eat. So feeding in crowds provides most subadults with an immediate advantage: eating more, especially in the presence of territorial adults. As already indicated, sharing also provides a long-term advantage: reducing the patchiness of an otherwise temporally unreliable food resource.

What's in a Yell?

Delving deeper into the mechanism behind food sharing by ravens steered us toward sounds. Ravens possess a large and varied vocal repertoire, which might include recruitment calls. Many of the calls advertise territories or attract mates, but we concentrated on calls that ravens give when crowds feed. The first call, a plaintive "yell," quickly recruits nearby ravens when a recording of it is played in the wild. Given that a yell attracts other ravens and that givers of the call gain feeding privileges (access to defended meat), yelling should lead to a demonstrated advantage. But what causes the yelling behavior?

Manipulating the food available to ravens in our aviary showed that yelling increases as a function of hunger when a subadult sees food. A yell resembles other food-related calls, including one made by young birds when their parents come near them and another call that a female makes when she begs for her mate to feed her while she sits on a nest. In young birds that are out of the nest, a juvenile's yell tells a parent where to find the juvenile,

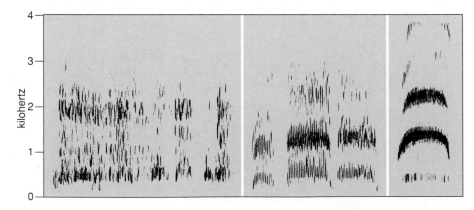

Figure 7. Calls made by juvenile ravens can lead to eating. Before fledging, ravens beg their parents for food with a raspy call, depicted in a sonogram *(left)*. Soon after fledging, a raven's beg possesses more distinct tones *(middle)*. A couple of months after fledging, a raven's beg develops into a so-called yell *(right)* that attracts other juveniles to a discovered food bonanza.

a

b

c

d

Figure 10. Sharing by ravens depends largely on communication. Although resident adults try to defend a carcass *(a)***, a lone juvenile (identified by a** *yellow* **wing tag) will eventually discover it** *(b)***. That juvenile relays the location of the carcass to a roost of other ravens** *(c)***, and the group flies to the carcass and overthrows the resident adults** *(d)***. That cooperation helps young ravens eat consistently, even when food runs scarce.**

moved the most dominant raven, then the second-most dominant one did the yelling. Removing the second in line stimulated yelling by the third, and so on. So hunger stimulates yelling, and socially superior ravens suppress it. A lone vagrant does not yell near a territorial male for fear of being beaten. Dominant ravens in a crowd of vagrants do yell, which tells other vagrants that they might join the feeding with little interference from defending territorial adults.

Subadult ravens make another call that also attracts other birds. In our aviary, when a subadult approached adults with food, the adults attacked, and the subadult made a begging call that attracted the rest of the subadults. Playing recorded begs also attracted subadults. Begging calls also reduce adult aggression, probably because an adult wants to limit the calling that recruits other subadults.

Although both yelling and begging attract other ravens, calls near a carcass attract only nearby individuals. When a carcass is first discovered and few vagrants happen to be in the area, yelling and begging would not attract the crowds that gather. In fact, when we

Figure 8. Kettle of young ravens broadcasts an impending move to a new roost. Most days, ravens return to their nearby roost after feeding at a carcass and settle in for the night. After consuming all of a carcass, however, ravens return to their roost and circle over it, soaring as high as 2,000 feet. Then the ravens fly to a new roost, presumably near a new carcass.

as well as which one is hungry. In addition, juvenile yells tell other members of the same brood (after they have left the nest) where their parents and food can be found. The juvenile yell persists for months or years after leaving the nest, presumably because it continues to provide food to a caller. In subadults, however, the call attracts other vagrants, not parents.

One might expect that the loudest yells would come from a lone vagrant located near adults with food. In our aviary, however, lone subadults did not yell when they saw meat near the adults. It turns out that social status affects yelling. Even among subadult vagrants, the most dominant bird does most of the yelling, and it suppresses yelling in others. If we re-

Figure 9. Postures and feather configurations portray a bird's status. A raven at an uncontested food source *(a)* holds its head up and keeps its feathers smooth across its head. A raven first approaching food *(b)* lowers its head. A vagrant at an adult-protected carcass *(c)* keeps its head up and the feathers on its head fluff out. When juveniles swamp a carcass, a resident adult performs a dominance display *(d)*, which includes erect posture, raised bill, raised earlike feathers and fluffed-out throat and leg feathers.

broadcast yells in a forest, we did not attract any ravens, presumably because their low density reduces the chance of being within a couple of miles of a bird.

Roost Reporting

Nocturnal roosts provide the primary crowd-forming capability of ravens. Subadult ravens often sleep in communal roosts that form at dusk in pine groves that lie within a few miles of a food bonanza. In general, most or all of the ravens from a roost leave *en masse* at dawn and fly directly to the food. Throughout the day, ravens feed at a carcass and also disperse, possibly looking for other carcasses, but only feeding on one. In the evening, the ravens return to a roost, coming in from many directions. According to our radio-tagging studies, ravens often join different roosts on different nights. After depleting a carcass, the ravens either disperse or move on to another source of food.

In 1971 the Israeli biologist Amotz Zahavi proposed that communal bird roosts serve as information centers, and that hypothesis spawned a long debate in the ornithological literature. Our results prove that they do. When we released naive ravens (which had been held in captivity for at least several weeks) in the evening and near a roost, they immediately joined the strangers. The following morning, the naive ravens appeared at a carcass that the roost birds were eating. By contrast, other naive ravens that were released in the evening and at equal distance from a carcass but not in the vicinity of a roost did not appear at the carcass. These results suggest that wandering ravens need merely locate a roost to be led directly to food.

Carcass size probably regulates roost size. If a deer carcass, for example, offers room for 15 feeding birds, then additional birds would gain little access without fighting constantly, and so they move on. On the other hand, if a moose carcass provides places for 40 ravens, then the feeding crowd grows, as does the roost. In Maine, roosts usually consisted of less than 50 birds, about the limit that we observed feeding at any time at moose or other large carcasses that we provided. In the western United States, where ravens eat insects and grain in open rangeland, relatively permanent raven roosts of more than 1,000 individuals exist.

Although most ravens feed at a carcass for only a short time, often not returning again until days or even weeks later, the number of birds feeding on a carcass remains high, suggesting that many birds know of a carcass's location. In addition, ravens probably know the location of more than one carcass at a time. So if members of a roost know about several different potential feeding sites, why do all the ravens in a roost go to the same carcass?

Soaring

Several observations suggest that roost relocation depends on consensus. We routinely climbed to the tops of tall spruce trees for the panoramic view necessary to watch roost formation in late afternoon, from two hours before dusk until dark. During 328 nights of roost watching, we observed 72 cases in which from 3 to 103 ravens circled over a roost, and then the whole roost rose out of the trees and all of the birds disappeared into the distance. In addition, we also saw a single radio-tagged bird discover a fresh carcass, and then return the following dawn with 20 or more followers. (Such observations are rare because as long as ravens know of a feeding site they have no reason to go to a new site even when a roost member discovers one.)

Spectacular soaring displays usually accompany roost moves. When a carcass has been nearly cleaned up, the ravens return to their old roost in the late afternoon or evening, but instead of flying directly into the trees to roost, as they usually do, the ravens ascend high into the air, sometimes to 2,000 feet or more, and fly in large "kettles." A kettle consists of birds flying noisily, diving and tumbling. The displays last from 15 minutes to more than two hours, and neighboring ravens keep joining in, making the soaring crowd grow. Finally, the growing aggregation stops circling, and then the birds fly in long lines, all traveling in the same direction as they disappear over the horizon. The next day, the ravens do not return to their old carcass, because the new roost probably lies near another feeding site. We have also observed this sequence from the other direction. After putting out bait, we may see only a pair of resident adults for many days. One evening, though, soaring ravens will settle nearby and begin feeding the next day. The resident adults will be swamped, no longer able to defend the meat against vagrants.

Almost all of the hundreds of carcasses that we have provided in Maine over the last 11 years were ultimately shared by crowds of ravens. Birds from the surrounding hundreds of square miles eventually participate in eating a large carcass. Working with Delia Kaye, Kristin Schaumburg and Ted Knight, for example, we radio-tagged 10 birds and then attempted to find them on a daily and nightly basis for two months. Most of those birds ranged over more than 1,000 square miles. In another experiment, we spread 10 carcasses over a linear distance of 30 miles, and nine of the 10 baits were eaten in turn by the marked birds and others.

Our data show that carcass-sharing behavior by ravens did not evolve because of altruism acting through intelligence and foresight, or generosity. Instead, ravens share because their system serves the common good by harnessing self-interest, not suppressing it. That combination of self-interest and common good gives the common raven—in the forests of New England and presumably elsewhere—a large edge over all other species in harvesting a rich resource of food, which is not available on a steady basis to any other bird. Surprisingly, harnessing the most selfish of motivations in an extremely aggressive species creates amazing cooperation for the common good.

Bibliography

Heinrich, B. 1988. Winter foraging at carcasses by three sympatric corvids with emphasis on recruitment by the raven, *Corvus corax*. *Behavioral Ecology and Sociobiology* 23:141–156.

Heinrich, B. 1989. *Ravens in Winter.* New York: Summit Books.

Heinrich, B. 1993. A birdbrain nevermore. *Natural History* 102:51–56.

Heinrich, B. 1994. Does the early bird get (and show) the meat? *The Auk* 111:764–769.

Heinrich, B., D. Kaye, T. Knight and K. Schaumburg. 1994. Dispersal and association among a "flock" of common ravens, *Corvus corax. The Condor* 96:545–551.

Heinrich, B., and J. M. Marzluff. 1991. Do common ravens yell because they want to attract others? *Behavioral Ecology and Sociobiology* 28:13–21.

Heinrich, B., J. M. Marzluff and C. S. Marzluff. 1993. Ravens are attracted to the appeasement calls of discoverers when they are attacked at defended food. *The Auk* 110:247–254.

Marzluff, J. M., B. Heinrich and C. S. Marzluff. (in press). Raven roosts are mobile information centers. *Animal Behavior.*

Parker, P. G., T. A. Waite, B. Heinrich and J. M. Marzluff. 1994. Do common ravens share ephemeral food resources with kin? DNA fingerprinting evidence. *Animal Behavior* 48:1085–1093.

Making Decisions in the Family: An Evolutionary Perspective

The complex social interactions in a family of white-fronted bee-eaters are governed by some simple rules of reproductive success

Stephen T. Emlen, Peter H. Wrege and Natalie J. Demong

The family has been the fundamental social unit throughout much of human evolutionary history. For countless generations, most people were born, matured and died as members of extended families. However, human beings are not the only animals that form such social structures. Some of the most outstanding examples can be found among birds, of whom nearly 300 species form social bonds that are unquestionably recognizable as family units. In most cases, the family appears to play a crucial role in the socialization and survival of the individual.

The significance of the family to the development of the individual is not lost on biologists, who are inclined to ask whether certain social interactions between family members might be better understood in an evolutionary framework. Given the intensity of the interactions within a family, it is natural to expect that natural selection has shaped many of the behaviors that emerge. Could the same forces that act on birds act also on the human species? Such questions are controversial but compelling.

The evolutionary framework that is used to understand most social interaction is the theory of kin selection, for-malized by William D. Hamilton in 1964. Hamilton emphasized that individuals can contribute genetically to future generations in two ways: directly, through the production of their own offspring, and indirectly, through their positive effects on the reproductive success of their relatives. This is because a relative's offspring also carry genes that are identical to one's own by virtue of common descent. The closer the genetic relationship, the greater the proportion of shared genes. The sum of an individual's direct and indirect contributions to the future gene pool is his or her inclusive fitness.

Because of this genetic relatedness, the social dynamics of family life is expected to differ in significant ways from the dynamics of other types of group living. The degree of kinship is predicted to influence the types of behavior exhibited among individuals. All else being equal, closely related individuals are expected to engage in fewer actions that have detrimental reproductive consequences for one another, and more actions with beneficial reproductive consequences. Although we expect significant amounts of cooperation within families, we must also recognize that not all familial interactions will be harmonious. Kinship may temper selfish behavior, but it does not eliminate it. Individuals will often differ in their degrees of relatedness to one another, in their opportunities to benefit from others, and in their abilities to wield leverage over others. These variables should predict the contexts of within-family conflicts, the identity of the participants and even the probable outcomes.

Human beings are notoriously difficult subjects for such studies because so much of our behavior is sculpted by cultural forces. In contrast, family-dwelling birds provide excellent opportunities for testing evolutionary predictions about social interactions among relatives. They have a large repertoire of complex social behavior, yet they have few culturally transmitted behaviors that might confound the analysis. They are a natural system in which to search for fundamental biological rules of social interaction.

It is in this light that we spent eight years studying the white-fronted bee-eaters at Lake Nakuru National Park in Kenya. Our original motivation was to study the altruistic behavior of these birds, in particular their tendency to help others at the nest. We came to realize, however, that the birds simultaneously engaged in a number of selfish behaviors as well. Indeed, the birds displayed a wide range of subtle tactics, some mutually beneficial but others clearly exploitative.

An Extended Family

In biological terms, a family exists when offspring continue to interact with their parents into adulthood. This distinguishes families from temporary child-rearing associations in which young members disperse from their parents when they reach sexual maturity. We can further narrow the definition by stipulating that the parents must maintain a preferential social and sexual bond with each other. The white-fronted bee-eaters of Kenya fulfill these qualifications.

Indeed, the heart of the bee-eater society is the extended family, a multigenerational group consisting of 3 to 17 individuals. A typical family contains two or three mated pairs plus a small assortment of single birds (the unpaired and the widowed). A young bee-eater matures in a group of close relatives,

Stephen Emlen is a professor of animal behavior at Cornell University. His research focuses on cooperation and conflict in animal societies and on animal mating systems. He has also worked on the orientation, navigation and acoustic communication of birds. Peter Wrege is a research associate at Cornell where he received his Ph.D. in 1980 for his studies on the social foraging strategies of the white ibis. Natalie J. Demong is a freelance writer and photographer who specializes in avian field studies. Emlen's address: Department of Neurobiology and Behavior, Mudd Hall, Cornell University, Ithaca, NY 14853-2702. Internet: ste1@cornell.edu

Figure 1. White-fronted bee-eaters of Kenya provide a culture-free animal model for studying the complex dynamics within a group of closely related individuals—a family unit. Colorful tags on each member of a family allow the authors to document the interactions between specific birds. Such studies reveal that bee-eaters make sophisticated decisions based on the status and genetic relatedness of the individual with which they interact. (Photograph courtesy of Marie Read.)

and most continue to interact with parents, siblings, grandparents, uncles, aunts, nephews and nieces into adulthood. Families can even include steprelatives (stepparents and half-siblings) when individuals remate after the death or divorce of a partner. As a result, bee-eater families often have very complex genealogies.

About 15 to 25 families (100 to 200 birds) roost and nest together in a colony. The nests are excavated in sandy cliff faces where the birds dig meter-long tunnels that end in enlarged nesting chambers. Late in the afternoon all bee-eaters congregate at their colony to socialize and roost.

During pair formation, one member leaves its own family and moves to that of the other. This dispersal rule reduces the likelihood of within-family pairings; indeed, we have never witnessed an in-

cestuous pairing among bee-eaters. As a consequence, the resident member of the pair continues to live in a network of close genetic kin. As in most species of birds, it is the bee-eater females that usually disperse. A paired female becomes socially integrated into her mate's family, but the genetic kinship links are lacking. Unrelated females are the functional equivalents of "in-laws."

Once paired, bee-eaters are socially monogamous, exhibiting high mate fidelity over years. Divorce rates are low, with the effect that most individuals remain paired to the same partner for life. Both sexes share equally and heavily in all aspects of parental care.

In many respects, the social structure of bee-eaters has similarities to the supposed organization of ancestral human beings, who are thought to have formed long-term pair bonds, who lived in vil-

lages consisting of several extended-family groups, and whose families included both related and unrelated (in-law) members.

Helping Whom?

The most dramatic aspect of bee-eater reproductive behavior is the phenomenon of helping at the nest. Helpers play a major role in almost every aspect of nesting except copulation. Even before breeding begins, helpers aid in digging the nest chamber, a task that may take 10 to 14 days. Helpers also bring food to breeding females during the week in which they are energetically burdened by egg production. After the eggs have been laid, helpers of both sexes undergo physiological changes, enabling them to incubate the clutch. Helpers will defend the young birds for weeks after they are hatched and for several weeks after

Figure 2. Colony of white-fronted bee-eaters in the face of a sandy cliff may contain as many as 25 families, or 200 birds. The birds excavate a nesting chamber at the end of a meter-long tunnel in the wall. The colony serves as a year-round site for nesting and roosting. (Photograph courtesy of Natalie Demong.)

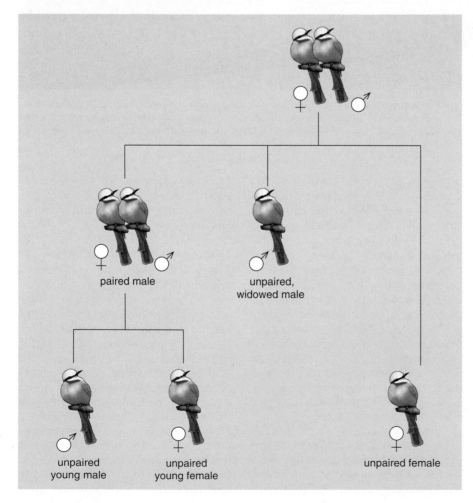

paired male

unpaired, widowed male

unpaired young male

unpaired young female

unpaired female

they are fledged. Helping usually ceases only when the young are completely self-sufficient.

By far the most significant component of helping is providing food for the young. Because the abundance of the bee-eaters' staple food (flying insects) varies unpredictably, the extraparental helpers can have a major effect on a pair's breeding success. In our study, one-half of all nestlings died of starvation before leaving the nest. However, the presence of even a single helper reduced the starvation losses to the point of doubling the fledgling success of an unaided pair!

Because bee-eaters tend to provide aid only to family members (genetic relatives), helpers play a major role in the reproductive success of their nondescendent kin. This means that helpers are indirectly increasing their own inclusive fitness. Interestingly, bee-eater helpers gain no measurable direct benefits from helping. In many other cooperative breeders (species with helpers at the nest or den) the experience of helping often translates into increased personal reproductive success later in life. This is true if the act of helping increases the likelihood that a helper will become a breeder in the future or if helping provides a better breeding slot. It is also true if the experience of helping makes one a better parent in the future. None of these personal benefits accrued to the bee-eater helpers at Nakuru; their helping behavior appears to be maintained entirely through kin selection.

If the major benefit the helpers accrue is through kin selection, bee-eaters should be sensitive to their degree of kinship to different family members. This is indeed the case. When a bee-eater faced the choice of helping one of several relatives, the helper chose to aid the most closely related breeding pair in over 90 percent of the cases (108 of 115).

Kin-selection theory also helps to explain why nearly half (44 percent) of all bee-eaters neither breed nor provide

Figure 3. Extended family of white-fronted bee-eaters may contain three or more generations of birds. Males remain in their natal family after taking a mate and are surrounded by close genetic relatives throughout most of their lives. When females pair they leave their natal group to live with their mate's family and consequently are not closely related to the birds in their new home. The difference of living with or without genetic relatives is associated with striking differences in the social interactions of paired males and females.

help in any given year. There is little profit in helping distant kin. Indeed, most nonhelpers are individuals with no close relatives in their social group. The largest subset of nonhelpers are the females who separated from their own families at the time of pairing. Helping does not increase the inclusive fitness of such females until they have raised fully grown (and breeding) offspring of their own. At this point, they again become helpers, selectively aiding their breeding sons to produce grand-offspring.

On the other hand, the benefits of helping close kin also explain instances in which birds whose own nesting attempts fail, change roles and addresses and become helpers at nests of other breeders in the family. Through such redirected helping, they can recoup much of their lost inclusive fitness. This "insurance" option is typically available only to the males, since they are more likely to be surrounded by close genetic relatives. As predicted, the vast majority (90 percent) of redirected helping involves males. Although females typically relocate to the new nesting chamber with their mate, they rarely participate in rearing unrelated young. The contrasting behaviors of the male and the female are especially striking in the light of all the stimuli—eggs, incubating adults, begging nestlings and attending adults feeding the nestlings—that

would seemingly induce the female to help at the nest.

Coercion by Parents

Since helpers have a large positive effect on nesting success, their services are a valuable resource in a bee-eater family. As a result, we would expect some competition among breeders for a helper's services, and even occasional conflicts between breeders and potential helpers over whether the latter should help. In some instances, helping at the nest might be forcefully "encouraged."

Bee-eaters do, in fact, engage in seemingly coercive behaviors that result in the disruption of nesting attempts of subordinate birds and their subsequent recruitment as helpers at the nest of the disrupter. Older birds will repeatedly interfere with the courtship feeding of a newly formed pair and block the pair from gaining access to its nesting chamber. Both actions increase the probability that the harassed pair will fail to initiate breeding and that the kin-related subordinate bird will help at the nest of the older bird.

The surprise is that the harassing birds are close genetic relatives of the pair they disrupt. Indeed, parents (mostly fathers) are the most frequent harassers; they disrupt the breeding attempts of their own sons. Over half (54 percent) of one-year-old sons whose

parents are breeding fail to breed themselves apparently because they are successfully recruited. This proportion drops as the sons become older and gain in dominance status. By the time sons are three years old, they are practically immune to coercion attempts.

The existence and the resolution of this conflict become understandable when we consider the relatively large net fitness benefit to the breeder and the small net cost to the potential helper when the latter is a son. For one thing, a son is equally related to his own offspring and his parents' offspring (which are his full siblings, provided that no cuckoldry or parasitic egg dumping has occurred). Since an unaided breeder (such as a subordinate son) produces only slightly more young on his own than he does if he contributes as a helper at another's nest, the genetic cost of the tradeoff is minimal to him. Sons apparently do not resist, because the fitness benefits of the two options are nearly equal for them. In contrast, the parents gain considerably more genetic fitness for themselves by using their son to help them increase the production of their own offspring (each of whom shares one-half of a parent's genes by descent) than they would if their son bred and produced grand-offspring (each of whom shares only one-quarter of a grandparent's genes by descent). In

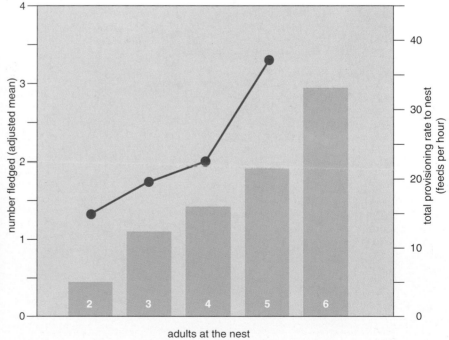

Figure 4. Feeding of juveniles by adult members of a family is crucial to the survival of the younger birds. The number of adults at the nest *(right)* **affects the rate at which juveniles are fed** *(orange line),* **which is closely associated with the number of birds that survive to fledgling status** *(green bars).* **Here adults bring food to nestlings waiting within the tunnels** *(left).* **(Photograph courtesy of Natalie Demong.)**

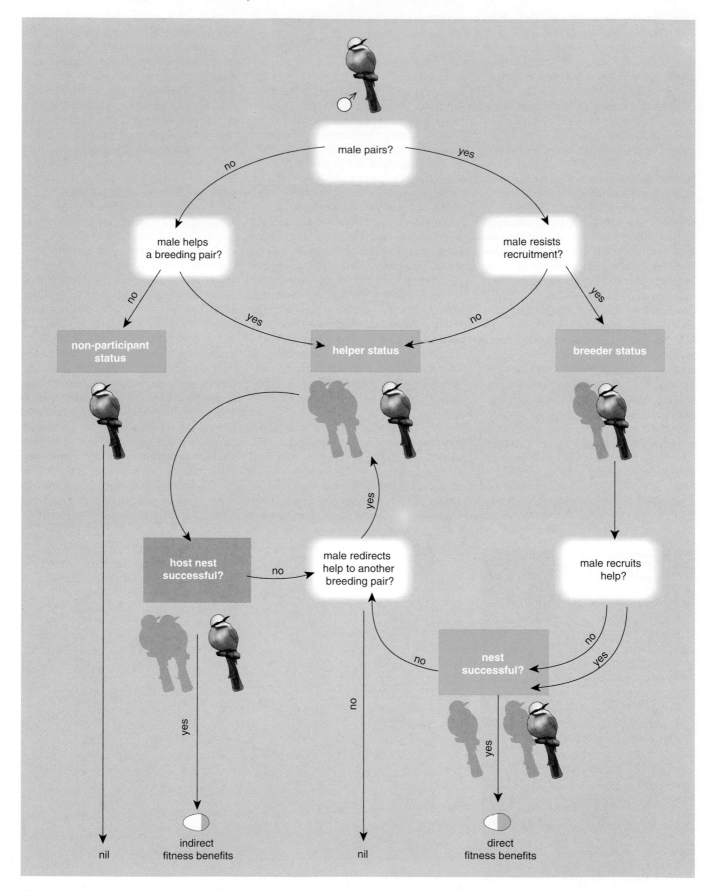

Figure 5. Male bee-eater faces a number of decisions every breeding season that influence his direct and indirect reproductive fitness. A male gets *direct* fitness benefits (corresponding to 0.5 units for each offspring produced, *green half egg*) by acquiring a mate and tending a successful nest. A male gets *indirect* fitness benefits (varying from near zero to 0.5 for each offspring produced, *green quarter egg*) by helping a close relative raise young, rather than breeding on his own. At each decision point the male generally chooses the option that maximizes his inclusive fitness.

this light, the harassment of the son by the parents makes evolutionary sense.

Other members of the family find themselves in a very different situation. Although a breeder will always gain by recruiting a helper, the cost to the helper increases dramatically when he or she is more distantly related to the harasser. Potential helpers who are distantly related to the harasser should, and do, show much greater resistance to recruitment attempts. An older dominant bird can exert leverage over a younger subordinate, but only to a point. It is not surprising then that harassers preferentially select the youngest, most closely related male family members as their targets.

The Female's Options

Since female bee-eaters break the social bonds with their natal families when they pair, their choice of reproductive options differs from those of male bee-eaters. For one thing, they largely forfeit the ability to obtain indirect benefits by helping. Unlike her mate, a female's inclusive fitness (after pairing) depends almost entirely on her success in breeding.

Since a female bee-eater lives with her mate's family, her breeding success is strongly affected by the composition and social dynamics of his family. The likelihood that the new pair will have helpers of its own, or will be able to breed unharassed by others, depends on the male's social and genealogical position within his family. We would expect females to incorporate social components of male quality in their mating choices. Females should pay attention to the prospective mate's social dominance and to the nature of his kin, who may be potential helpers or harassers.

These predictions have been confirmed. Widowed or divorced, older males with offspring of their own were nearly twice as likely to become paired as were young males with older close relatives. The older males were more likely to provide the pairing female with helpers at her initial nesting, whereas younger males were more likely to have their initial nesting disrupted.

Unpaired females who postpone the decision to take a mate retain the option of gaining indirect benefits from helping members of their natal family. Females with close breeding kin should be more likely to remain single. Again, this prediction was borne out: Females with both parents breeding were nearly twice

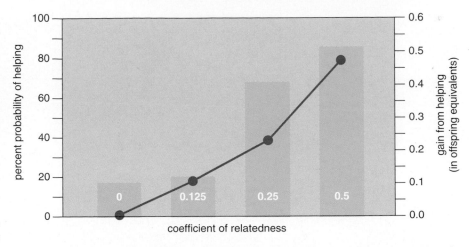

Figure 6. Helper's gain in indirect reproductive benefits *(orange line)* is proportional to the degree of genetic relatedness between the helper and the juvenile being helped. Not surprisingly, the degree of genetic relatedness is a strong predictor of the probability *(green bars)* that one bee-eater will help another.

as likely to remain single as were females with only distantly related breeders in their family.

Females appear to be making a very sophisticated assessment of their options. They act as if they compare the expected benefits of helping versus breeding. We compared the females' actual decisions to those predicted on the basis of the expected benefits given their circumstances (the identity of their breeding natal relatives and the status of their chosen mate within his family). We found that more than 90 percent of the females (67 of 74 cases) behaved as our model predicted. They paired when a potential mate was in a social position

that provided a net increase in their expected inclusive fitness benefits, but they remained in their natal families when their benefits were greater as unpaired helpers. For many females it is better to delay breeding for a season than to accept a mate of poor social standing.

After pairing, a female bee-eater is faced with another series of reproductive choices. If she succeeds in mating with a male in good standing, her problems are solved. But what options remain if her nesting attempts end in failure? Returning to her natal family to help at the nest seems to be an obvious choice, but we have seen this be-

Figure 7. Aggressive interactions take place when dominant bee-eaters attempt to recruit subordinate relatives to help raise the aggressor's offspring, rather than permit the relatives to breed on their own. Most coercive interactions take place between a father and a young son. The existence and the outcomes of such conflicts can be predicted on the basis of the reproductive benefits to the individuals. (Photograph courtesy of Natalie Demong.)

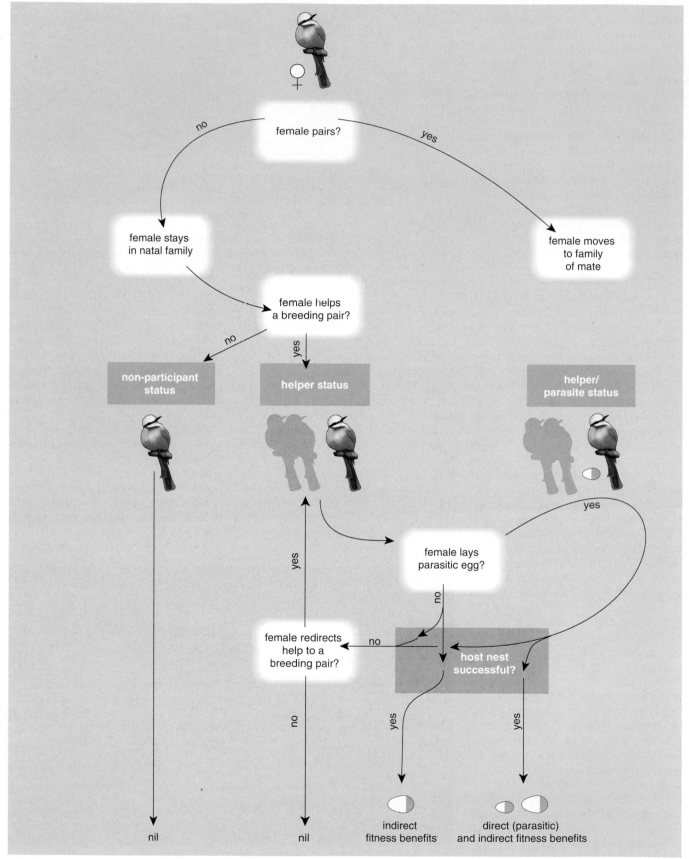

Figure 8. Female bee-eaters that choose not to mate during a breeding season stay in the natal family group. If they choose not to help during a breeding season, their net fitness benefit is nil. If they choose to help at a relative's nest, they receive *indirect* fitness benefits. Occasionally an unpaired female may copulate with a neighboring paired male and then return to the natal nest. If such a female lays a fertile egg in the nest of the relative she is helping she will receive both *direct (small, pink half egg)* and *indirect* fitness benefits (*large, pink quarter egg*).

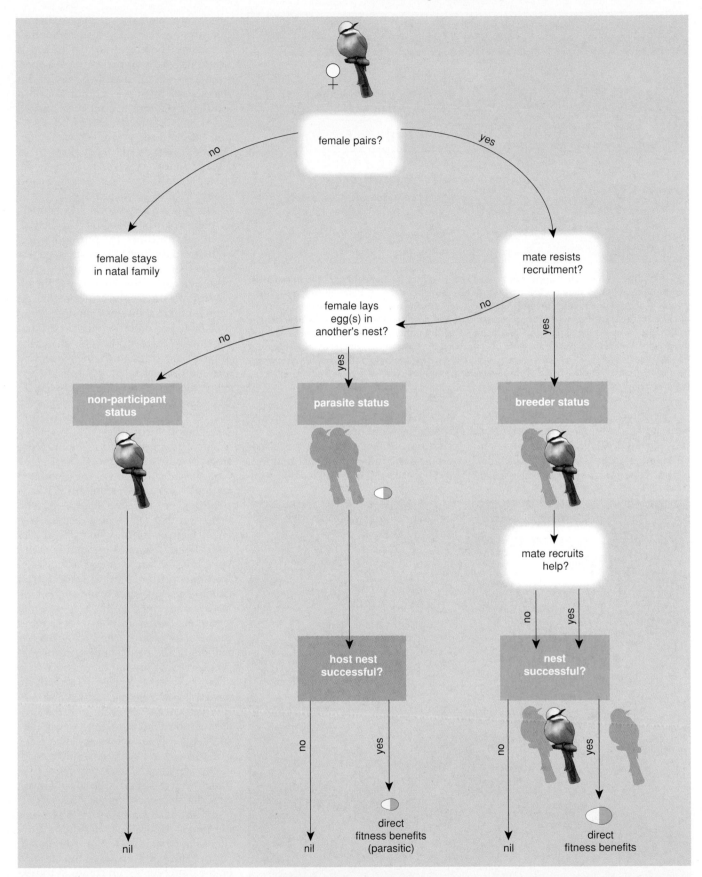

Figure 9. Female bee-eaters that take a mate leave the natal group to live with the male's family group. If the female's mate resists recruitment by his relatives, the pair can establish their own nest and receive *direct* fitness benefits *(large egg)* by having offspring. If the female's mate does not resist recruitment, a female has the option of parasitically laying her egg(s) in another's nest and still gaining direct fitness benefits *(small egg)*. Parasitism is a relatively frequent tactic, with about 16 percent of nests containing a parasitic egg. If the female's mate is recruited, the female usually remains in her mate's family group but does not help raise the offspring of her mate's relatives.

factors in a female's decision whether to mate or stay in her natal group	
concerns within natal family	• kinship relationships to breeders • natal group size
concerns within mate's family	• mate's kinship relationships • mate's family group size • mate's age • mate's relative dominance poisition

Figure 10. Female bee-eaters appear to make a very sophisticated assessment of their potential fitness benefits when they decide whether to take a mate or stay at home during a breeding season. A female must weigh the indirect fitness benefits she might gain by helping at the nest of close relatives in her family group compared to the potential for direct fitness benefits she could acquire by taking a mate and producing offspring of her own. She must assess whether the potential mate can recruit help from his family on the basis of his genetic relationships and his dominance status, or whether he is likely to be recruited to help at the nest of other members of his family. On the basis of these factors the authors were able to correctly predict the females decision in 67 of 74 cases.

havior only a handful of times in eight years of studying these birds. We can only speculate that returning home entails some hidden costs. One possibility is that a prolonged separation from one's mate increases the risk of dissolving the pair bond.

It turns out that a female who fails at nesting has another option. If nest failure takes place while she is still at the

Figure 11. Breeding female actively removes eggs laid by parasitic females before she initiates her own clutch. A successful parasite must overcome the host's defenses *and* lay her eggs within the two- or three-day period that the host lays her own eggs. (Photograph courtesy of Marie Read.)

egg-laying stage, she can deposit her remaining egg(s) in the nest of another bee-eater. The large number of active nests in a colony provides ample opportunity for such parasitic behavior. Indeed, parasitism was common in our study populations: About 16 percent of bee-eater nests were parasitized, and 7 percent of all eggs were laid by foreign females. Despite its frequency, parasitism is a low-yield tactic: Parasitic females usually lay only one egg, and many of these do not survive.

The reproductive costs and benefits of parasitism have resulted in behavioral adaptations by parasites and counter-adaptations by hosts. Breeders and helpers take turns guarding their nests against all trespassers, and breeding females actively remove foreign eggs found in their nest chambers until they have laid their own first egg. Parasites must locate a potential host at the appropriate stage in the nesting cycle and gain access to the chamber when there is a lapse in the host's defenses. Eggs laid too early will be removed, whereas eggs laid too late will fail to hatch before incubation ends.

There is an interesting twist to the story of parasitism among bee-eaters: Not all parasitic females are unrelated to their hosts. About one-third of the parasites are unpaired daughters who were assisting at the nest of their parents. The parasitic daughter actively defends the nest against nonfamily members, but she slips an egg of her own into her mother's (or stepmother's) clutch. In one instance, a daughter removed one of her mother's eggs before laying her own in its place. These intrafamilial parasites remain active as helpers at the nest, sharing in incubation and providing food.

How are these females fertilized? We have watched a few such daughters closely: They actively trespassed onto the territory of a neighboring family, where they solicited a copulation with a paired male! Thus their eggs are not the result of an incestuous mating. Rather, parasitism by a daughter appears to be a tactic involving a specific series of complex behaviors.

Intrafamilial parasitism offers a single female the option of achieving *direct* fitness benefits in addition to the *indirect* benefits gained by helping. However, the daughter's gain comes at the expense of the parent. It is not clear whether the parents tolerate their daughter's egg dumping to retain her as a helper or whether the daughter is surreptitiously taking advantage of her parents. In either case, the existence of this form of parasitism underscores the flexibility of the bee-eaters' reproductive options and the subtle conflicts that take place in this species.

Conclusion

The tactics that individuals use in their interactions with one another have only recently become the subject of evolutionary analysis. This is because the expression of social tactics is very plastic: Most organisms can adopt a variety of roles according to the situation and the identity of the other participants. Early workers found it difficult to reconcile this plasticity with the view that specific genes literally determine specific behaviors. It is now recognized that natural selection can operate on the *decision-making process* itself.

As long as there is heritable variation in the decision rules that the birds use, natural selection will favor variants that result in the expression of situation-dependent behaviors that maximize the inclusive fitness of the actor. One of the pioneers of this approach, Robin Dunbar of the University College of London speaks heuristically of organisms as "fitness maximizers." They make decisions based on their ability to assess the costs and benefits of the options available to them.

We have observed that bee-eaters behave as if they assess the relative costs and benefits of pursuing different options in very complex social situations. Gender, dominance and kinship all influence the fitness tradeoffs of the various tactical alternatives available to bee-eaters. Knowing these variables allows us to predict with considerable accuracy whether an individual will attempt to breed, whether it will help at a nest and whom it will help. We can also ascertain whether a bird will be harassed and whether harassment will be successful. Differences in the behavior of genetic and nongenetic members in an extended family group would remain mysterious if it were not for the explanatory power of inclusive-fitness theory.

Gender, dominance and kinship should be important predictors of family dynamics in any species that exhibits long-term pair bonding, sex-biased dispersal and interactions where one family member can influence the reproductive success of another. Cases of breeding harassment and even reproductive suppression are common features of many species that live in family-based societies. Analyzing the fitness consequences of such behavior from the perspectives of the various participants provides an evolutionary framework for understanding such social dynamics.

Can we learn anything about the dynamics in a human family from the behavior of the bee-eaters? More than any other species, the behavior of human beings is shaped by culture. The rewards and punishments that accompany human social actions are largely determined by society. The currency human beings use in assessing the costs and benefits of a particular tactic is no longer solely based on reproductive fitness. But this does not mean that we do not possess a set of behavioral predispositions based on flexible decision rules that were adaptive in our evolutionary past. Such tendencies would have been molded during our long history of living in extended family groups. It is these underpinnings that surface more clearly in animal studies of family-dwelling species.

A small but growing number of psychologists and anthropologists are incorporating an evolutionary perspective into their studies of human families. Investigation of the roles of nonparental family members in childrearing has focused on the role of siblings (especially the mother's brother) and grandparents as human analogues of helpers at the nest. Martin Daly and Margo Wilson of McMaster University have studied the effects of relatedness (parent versus stepparent) on child abuse. Robert Trivers, now of Rutgers University, has looked at the theoretical basis for parent-offspring conflict. Trivers and his colleague Dan Willard have proposed an evolutionary hypothesis to explain why some parents invest unequally in their sons and daughters.

We believe that an evolutionary framework has great potential for increasing our understanding of the social dynamics of family-based societies. By focusing on the fitness consequences of different actions to different individuals, it provides a functional explanation for why particular behavioral predispositions may have evolved. It also provides a theoretical basis for predicting the social roles that different individuals will adopt under differing circumstances. We fully expect that the same general variables found to be important predictors of bee-eater behavior—gender, dominance and kinship—will be important predictors of cooperation, conflict and the resolution of conflict, in most other social species, including human beings. We expect that the incorporation of this Darwinian approach into the social sciences will provide a valuable additional perspective to our understanding of human family interactions.

Bibliography

Betzig, L. L., M. Borgerhoff Mulder and P. Turke. 1988. *Human Reproductive Behaviour: A Darwinian Perspective*. Cambridge: Cambridge University Press.

Daly, M., and M. Wilson. 1985. Child abuse and other risks of not living with both parents. *Ethology and Sociobiology* 6:197–210.

Daly, M., and M. Wilson. 1987. Evolutionary psychology and family violence. In *Sociobiology and Psychology*, ed. C. Crawford, M. Smith and D. Krebs. Hillsdale, New Jersey: Lawrence Erlbaum Associates.

Dunbar, R. 1989. *Reproductive Decisions: An Economic Analysis of Gelada Baboon Social Strategies*. Princeton: Princeton University Press.

Emlen, S. T. 1991. The evolution of cooperative breeding in birds and mammals. In *Behavioural Ecology: An Evolutionary Approach*, ed. J. Krebs and N. Davies, pp. 301–337. Blackwell Scientific Publishers.

Emlen, S. T. 1994. Benefits, constraints and the evolution of the family. *Trends in Ecology and Evolution* 9:282–285.

Emlen, S. T., and P. H. Wrege. 1986. Forced copulations and intra-specific parasitism: Two costs of social living in the white-fronted bee-eater. *Ethology* 71:2–29.

Emlen, S. T., and P. H. Wrege. 1988. The role of kinship in helping decisions among white-fronted bee-eaters. *Behavioral Ecology and Sociobiology* 23:305–315.

Emlen, S. T., and P. H. Wrege. 1989. A test of alternate hypotheses for helping behavior in white-fronted bee-eaters. *Behavioral Ecology and Sociobiology* 25:303–319.

Emlen, S. T., and P. H. Wrege. 1991. Breeding biology of white-fronted bee-eaters at Nakuru: The influence of helpers on breeding success. *Journal of Animal Ecology* 60:309–326.

Emlen, S. T., and P. H. Wrege. 1992. Parent-offspring conflict and the recruitment of helpers among bee-eaters. *Nature* 356:331–333.

Emlen, S. T., and P. H. Wrege. 1994. Gender, status and family fortunes in the white-fronted bee-eater. *Nature* 367:129–132.

Hamilton, W. D. 1964. The genetical evolution of social behaviour. *Journal of Theoretical Biology* 7:1–52.

Hegner, R. E., S. T. Emlen and N. J. Demong. 1982. Spatial organization of the white-fronted bee-eater. *Nature* 296:702–703.

Smith, E. A., and B. Winterhalder. 1991. *Ecology, Evolution and Human Behavior*. New York: Aldine de Gruyter.

Trivers, R. L. 1974. Parent-offspring conflict. *American Zoologist* 14:249–264.

Trivers, R. L., and D. E. Willard. 1973. Natural selection of parental ability to vary the sex ratio of children. *Science* 179:90–92.

Wrege, P. H., and S. T. Emlen (1994). Family structure influences mate choice in white-fronted bee-eaters. *Behavioral Ecology and Sociobiology* 35:185–191.

Naked Mole-Rats

Like bees and termites, they cooperate in defense, food gathering and even breeding. How could altruistic behavior evolve in a mammalian species?

Rodney L. Honeycutt

Biological evolution is generally seen as a competition, a contest among individuals struggling to survive and reproduce. At first glance, it appears that natural selection strongly favors those who act in self-interest. But in human society, and among other animal species, there are many kinds of behavior that do not fit the competitive model. Individuals often cooperate, forming associations for their mutual benefit and protection; sometimes they even appear to sacrifice their own opportunities to survive and reproduce for the good of others. In fact, apparent acts of altruism are common in many animal species.

It is easy to admire altruism, charity and philanthropy, but it is hard to understand how self-sacrificing behavior could evolve. The evolutionary process is based on differences in individual fitness—that is, in reproductive success. If each organism strives to increase its own fitness, how could natural selection ever favor selfless devotion to the welfare of others? This question has perplexed evolutionary biologists ever since Charles Darwin put forth the concepts of natural selection and individual fitness. An altruistic act—one that benefits the recipient at the expense of

Rodney L. Honeycutt is an associate professor in the Department of Wildlife and Fisheries Sciences and a member of the Faculty of Genetics at Texas A&M University. He began his research on the genetics and systematics of African mole-rats in 1983 during his tenure as assistant professor of biology at Harvard University and assistant curator of mammals at Harvard's Museum of Comparative Zoology. He is interested in the evolution and systematics of mammals and has taught courses in mammalian biology for the past seven years. His research has taken him to regions of Africa, South America, Central America and Australia. Address: Department of Wildlife and Fisheries Sciences, Texas A&M University, 210 Nagle Hall, College Station, TX 77843.

the individual performing the act—represents one of the central paradoxes of the theory of evolution.

In seeking to explain this paradox, biologists have focused their attention on the social insects—ants, bees, wasps and termites. These species exhibit an extreme form of what has been called reproductive altruism, whereby individuals forgo reproduction entirely and actually help other individuals reproduce, forming entire castes of sterile workers. Since reproductive success is the ultimate goal of each player in the game of natural selection, reproductive altruism is a remarkable type of self-sacrifice.

Helping behavior is common in vertebrate societies as well, and some species cooperate in breeding. But until recently there did not appear to be a close vertebrate analogue to the extreme form of altruism observed in social insects. Such a society may now have been found in the arid Horn of Africa, where biologists have been studying underground colonies of a singularly unattractive but highly social rodent.

The naked mole-rat, *Heterocephalus glaber*, appears to be a eusocial, or truly social, mammal. It fits the classical definition of eusociality developed by Charles Michener (1969) and E. O. Wilson (1971), who extensively studied the social insects. In the burrow colonies of naked mole-rats there are overlapping adult generations, and as in insect societies brood care and other duties are performed cooperatively by workers or helpers that are more or less nonreproductive. A naked mole-rat colony is ruled, as is a beehive, by a queen who breeds with a few select males. Furthermore, the other tasks necessary to underground life—food gathering, transporting of nest material, tunnel expansion and cleaning and defense against predators—appear to

be divided among nonreproductive individuals based on size, much as labor in insect societies is performed by the sterile worker castes.

The naked mole-rat is not the only vertebrate that can be described as eusocial, but no other vertebrate society mimics the behavior of the eusocial insects so closely. The fact that highly social behavior could evolve in a rodent population suggests that it is time to reexamine some old theories about how eusocial behavior could come into being—theories that were based on the characteristics of certain insects and their societies. In the past decade, since Jennifer U. M. Jarvis first revealed the unusual social structure of a naked mole-rat colony, a number of biologists have been at work considering how a eusocial rodent could evolve. I shall discuss the state of that work briefly here, examining what is known about the naked mole-rat's ecology, behavior and evolution and about altruistic animal societies.

Introducing the Naked Mole-Rat

The naked mole-rat is a member of the family Bathyergidae, the African mole-rats—so named because they resemble rats but live like moles. Many rodents burrow and spend at least part of their life underground; all 12 species of Bathyergidae live exclusively underground, and they share a set of features that reflect their subterranean lifestyle and that demonstrate evolutionary convergence, the independent development of similar characteristics. Like the more familiar garden mole, a mole-rat has a stout, cylindrical body, a robust skull, eyes that are small or absent, reduced external ears, short limbs, powerful incisors and sometimes claws for digging, and a somewhat unusual physiology adapted to the difficulties of life underground, including a burrow

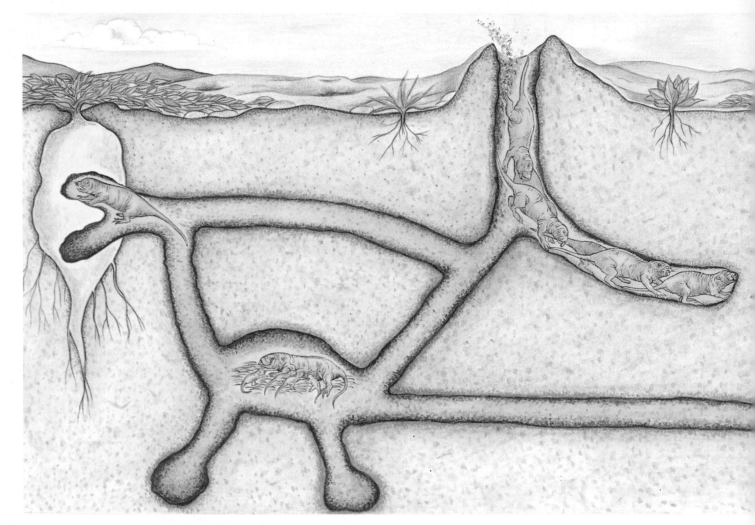

Figure 1. Burrow system built by naked mole-rats beneath the East African desert illustrates the complex social organization that makes the subterranean species unusual. Reproduction in a naked mole-rat colony, which usually has 70 to 80 members, is controlled by a queen, the only breeding female, shown here nursing newborns in a nest chamber. Digging tunnels to forage for food is one of the functions of

atmosphere high in carbon dioxide. All Bathyergidae species are herbivorous, and all but one sport fur coats.

Field biologists who encountered naked mole-rats in the 19th century thought that these small rodents—only three to six inches long at maturity, with weights averaging 20 to 30 grams— were the young of a haired adult. But subsequent expeditions showed that adult members of the species are hairless except for a sparse covering of tactile hairs. Oldfield Thomas, noting wide variations in the morphological characteristics of the naked mole-rats, identified what he thought were several species. *H. glaber* is currently considered a single species, within which there is great variation in adult body size.

Naked mole-rats inhabit the hot, dry regions of Ethiopia, Somalia and Kenya. Like most of the Bathyergidae species, they build elaborate tunnel systems. The tunnels form a sealed, compartmentalized system interconnecting nest sites, toilets, food stores, retreat routes and an elaborate tunnel system allowing underground foraging for tubers *(Figure 1)*. Like the morphology of the animals, the tunnel system is an example of convergent evolution, being similar to those of the other mole-rats in its compartmentalization, atmosphere and more or less constant temperature and humidity. Naked mole-rats subsist primarily on geophytic plants (perennials that overwinter in the form of bulbs or tubers), which are randomly and patchily distributed. The mole-rats forage broadly by expanding their burrows, but their distribution is limited by food supply and soil types. Like most rodents that live underground, they are not able to disperse over long distances.

The tunnel systems of naked mole-rats can be quite large, containing as many as two miles of burrows. The average colony is thought to have 70 to 80 members. In order to study the social organization of the naked mole-rats, biologists have had to devise ways to capture whole colonies and recreate their burrow systems in the laboratory. This is not an easy task, but it is possible because the rodents have a habit of investigating opened sections of their burrow systems and then blocking them. One can create an opening, then capture the naked mole-rats as they come to seal it. Cutting off their retreat requires quick work with a spade, hoe or knife, and the procedure must be repeated in various parts of the tunnel system in order to retrieve an entire colony. A carefully reconstructed colony can survive quite well in captivity, and naked mole-rats are beginning to become an attraction at zoos.

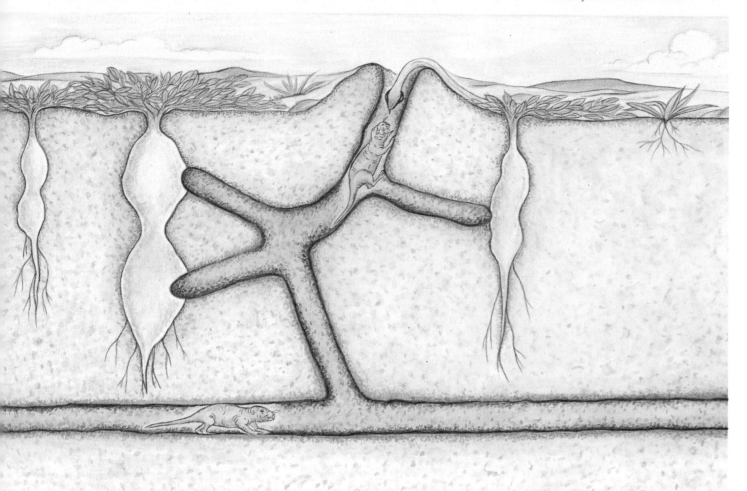

nonreproductive workers, which often form digging teams; one individual digs with its incisors while others kick the dirt backward to a mole-rat that kicks it out of the tunnel. The molehills or "volcanoes" formed in this way are plugged to create a closed environment and deter predators such as the rufous-beaked snake. Tubers and bulbs are the naked mole-rats' food source.

Most African mole-rats excavate by digging with their large incisors, removing the dirt from the burrow with their hind feet. The digging behavior of naked mole-rats, which are most active during periods when the soil in their arid habitat is moist, appears to be unlike that of the other mole-rats in two respects. First, instead of plugging the surface opening to a tunnel during excavation, the naked mole-rats "volcano," kicking soil through an open hole to form a tiny volcano-shaped mound. When excavation is complete, the tunnel is plugged to form a relatively airtight, watertight and predator-proof seal *(Figure 4)*. Second, naked mole-rats have been observed digging cooperatively in a wonderfully efficient arrangement that resembles a bucket brigade. One animal digs while a chain of animals behind move the dirt backward to an

Figure 2. Wrinkled, squinty-eyed and nearly hairless, the first naked mole-rats found by biologists were thought to be the young of a haired adult. The rodents are just three to six inches long at maturity, although there is great variation in body size within each colony. Other morphological features reflect the fact that the naked mole-rats live entirely underground: small eyes, two pairs of large incisors for digging, and reduced external ears. (Except where noted, photographs courtesy of the author.)

Figure 3. Habitat of the naked mole-rats is hot, dry and dotted with patches of vegetation. Visible in the foreground of this photograph, taken in Kenya, are the molehills formed by the rodents.

Figure 4. "Volcanoes" formed when naked mole-rats kick sand out of a tunnel, then plug the opening, make the animals' burrows easy to find. Naked mole-rats are most vulnerable to predators while forming volcanoes; the activity often attracts the attention of snakes.

animal at the end, which kicks the dirt from the burrow. One 87-member colony was seen to remove about 500 kilograms of soil per month by this process. Another colony of similar size moved an estimated 13.5 kilograms in an hour—about 380 times the mean body weight of a naked mole-rat. A team kicking dirt through a surface opening is vulnerable to attack from snakes; the mounds also make *H. glaber*'s colonies easy for scientists to find.

Naked mole-rats are long-lived animals and prolific breeders. Several individuals caught in the wild are surviving after 16 years in captivity; two of these are females that still breed. In captive colonies females have produced litters as large as 27, and in wild populations litter sizes can be as high as 12. The naked mole-rat breeds year-round, giving birth about every 70 to 80 days. This fecundity is unusual among the Bathyergidae. The other highly social species of African mole-rat, *Cryptomys damarensis*, is also a year-round breeder but produces smaller litters, with an average size of five.

The major threat to the longevity of a naked mole-rat, and probably to all of the mole-rats, is predation. On at least two occasions I have encountered the rufous-beaked snake in a mole-rat burrow; one snake had three mole-rats in its stomach. Similar field observations have been made by other investigators. Encounters between mole-rats and snakes in the laboratory suggest that avoidance may not be the mole-rat's only strategy against predators; individuals have also been seen attacking the predator in their defense of the colony.

The naked mole-rat's closest relatives are the 11 other species in the Bathyergidae, which are all of exclusively African origin and distribution (*Figure 5*). It has been difficult to determine which of the 32 other rodent families shares a common ancestry with the Bathyergidae, but a consensus arising from recent studies places the family in the rodent suborder Hystricognathi, which includes caviomorph rodents from the New World—porcupines, guinea pigs and chinchillas—and porcupines and cane rats from the Old World. The naked mole-rat is the most divergent species within the Bathyergidae, its evolutionary branch splitting off at the base of the family's phylogenetic tree (*Figure 6*).

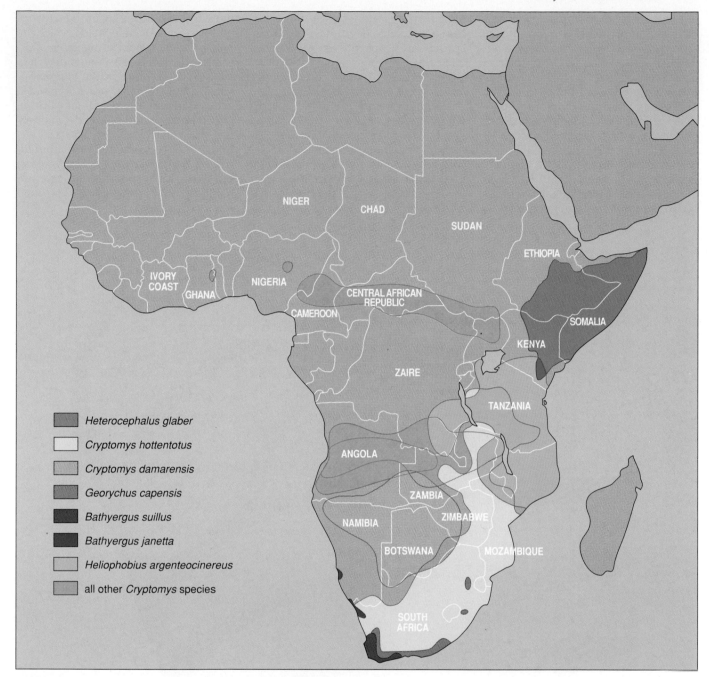

Figure 5. Geographic range of the naked mole-rat, *Heterocephalus glaber*, is limited to the hot, dry region called the Horn of Africa—parts of Ethiopia, Kenya and Somalia. On the map are shown the areas inhabited by other species of African mole-rats. All species in the family Bathyergidae live entirely underground. Most are solitary or colonial; the other species with a highly developed social structure, *Cryptomys damarensis*, is found in Southern Africa.

How Do Altruistic Societies Evolve?

Darwin called the development of sterile castes in insect societies a "special difficulty" that initially threatened to be fatal to his theory of natural selection. His solution to the problem was surprisingly close to current hypotheses based on genetic relatedness, even though he did not have a knowledge of genetics. Darwin suggested that traits, such as helping, that were observed in sterile form could survive if individuals that expressed the traits contributed to the reproductive success of those individuals that had the trait but did not express it.

Today the notion of *inclusive fitness* forms the foundation for theories about how reproductive altruism might evolve. The idea arose in 1964 from William Hamilton's remarkable genetic studies of the Hymenoptera, the insect order that includes the social ants, bees and wasps. Hamilton showed that if the genetic ties within a generation are closer than the ties between generations, each member of the generation might be motivated to invest in a parent's reproductive success rather than his or her own. Inclusive fitness is a combination of one's own reproductive success and that of close relatives.

In the Hymenoptera, Hamilton found an asymmetric genetic system that could contribute to the development of reproductive altruism by giving individuals chances to maximize their inclusive fitness without reproducing. Hymenopteran males arise from unfertilized eggs and thus have only one set of chromosomes (from the mother); females have one set from each parent. The males are called haploid, the females diploid, and this system of sex determination is referred to as *haplodiploidy (Figure 9)*. The daughters of a monogamous mother share identical genes from their father and half their mother's genes; they thus have three-quarters of their genes in common. A female who is more closely related to her sister than to her mother or her offspring can propagate her own genes most effectively by helping create more sisters. Sterile workers in hymenopteran insect colonies are all female.

Hamilton's work prompted a flurry of interest in genetic asymmetry, but he and others recognized that it was not a general explanation for how eusocial societies might evolve. There are many limitations; for instance, multiple matings by females reduce the closeness of relationships between sisters, and it is hard to explain the incentives for females to tend juvenile males, which are not as closely related as are sisters. Furthermore, although eusociality has evolved more times in the Hymenoptera than in any other order, it has also evolved in parts of the animal world in which both sexes are diploid— namely Isoptera, which includes the social termites, and Rodentia, the order that includes the naked mole-rat. Finally, there are many arthropod species that are haplodiploid and have not developed highly social behavior.

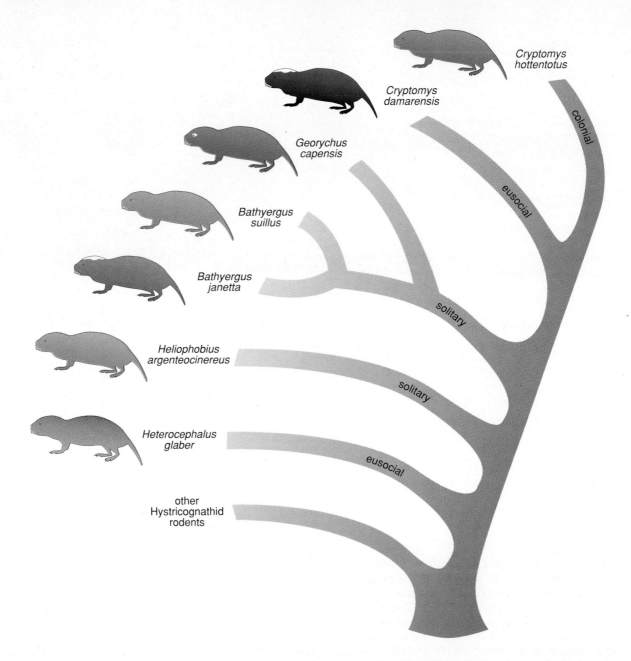

Figure 6. Phylogenetic tree for the family Bathyergidae, the African mole-rats, shows that the two eusocial, or truly social, species are quite divergent. Among other rodents, the suborder Hystricognathi, which includes porcupines, guinea pigs and chinchillas, appears to have the closest genetic link with the African mole-rats. Although there is much similarity among the Bathyergidae species in their physiological characteristics and their subterranean lifestyle, the phylogenetic distance between the eusocial species of mole-rats suggests that complex social behavior evolved separately in the two cases.

There is another way that close kinship might develop among the members of a generation, and it is considered a possible explanation for the evolution of the termite and naked mole-rat societies. Several generations of inbreeding could result in a higher degree of relatedness among siblings than between parents and offspring *(Figure 10)*. When male and female mates are unrelated, but each is the product of intense inbreeding, their offspring can be genetically identical and might be expected to stay and assist their parents for the same reasons set forth in the haplodiploid model. The inbreeding model was developed by Stephen Bartz in 1979 to explain the development of eusocial behavior in termites, which live in a contained and protected nest site conducive to multigenerational breeding.

Genetics alone cannot provide a comprehensive explanation for the evolution of eusociality. Other possible explanations, especially relevant to termites and vertebrate helpers, lie in combinations of ecological and behavioral factors. These factors perhaps provided preconditions or starting points for the eventual evolution of a eusocial lineage or species. The best way to understand the development of eusociality may be to consider the costs and benefits associated with remaining in the natal group and helping, as compared to the costs and benefits of dispersing and breeding.

Probably one of the most important preconditions for the development of eusociality is parental care in a protected nest, where offspring are defended against predators and provided with food. If there is a high cost associated with dispersal—in terms of restricted access to food, lack of breeding success or increased vulnerability to predators—then there may be an incentive for juveniles to remain in the protected nest and become helpers. Helpers that remain in the nest for multiple generations may forgo reproduction indefinitely as a consequence of maternal manipulation.

The short-term benefits of group living seem to accrue mainly to those individuals who are reproducing, since they benefit from the help others provide with defense and obtaining food. In fact, there is a correlation between the breeder's reproductive fitness and the number of helpers in cooperatively breeding vertebrate species. Thus the long-term effect of helping may be an

Figure 7. Catching naked mole-rats requires some understanding of their behavior. Mole-rat catchers create an opening from the surface to a burrow, which is normally kept sealed by the animals, and wait quietly for a mole-rat to investigate. A spade, hoe, pick or knife blade is driven quickly into the tunnel to block the mole-rat's escape. (Photograph courtesy of Stan Braude, University of Missouri at St. Louis.)

Figure 8. Captive naked mole-rats, carrying identifying tattoos, adapt well to being placed together in bins, apparently because the highly social animals tend to huddle together for warmth in their burrows in the wild. These rodents are part of Jennifer U. M. Jarvis's collection at the University of Cape Town.

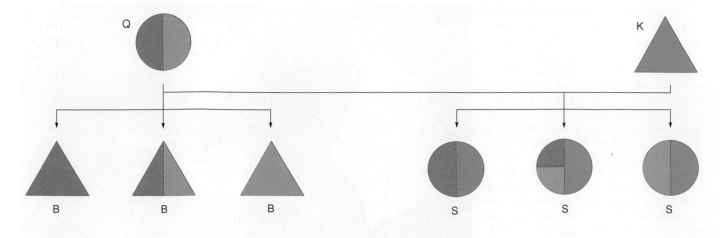

degrees of relatedness in haplodiploid species

	daughter	son	mother	father	sister	brother
female	$\frac{1}{2}$	$\frac{1}{2}$	$\frac{1}{2}$	$\frac{1}{2}$	$\frac{3}{4}$	$\frac{1}{4}$
male	1	0	1	0	$\frac{1}{2}$	$\frac{1}{2}$

Figure 9. Haplodiploidy, an asymmetric genetic system, is thought to contribute to the development of reproductive altruism in ants, bees and wasps—species with intricate social systems that include sterile castes of workers. In a haplodiploid species, males *(triangles)* arise from unfertilized eggs and have only one set of chromosomes, whereas females *(circles)* have one set of chromosomes from each parent. The relatedness between sisters—the fraction of their genes that are shared—is thus greater than the relatedness of mother and daughter *(bottom panel)*. William D. Hamilton hypothesized that females seeking to increase their inclusive fitness—a combination of their own reproductive success and that of close relatives—might in a haplodiploid species become helpers, advancing the continuation of their own genetic heritage by helping with the reproduction of sisters rather than their own offspring. Although haplodiploidy is not considered a full explanation of how eusocial behavior would evolve in ants, bees and wasps, it is notable that most species in which reproductive altruism has evolved are haplodiploid, and that the sterile workers among the haplodiploid insects are all female. In this illustration, the parents are labeled *Q* and *K* and the offspring *S* and *B*, following the scheme in Figure 10; for simplicity, the effects of any recombination of genes are not depicted.

increase in inclusive fitness for the helpers. This may prove to be a very important consideration in species where the probability of a dispersing individual procuring a nest site and eventually breeding is extremely low.

Naked Mole-Rat Society
In some ways the social organization observed in naked mole-rat colonies is more akin to the societies of the social insects than to the social organization of any other vertebrate species. In other respects, mole-rats are unique and may always remain a bit of a mystery.

Some similarities between naked mole-rat societies and the insect societies are striking. A naked mole-rat colony, like a beehive, wasp's nest or termite mound, is ruled by its queen or reproducing female. Other adult female mole-rats neither ovulate nor breed. The queen is the largest member of the colony, and she maintains her breeding status through a mixture of

behavioral and, presumably, chemical control. She is aggressive and domineering; queenly behavior in a naked mole-rat includes facing a subordinate and shoving it along a burrow for a distance. Queens have been long-lived in captivity, and when they die or are removed from a colony one sees violent fighting among the larger remaining females, leading to a takeover by a new queen.

Most adult males produce sperm, but only one to three of the larger males in a colony breed with the queen, who initiates courtship. There is little aggression between breeding males, even upon removal of the queen. The queen and breeding males do not participate in the defense or maintenance of the colony; instead, they concern themselves with the handling, grooming and care of newborns.

Eusocial insect societies have a rigid caste system, defined on the basis of distinctions in behavior, morphology

and physiology. Mole-rat societies, on the other hand, demonstrate behavioral asymmetries related primarily to reproductive status (reproduction being limited to the queen and a few males), body size and perhaps age. Smaller nonbreeding members, both male and female, seem to participate more in gathering food, transporting nest material and clearing tunnels. Larger nonbreeders are more active in defending the colony and perhaps in removing dirt from the tunnels. Jarvis has suggested that differences in growth rates may influence the length of time that an individual performs a task, regardless of its age.

Naked mole-rats, being diploid in both sexes, do not have an asymmetric genetic system such as haplodiploidy. As Bartz has proposed for termites, inbreeding in naked mole-rats may create a genetic asymmetry that mimics the result of haplodiploidy. There is genetic evidence suggesting that naked mole-rats are highly inbred within colonies and even between colonies in a local area. An important part of the question about breeding within and between colonial groups cannot be answered, however, since there is very little information on how mole-rat colonies are established. This makes it difficult to evaluate the naked mole-rats using Bartz's model of inbreeding and eusociality in termites.

Still, among the eusocial insects termites offer the closest comparison with the naked mole-rats. Termites are the only eusocial insects outside the Hymenoptera, and all termites are diploid, with two sets of chromosomes. Worker groups include nonreproductive males and females, and they perform primarily tasks associated with maintaining and defending the colony. The queen termite is more passive than a naked mole-rat queen and uses chemical control. Termite colonies are much larger, sometimes having more than 10,000 workers, and the definition of castes is more rigid.

The naked mole-rat cannot be considered the only eusocial vertebrate species, but it does represent the most advanced form of vertebrate eusociality and the one most analogous to eusociality in insects. Helping or cooperative breeding has evolved many times in vertebrates, and in many of those species the social system includes both a small number of reproducing individuals (usually a dominant breeding pair) and several nonbreeding individuals (males and females), representing offspring from previous years, that serve as helpers or alloparents. As in naked mole-rats, these nonbreeders participate in foraging for food, care of young and defense against predators. Unlike naked mole-rats, most cooperatively breeding vertebrates (an exception being the wild dog, *Lycaon pictus*) are dominated by a pair of breeders rather than by a single breeding female. The division of labor within a social group is not as pronounced in other vertebrates, and the colony size is much smaller. In addition, mating by subordinate females in many social vertebrates is not totally suppressed, whereas in naked mole-rat colonies subordinates are not sexually active, and many may never breed.

Several ecological and behavior factors may have facilitated the evolution of eusociality in naked mole-rats. Richard Alexander, Katharine Noonan and Bernard Crespi (1991) have sug-

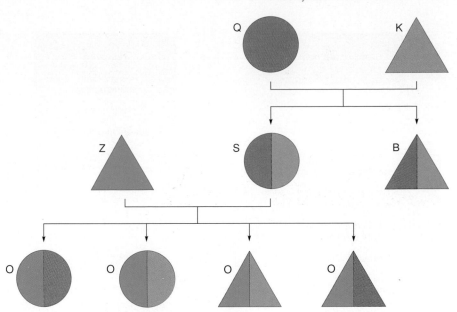

Figure 10. Genetic asymmetry can be produced by cycles of inbreeding and outbreeding in a way that may encourage the evolution of reproductive altruism. Stephen Bartz developed a genetic model to explain how complex social behavior could have evolved in termites living within the confines of a bark-covered chunk of rotting wood. Bartz's hypothesis begins with the mating of a male and a female who are unrelated but are each the product of intense inbreeding (the "queen" and "king," or *Q* and *K, above*), so that for each, both halves of the genotype are essentially identical. The products of this union *(S* and *B)* are essentially identical and therefore more related to one another than to their parents; this genetic asymmetry is thought to encourage helping behavior in both sexes because each sibling can increase its inclusive fitness by assisting in the creation of brothers and sisters. If one of the offspring mates with a similarly inbred but unrelated individual, as in the case of *S* and *Z*, the new parents and the new offspring *(O)* are less closely related than than are the original siblings, *S* and *B*. The result mimics the close ties between siblings that are produced by haplodiploidy *(Figure 9)*, but the genetic asymmetry disappears in subsequent generations unless specific patterns of inbreeding and outbreeding are followed. (Adapted from Bartz 1979.)

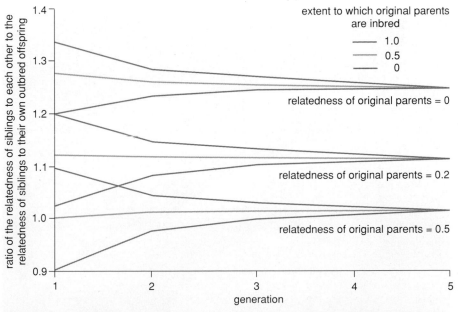

Figure 11. Brother-sister incest might perpetuate the genetic asymmetry shown in Figure 9 over several generations. On this graph, a ratio of relatedness greater than 1.0 means that siblings are more related to one another than to their offspring and are therefore encouraged to become helpers rather than breeders. It is evident that helping behavior is most encouraged when the original parents are highly inbred but unrelated; brother-sister mating makes the inbreeding of the parents unimportant after a few generations, but the importance of the relatedness of the original parents persists. A similar pattern might have contributed to the evolution of social behavior in the confined quarters in which naked mole-rats breed. (From Bartz 1979.)

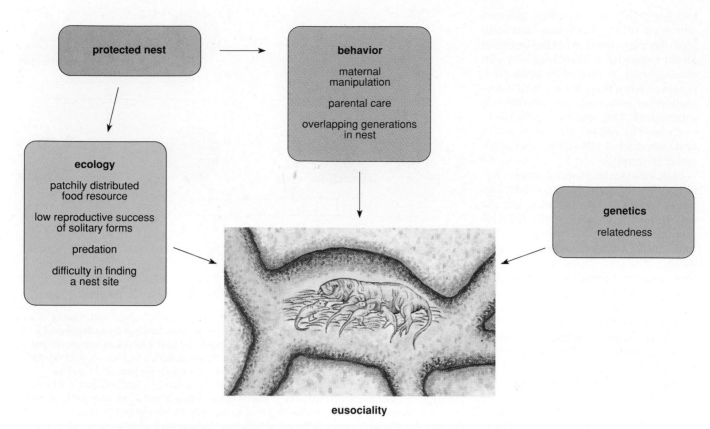

eusociality

Figure 12. Genetic, ecological and behavioral factors probably combine to promote the development of eusociality in an animal species. All of the factors shown above may have been important in the evolution of the complex social organization of naked mole-rat colonies. Inbreeding in the rodents' underground burrows may have created a high degree of genetic inbreeding within generations, promoting helping behavior. In the ecosystem inhabited by the naked mole-rats, the costs associated with leaving the nest may be high compared to the benefits of group living. Behavioral patterns such as parental care may also have predisposed the species to large-scale group living. A closed burrow system that provides a protected nest environment might be crucial in tipping the ecological and behavioral balance toward organized group living and cooperative breeding.

Figure 13. Ruler of a famous group of naked mole-rats, the queen of Jennifer U. M. Jarvis's captive colony at the University of Cape Town in South Africa is distinguished from her subjects by her large size. She is still breeding after 16 years of captivity. Jarvis's colony served as the basis for the first description of eusocial behavior in a mammal.

gested that the subterranean niche shared by termites and naked mole-rats may be an important precursor for the evolution of eusociality. Life underground provides relative safety from predators and access to a readily available food source that does not require exit from the underground chamber. It also offers an expandable living place that can accommodate a large group.

Since the naked mole-rats share their subterranean niche with the other mole-rat species, it is interesting to speculate about why eusocial behavior has or has not evolved among the other Bathyergidae. Of the other 11 species, all are solitary but two—one of which, *Cryptomys damarensis*, may be termed eusocial. *C. damarensis* is not a particularly close relative of the naked mole-rat; whereas *H. glaber* diverged at the base of the family phylogenetic tree, *C. damarensis* is a distantly related and much more recent species, suggesting that complex social behavior in the two species evolved quite separately *(Figure 6)*. Another species in the *Cryptomys* genus, *C. hottentotus*, is the only other social member of the family; its small colonies (two to 14 members) have less well-developed social structures and vary in size and organization.

C. damarensis colonies are somewhat smaller than those of the naked mole-rat, having eight to 25 members. They also include a single reproductive female and one or more reproductive males and exhibit a division of labor among reproductive individuals based on size. One important difference between the two species has been suggested: *C. damarensis* colonies appear to be less stable over time, and the effects of multigenerational group living and inbreeding may be less pronounced. The significance of the presumed differences is a matter that will require further study because the dynamics associated with the duration of colonies and the founding of new colonies in wild naked mole-rats are not well understood. For instance, new colonies presumably are formed from existing colonies by budding, or fissioning, but the frequency of this event and its causes are not known.

The features of the subterranean niche may supply part of the explanation of social living in mole-rats, even though many solitary, non-social species of rodents in the Bathyergidae and other families occupy a similar niche. Restricted access to food and an unpredictable environment may also provide clues to the evolution of eusociality in both naked mole-rats and *C. damarensis* because as resources become more difficult to find, the energetic cost associated with finding them increases. Several authors have suggested that cooperation in food foraging and communal living might be promoted by the patchy distribution of the food source.

There is no simple explanation for the evolution of eusociality, and the hypotheses that fit the naked mole-rats and the other social species should not be considered mutually exclusive. Reproductive altruism is more likely to occur among genetically related individuals, but relatedness is not a sufficient explanation. Each eusocial species has a unique combination of life-histo-

Figure 14. *Cryptomys damarensis* is an African mole-rat distantly related to the naked mole-rats but sharing many kinds of eusocial behavior. Its somewhat larger colonies appear to be less stable over time. The species is found in Southern Africa.

ry characteristics associated with both its ecology and its behavior, and some or perhaps all of these characteristics may have predisposed a particular species for group living and cooperative breeding. The fact that various factors can work together in the development of eusociality may provide the ultimate explanation for the novelty, and therefore the mystery, of each example of eusocial behavior.

Bibliography

Alexander, R. D., K. M. Noonan and B. J. Crespi. 1991. The evolution of eusociality. In *The Biology of the Naked Mole-Rat*, ed. P. W. Sherman, J. U. M. Jarvis and R. D. Alexander, 3–44. Princeton, N.J.: Princeton University Press.

Allard, M. W., and R. L. Honeycutt. 1992. Nucleotide sequence variation in the mitochondrial 12S rRNA gene and the phylogeny of African mole-rats (Rodentia: Bathyergidae). *Molecular Biology and Evolution* 9 (in press).

Andersson, M. 1984. The evolution of eusociality. *Annual Review of Ecology and Systematics.* 15:165–189.

Bartz, S. H. 1979. Evolution of eusociality in termites. *Proceedings of the National Academy of Sciences (U.S.A.)* 76:5764–5768.

Bennett, N. C., and J. U. M. Jarvis. 1988. The social substructure and reproductive biology of colonies of the mole-rat, *Cryptomys damarensis* (Rodentia, Bathyergidae). *Journal of Mammalogy.* 69:293–302.

Brown, J. L. 1987. *Helping and Communal Breeding in Birds.* Princeton, N. J.: Princeton University Press.

Emlen, S. T. 1991. Evolution of cooperative breeding in birds and mammals. In *Behavioral Ecology: An Evolutionary Approach*, 3rd edition, ed. J. R. Krebs and N. B. Davies, 301–337. Palo Alto, Calif.: Blackwell Scientific Publications.

Genelly, R. E. 1965. Ecology of the common mole-rat (*Cryptomys hottentotus*) in Rhodesia. *Journal of Mammalogy* 46:647–665.

Hamilton, W. D. 1964. The genetical evolution of social behavior. *Journal of Theoretical Biology* 7:1–52.

Jarvis, J. U. M. 1981. Eusociality in a mammal: Cooperative breeding in naked mole-rat colonies. *Science* 212:571–573.

Macdonald, D. W., and P. D. Moehlman. 1982. Cooperation, altruism, and restraint in the reproduction of carnivores. In *Perspective in Ethology*, Vol. 5, ed. P. P. G. Bateson and P. H. Klopfer, 433–467. New York: Plenum Press.

Michener, C. D. 1969. Comparative social behavior of bees. *Annual Review of Entomology* 14:277–342.

Reeve, H. K., D. F. Westneat, W. A. Noon, P. W. Sherman and C. F. Aquadro. 1990. DNA "fingerprinting" reveals high levels of inbreeding in colonies of the eusocial naked mole-rat. *Proceedings of the National Academy of Sciences (U.S.A.).* 87:2496–2500.

Trivers, R. 1985. *Social Evolution.* Menlo Park, Calif.: The Benjamin/Cummings Publishing Company, Inc.

Wilson, E. O. 1971. *The Insect Societies.* Cambridge, Mass.: Harvard University Press.

The Honey Bee Colony as a Superorganism

Thomas D. Seeley

In an essay titled "The Architecture of Complexity," the economist Herbert A. Simon (1962) presented a parable about two watchmakers. Although both craftsmen built fine watches and both received frequent calls from customers placing orders, one, Hora, grew richer while the other, Tempus, became poorer and eventually lost his shop. This difference was traced to different methods used in assembling the watches, which in both cases consisted of 1,000 parts. Tempus's procedure was such that if he had a watch partially assembled and then had to put it down—to take an order, for example—it fell apart and had to be reassembled from scratch. Hora's watches were no less complex than those of Tempus but were designed so that he could put together stable subassemblies of about ten parts each. Ten of the subassemblies would, in turn, form a larger and also stable subassembly, and ten of the latter subassemblies constituted a complete watch. Thus each time Hora answered his phone he sacrificed only a small part of his labors and consequently was far more successful than Tempus at finishing watches.

The lesson of this story is that complex systems most likely arise through a sequence of stable subassemblies, with each higher-level unit being a nested hierarchy of lower-level units. This is certainly the path followed in the evolution of life (Margulis 1981; Bonner 1988). The biological hierarchy of functionally organized units consists of macromolecules within prokaryotic cells, prokaryotic cells within eukaryotic cells, eukaryotic cells within organisms, and, in certain species, organisms within thoroughly unified societies which have been called superorganisms (Wheeler 1928; Wilson 1971). To explain why natural selection has favored the formation of ever larger units of life, Richard Dawkins (1982) pointed out that all functional units above the level of the genes can be viewed as "vehicles" built by the genes to enhance their survival and reproduction, and that larger and more complex vehicles have evi-

Natural selection has made the colony a vehicle for the survival of genes

dently proved superior to smaller and simpler vehicles in certain ecological settings. By virtue of its greater size and mobility and other traits, a multicellular organism is sometimes a better gene-survival machine than is a single eukaryotic cell (Bonner 1974). Likewise, the genes inside organisms sometimes fare better when they reside in an integrated society of organisms rather than in a single organism, because of the superior defensive, feeding, and homeostatic abilities of functionally organized groups (Alexander 1974; Wilson 1975).

What is especially puzzling about the evolution of life is how each of the transitions to a higher level of biological organization was achieved. Individual units, each honed by natural selection to be a successful, free-living entity, must have begun somehow to interact cooperatively, eventually evolving into a larger, tightly integrated unit composed of mutually interdependent parts. The details of how this happened in the origin of prokaryotic cells or the advent of multicellular organisms are particularly obscure because in both instances the evolution of separate parts into integrated wholes has progressed so far that the original components have become altered beyond recognition. Furthermore, the integration of cells and organisms is so far advanced that it is difficult to see how the original cells or multicellular organisms were built.

The situation is quite different for the transition from organism to superorganism. This transition began relatively recently and indeed can possibly be viewed as the current frontier in the evolution of biological organization. Whereas the origin of prokaryotic cells occurred 3,500 million years ago, the advent of eukaryotic cells took place 1,300 million years ago, and multicellular organisms arose 700 million years ago (Margulis 1981), superorganism-grade insect societies began to appear only about 100 million years ago and presumably are still taking shape (Burnham 1978). Therefore, it is perhaps not surprising that even in the most advanced insect societies, such as army ants, fungus-growing termites, or honey bees, the differentiation and integration of a society's members have not reached the point at which each member's original nature has been erased. A colony of honey bees, for example, functions as an integrated whole and its members cannot survive on their own, yet individual honey bees are physically independent and closely resemble in physiology and morphology the solitary bees from

Thomas D. Seeley, an associate professor of animal behavior at Cornell University, received his doctorate at Harvard University, where he was a Junior Fellow. His main interest is in the design and evolution of insect societies, and he has written a summary of this subject for honey bees, Honeybee Ecology *(Princeton University Press, 1985). The area of research described in this article will be reviewed more fully in a book to be published by Cornell University Press. Address: Section of Neurobiology and Behavior, Seeley G. Mudd Hall, Cornell University, Ithaca, NY 14853.*

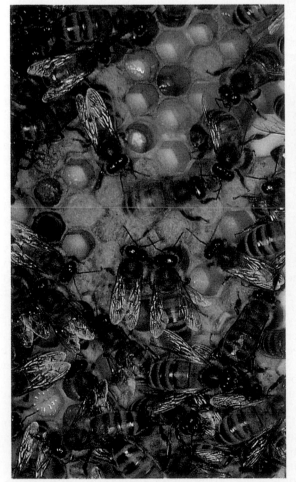

Figure 1. Signals are stimuli that convey information and have been molded by natural selection to do so; cues are stimuli that contain information but have not been shaped by natural selection specifically to convey information. Information can also pass between the members of a colony indirectly, through any component of their shared environment. Shown at the left are bees following another bee performing waggle dances, which are elaborate signals that indicate with precision the distance and direction of rich patches of flowers. In the top photograph, a food-storer bee (*left*) is unloading nectar from a forager. The delay a forager experiences before she can pass off her nectar is a cue that indicates the colony's nutritional status. In the photograph above, bees are fanning their wings in order to expel warm, moist air from their hive. The effect of this fanning—a cooler, drier atmosphere inside the hive—conveys information to other bees about the colony's need for ventilation. (Photos courtesy of P. K. Visscher, *left and top*, and S. Camazine, *above*.)

which they evolved (Fig. 1). In a colony of honey bees two levels of biological organization—organism and superorganism—coexist with equal prominence. The dual nature of such societies provides us with a special window on the evolution of biological organization, through which we can see how natural selection has taken thousands of organisms that were built for solitary life and merged them into a superorganism.

Is it a superorganism?

The term "superorganism" was coined by William Morton Wheeler (1928) to denote insect societies that possess features of organization analogous to the physiological processes of individual organisms. These include advanced social insects like army ants, leaf-cutter ants, fungus-growing termites, stingless bees, and honey bees. Although sociologists dealing with insects have used the superorganism concept more as a heuristic device than as a category of societal complexity (Lüscher 1962; Southwick 1983), recent insights into the logic of natural selection support the use of this term in a manner close to Wheeler's original intent (Hull 1980; Dawkins 1982; Wilson and Sober 1989). It seems correct to classify a group of *organisms* as a superorganism when the organisms form a cooperative unit to propagate their genes, just as we classify a group of *cells* as an organism when the cells form a cooperative unit to propagate their genes. By this definition, most groups of organisms are not perfect superorganisms because there is usually intense intragroup conflict when members compete for reproductive success (Trivers 1985). Indeed, in many species of social insects the female members of a colony (queens and workers) fight over who will lay eggs (West-Eberhard 1981; Bourke 1988). In the most advanced species of social insects, however, there appears to be little if any conflict within colonies, so that these colonies do represent superorganisms.

How complete is the cooperation in a honey bee colony, and thus to what extent is a colony of honey bees truly a superorganism? The best way to answer these questions is to determine the degree of congruence in the genetic interests of a colony's members. Consider the typical situation of a colony comprising one queen and some 20,000 workers, all daughters of the queen. At first glance, it might seem that there will be tremendous divergence of genetic interests within the colony. As a result of sexual reproduction, the queen's genotype does not match that of her workers; furthermore, although the workers are all offspring of the queen, because of segregation and recombination of the queen's genes during meiosis and because the queen has mated with ten or more males (Page 1986), the workers possess substantially different genotypes.

A closer look, however, reveals several features of the biology of honey bees that indicate a close alignment of genetic interests among the members of a colony, despite these genetic differences (Ratnieks 1988). Although worker bees possess ovaries and will lay eggs to produce sons if they lose their queen (Page and Erickson 1988), in the presence of the queen, workers engage in essentially no direct, personal reproduction. Workers cannot mate, so their only possible avenue of direct reproduction is through haploid sons from unfer-

tilized eggs. A recent study in which the extent of worker reproduction in colonies with queens was measured using genetic markers to distinguish drones from queen-laid and worker-laid eggs, reported that only one in one thousand drones in a colony is the offspring of workers (Visscher, in press). This means that as long as the queen is present there is a reproductive bottleneck in which every individual's gene propagation occurs virtually exclusively through a common pathway—the reproductive offspring (queens and drones) of the mother queen. This situation promotes strong cooperation among the queen and all workers; ultimately each worker focuses her efforts on the welfare and reproductive success of one individual, the queen.

This reproductive bottleneck does not, however, indicate that a perfect alignment of the genetic interests of a colony's members has evolved. The workers in colonies with queens may still disagree over which eggs should be reared into queens when it is time to produce new queens. This potential conflict of interest traces to the multiple mating of honey bee queens, which produces a set of patrilines within each colony. Because workers share three times as many genes with full-sister queens (same patriline) as with half-sister queens (different patriline), they are expected to prefer that queens produced in a colony be their full sisters. Over the last few years several investigators have searched for intracolony competition during queen rearing, and a growing body of evidence indicates that some patrilines within a colony do achieve a small bias in their favor (Noonan 1986; Visscher 1986; Page et al. 1989). However, all studies that have reported preferential rearing of more closely related queens involved somewhat artificial test conditions, such as transfers of larval queens between colonies or use of colonies containing only two or three instead of the normal number (ten or more) of patrilines. It may be that even the slight bias in queen rearing observed in these studies is greater than what occurs under natural conditions (Hogendoorn and Velthuis 1988).

Given the bottleneck for gene propagation and the strong indication that workers have nearly equal genetic stakes in a colony's production of reproductives (due to meiosis in the queen, together with little patriline bias in queen rearing), we can conclude that the genetic interests of the workers in a colony led by a queen are nearly, though not perfectly, congruent. Furthermore, we know that the mother queen and the workers have evolved similar interests in matters such as who lays the eggs that produce the colony's drones, the ratio of the colony's investment in queens and drones, and the timing of replacement of the queen (Seeley 1985; Ratnieks 1988). Thus it appears that there is minimal conflict within honey bee colonies as long as the mother queen is present. Therefore, we may conclude that honey bee colonies containing queens are nearly true superorganisms.

This conclusion, based on analyses of the genetic interests of a colony's members, is reinforced by the picture of pervasive cooperation which has emerged from analyses of colony functioning. In choosing a nest site, building a nest, collecting food, regulating the nest temperature, and deterring predators, a honey bee colony containing a queen resembles a smoothly running machine in which each part always contributes to the efficient operation of the whole (Seeley 1985; Winston 1987). As we will see, in a normal honey bee colony, food, information, and aid appear to pass freely among the members in ways that apparently promote the economic success of the whole colony.

It should be very revealing, and at most only slightly misleading, to view a honey bee colony as an integrated biological machine that promotes the success of the colony's genes. Given this perspective, the outstanding biological question becomes: How did evolution take a large number of organisms built for solitary life and forge them into a single vehicle of gene survival? The answer to this question has two parts. One concerns the ultimate forces of natural selection, which caused the evolution of unified colonies; the other involves the proximate mechanisms by which colonies function as integrated wholes. This article focuses on the second half of the answer. The key to understanding this aspect of the puzzle involves understanding the flow of information within colonies. Coordination in any complex system depends upon each part having access to appropriate information at the right time and place (Wiener 1961). Coherence implies communication.

Architecture of information flow

Coordination of the activities in a honey bee colony arises without any centralized decision making. There is no evidence of an information and control hierarchy, with some individuals taking in information about the colony, deciding what needs to be done, and issuing commands to other individuals who then perform necessary tasks. As the biblical King Solomon observed, there is "neither guide, overseer, nor ruler." In particular, it is clear that the queen does not supervise the activities of her workers. She does emit a chemical signal, the queen-substance pheromone, which plays a role in regulating the colony's production of additional queens (Free 1987), but this signal cannot provide comprehensive supervision of the activities of the tens of thousands of workers in a colony.

A colony's coherence depends instead upon the ability of its members to circulate throughout the hive, gather information about the colony's needs, and adjust the supply of their labor to the demands they sense. This idea was suggested in the early 1950s by Martin Lindauer (1952), who painstakingly followed individual workers within colonies living in glass-walled observation hives. He learned that the bees devote about 30% of their time to walking about the nest, and that this patrolling is punctuated by bouts of activity in a wide variety of tasks (Fig. 2). A typical 30-minute segment from Lindauer's records reveals the following behavior for a seven-day-old bee: patrolling, shaping comb, patrolling, feeding young brood, cleaning cells, patrolling, shaping comb, eating pollen, resting, patrolling, shaping comb. The task performed at any given moment presumably depends upon the specific labor need sensed by the bee.

Why are honey bee colonies organized in this way? Decentralized control is possibly superior to centralized control for bees. Systems with decentralized control generally have faster responses to local stresses than

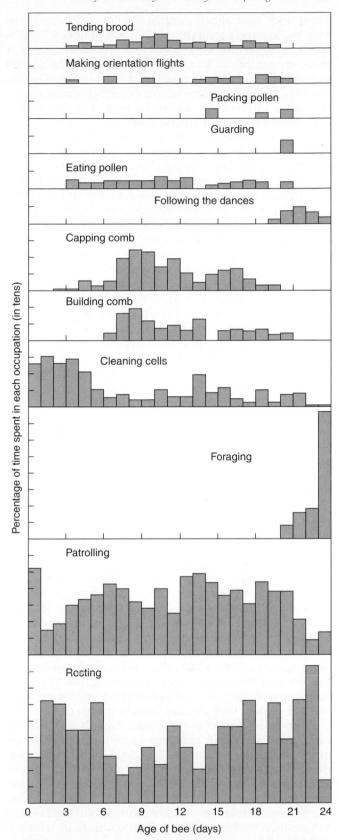

Figure 2. In the course of her life, a worker honey bee performs a variety of tasks. As the distribution above indicates, she can perform several different tasks on any given day and at any given age. She thereby behaves flexibly, responding to different needs encountered in the hive. The large amount of time spent patrolling is evidently related to the gathering of information—through cues, signals, and the shared environment—about the colony's labor needs. (After Lindauer 1952.)

those with centralized control (Miller 1978), and this may be extremely valuable to colonies of honey bees. A colony may be able to respond to a predator's attack at the nest entrance or to a temperature rise in the central broodnest much more quickly if the workers at the trouble sites can perform corrective actions immediately than if information has to be sent to a supervisor, who would then issue instructions.

A perhaps more likely explanation of decentralized control is not that it is superior to centralized control, but that it is the best the bees can do given the limited communication processes that have developed through evolution. As we will see, the mechanisms of communication in colonies of social insects are rather rudimentary, at least relative to what exists in human organizations or in multicellular organisms. Colonies of army ants, fungus-growing termites, honey bees, and other superorganisms have yet to invent anything like a mail system, telephone, or computer network. Such technologies make it possible for information to flow rapidly and efficiently between the different parts of a human organization with centralized control. One piece of evidence in support of this second hypothesis is that at the level of organization just below the superorganism, the multicellular organism, where a sophisticated intercellular communication system has evolved, there exists centralized control, with the brain taking in information about the whole organism and issuing commands to cells of the body. Whatever the underlying reason, the fact of decentralized control in honey bee colonies tells us that understanding how colonial coordination arises follows from understanding how each worker acquires information about her colony's needs.

Pathways of information

To what do worker bees respond when patrolling? The answer to this question is complex because evolution has been highly opportunistic in building pathways for information in honey bee colonies. It has shaped the workers so that they are sensitive to virtually all variables and stimuli that contain useful information: the temperature of the nest interior, the degree of crowding at a food source, the moistness of larvae, the recruitment dances of nestmates, the shape of a beeswax cell, the odor of dead bees. Furthermore, given the close alignment of the genetic interests of a colony's workers, we can expect that natural selection has molded the workers to be skilled at generating signals for information transfer. Within colonies there are various tappings, tuggings, shakings, buzzings, strokings, wagglings, crossing of antennae, and puffings and streakings of chemicals, all of which seem to be communication signals. The result is that within a honey bee colony there exists an astonishingly intricate web of information pathways, the full magnitude of which is still only dimly perceived.

Information can flow between colony members in two ways: directly, through signals and cues, as we will discuss below, or indirectly, through some component of the shared environment. An example of the latter process is the transfer of information through the process of comb building. The construction of a particular cell in a beeswax comb may involve several bees, yet these bees never need to come together and exchange information directly. The building activities can be completely and efficiently coordinated by information embodied in the structure of the partially completed cell. Thus one bee might begin a cell wall by depositing a small ridge of beeswax; a second bee might finish sculpting the wall, guided by the shape of the wax ridge left by the first bee. Another example of information flow through the shared environment is thermoregulation of the nest. A colony maintains the central broodnest at 34 to 36°C in the face of ambient temperatures that may range from −20 to 40°C. The coordinated heating and cooling of a nest occurs automatically: each bee responds to the temperature of her immediate environment by appropriately heating it (by making intense isometric contractions of her flight muscles) or cooling it (by fanning her wings to draw cooler air into the area) (Heinrich 1985). In effect, the temperature of the air and comb inside a hive provides a communication network regarding the colony's heating and cooling needs.

Several authors have expressed the concept of information flow through the shared environment in social insect colonies. These include Pierre-Paul Grassé (1959), who coined the term "stigmergy" to explain coordination in nest construction by termites, and Charles D. Michener (1974), who pointed out that "indirect social interactions," such as transfers of information through the food stored in the nest, are an important integration mechanism in colonies of social bees. Future studies of information flow in social insect colonies may reveal that more information is transmitted indirectly than directly. The use of the shared environment as a communication pathway has certain attractions, including easy asynchronous transfer of information between individuals and virtually automatic transfer of information between any two individuals sharing some portion of the nest environment. It also has the important feature whereby information can pass from a group to an individual whenever an individual responds to the environmental effects of a group. Because the process of integration of a group is largely a matter of information flow from group to individual, it may be that information flow through the shared environment has been natural selection's principal technique of integration in building superorganisms.

Signals and cues

There are two types of direct communication channels: signals and cues (Lloyd 1983). Signals are stimuli that convey information and have been shaped by natural selection to do so, whereas cues are stimuli that contain information but have not been shaped by natural selection specifically to convey information. Cues carry information only incidentally. The distinction between signals and cues deserves emphasis because studies of information flow in social insect colonies have tended to overlook cues and have focused instead on conspicuous visual, tactile, acoustical, and chemical signals (reviewed by Wilson 1971; Hölldobler 1977). The emphasis on signals reflects the fact that information transfer via signals is relatively easily detected by humans because in the mutualistic setting of a social insect colony natural selection will have shaped signals to be powerful and unambiguous carriers of information (Markl 1985). In contrast, information transfer via cuing will usually be subtle; cues are simply by-products of behaviors performed for reasons other than communication.

The famous dance language, through which a bee can inform her nestmates of the direction and distance of a rich food source, is a classic example of a signal (von Frisch 1967). Given the richness of the information and the high precision of encoding the information in the dances, there is no doubt that the dance language has been intensively molded by natural selection for the efficient transfer of information.

One example of a cue involves the regulation of a colony's choosiness among nectar sources in relation to its nutritional status. When a colony is well nourished, its foragers exploit only highly profitable patches of flowers, but if the colony is near starvation, the foragers exploit both high-

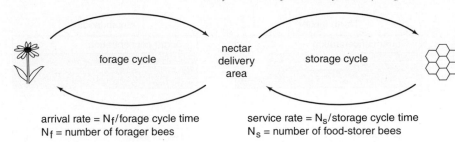

arrival rate = N_f/forage cycle time
N_f = number of forager bees

service rate = N_s/storage cycle time
N_s = number of food-storer bees

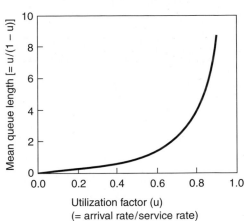

Figure 3. The process of collecting nectar in honey bee colonies is shown schematically as two separate but intersecting forage and storage cycles. In the forage cycle, foragers collect nectar from flowers, bring it back to the hive, and then return to the flowers to gather more nectar. The storage cycle takes place entirely within the hive as food-storer bees in the delivery area (just inside the entrance) unload the fresh nectar from foragers, transport it deep inside the hive to the honey comb for storage, and then crawl back to the delivery area. The amount of time a nectar forager has to wait to begin unloading to a food-storer bee is a cue that indicates the nutritional status of the colony.

ly profitable and less profitable flower patches. This raises the question of how foragers stay informed about their colony's nutritional status. My research (Seeley 1989) has recently confirmed a hypothesis originally proposed by Lindauer (1948) that the delay a returning forager experiences before she can unload her nectar to a food-storer bee suggests to the forager the colony's nutritional status. (The food storers are bees who are slightly younger than the foragers and who specialize in receiving the fresh nectar, concentrating it into honey, and storing it in the honeycombs.) If a forager can find a food-storer bee within approximately 15 seconds of entering the hive, then she knows that there is little nectar coming into the hive and little honey stored in the hive—her colony is approaching starvation. But if a forager has to wait more than about 15 seconds (and as much as 100 seconds or more), then she knows that either there is much nectar currently being gathered or there is much honey already stored in the hive—either way, her colony is well nourished.

The link between waiting time and the colony's nutritional status is shown schematically in Figure 3. Nectar collection involves two cycles, a forage cycle and a storage cycle, which intersect at the point of nectar transfer from foragers to food storers. This sort of system, in which there is a stream of individuals in one group arriving at a location to be serviced by individuals in a second group, is quite common. It occurs at toll booths along highways, at the service windows of banks, and at the checkout counters of supermarkets. A critical variable of all such systems is the utilization factor, U, which is the ratio of the rate of arrival of individuals needing servicing and the rate at which the servers can provide service. The mathematical theory of queues (Morse 1958) reveals that the average length of the waiting line, Q, that an arriving individual will face is a simple function of the utilization factor: $Q = U / (1 - U)$. Thus in the

case of honey bees, if U is low—say less than 0.5—then the average queue size does not exceed one, and nectar unloading proceeds with little delay. If U reaches 0.8, the wait becomes appreciable, with an average waiting line of four individuals. Any further rise in the utilization factor entails a disproportionately sharp increase in the length of the waiting line.

Consider the case of a colony near starvation. Little nectar is being gathered, so the arrival rate of foragers is low and there is abundant empty storage comb, enabling the food storers to complete each storage cycle quickly (usually in less than ten minutes). The effect of these conditions is a low utilization factor, and a negligible waiting time for returning foragers. Now consider a well-nourished colony whose hive is brimming with honey. In this situation there is little empty storage comb, which can cause the food storers to take 40 minutes or more to complete a storage cycle, so the service rate is low. The utilization factor is therefore high and the waiting time to begin unloading is long. Experiments have demonstrated that the foragers do indeed respond to the waiting time in determining their colony's nutritional status (Seeley 1989). The critical test involved removing most of the food-storer bees from a colony with little honey in its hive, thereby reducing the service rate and so increasing the waiting time to begin unloading. The colony's foragers were then observed to cease recruiting nestmates to a feeder with a concentrated sucrose solution. These bees behaved as if their hive were packed with honey, although the combs were nearly empty!

Coordinated group action

Let us now consider an example that illustrates how natural selection has linked multiple pathways of information flow to achieve an impressive feat of coordination at the colony level. As flower patches bloom and wither

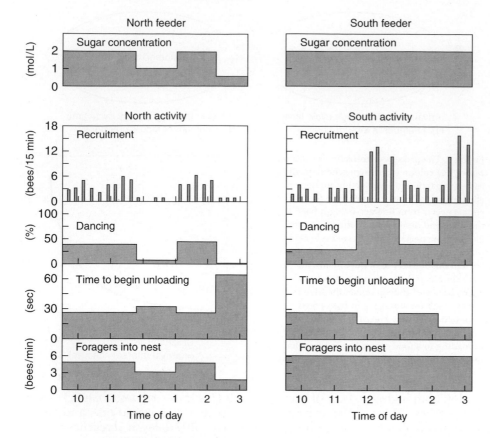

Figure 4. In a six-hour experiment, the quality of one food source (north feeder) was decreased, increased, and decreased again while the quality of a second source remained fixed. When the food source in the north deteriorated, the bees promptly increased recruitment of foragers to the food source in the south. This can be explained by the following scenario: When the north food source declined, the bees foraging there lowered their foraging rate, which depressed the total arrival rate of foragers at the hive. This caused the foragers from the south food source to experience a shorter wait when unloading nectar. This in turn stimulated these foragers to increase their production of recruitment signals, thereby producing a rise in the number of recruits at the south feeder. (After Seeley 1986.)

across the countryside, a hive of bees continuously adjusts the distribution of its foragers among the patches to track the changes in foraging opportunities (Visscher and Seeley 1982). This process is depicted in Figure 4, which shows how a colony foraging from two experimental food sources deftly altered its behavior following changes in the profitability of one of the sources.

From mid-morning to shortly before noon, the two feeders, north and south of the hive, offered equal sucrose solutions, and the colony exploited each at the same moderate level. Loaded foragers returned to the hive from each feeder at a rate of six bees per minute, and approximately four recruits joined the work force for each feeder every 15 minutes. (All recruits were captured, so each feeder's work force actually remained stable.) Then the sucrose solution of the north feeder was diluted from a two- to a one-molar concentration. The colony responded within 15 minutes, shifting to a strongly asymmetrical pattern in which the rate of forager visits to the feeder with the dilute solution fell by nearly 50%, whereas the rate of visits to the other feeder held steady. The rate of recruitment of additional workers to the feeder with the dilute solution dwindled to zero, while the rate of recruitment to the other feeder rose threefold.

The most intriguing aspect of this reallocation of foraging effort is that it involved behavioral changes by bees foraging at *both* food sources. Foragers at both sources were distinctively labeled, and each feeder was monitored for bees from the other feeder; no cross visits were observed. Information about the waning food source must have passed from one group of foragers to the other; only with this information could the foragers at the steady food source have known to boost their dancing.

The multistage pathway of this information transfer can be traced using a flow diagram. Following the dilution of the north food source, the north foragers virtually ceased dancing, which shut off recruitment to the north, and they halved their foraging tempo. These adjustments reduced the arrival rate of foragers from the north, thereby diminishing the colony's total rate of arrivals at the hive. This lowered the colony's utilization factor and nearly halved the period during which south foragers had to wait to begin unloading—13 seconds instead of 22 seconds. (The waiting time for the north foragers rose somewhat—to 28 seconds instead of 22 seconds—because these bees no longer tried to make contact with food storers immediately upon entering the hive.) The drop in waiting time for south foragers stimulated them to recruit foragers more vigorously; the percentage of the south foragers dancing upon returning to the hive soared from 30 to 90% (Fig. 4). In sum, the colony responded to a decline in one part of its food-source array with adjustments in the foraging effort throughout the array, and this coordination involved information passing between four groups of bees (north foragers, food storers, south foragers, recruits) via a combination of signals and cues.

Building integration

This article began by noting what is perhaps the single most important question raised by colonies of honey bees and other advanced social insects: How did evolution take a large number of organisms built for solitary life and forge them into a unified, group-level vehicle of gene survival? With respect to the mechanisms of integration, the solution to this puzzle lies in how information flows among the members of a colony, enabling them to coordinate joint efforts in nest construction, thermoregulation, social foraging, and other colony-level adaptations. The message that is emerging from studies of integrative phenomena in honey bee colonies is that much of the intricate orchestration of a colony's members is achieved through surprisingly rudimentary information transfer. Traditionally, studies of communication in the social insects have emphasized sophisticated and conspicuous communication processes that involve signals honed by natural selection, such as the dance language behavior. There is no question that these processes are important. Nevertheless, I predict that the relatively subtle communication mechanisms of cues and the shared environment will prove even more important than the more obvious signals. If so, then the impressive feats of internal coordination shown by superorganisms will often prove to be built of rather humble devices. This should not surprise us, for as Colin Pittendrigh (1958) so nicely put it, adaptive organization is "a patchwork of makeshifts pieced together, as it were, from what was available when opportunity knocked, and accepted in the hindsight, not the foresight, of natural selection."

References

Alexander, R. D. 1974. The evolution of social behavior. *Ann. Rev. Ecol. Syst.* 5:325–83.

Bonner, J. T. 1974. *On Development.* Harvard Univ. Press.

———. 1988. *The Evolution of Complexity.* Princeton Univ. Press.

Bourke, A. F. G. 1988. Worker reproduction in the higher eusocial Hymenoptera. *Quart. Rev. Biol.* 63:291–311.

Burnham, L. 1978. Survey of social insects in the fossil record. *Psyche* 85: 85–133.

Dawkins, R. 1982. *The Extended Phenotype.* Freeman.

Free, J. B. 1987. Pheromones of Social Bees. Cornell Univ. Press.

Grassé, P.-P. 1959. La reconstruction du nid et les coordinations interindividuelles chez *Bellicositermes natalensis* et *Cubitermes* sp. La théorie de la stigmergie: Essai d'interprétation du comportement des termites constructeurs. *Insectes Sociaux* 6:41–83.

Heinrich, B. 1985. The social physiology of temperature regulation in honeybees. In *Experimental Behavioral Ecology and Sociobiology,* ed. B. Hölldobler and M. Lindauer, pp. 393–406. Sinauer.

Hogendoorn, K., and H. H. W. Velthuis. 1988. Influence of multiple mating on kin recognition by worker honeybees. *Naturwissenschaften* 75:412–13.

Hölldobler, B. 1977. Communication in social Hymenoptera. In *How Animals Communicate,* ed. T. A. Sebeok, pp. 418–71. Indiana Univ. Press.

Hull, D. L. 1980. Individuality and selection. *Ann. Rev. Ecol. Syst.* 11: 311–32.

Lindauer, M. 1948. Über die Einwirkung von Duft- und Geschmacksstoffen sowie anderer Faktoren auf die Tänze der Bienen. *Zeitschrift für vergleichende Physiologie* 31: 348–412.

———. 1952. Ein Beitrag zur Frage der Arbeitsteilung im Bienenstaat. *Zeitschrift fur vergleichende Physiologie* 34:299–345.

Lloyd, J. E. 1983. Bioluminescence and communication in insects. *Ann. Rev. Entomol.* 28:131–60.

Lüscher, M. 1962. Sex pheromones in the termite superorganism. *Gen. Comp. Endocrinol.* 2:615.

Margulis, L. 1981. *Symbiosis in Cell Evolution.* Freeman.

Markl, H. 1985. Manipulation, modulation, information, cognition: Some of the riddles of communication. In *Experimental Behavioral Ecology and Sociobiology,* ed. B. Hölldobler and M. Lindauer, pp. 163–94. Sinauer.

Michener, C. D. 1974. *The Social Behavior of the Bees.* Harvard Univ. Press.

Miller, J. G. 1978. *Living Systems.* McGraw-Hill.

Morse, P. M. 1958. *Queues, Inventories, and Maintenance.* Wiley.

Noonan, K. C. 1986. Recognition of queen larvae by worker honey bees (*Apis mellifera*). *Ethology* 73:295–306.

Page, R. E., Jr. 1986. Sperm utilization in social insects. *Ann. Rev. Entomol.* 31:297–320.

Page, R. E., Jr., and F. H. Erickson, Jr. 1988. Reproduction by worker honey bees (*Apis mellifera* L.). *Behav. Ecol. Sociobiol.* 23:117–26.

Page, R. E., Jr., G. E. Robinson, and M. K. Fondrk. 1989. Genetic specialists, kin recognition and nepotism in honey-bee colonies. *Nature* 338:576–79.

Pittendrigh, C. S. 1958. Adaptation, natural selection, and behavior. In *Behavior and Evolution,* ed. A. Roe and G. G. Simpson, pp. 390–416. Yale Univ. Press.

Ratnieks, F. L. W. 1988. Reproductive harmony via mutual policing by workers in eusocial Hymenoptera. *Am. Naturalist* 132:217–36.

Seeley, T. D. 1985. *Honeybee Ecology.* Princeton Univ. Press.

———. 1986. Social foraging by honey bees: How colonies allocate foragers among patches of flowers. *Behav. Ecol. Sociobiol.* 19:34–54.

———. 1989. Social foraging in honey bees: How nectar foragers assess their colony's nutritional status. *Behav. Ecol. Sociobiol.* 24:181–99.

Simon, H. A. 1962. The architecture of complexity. *Proc. Am. Philosoph. Soc.* 106:467–82.

Southwick, E. E. 1983. The honey bee cluster as a homeothermic superorganism. *Comp. Biochem. Physiol.* 75A:641–45.

Trivers, R. 1985. *Social Evolution.* Benjamin/Cummings.

Visscher, P. K. 1986. Kinship discrimination in queen rearing by honey bees (*Apis mellifera*). *Behav. Ecol. Sociobiol.* 18:453–60.

———. In press. A quantitative study of worker reproduction in honey bee colonies. *Behav. Ecol. Sociobiol.*

Visscher, P. K., and T. D. Seeley. 1982. Foraging strategy of honeybee colonies in a temperate deciduous forest. *Ecology* 63:1790–1801.

von Frisch, K. 1967. *The Dance Language and Orientation of Bees.* Harvard Univ. Press.

West-Eberhard, M. J. 1981. Intragroup selection and the evolution of insect societies. In *Natural and Social Behavior,* ed. R. D. Alexander and D. W. Tinkle, pp. 3–17. Chiron.

Wheeler, W. M. 1928. *The Social Insects: Their Origin and Evolution.* Kegan Paul, Trench, and Trubner.

Wiener, N. 1961. *Cybernetics: Or Control and Communication in the Animal and the Machine.* MIT Press.

Wilson, D. S., and E. Sober. 1989. Reviving the superorganism. *J. Theor. Biol.* 136:337–56.

Wilson, E. O. 1971. *The Insect Societies.* Harvard Univ. Press.

———. 1975. *Sociobiology.* Harvard Univ. Press.

Winston, M. L. 1987. *The Biology of the Honey Bee.* Harvard Univ. Press.

The Essence of Royalty: Honey Bee Queen Pheromone

A queen bee induces thousands of worker bees to submit to her hegemony by secreting a potent chemical blend, which is spread by physical contact throughout the colony

Mark L. Winston and Keith N. Slessor

Beekeeping, like most skills, has developed its own shorthand language, by which a word or simple phrase can evoke complex concepts that quickly are understood by those who practice the craft. One such word is "queenright," meaning that a colony's queen is present and thousands of her worker offspring are collaborating diligently on communal tasks. Hidden in this beekeeper jargon, however, lies the most fundamental mystery of social insects: the mechanisms of coordination and integration that mediate the tasks of thousands of individuals into the smoothly functioning unit of the colony. In this article, we shall describe some new findings concerning one aspect of these integrative mechanisms, the honey bee queen's pheromonal influence over worker bees. The discovery of the nature and function of this "essence of royalty" has provided profound insights into the functioning of social-insect colonies, and also has led to some significant commercial applications for crop pollination and beekeeping.

Mark L. Winston, who earned B.A. and M.Sc. degrees at Boston University and his Ph.D. at the University of Kansas, is the author of two recent books about honey bees, The Biology of the Honey Bee *(1987) and* Killer Bees: The Africanized Honey Bee in the Americas *(1992), both published by Harvard University Press. Keith N. Slessor, who received his Ph.D. at the University of British Columbia, is a native British Columbian who finds organic chemistry an excellent medium for both teaching and exploration, and regards it as a challenge to decipher the communication skills of economically important insects. Winston and Slessor are professors at Simon Fraser University, Burnaby, B.C. V5A 1S6, Canada, in the departments of Biological Sciences and Chemistry/Biochemistry, respectively.*

The study of social-insect pheromones, or sociochemistry, is not a new discipline. Many pheromones are known throughout the social insects that are used in myriad tasks such as alarm behavior, orientation, trail marking, nest recognition, mate attraction and others (Bradshaw and Howse 1984; Duffield, Wheeler and Eickwort 1984; Free 1987; Hölldobler and Wilson 1990; Howse 1984; Wilson 1971; Winston 1987). The honey bees, for example, are estimated to produce at least 36 pheromones, which together constitute a chemical language of some intricacy. However, virtually all of the known chemicals are secretions that "release," or elicit, a specific behavior. The queen sociochemicals belong to another class, called primer pheromones, that exercise a more fundamental level of control by mediating worker and colony reproduction and influencing broad aspects of foraging and other behaviors. Further, they can be both stimulatory and inhibitory, depending on the specific function. The existence of primer sociochemicals is well known, but their identification and synthesis have proved difficult.

Indeed, the only primer pheromone that has been identified comes from the queen honey bee (*Apis mellifera* L.); it is secreted by the mandibular glands, a paired set of glands on either side of the queen's head (Figure 2). Here we discuss the research that led to the pheromone's identification, new findings concerning its effects on worker bees and colony functions, transmission mechanisms within colonies, and some applications for crop pollination and beekeeping.

Figure 1. Retinue response is a stereotyped honey bee behavior by which various pheromones are transferred from the queen to the worker bees. In the photograph on the opposite page workers turn to face a queen, touching her with their antennae and licking her. By this means they pick up the pheromone that she exudes on her body. For the next half hour, the workers act as dispersers of pheromone, travelling throughout the nest and contacting other workers more frequently than normal. It is the pheromone itself—including the mixture of substances secreted by the queen's mandibular glands—that induces the retinue response. The effectiveness of the pheromone is demonstrated by experiments in which worker bees form a retinue around a glass lure coated with mandibular-gland extract, as in the photograph above. The response has formed the basis of a bioassay for synthetic pheromones. (Photograph opposite courtesy of Kenneth Lorenzen; photograph above by Keith N. Slessor.)

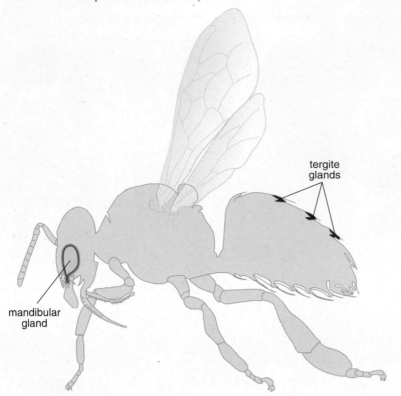

Figure 2. Two sets of glands in the queen honey bee are thought to secrete primer pheromones, the class of pheromones that regulate reproductive behavior in the colony. The best-studied primer pheromones are those produced in the mandibular glands on either side of the head. Each gland is connected to the mandible by a duct; a valve allows the bee to regulate the discharge of secretions. The queen may also synthesize primer pheromone in the tergite glands, which are large subepidermal complexes of glandular cells near the rear of some of the tergites, or dorsal plates. The functions of the secretions from the tergite glands have not been definitively established.

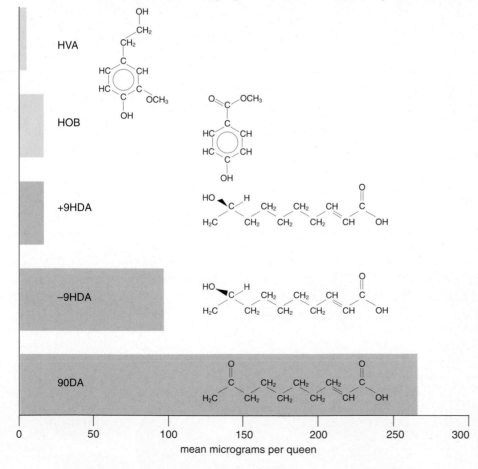

Early Research

For more than 30 years it has been recognized that queen pheromones mediate many colony activities. The pheromones were known to inhibit the rearing of new queens and the development of ovaries in worker bees; they were also observed to attract workers to swarm clusters, stimulate foraging, attract workers to a queen and allow workers to recognize their own queen.

Surprisingly, only two active substances had been identified. These two most abundant compounds in queen mandibular pheromone are the organic acids (E)-9-keto-2-decenoic acid, or 9ODA, which was identified in 1960 (Callow and Johnston 1960, Barbier and Lederer 1960) and (E)-9-hydroxy-2-decenoic acid, or 9HDA, which was identified in 1964 (Callow, Chapman and Paton 1964; Butler and Fairey 1964).

Experiments with the synthetic acids, however, showed that they did not fully duplicate the effects of mandibular extracts or of the queen's presence (reviewed by Free 1987; Winston 1987). Although analysis of crushed bee heads produced laundry lists of chemicals, none was shown on closer examination to be active, and progress stalled.

One reason it took so long to identify queen mandibular pheromone is that bioassays for primer pheromones are difficult and time-consuming. The potency of candidate releaser pheromones can be evaluated by relatively simple means. Orientation pheromones, for example, can be evaluated by counting the number of bees entering a hive after pheromone is deposited at the hive opening. Demonstrating that a synthetic blend or a natural extract is fully

Figure 3. Queen mandibular pheromone is a complex of five substances. Shown here is the average composition of one queen equivalent (Qeq) of mandibular pheromone obtained by analyzing mandibular-gland extracts from 55 queens. (The composition of the pheromone changes over the queen's life span and varies considerably even among mature, mated queens.) A queen equivalent is the average amount of pheromone in a queen's glands at a given time. The most abundant ingredient is (E)-9-keto-2-decenoic acid (9ODA). Another decenoic acid has two optical isomers, both of which must be present to achieve full activity. The isomers are designated (R,E)-(–)-9-hydroxy-2-decenoic acid (–9 HDA) and (S,E)-(+)-9-hydroxy-2-decenoic acid (+9 HDA). The remaining two ingredients are aromatic compounds: methyl *p*-hydroxybenzoate (HOB) and 4-hydroxy-3-methoxyphenylethanol (HVA).

equivalent to a primer pheromone, however, may require a battery of bioassays, some of which may take months to complete. To further complicate matters, there is often disagreement over a primer's functions.

The chemical composition of queen mandibular pheromone and its biological effects were not the only controversial topics. Even the means by which the pheromone reaches worker bees was unknown. Is the pheromone transmitted when the workers exchange food? Does it volatilize and spread through the air? Or is it spread by body contact? Here the main obstacle was the minute amount involved: A queen produces less than four ten-thousandths of a gram per day, and a worker bee is able to sense a ten-millionth of the queen's daily production.

Discovery

One of the stereotyped behaviors thought to be elicited by queen pheromones is the retinue response. Worker bees turn toward the queen, forming a dynamic retinue around her. About 10 workers at a time contact her with their antennae, forelegs or mouthparts for periods of minutes. Then the workers leave the queen, groom themselves and move through the nest, making frequent reciprocated contacts with other workers for the next half-hour (Allen 1960; Seeley 1979).

Bioassays involving retinue behavior had given particularly confusing results. Mandibular extracts seemed to be important to worker recognition of the queen and attraction to her. However, the most abundant component of the glands, 9ODA, did not by itself elicit retinue behavior. To add to the confusion, bees had been shown to exhibit the retinue response to queens whose mandibular glands had been surgically removed.

This is how matters stood in 1985 when, quite by accident, we put a glass lure coated with queen mandibular extract next to some stray workers on a laboratory bench. To our surprise, the workers formed a retinue around the lure. Clearly the extract had properties that its major known constituent did not. This was exciting because the retinue response is easily recognizable and takes place immediately. For both reasons it might serve as the basis for a practical bioassay for mandibular pheromone.

Indeed we were able to develop a bioassay based on the retinue response that was a particularly sensitive indica-

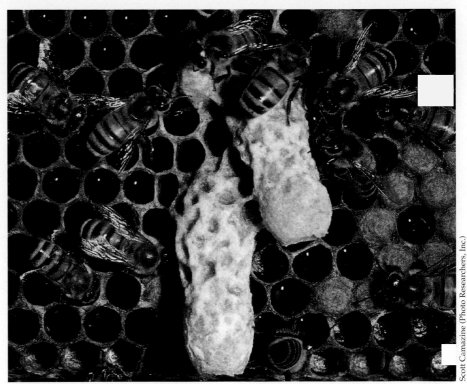

Scott Camazine (Photo Researchers, Inc.)

Figure 4. Queen rearing is one of the behaviors inhibited by queen mandibular pheromone. Queens are reared in large cells that hang from the comb. Workers, in contrast, are reared in cells that are capped at the level of the comb. The first sign of queen rearing is the construction of new cups suitable for rearing queens. Most eggs in these cups are laid there by the queen, but workers sometimes move fertilized eggs or very young larvae from worker cells into queen cups. Once the eggs hatch, the larvae are fed a specialized diet that induces them to develop into queens. As the larval queens grow, the cells are elongated downward. They are finally sealed at the end of the larval feeding period.

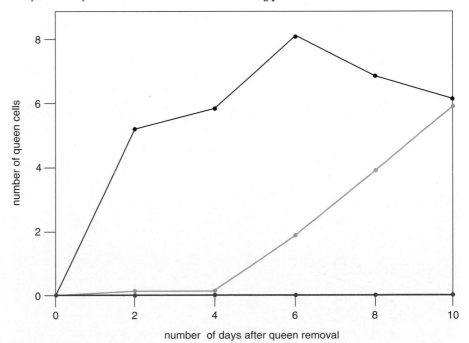

Figure 5. Queen mandibular pheromone suppresses queen rearing. This effect was demonstrated by comparing the number of queen cells in queenless colonies receiving various dosages of synthetic mandibular pheromone with the number of cells in queenright colonies. Shown here are three cases: a queenless colony receiving a glass slide coated only with the solvent used in preparing pheromone *(black)*; a queenless colony receiving one Qeq of synthetic pheromone *(gold)*; and a queenright colony *(red)*. Brood reared more than six days after the queen was removed would have developed into queens of inferior quality or into workers rather than queens.

tor of pheromonal activity. In a series of experiments we used this bioassay to evaluate complete or fractionated mandibular extracts. Interestingly, fractions that contained a single substance were relatively inactive. We began to see retinue behavior around the lure only when we combined certain fractions. Eventually, we identified the components of the active fractions and devised a synthetic blend of five components that duplicated the activity of the natural mandibular secretion (Figure 3) (Kaminski et al. 1990, Slessor et al. 1988).

The ingredients include two forms of 9HDA. This substance is chiral: It has a left-handed and a right-handed form. Earlier, we had demonstrated that one form, or enantiomer, is significantly more effective in evoking some behavioral responses than the other (Winston et al. 1982). We had also determined that one enantiomer predominates in the natural pheromone (Slessor et al. 1985). But it turned out that both enantiomers also have to be present to evoke the full retinue response.

The remaining missing ingredients were two small, aromatic compounds. One of the aromatic compounds, methyl p-hydroxybenzoate, or HOB, had been identified earlier (Pain, Hugel and Barbier 1960) but had not been considered to be active as a pheromone. The other aromatic molecule, 4-hydroxy-3-methoxyphenylethanol, or HVA, was first identified in the course of our experiments.

The pheromonal activity of the aromatic molecules was quite unexpected. Most insect pheromones are aliphatic compounds, built up out of straight or branched chains of carbon atoms; they are derived from fatty acids or terpenes. The decenoic acids in mandibular gland pheromone fit this description, for example. The aromatics, which feature closed rings of carbon atoms, are a different class of chemical entirely. Indeed we probably found them only because we reversed the customary experimental procedure. Instead of identifying substances in gland extracts and then screening them for pheromonal activity, we screened for activity and then identified the compounds we had isolated.

Our research into the mandibular gland's composition and activity cleared up several points of confusion. First, the complete ensemble of molecules is necessary for full efficacy. Removing any of the components reduced the pheromone's effect by as much as 50 percent, and individual components are inactive when they are tested alone. This is why the pheromone's major component by itself had failed to elicit retinue behavior.

Second, although we managed to duplicate the effects of mandibular secretions, we should emphasize that we did not duplicate the queen's complete arsenal of pheromonal effects. The mandibular glands are not the only site of production for queen pheromones; they are also probably secreted by the tergite glands, which are located in some of the membranes between the queen's abdominal segments (Renner and Baumann 1964, Velthuis 1970). Moreover, the mandibular and other pheromones may have closely linked or overlapping effects, which may explain why demandibulated queens still evoked the retinue response.

Third, the pheromonal blend a queen produces varies with her age. A newly emerged virgin queen has virtually no pheromone in her glands, but when she begins to mate some six days later, she is secreting the decenoic acids, and her glands contain about half the amount of acid a mated queen's contain. The full blend, includ-

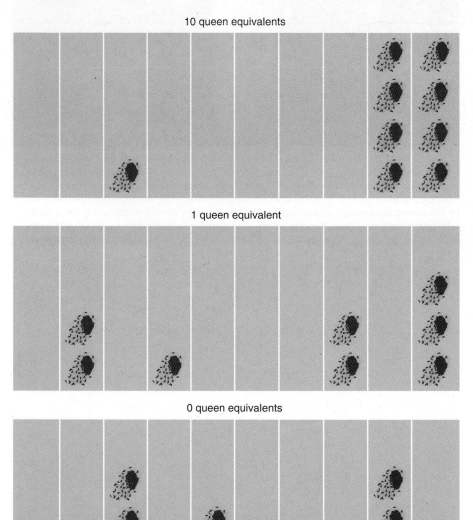

Figure 6. Queen mandibular pheromone plays a role in the timing of swarming. Experiments in which supplemental pheromone was administered on lures to queenright colonies suggest that hives swarm when congestion causes the levels of pheromone reaching workers to fall below a threshold. In the experiments summarized here, 10 Qeq of supplemental pheromone significantly delayed swarming, but one Qeq of supplemental pheromone did not. Other experiments suggest that a lower dose is effective if the pheromone is well dispersed.

Figure 7. Attracting flying workers to swarm clusters is another function of queen mandibular pheromone. The fence posts in this photograph were baited with lures containing one Qeq of synthetic pheromone. The synthetic pheromone was almost as effective as a queen in attracting workers. (Photograph courtesy of K. Naumann.)

ing the aromatic compounds, is secreted only after a queen has finished mating and begins to lay eggs (Slessor et al. 1990). This variation in the composition of the pheromone with the queen's age also may help to explain differing reports of effectiveness.

Finally, the mandibular pheromone is effective over a wide range of dosages. We defined the average amount of pheromone found in the glands of a mated queen to be one queen equivalent (Qeq). Worker bees respond to dosages as low as 10^{-7} Qeq, or a ten-millionth of her daily production (Kaminski et al. 1990, Slessor et al. 1988). This finding was particularly interesting in relation to hypotheses about pheromone transmission.

Functions

Once we were satisfied that we had completely identified queen mandibular pheromone, we began to test the effects of the pheromone on colony functions. One of the first pheromone-mediated behaviors we examined was the inhibition of queen rearing. This function can be easily demonstrated by removing a queen from her colony; the workers become agitated after half an hour and begin rearing new queens within 24 hours. Beekeepers call this

"emergency queen rearing," because failure to promptly rear a new queen results in the death of the colony.

To rear queens, workers elongate the cells around a few newly hatched larvae (Figure 4). These larvae are fed a highly enriched food called royal jelly that adult workers produce in their brood-food glands. The specialized diet directs larval development away from the worker pathway and onto the queen pathway. Brood older than six days (three days as eggs and three as larvae) lose their ability to develop into queens. Since no new eggs are laid after the queen is lost, workers have only six days to begin queen rearing before the colony loses its ability to produce a new, functional queen.

We examined the role of mandibular pheromone in emergency queen rearing by comparing queen rearing in queenright colonies, queenless colonies and queenless colonies receiving daily doses of synthetic mandibular pheromone (Winston et al. 1989, 1990). The results were dramatic. When queenless colonies received one Qeq or more of pheromone per day, they made almost no attempt to rear new queens for four days after the queen was removed. Even six days after queen loss, there was no significant difference between

the number of queen cells in queenless colonies treated with pheromone and the number in queenright control colonies (Figure 5). We concluded that the queen's mandibular pheromone is largely responsible for the inhibition of queen rearing in queenright colonies.

Colonies also rear new queens in preparation for reproductive swarming, the process of colony division in which a majority of the workers and the old queen leave the colony and search for a new nest. Left behind in the old colony are developing queens, some of which may issue with additional swarms once they emerge, and one of which will reign over the old nest once swarming is completed.

Swarming poses something of a puzzle because it does not occur unless new queens are being reared, but the old queen is present during the initial stages of queen rearing and is presumably still secreting her inhibitory chemicals (Butler 1959, 1960, 1961; Seeley and Fell 1981; Simpson 1958). One hypothesis is that the transmission of queen pheromones is slowed as colonies grow and become more congested prior to swarming. As the amounts of pheromone reaching workers diminish, the workers begin to escape the queen's control.

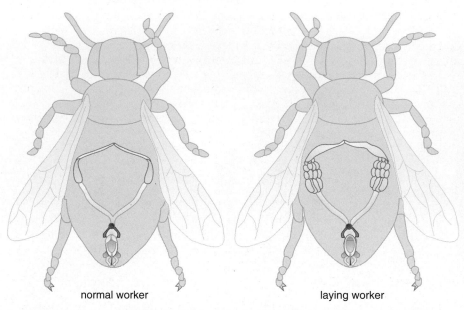

normal worker   laying worker

Figure 8. Suppression of worker ovary development, which had been thought to be a function of queen mandibular pheromone, is probably controlled by other primer pheromones, including substances produced by larval and pupal bees. Egg-laying is pre-eminently the queen's function but, under certain conditions, workers' ovaries also develop, and they can then lay eggs. In queenright colonies, a few workers have developed ovaries, but the workers lay few eggs unless the colony becomes queenless. If the queen is lost and the colony fails to rear a new queen, the workers with enlarged ovaries begin to lay eggs. Because workers lack the genital structures that allow the queen to mate and store sperm, any eggs they lay are unfertilized and usually develop into drones.

We tested the role of queen pheromone in reproductive swarming by supplementing the queen's normal secretion with additional, synthetic pheromone. Again, the pheromone had a significant effect on queen rearing. Colonies receiving 10 Qeq per day swarmed an average of 25 days later than control colonies (Figure ·6). The pheromone-treated colonies became extraordinarily crowded before they would swarm, indicating just how powerful a suppressant the queen pheromone can be.

For these experiments, the pheromone was presented on stationary lures, but queen bees move about in the nest. In another set of experiments, we tried to better mimic natural conditions by administering the pheromone as a spray. We found that a dose of one Qeq per day of supplemental pheromone was sufficient to suppress swarming. The fact that the better-dispersed spray was active at a much lower dose than the stationary lure supports the hypothesis that the slowing of pheromonal dispersal in congested colonies triggers swarming (Winston et al. 1991).

The queen mandibular pheromone plays two other roles in swarming as well: It attracts flying workers to the swarm cluster, and it stabilizes the

cluster, preventing workers from becoming restless and leaving before the swarm reaches a new nest site. We found that the synthetic mandibular pheromone is almost as effective as the queen in attracting workers to the cluster and in keeping them together, at least for the first few hours following cluster formation (Figure 7).

Next we examined the effect of mandibular pheromone on worker ovary development and egg laying. Both are known to be suppressed by queen pheromones. Indeed pheromonal suppression of worker reproduction is one of the hallmarks of advanced insect societies. Primitive social insects suppress nestmate reproduction primarily through dominance interactions. Many wasp and bumblebee queens, for example, seem to maintain dominance by biting and harassing the other females in the colony.

Honey bee workers have the potential to lay unfertilized eggs that can develop into males, or drones. However, some combination of pheromones secreted by the queen and the brood (the developing larvae or pupae) inhibits the maturation of the workers' ovaries, which usually remain small and nonfunctional (Figure 8) (Jay 1970, 1972). Nevertheless, even in queenright colonies a few workers manage to lay eggs

(Ratnieks and Visscher 1989, Visscher 1989) and many workers begin egg-laying two or three weeks after the queen and brood are removed (Winston 1987).

The ovary-suppressing substance produced by the queen was thought to be mandibular pheromone, particularly 9ODA (see reviews by Free 1987; Willis, Winston and Slessor 1990; Winston 1987). To test this hypothesis, we removed queens from groups of colonies and applied various doses of mandibular pheromone to the colonies for the next 43 days. Some colonies received daily doses as high as 10 Qeq. To our surprise, workers in queenless, pheromone-treated colonies developed ovaries at the same rate as workers in queenless, pheromone-free colonies (Willis, Winston and Slessor 1990). Apparently queen mandibular pheromone is not involved in the suppression of worker ovary development. Another primer pheromone must be responsible for this effect, possibly the unidentified queen tergite pheromone in combination with brood pheromone.

A final set of experiments demonstrated that queen mandibular pheromone is not just an inhibitor. In addition to suppressing reproductive activity, it can stimulate foraging and brood rearing. To establish this, we applied pheromone to newly founded, queenright colonies and counted the number of foragers, the size of their nectar and pollen loads and the extent of brood rearing (Figure 9). The pheromone did not increase overall foraging activity, but colonies that received one Qeq of supplemental pheromone daily had more pollen foragers, and those foragers carried heavier pollen loads. Pheromone applications increased the amount of pollen entering the nest by 80 percent, and pheromone-treated colonies reared 18 percent more brood, possibly due to the increased amounts of protein-rich pollen being brought into those colonies (Higo et al. 1992).

Transmission

The set of experiments we have just described demonstrated that queen mandibular pheromone has a wide range of functions, but we still did not know how much is produced and secreted by the queen daily or how these important compounds are transmitted to workers. The chemical components of the pheromone are not particularly volatile, and so they probably cannot spread throughout the colony by diffusion alone. The number of workers in an established colony and the rapidity with which the pheromone is lost make it unlikely that the queen is the sole disseminator of the pheromone. Some experiments had suggested that retinue workers act as messengers, transmitting the queen's pheromones to other workers when they touch antennae or mouthparts (Juška, Seeley and Velthuis 1981; Seeley 1979). Pheromone could not be found on worker bees, however, possibly because it was present in concentrations below the limit of detection of the instruments then in use.

We were able to follow pheromone secretion and transmission using the more sensitive chromatographic and spectroscopic equipment now available, as well as radiolabelled pheromone provided by Glenn Prestwich and Francis Webster of the State University of New York at Stony Brook and Syracuse (Webster and Prestwich 1988). Quantitative measurements of rates of production, transfer and loss allowed us to construct a mathematical model of pheromone dispersal. Our results essentially confirm the messenger-bee hypothesis, but they raise many interesting questions as well (Figure 10) (Naumann et al. 1991).

We first determined that a queen typically has five micrograms, or 0.001 Qeq, of pheromone on her body at any one time. To determine how much a queen secretes daily, we removed queens from colonies and measured the pheromone that built up on their cuticles. It turned out that a queen secretes between 0.2 and 2 Qeq of pheromone per day. These results fit well with our within-colony function studies, where pheromonal effects typically appeared at a dosage of about 1 Qeq.

We then quantified the processes by which pheromone is transferred or lost. For example, we allowed workers to contact the queen for different periods and then measured the amount of pheromone they had picked up. We found that each process could be approximated by a first-order rate equation. In other words, the amount of pheromone transferred during any short interval is proportional to the quantity present at the source during that interval; the constant of proportionality is called the rate constant. The first-order equation yields an exponential loss of material at the source; the "half life" is inversely proportional to the rate constant. We calculated the rate constants for all of the transmission pathways from our experimental

data and then used the value we had obtained for the queen's daily secretion and the rate equations to determine the amount of pheromone that would reach worker bees each day.

Our confidence in the model was bolstered when it predicted rates of pheromone transfer that are consistent with bee behavior. For example, the model predicts a flux of about one Qeq of pheromone through the nest per day; in our function experiments, a similar dose was typically the most active. The model also predicts that the amount of pheromone passed between workers will drop below the detectable level (10^{-7} Qeq) about 30 minutes after it is picked up from the queen, which is about how long it takes for workers to become restless after the queen is removed. Finally, the model predicts that pheromone deposited by the queen in the wax comb

are lickers, but they pick up over half the queen's pheromonal secretion. Each antennating worker picks up much less pheromone and passes on much less in subsequent contacts. Why some bees lick and others antennate is not known, nor do we understand the implications of this dual transmission system for the colony's social structure.

The wax comb also plays a role in the transfer of pheromone, although the queen deposits only 1 percent of her production there. Workers pick up a little of the pheromone the queen deposits. The remaining pheromone is probably released slowly over the next few hours. The slow release may explain why empty comb that has had brood in it is attractive to adult workers. The queen spends most of her time in the brood area, and her pheromonal footprints would be concentrated there.

Figure 9. Pollen foraging is yet another behavior influenced by queen mandibular pheromone, but in this case the pheromone stimulates rather than suppresses the behavior. A bee transports pollen in corbiculae, or pollen baskets, on the outer surfaces of the hind legs. The photograph shows a bee with heavily laden pollen baskets. Under some circumstances, queen mandibular pheromone stimulates workers to collect more pollen. (Photograph courtesy of Kenneth Lorenzen.)

will become undetectable in a few hours, which is about how long it takes the bees to begin rearing new queens.

The retinue workers do indeed remove the greatest fraction of the queen's pheromonal production. Not all bees in the retinue pick up and transfer the same amount of pheromone, however. There are two types of retinue bees, which we call licking and antennating messengers (Figure 11). The lickers touch the queen with their tongues, forelegs and mouthparts. The antennators brush the tips of their antennae lightly and quickly over her body. Only about 10 percent of the bees

One of our more surprising findings is that pheromone is lost primarily through internalization. The queen herself internalizes some 36 percent of the pheromone she produces. She swallows some of it, some is adsorbed on or bound to her cuticle, and some moves through the cuticle into the blood (hemolymph) system. What purpose internalization serves is unclear. Even if the internalized pheromone retains the same chemical identity, it is no longer available to the colony. It seems odd that such inefficient use is made of the queen's pheromone secretion. It is possible that the model somewhat overesti-

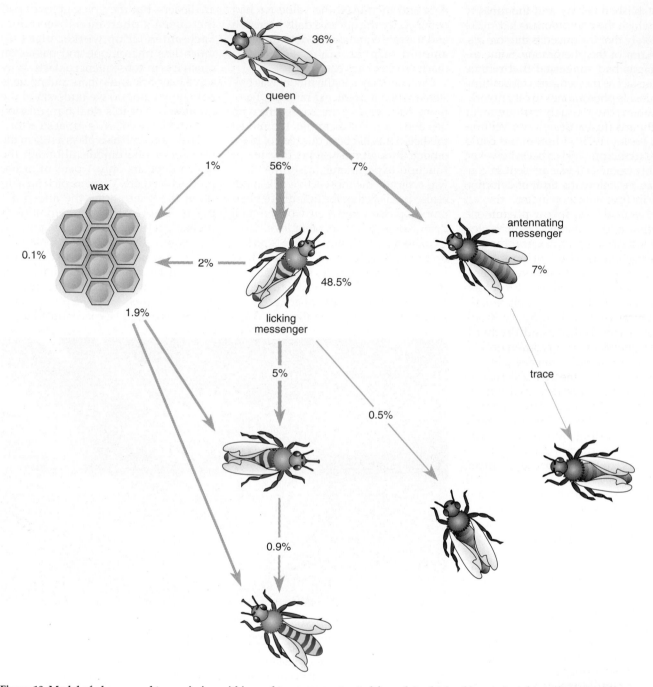

36%

queen

1%   56%   7%

wax   antennating
                                                        messenger

0.1%   2%   7%

                        48.5%

                        licking
            1.9%   messenger

                        5%

                                            0.5%   trace

                        0.9%

Figure 10. Model of pheromonal transmission within a colony was constructed from data obtained in a series of experiments with radiolabelled 9ODA as part of the synthetic pheromone blend. The pheromone is transmitted primarily by messenger bees, which gather it from the queen either by licking her body or by stroking with the antennae. Some pheromone is also deposited by the queen on the wax comb and picked up by worker bees. Only one bee in 10 is a licker, but the lickers gather much more pheromone than antennating bees and transfer more in subsequent contacts. For unknown reasons, large amounts of pheromone are internalized both by the queen herself and by the messenger bees. The numbers give the percentage of the queen's daily production that is transmitted in the indicated directions.

mates the queen's internalization of pheromone, since this was the only process that did not seem to conform to a first-order rate equation.

On the other hand, the messenger bees also internalize pheromone at a surprising rate (Figure 12). The licker bees swallow 40 percent of the pheromone they pick up. Most of the rest is quickly transferred from the worker's head to her abdomen by a combination of passive transport and active grooming (Naumann 1991). There the pheromone passes into and through the cuticle. Some of it ends up in the blood system within 30 to 60 minutes, and some remains bound to or adsorbed on the cuticle. In short, between the queen and the workers, nearly all of the pheromone is eventually internalized.

The mechanisms by which worker bees perceive the pheromone are not well understood. Sensory structures on a bee's antennae allow it to detect some odors at a concentration equal to a tenth or a hundredth of the concentration people can detect. The arrival of odorant at receptor cells inside the antennae causes associated nerve cells to fire, and the firing can initiate further behavioral and physiological changes. In most insects pheromones act by similar mechanisms.

But what about the large amounts of pheromone that are internalized? One hypothesis is that once the pheromone is translocated across the cuticle, its constituents or their breakdown products become active as hormones rather than as pheromones. We are only beginning to explore this intriguing hypothesis.

It also is unclear why the queen produces such a large amount of mandibular pheromone. The answer may be rooted in queen-worker conflict and the evolution of queen dominance in highly social insects. As we have mentioned, there are almost no dominance interactions in a normally functioning honey bee colony. Instead the queen controls the workers entirely through her sophisticated arsenal of pheromones. Perhaps over evolutionary history the workers and queen engaged in a kind of chemical arms race, where the workers began to break down the queen's compounds more rapidly and the queen responded by secreting more pheromone. Thus the swallowing and absorbing of pheromone by workers that we find so puzzling may be an attempt to catabolize the queen's pheromonal dominance and escape her control. Although this idea is speculative, it is certainly plausible. Indeed, this concept is reminiscent of most families and societies, which exhibit a complex blend of cooperation and conflict, with some objectives in common but with each individual also having its own goals.

Commercial Applications

We have been investigating commercial applications of queen pheromone for beekeeping and crop pollination. Several beekeeping applications for queen mandibular pheromone are already close to commercial realization. For example, we have shown that the pheromone allows packages of worker bees to be shipped without queens (Naumann et al. 1990). Beekeepers often establish a colony by buying a kilogram of workers and a queen in a wire package, but queens are typically produced by beekeepers who specialize in this art, and so they may come from a different source. The pheromone's ability to delay swarming also might be useful in bee management, since frequent swarming reduces honey production and can threaten colony survival. The pheromone's ability to stimulate pollen foraging and brood rearing might have commercial importance as well, since substantial increas-

Figure 11. Licking and antennating behavior are both part of the retinue response; thus actions induced by the pheromone are also responsible for its dissemination through the colony. In the upper photograph at least one worker bee can be seen licking the queen's abdomen; in the lower photograph several workers make antennal contact. (Photographs courtesy of Kenneth Lorenzen.)

es in colony growth rates can be achieved by these means. Finally, we are investigating the use of mandibular pheromone as an attractant for swarms. Such an attractant would be useful for the monitoring and control of honey bee diseases and the Africanized bee (the so-called "killer bee").

The most significant application of queen mandibular pheromone, however, may be in assisting crop pollination. Managed crop pollination is economically the most important function of bees. Colonies of bees are moved to blooming crops in order to provide enough bees to ensure good pollination and seed set. Beekeepers receive up to $45 per colony for this service, which is essential for the production of many fruits and vegetables. Not only does pollination affect yield, it also is often closely associated with the quality of fruits, vegetables and seeds. Each year the honey bee pollinates crops worth at least $1.4 billion in Canada and $9.3 billion in the United States

(Robinson, Nowogrodzki and Morse 1989; Scott and Winston 1984). Although hundreds of thousands of bee colonies are moved to crops each year, many crops are still inadequately pollinated. Reduced yields, occasional crop failures and lowered crop quality can all be attributed to this problem (Free 1970; Jay 1986; McGregor 1976; Robinson, Nowogrodzki and Morse 1989).

Since queen pheromone is highly attractive to flying workers, we thought it possible that pollination could be improved by spraying crops with a dilute blend of pheromone. To test this hypothesis, we sprayed blocks of trees or berries with various concentrations of pheromone and monitored the number of workers attracted to the blocks and the crop yields. In almost all of the experiments, up to twice the number of bees visited treated blocks compared to untreated ones.

The effect of pheromone spraying on yield was more variable. In preliminary trials with apple trees we found no increases in yield or improvements in fruit quality. However, pear trees sprayed with pheromone produced fruit that was heavier by an average of 6 percent, which translated to an increase of 30 percent in profits, or about $1,055 per hectare once spraying costs were deducted. Cranberries treated with pheromone yielded about 15 percent more berries by weight, which increased net returns by $4,465 per hectare averaged over two years (Currie, Winston and Slessor 1992; Currie, Winston, Slessor and Mayer 1992). In some cases there was no improvement, whereas in others the yield increase was much greater than these average values. Clearly, if we are able to exercise in the farmer's field the kind of sociochemical control the queen exercises in the nest, we can dramatically expand the economic value of this already beneficial social insect.

Acknowledgments

We are grateful to all of our collaborators in the research this article describes, including: J. H. Borden, S. J. Colley, R. W. Currie, H. A. Higo, L.-A. Kaminski, G. G. S. King, K. Naumann, T. Pankiw, E. Plettner, G. D. Prestwich, F. X. Webster, L. G. Willis and M. H. Wyborn. We would also like to acknowledge the technical assistance of P. LaFlamme. Funding for the research came from the Natural Sciences and Engineering Research Council of Canada, the Science Council of British Columbia, the Wright Institute and Simon Fraser University.

Bibliography

Allen, M. D. 1960. The honeybee queen and her attendants. *Animal Behavior* 8:201–08.

Barbier, J., and E. Lederer. 1960. Structure chimique de la "substance royale" de la reine d'abeille (*Apis mellifica*). *Comptes Rendus des Séances de L'Académie de Science* (Paris) 251:1131–35.

Bradshaw, J. W. S., and P. E. Howse. 1984. Sociochemicals of ants. In: *Chemical Ecology of Insects*, W. J. Bell and R. T. Carde (eds). Sunderland, Mass.: Sinauer.

Butler, C. G. 1959. Queen substance. *Bee World* 40:269–75.

Butler, C. G. 1960. The significance of queen substance in swarming and supersedure in honey-bee (*Apis mellifera*) colonies. *Proceedings of the Royal Entomological Society* (London) (A) 35:129–32.

Butler, C. G. 1961. The scent of queen honey bees (*Apis mellifera*) that causes partial inhibition of queen rearing. *Journal of Insect Physiology* 7:258–64.

Butler, C. G., and E. M. Fairey. 1964. Pheromones of the honey bee: biological studies of the mandibular gland secretion of the queen. *Journal of Apicultural Research* 3:65–76.

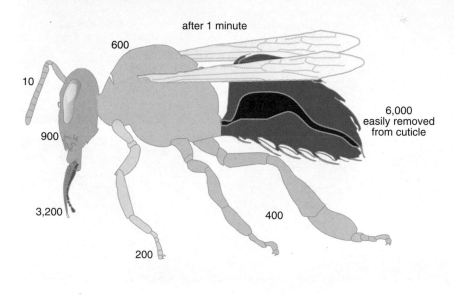

after 1 minute

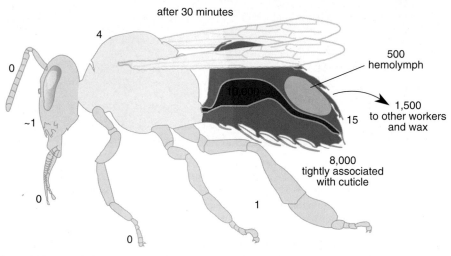

after 30 minutes

Figure 12. Maps of pheromone on a licking messenger bee illustrate one of the mysteries of pheromone transmission: Most of the pheromone is quickly internalized. The distribution of the pheromone is shown one minute (*upper diagram*) and 30 minutes (*lower diagram*) after the bee contacted the queen. To quantify the internalization of the pheromone by worker bees, 20,000 picograms of radiolabelled 9ODA (the amount a worker typically picks up in her contact with the queen) was applied topically to individual bees. Each bee was then isolated, so that the pheromone could not be transferred to other workers. After a delay, the bee was washed with methanol to dissolve any pheromone on the body surface. The crop and gut were then excised. The amounts of 9ODA in the washes, excised organs and the corpse were determined by liquid scintillation counting. The corpse scintillation count indicated how much pheromone had passed into the body or was so closely associated with the cuticle that it could not be removed by methanol. The pheromone depicted as being external remains available to the colony; it can be picked up by other workers or deposited on the wax comb. The internalized pheromone is effectively removed from circulation.

Callow, R. K., J. R. Chapman and P. N. Paton. 1964. Pheromones of the honey bee: Chemical studies of the mandibular gland secretion of the queen. *Journal of Apicultural Research* 3:77–89.

Callow, R. K., and N. C. Johnston. 1960. The chemical constitution and synthesis of queen substances of honey bees (*Apis mellifera* L.). *Bee World* 41:152–3.

Currie, R. W., M. L. Winston and K. N. Slessor. 1992. Impact of synthetic queen mandibular pheromone sprays on honey bee pollination of berry crops. *Journal of Economic Entomology* (in press).

Currie, R. W., M. L. Winston, K. N. Slessor and D. F. Mayer. 1992. Effect of synthetic queen mandibular pheromone sprays on pollination of fruit crops by honey bees. *Journal of Economic Entomology* (in press).

Duffield, R. M., J. W. Wheeler and G. C. Eickwort. 1984. Sociochemicals of bees. In: *Chemical Ecology of Insects*, W. J. Bell and R. T. Carde (eds). Sunderland, Mass: Sinauer.

Free, J. B. 1970. *Insect Pollination of Crops*. London: Academic Press.

Free, J. B. 1987. *Pheromones of Social Bees*. Ithaca: Cornell University Press.

Higo, H. A., S. J. Colley, M. L. Winston and K. N. Slessor. 1992. Effects of honey bee queen mandibular gland pheromone on foraging and brood rearing. *Canadian Entomologist* 124:409–418.

Hölldobler, B., and E. O. Wilson. 1990. *The Ants*. Cambridge, Mass.: Harvard University Press.

Howse, P. E. 1984. Sociochemicals of termites. In: *Chemical Ecology of Insects*, W. J. Bell and R. T. Carde (eds). Sunderland, Mass: Sinauer.

Jay, S. C. 1970. The effect of various combinations of immature queen and worker bees on the ovary development of worker honey bees in colonies with and without queens. *Canadian Journal of Zoology* 48:168–73.

Jay, S. C. 1972. Ovary development of worker honeybees when separated from worker brood by various methods. *Canadian Journal of Zoology* 50:661–4.

Jay, S. C. 1986. Spatial management of honey bees on crops. *Annual Review of Entomology* 31:49–66.

Juška, A., T. D. Seeley and H. H. W. Velthuis. 1981. How honeybee queen attendants become ordinary workers. *Journal of Insect Physiology* 27:515–19.

Kaminski, L.-A., K. N. Slessor, M. L. Winston, N. W. Hay and J. H. Borden. 1990. Honey bee response to queen mandibular pheromone in a laboratory bioassay. *Journal of Chemical Ecology* 16:841–49.

McGregor, S. E. 1976. *Insect Pollination of Cultivated Crop Plants*. Agricultural Handbook No. 496. Washington, D.C.: U.S. Department of Agriculture.

Naumann, K. 1991. Grooming behaviors and the translocation of queen mandibular gland pheromone on worker honey bees. *Apidologie* 22:523–531.

Naumann, K., M. L. Winston, M. H. Wyborn and K. N. Slessor. 1990. Effects of synthetic honey bee (Hymenoptera: Apidae) queen mandibular gland pheromone on workers in packages. *Journal of Economic Entomology* 83:1271–75.

Naumann, K., M. L. Winston, K. N. Slessor, G. D. Prestwich and F. X. Webster. 1991. The production and transmission of honey bee queen (*Apis mellifera* L.) mandibular gland pheromone. *Behavioral Ecology and Sociobiology* 29:321–32.

Pain, J., M.-F. Hugel and M. C. Barbier. 1960. *Comptes Rendus des Séances de L'Académie de Science* (Paris) 251:1046–8.

Ratnieks, F. L. W., and P. K. Visscher. 1989. Worker policing in the honey bee. *Nature* 342:796–97.

Renner, M., and M. Baumann. 1964. Uber komplexe von subepidermalen drusenzallen (Duftdrusen?) der bienenkonigin. *Naturwissenschaften* 51:68–9.

Robinson, W. S., R. Nowogrodzki and R. A. Morse. 1989. The value of honey bees as pollinators of U.S. crops. *American Bee Journal* 129:411–23, 477–87.

Scott, C. D., and M. L. Winston. 1984. The value of bee pollination to Canadian apiculture. *Canadian Beekeeping* 11:134.

Seeley, T. D. 1979. Queen substance dispersal by messenger workers in honey bee colonies. *Behavioral Ecology and Sociobiology* 5:391–415.

Seeley, T. D., and R. D. Fell. 1981. Queen substance production in honey bee (*Apis mellifera*) colonies preparing to swarm. *Journal of the Kansas Entomological Society* 54:192–96.

Simpson, J. 1958. The factors which cause colonies of *Apis mellifera* to swarm. *Insectes Sociaux* 5:77–95.

Slessor, K. N., G. G. S. King, D. R. Miller, M. L. Winston and T. L. Cutforth. 1985. Determination of chirality of alcohol or latent alcohol semiochemicals in individual insects. *Journal of Chemical Ecology* 11:1659–67.

Slessor, K. N., L.-A. Kaminski, G. G. S. King, J. H. Borden and M. L. Winston. 1988. Semiochemical basis of the retinue response to queen honey bees. *Nature* 332:354–56.

Slessor, K. N., L.-A. Kaminski, G. G. S. King and M. L. Winston. 1990. Semiochemicals of the honey bee queen mandibular glands. *Journal of Chemical Ecology* 16:851–60.

Velthuis, H. H. W. 1970. Queen substances from the abdomen of the honey bee queen. *Zeitschrift fuer vergleichende Physiologie* 70:210–22.

Visscher, P. K. 1989. A quantitative study of worker reproduction in honey bee colonies. *Behavioral Ecology and Sociobiology* 25:247–54.

Webster, F. X., and G. D. Prestwich. 1988. Synthesis of carrier-free tritium-labelled queen bee pheromone. *Journal of Chemical Ecology* 14:957–62.

Willis, L. G., M. L. Winston and K. N. Slessor. 1990. Queen honey bee mandibular pheromone does not affect worker ovary development. *Canadian Entomologist* 122:1093–99.

Wilson, E. O. 1971. *The Insect Societies*. Cambridge, Mass.: Harvard University Press.

Winston, M. L., K. N. Slessor, M. J. Smirle and A. A. Kandil. 1982. The influence of a queen-produced substance, 9HDA, on swarm clustering behavior in the honey bee *Apis mellifera* L. *Journal of Chemical Ecology* 8:1283–88.

Winston, M. L. 1987. *The Biology of the Honey Bee*. Cambridge, Mass.: Harvard University Press.

Winston, M. L., K. N. Slessor, L. G. Willis, K. Naumann, H. A. Higo, M. H. Wyborn and L.-A. Kaminski. 1989. The influence of queen mandibular pheromones on worker attraction to swarm clusters and inhibition of queen rearing in the honey bee (*Apis mellifera* L.). *Insectes Sociaux* 36:15–27.

Winston, M. L., H. A. Higo and K. N. Slessor. 1990. Effect of various dosages of queen mandibular gland pheromone on the inhibition of queen rearing in the honey bee (Hymenoptera: Apidae). *Annals of the Entomological Society of America* 83:234–38.

Winston, M. L., H. A. Higo, S. J. Colley, T. Pankiw and K. N. Slessor. 1991. The role of queen mandibular pheromone and colony congestion in honey bee (*Apis mellifera* L.) reproductive swarming. *Journal of Insect Behavior* 4:649–659.

INDEX